OPTICAL CHARACTER RECOGNITION • *Shunji Mori, Hirobumi Nishida, and Hiromitsu Yamada*

ANTENNAS FOR RADAR AND COMMUNICATIONS: A POLARIMETRIC APPROACH • *Harold Mott*

INTEGRATED ACTIVE ANTENNAS AND SPATIAL POWER COMBINING • *Julio A. Navarro and Kai Chang*

FREQUENCY CONTROL OF SEMICONDUCTOR LASERS • *Motoichi Ohtsu (ed.)*

SOLAR CELLS AND THEIR APPLICATIONS • *Larry D. Partain (ed.)*

ANALYSIS OF MULTICONDUCTOR TRANSMISSION LINES • *Clayton R. Paul*

INTRODUCTION TO ELECTROMAGNETIC COMPATIBILITY • *Clayton R. Paul*

INTRODUCTION TO HIGH-SPEED ELECTRONICS AND OPTOELECTRONICS • *Leonard M. Riaziat*

NEW FRONTIERS IN MEDICAL DEVICE TECHNOLOGY • *Arye Rosen and Harel Rosen (eds.)*

ELECTROMAGNETIC PROPAGATION IN MULTI-MODE RANDOM MEDIA • *Harrison E. Rowe*

NONLINEAR OPTICS • *E. G. Sauter*

InP-BASED MATERIALS AND DEVICES: PHYSICS AND TECHNOLOGY • *Osamu Wada and Hideki Hasegawa (eds.)*

DESIGN OF NONPLANAR MICROSTRIP ANTENNAS AND TRANSMISSION LINES • *Kin-Lu Wong*

FREQUENCY SELECTIVE SURFACE AND GRID ARRAY • *T. K. Wu (ed.)*

ACTIVE AND QUASI-OPTICAL ARRAYS FOR SOLID-STATE POWER COMBINING • *Robert A. York and Zoya B. Popović (eds.)*

OPTICAL SIGNAL PROCESSING, COMPUTING AND NEURAL NETWORKS • *Francis T. S. Yu and Suganda Jutamulia*

SiGe, GaAs, and InP HETEROJUNCTION BIPOLAR TRANSISTORS • *Jiann Yuan*

SiGe, GaAs, and InP Heterojunction Bipolar Transistors

SiGe, GaAs, and InP Heterojunction Bipolar Transistors

JIANN S. YUAN
University of Central Florida

A WILEY INTERSCIENCE PUBLICATION
JOHN WILEY & SONS, INC.
NEW YORK / CHICHESTER / WEINHEIM / BRISBANE / SINGAPORE / TORONTO

This book is printed on acid-free paper. ♾

Published simultaneously in Canada.

Library of Congress Cataloging in Publication Data:

Yuan, Jiann S., 1959–
SiGe, GaAs, and InP heterojunction biopolar transistors / Jiann S. Yuan
p. cm. — (Wiley series in microwave and optical engineering)
Includes index.
ISBN 0–471–19746–7 (cloth : alk. paper)
1. Bipolar transistors. 2. Junction transistors. I. Title. II. Series.
TK7871.96.B55Y83 1999
621.3815′28—dc21 98–38194

Printed in the United States of America.

10 9 8 7 6 5 4 3 2 1

To my wife, Huili, and our sons, Michael and Jerry

Contents

Preface xiii

About the Author xvii

1 Introduction **1**

References 4

2 Material Properties and Technologies **6**

2.1 SiGe and Group III/V Compound Semiconductors 6

2.1.1 Bandgaps and Lattice Constants 6

2.1.2 Velocity Overshoot 8

2.1.3 Bandgap Discontinuity 10

2.1.4 Bandgap Narrowing 11

2.1.5 Strained Layer and Critical Thickness 14

2.1.6 Electron Mobility 16

2.1.7 Hole Drift Mobility 20

2.2 Heterojunction Technologies 24

2.2.1 Vapor-Phase Epitaxy 25

2.2.2 Molecular Beam Epitaxy 27

2.2.3 Gas-Source MBE and Metal-Organic MBE 27

2.3 Device Fabrication 32

2.3.1 SiGe HBTs 32

2.3.2 AlGaAs/GaAs HBTs 36

2.3.3 InP HBTs 40

References 50

Problems 53

3 DC Performance **54**

3.1 General Structures and Steady-State Behavior 54

3.1.1 Electron and Hole Currents 54

3.1.2 Abrupt and Graded Heterojunctions 55

3.1.3 Undoped Setback Layer 59

3.1.4 Graded-Base HBTs 61

3.1.5 Double Heterojunctions 62

3.1.6 Electron Quasi-Fermi Level Splitting 65

3.1.7 Collector–Emitter Offset Voltage 67

3.1.8 Early Voltage 69

3.1.9 Bias-Dependent Base Resistance 71

3.1.10 High Injection Barrier Effect 72

3.2 SiGe Heterojunction Bipolar Transistors 76

3.2.1 Current Gain and Early Voltage Product 76

3.2.2 Temperature-Dependent Current Gain 79

3.2.3 Current Gain Roll-off in Graded SiGe Base 81

3.2.4 Early Voltage, Including Recombination in the SiGe Base 82

3.2.5 Inverse Base Width Modulation Effect 89

3.3 III/V Compound Heterojunction Bipolar Transistors 92

3.3.1 Self-Heating Effect 92

3.3.2 Recombination Currents 97

3.3.3 Temperature-Dependent Current Gain of AlGaAs/GaAs HBTs 99

3.3.4 Temperature-Dependent Current Gain of InP-Based HBTs 103

References 105

Problems 107

4 RF and Transient Performance **109**

4.1 General Device Behavior 109

4.1.1 Output Conductance 109

4.1.2 Transconductance 109

4.1.3 Heterojunction Junction Capacitance 109

4.1.4 Base Transit Time 111

4.1.5 Collector–Base Space-Charge-Layer Delay 112

4.1.6 Cutoff Frequency 119

4.1.7 Maximum Frequency of Oscillation 120

4.1.8 NPN Versus PNP on RF Performance 121

4.1.9 Collector-Up Versus Collector-Down Influence on RF Performance 124
4.1.10 Noise 127
4.1.11 S-Parameters 133
4.1.12 Turn-off Transient 138
4.2 Silicon–Germanium Heterojunction Bipolar Transistors 143
4.2.1 Effect of Ge Profiles on g_o and τ_B 143
4.2.2 Effect of Ge Profiles on f_T and $f_{\max}$ 149
4.2.3 Effect of Inverse Base Width Modulation on τ_B and τ_C 152
4.2.4 Transconductance Degradation at High Current Densities and Low Temperatures 155
4.2.5 Ge and Collector Doping Profile Design to Improve the Clipping Effect 156
4.3 III/V Compound Heterojunction Bipolar Transistors 161
4.3.1 Emitter Delay 161
4.3.2 AlGaAs and InGaAs Graded Bases 163
4.3.3 Heterojunction Capacitance, Including the Composition Grading and a Setback Layer 166
4.3.4 Thermal Effects on f_T and $f_{\max}$ 170
References 172
Problems 176

5 HBT Modeling 177

5.1 Silicon–Germanium HBT Models 177
5.1.1 Analytical Collector Current Equation 177
5.1.2 Generalized Integral Charge-Control Relation for SiGe HBTs 179
5.1.3 Base Current of SiGe HBTs 181
5.1.4 High Current Operation 183
5.2 III/V Compound HBT Models 185
5.2.1 Thermionic-Field-Diffusion Model 187
5.2.2 Grinberg–Luryi Physics-Based Collector Current Model 188
5.2.3 New Charge-Control Model 191
5.2.4 Base Recombination Currents 193
5.2.4.1 Space-Charge-Region Recombination Currents 193
5.2.4.2 Surface Recombination Currents 198
5.2.4.3 Quasi-Neutral Recombination Currents 200

5.2.5 Analytical Collector Current Model, Including the Self-Heating Effect 200
5.2.6 Compact Gummel–Poon Model, Including the Self-Heating Effect 204
5.3 Large- and Small-Signal Models for RF Applications 209
5.4 Parameter Extraction 215
References 224
Problems 227

6 Heterojunction Device Simulation **228**
6.1 Boltzmann Transport Equation 228
6.2 Monte Carlo Simulation 232
6.3 Drift and Diffusion Equations 244
6.4 Hydrodynamic Equations 250
6.5 Transistor Design Using Heterojunction Device Simulation 254
6.6 Multiemitter Simulation 268
References 277
Problems 279

7 Breakdown and Thermal Instability **281**
7.1 Avalanche Breakdown 281
7.1.1 Reverse Base Current Phenomenon 285
7.1.2 Nonlocal Avalanche Effect 291
7.1.3 Influence of the Base Thickness on the Collector Breakdown 294
7.1.4 Avalanche Effect on the Collector-Base Junction Capacitance 296
7.1.5 Avalanche Effect on the Output Conductance 298
7.1.6 Breakdown and Speed Considerations in InGaAs HBTs 300
7.2 Thermal Instability 304
7.2.1 Emitter Collapse Phenomenon 306
7.2.2 Relation Between Emitter Collapse and Avalanche Breakdown 307
7.2.3 InP HBT Thermal Instability 309
7.2.4 Modeling the Emitter Collapse Loci 312
7.3 Design In Thermal Stability 314
7.3.1 Emitter Ballasting Resistors 314
7.3.2 Emitter Thermal Shunt 317
7.3.3 Base Ballasting Resistors 319
References 323
Problems 325

8 Reliability **327**

8.1 Electrical and Thermal Overstress 327
8.1.1 Forward- and Reverse-Bias Stress Effects 330
8.1.2 Thermal Overstress 335
8.1.3 Burn-in 338
8.2 Process-Related Reliability Issues 340
8.2.1 Base Dopant Out-diffusion 340
8.2.2 Sensitivity of Emitter–Base Junction Design 345
8.2.3 Influence of Dislocations on the Transistor Current Gain 346
8.2.4 Effect of Passivation on InAlAs/InGaAs HBTs 348
8.2.5 Effect of Hydrogen Out-diffusion in InGaP/GaAs HBTs 349
8.3 Hot Carrier Behavior 351
8.4 Radiation Effects 357
8.4.1 Si-Based Bipolar Transistors 358
8.4.1.1 Oxide Trapped Charge and Excess Base Current 362
8.4.1.2 Low-Dose-Rate Radiation 363
8.4.1.3 Implications for Circuit Behavior 364
8.4.2 GaAs- and InP-Based Bipolar Transistors 366
8.4.2.1 Total Dose Effects 366
8.4.2.2 Transient Radiation Effects 368
References 369
Problems 373

9 RF and Digital Circuits for Low-Voltage Applications **375**

9.1 Low-Voltage Applications 376
9.2 Wideband Amplifiers 376
9.3 RF Power Amplifiers 380
9.3.1 Power-Added Efficiency 381
9.3.2 Impact of Device Parameters on the HBT Large-Signal Gain 383
9.3.3 Heterojunction Bipolar Transistor Design for Power Applications 385
9.3.4 Class E Power Amplifiers 390
9.3.5 Third-Order Intermodulation 397
9.3.6 Self-Linearizing Technique for the L-Band HBT Power Amplifier 401
9.4 Low-Noise Amplifiers 404
9.5 HBT Oscillators 411
9.5.1 Modeling the Bipolar Phase Noise 416

9.6 Analog Multipliers 422
9.7 A/D Converters 423
9.8 Diode-HBT Logic with ECL/CML Circuits 428
9.8.1 Gate Delay Versus Power 433
9.8.2 Figure of Merit for CML 435
9.8.3 Figure of Merit for ECL 439
9.8.4 SiGe Digital Circuit Performance 441
9.9 Phototransistors 445
9.10 Photoreceivers 450
References 457

Index **461**

Preface

Advances in wireless communication and information processing systems require implementation of very high performance electronic systems. In recent years, heterojunction bipolar transistors (HBTs) have emerged as a leading contender to satisfy these demands. The use of wide-bandgap heterojunction emitters allows design of HBTs with very high base doping levels, which significantly reduce base resistance and increase device speed. The low emitter–base turn-on voltage and device scaling to submicron dimensions significantly reduce power consumption in circuit operation. With the increasing demand placed on voice and data communications, transmitting, receiving, and processing information at high frequencies and high speeds using both microwave and optical means has become another area where heterojunction bipolar transistors have become increasingly important.

This book is intended for use by senior undergraduate or first-year graduate students in applied physics, electrical engineering, and materials sciences, and as a reference for engineers and scientists involved in semiconductor device research and development. It is assumed that the reader has already acquired a basic understanding of bipolar device operation. With this as a basic, the present book elaborates on the high-speed and radio-frequency (RF) aspects of heterojunction bipolar transistor performance. Many important aspects of SiGe-, GaAs-, and InP-based HBTs are presented in a unified manner.

In Chapter 1, the advantages of heterojunction bipolar devices for wireless telecommunication applications are introduced. Comparisons are made between HBTs, MESFETs, and HEMTs for high-speed and high-frequency operations.

In Chapter 2, SiGe and group III/V compound semiconductor properties such as bandgap versus lattice constant, strained and unstrained layers, critical thickness, bandgap narrowing and discontinuities, electron and hole mobilities, and velocity overshoot are presented. Heterojunction technologies using MOCVD and MBE are discussed. Step-by step processes of self-aligned Si/SiGe, AlGaAs/GaAs, and InP/InGaAs HBTs using heterojunction technologies are illustrated.

In Chapter 3, heterojunction-device dc performance is examined. The chapter begins with a description of general steady-state behavior for SiGe-, GaAs-, and

InP-based HBTs. Bandgap engineering to control the electron and hole currents of HBTs is introduced. Heterojunction systems such as abrupt heterojunction, graded heterojunction, undoped setback layer, graded base, and double heterojunctions are explained. Physical phenomena such as quasi-Fermi level separation, collector–emitter offset voltage, and high injection barrier effect are described. Unique issues such as the current gain and Early voltage product, current-gain roll-off in graded SiGe base at high currents, and inverse base width modulation effect for SiGe HBTs are also presented. The self-heating effect and temperature-dependent current gain of AlGaAs/GaAs- and InP-based HBTs are given in another section for a smooth presentation of several important subjects for different heterojunction technologies.

Chapter 4 illustrates the RF and transient performance of HBTs. General ac characteristics such as the output conductance, transconductance, and heterojunction capacitance are introduced, followed by base transit time and base–collector space-charge-layer delay for the determination of cutoff frequency and maximum frequency of oscillation. Device noise, S-parameters, and turn-off transient are also presented. For SiGe HBTs, effects of the Ge profiles on the output conductance, base transit time, cutoff frequency, and maximum frequency of oscillation are investigated. The transconductance clipping at high currents and the design of Ge and collector profiles to improve the clipping effect are introduced.

In Chapter 5 we examine HBT device modeling. Both physics-based and compact modeling schemes are investigated. The physics-based models include the thermionic-field-diffusion model, the Grinberg–Luryi physical model, Parikh–Lindholm new charge model, and Yuan–Ning analytical model. The compact models are Gummel–Poon-like models. Parameter extraction schemes for a Gummel–Poon model are introduced.

In Chapter 6 we present numerical heterojunction-device simulation for device design. The presentation begins with introduction of the Boltzmann transport equation. Monte Carlo simulation is then presented. Standard drift and diffusion equations are explained in detail. Hydrodynamic equations suitable for heterojunction device simulation are described. Device designs for multiemitter-finger HBTs are illustrated.

In Chapter 7 we discuss the avalanche breakdown and thermal instability of multiemitter- finger HBTs. Thermal instability, such as the emitter collapse phenomenon and its relation to avalanche breakdown, is explained. The different mechanisms triggering thermal instability in the GaAs and InP HBTs are explained. HBT designs using emitter ballast resistors, emitter thermal shunt, and base ballast resistors to improve thermal stability are given.

In Chapter 8 we discuss HBT reliability, such as electrical and thermal overstress. Process-related issues on HBT reliability are also discussed. Hot electron and radiation effects for Si-, GaAs-, and InP-based HBTs are presented.

In Chapter 9 we cover HBT RF and digital circuits, including wideband amplifiers, power amplifiers, low-noise amplifiers, oscillators, analog–digital converters, emitter-coupled logic, phototransistors, and photoreceivers. Key device and circuit parameters are examined. Many experimental data published in the literature are included.

This is the first book on the SiGe HBT of which we are aware. It is also the first book to treat InP, GaAs, and SiGe heterojunction bipolar transistors in a unified manner. The book has a balance between depth and breadth and covers a wide range of important subjects regarding HBTs. Other unique aspects of the book include (1) presentation of HBT dc and RF performance in a clear and organized way, (2) description of compact HBT models for RF applications, (3) inclusion of heterojunction device simulation for device design and research, (4) investigation of various important HBT reliability issues for the improvement of device lifetime, (5) illustration of HBT cryogenic and high-temperature behavior in dc and RF performance, and (6) presentation of HBT digital and RF circuits for low-voltage applications. Numerous up-to-date results from journals, conference proceedings, and books are cited. More than 400 figures are provided for illustrative purposes.

I hope that the book will serve as a complete reference and guide for heterojunction bipolar transistor research and development and stimulate further work in this exciting and expanding field.

JIANN S. YUAN
Electrical and Computer Engineering Department
University of Central Florida, Orlando, Florida
email: yuanj@pegasus.cc.ucf.edu

About the Author

Jiann S. Yuan received both the M.S. and Ph.D. degrees (1984 and 1988) from the University of Florida. He joined the faculty at the University of Central Florida in 1990 after one year of industrial experience at Texas Instruments, Inc., where he was involved with the 16-MB CMOS DRAM circuit design. Currently, he is an associate professor and director of the microelectronics CAD lab at the University of Central Florida.

Dr. Yuan has published more than 120 papers in refereed journals and conference proceedings in the area of semiconductor devices and circuits. He has also authored the book *Semiconductor Device Physics and Simulation*, published by Plenum in 1998. Since 1990 he has conducted many research projects funded by NSF, Motorola, Harris, Lucent Technologies, National Semiconductor, and Enterprise Florida. He has held consulting positions with Harris, NEC, and National Semiconductor. He serves regularly as a reviewer for the *IEEE Transactions on Electron Devices*, *IEEE Electron Device Letters*, and *Solid-State Electronics*. He has received many awards. These include the Distinguished Researcher Award, UCF, 1993 and 1996, Outstanding Engineering Educator Award, IEEE Orlando Section and Florida Council, 1993, and Teaching Incentive Program Award, UCF, 1995. He is listed in *Who's Who in American Education* and *Who's Who in Science and Engineering*.

Dr. Yuan is a senior member of IEEE and a member of Eta Kappa Nu and Tau Beta Pi. He is the chairman of the Electron Device Chapter in the IEEE Orlando Section, and the associate editor of the *International Journal of Modeling and Simulation*.

CHAPTER ONE

Introduction

Due to the phenomenal growth of applications such as computing and wireless communications, the microelectronics industry, has become a major economic force on the world scene with annual sales in excess of $100 billion and continuing growth with no leveling in sight. Competition in this field is fierce among semiconductor companies to maintain their competitiveness and to protect their market share. The explosive demand for low-power wireless communication systems has increased the pace of radio-frequency (RF) technology development. A low-cost, high-performance, highly integrated technology is essential to implementation of wireless communication systems on a chip and makes time to market minimal. The emerging SiGe and InP technologies have the advantage over Si and GaAs technologies of higher-speed performance at low power density.

The concept of the heterojunction bipolar transistor (HBT) was introduced by William Shockley in 1948 (U.S. patent 2,569,347). A detailed theory related to this device was developed by Kroemer in 1957 [1]. The great potential advantages of the heterostructure design over conventional homostructure design have long been recognized [1,2]. It was not until the early 1970s, though, that the technology evolved to build practical transistors of this kind. The situation began to change with the emergence of liquid-phase epitaxy (LPE) as a technology for group III and V compound semiconductor heterostructures. Since the mid-1970s, two promising technologies have appeared: metal-organic chemical vapor deposition (MOCVD) [3] and molecular beam epitaxy (MBE) [4]. Impressive results for MOCVD- and MBE-grown heterojunction bipolar transistors have been attained [5,6].

Both technologies are capable of growing epitaxial layers with high crystalline perfection and purity. Highly controlled doping levels up to 10^{19} cm^{-3} or more can be achieved, and highly controlled changes in the doping level are possible during growth with a minor adjustment in growth parameters. The doping may be changed either gradually or abruptly. Because of the relatively low growth temperatures, diffusion effects during growth are weak. With certain dopants, much more abrupt doping steps can be achieved than with any other techniques. Most important in the context of heterostructures is that it is possible in both techniques to change from one group III/V compound semiconductor to a different lattice-matched group III/V semiconductor with greater ease than in any of the other techniques.

Heterojunction bipolar transistors have a number of inherent advantages over silicon homojunction bipolar transistors: (1) high Early voltage due to higher base doping, which reduces base width modulation effect; (2) small base resistance as a result of higher base doping; (3) high electron mobility and velocity overshoot, reducing electron transit times; (4) low base–emitter junction capacitance since emitter doping is lowered; (5) reduced parasitic capacitance, owing to the use of a semi-insulating substrate; (6) reduced high injection effect in the base; and (7) greater radiation hardness.

Heterojunction bipolar transistors also have many intrinsic advantages over field-effect transistors (FETs) [7,8]:

- The key distances that govern electron transit time are established by epitaxial growth, not by lithography. This allows high cutoff frequency f_T with modest processing requirements.
- The entire emitter area conducts current, leading to high current-handling capability per unit area.
- The threshold voltage for current flow is governed by built-in junction potential, leading to well-matched characteristics.
- The transistor is well shielded from traps in the bulk and surface regions, contributing to low $1/f$ noise and the absence of trap-induced frequency dispersion behavior.
- Breakdown voltage is directly controllable by the epitaxial structures of the device.
- High transconductance g_m results from direct control over current flow by the input voltage, resulting in exponential input–output characteristics.
- A significant voltage amplification factor g_m/g_o is attainable due to high transconductance and low output conductance g_o.

Table 1.1 compares GaAs HBTs, MESFETs, and HEMTs RF and power performance. The HBT obviously has advantages in collector efficiency, power density, g_m/g_o ratio, noise figure, phase noise, IP3/P_{dc} ratio, gain–bandwidth product, and peak cutoff frequency. Relatively speaking, the GaAs MESFET is the least competitive of the three. However, the GaAs MESFET is a good candidate for large-scale MMIC integration applications.

InP-based heterojunction bipolar transistors grown on InP substrates are emerging as an alternative technology to GaAs-based HBTs. InP heterojunction devices offer numerous advantages over GaAs devices for high-speed, low-power analog, digital, and optoelectronic applications, due to the following superior properties [9,10]:

- A smaller bandgap, which reduces the turn-on voltage and minimizes power dissipation.

TABLE 1.1 Comparison of HBT, MESFET, and HEMT Performances[a]

Merit Parameter	MESFET	HEMT	HBT	Remark
f_T	L	M	H	HBT has the highest speed.
f_{max}	M	H	L	HEMT has the highest frequency.
Gain–bandwidth	M	H	H	HEMT is best for wideband at high frequency.
Noise figure	M	L	H	HEMT is best for low-noise amplifiers.
Phase noise	M	H	L	HBT is best for voltage-controlled oscillators.
g_m/g_o	L	M	H	HBT is best for higher linear amplifiers.
IP3/P_{dc}	H	M	H	MESFET is best at high frequency; HBT is best at low frequency.
V_{th} uniformity	L	M	L	HBT is best for analog LSI circuits.
Hysteresis	H	M	H	HBT is best for sampling/hold.
Collector efficiency	M	H	H	HBT is potentially the best power device.
Power density	M	M	H	HBT is potentially the best power device.

Source: After Ali and Gupta, Ref. 8 © Artech House.
[a] H, high; M, medium; L, low.

- A higher Γ–L valley separation, which gives pronounced velocity overshoot and shorter space-charge-layer transit time.
- Higher carrier mobilities, which translate into superior minority-carrier transport and bulk resistance.
- A lower surface recombination velocity, which reduces $1/f$ noise and surface recombination current density.
- Excellent specific contact resistance for nonalloyed ohmic contacts to n- and p-type InGaAs.
- Compatibility with 1.3 to 1.55-μm lightwave communication systems.

InP-based HBTs also have a higher thermal conductive substrate and are more radiation hard. Devices can be fabricated using several different combinations of epitaxial InP, GaInAs, and AlInAs lattice matched to InP. For the wide-bandgap emitter, AlInAs and InP have shown comparable device performance. To achieve the lowest turn-on voltage while maintaining a large valence-band discontinuity at the emitter–base junction, it is necessary to compositionally grade the base–emitter junction between the GaInAs base and the AlInAs or InP emitter. A compositionally graded base–emitter junction also improves the reliability of the HBT significantly with respect to current gain and turn-on voltage stability. For applications requiring higher base–collector breakdown voltages and lower output conductances, a wide-bandgap InP collector is the material of choice.

The recent boom in wireless consumer applications has emphasized the requirement for low-cost, highly integrated RF parts. Steady improvement in transistor performance and desire for higher levels of integration have led to increased use of silicon technology. Material systems in group V and heterojunction growth techniques that are compatible with silicon technology have been developed [11–13]. Advances in the growth of strained SiGe epitaxial layers on silicon substrates have enabled realistic large-scale fabrication of high-performance devices and circuits using SiGe HBTs and MOSFETs. With MOSFETs, there is growing interest in the p-channel SiGe MOSFET [14–16] in view of its performance advantage over Si p-MOS, which is currently the speed-limiting part of CMOS circuits. The significant valence-band offset allows accumulation of the hole charge at the Si/SiGe heterointerface, increased hole mobility arising from strain-induced band distortion in the alloy, and reduced surface scattering.

REFERENCES

1. H. Kroemer, "Theory of wide-gap emitter for transistor," *Proc. IRE*, **45**, 1535 (1957).
2. H. Kroemer, "Heterostructure bipolar transistors and integrated circuits," *Proc. IEEE*, **70**, 13 (1982).
3. R. D. Dupuis, L. A. Moudym, and P. D. Dapkus, "Preparation and properties of $Ga_{1-x}Al_xAs$ GaAs heterojunction grown by metal-organic chemical vapor deposition," *First Physics Conference on Gallium-Arsenide and Related Compounds*, **45**, 1 (1979).
4. A. Y. Cho and J. R. Arthur, "Molecular beam epitaxy," *Proc. Solid-State Chem.*, **10**, 157 (1975).
5. M. Konnzai, K. Katsukawa, and K. Takahashi, "(GaAl)As/GaAs heterojunction phototransistors with high current gain," *J. Appl. Phys.*, **48**, 4389 (1979).
6. H. Beneking and L. M. Su, "GaAlAs/GaAs heterojunction microwave bipolar transistor," *Electron. Lett.*, **16**, 41 (1980).
7. P. M. Asbeck et al., "Heterojunction bipolar transistors for microwave and millimeter-wave integrated circuits," *IEEE Trans. Microwave Theory Tech.*, **MTT-35**, 1462 (1987).
8. F. Ali and A. Gupta, eds., *HEMTs and HBTs: Devices, Fabrication, and Circuits*, Artech House, Norwood, MA (1991).
9. S. M. Sze, *High-Speed Semiconductor Devices*, Wiley, New York (1990).
10. B. Jalali and S. J. Pearton, eds., *InP HBTs: Growth, Processing, and Applications*, Artech House, Norwood, MA (1995).
11. S. S. Iyer, G. L. Patton, S. L. Delage, S. Tiwari, and J. M. C. Stork, "Silicon–germanium base heterojunction bipolar transistors by molecular beam epitaxy," *IEDM Tech. Dig.*, 874 (1987).

12. C. A. King, J. L. Hoyt, and J. F. Gibbons, "Bandgap and transport properties of $Si_{1-x}Ge_x$ by analysis of nearly ideal $Si/Si_{1-x}Ge_x/Si$ heterojunction bipolar transistors," *IEEE Trans. Electron Devices*, **ED-36**, 2093 (1989).

13. D. L. Harame, J. H. Comfort, J. D. Cressler, E. F. Crabbé, J. Y.-C. Sun, B. S. Meyerson, and T. Tice, "Si/SiGe epitaxial-base transistors: I. Materials, physics, and circuits," *IEEE Trans. Electron Devices*, **ED-42**, 455 (1995).

14. S. Verdonckt-Vanderbroek, E. F. Crabbé, B. S. Meyerson, D. L. Harame, P. J. Restle, J. M. C. Stork, and J. B. Johnson, "SiGe-channel heterojunction p-MOSFET's," *IEEE Trans. Electron Devices*, **ED-41**, 90 (1994).

15. M. Arafa, P. Fay, K. Ismail, J. O. Chu, B. S. Meyerson, and I. Adesida, "High speed p-type SiGe modulation-doped field effect transistors," *IEEE Electron Device Lett.*, **EDL-17**, 124 (1996).

16. M. Arafa, K. Ismail, J. O. Chu, B. S. Meyerson, and I. Adesida, "A 70-GHz f_T low operating bias self-aligned p-type SiGe MODFET," *IEEE Electron Device Lett.*, **EDL-17**, 586 (1996).

CHAPTER TWO

Material Properties and Technologies

In the first part of this chapter we present the material properties of SiGe, GaAs, and InP semiconductors, such as bandgaps, lattice constants, velocity-field characteristics, conduction- and valence-band discontinuities, heavy doping effect, and carrier mobilities. In the second part of the chapter we illustrate the semiconductor technologies and fabrication of SiGe/Si-, AlGaAs/GaAs-, and InP-based heterojunction bipolar transistors.

2.1 SiGe AND GROUP III/V COMPOUND SEMICONDUCTORS

2.1.1 Bandgaps and Lattice Constants

When Si epitaxial layers are grown on Si substrates, there is a natural matching of the crystal lattice, and high-quality single-crystal layer result. On the other hand, it is often desirable to obtain epitaxial layers that differ somewhat from the substrate. This can be accomplished easily if the lattice structure and lattice constant match for the two materials. For example, GaAs and AlAs both have the zincblende structure, with a lattice constant of about 5.65 Å. As a result, epitaxial layers of the ternary alloy AlGaAs can be grown on GaAs substrates with little lattice mismatch [1]. Similarly, GaAs can be grown on Ge substrate.

Figure 2.1 shows the energy bandgap as a function of lattice constant for several elemental and compound semiconductors. Since AlAs and GaAs have similar lattice constants, the ternary alloy AlGaAs has essentially the same lattice constant over the entire range of compositions from AlAs to GaAs. As a result, one can choose the composition x of the ternary compound $Al_xGa_{1-x}As$ to fit the particular device requirement and grow this composition on a GaAs wafer. The resulting epitaxial layer will be lattice matched to the GaAs substrate.

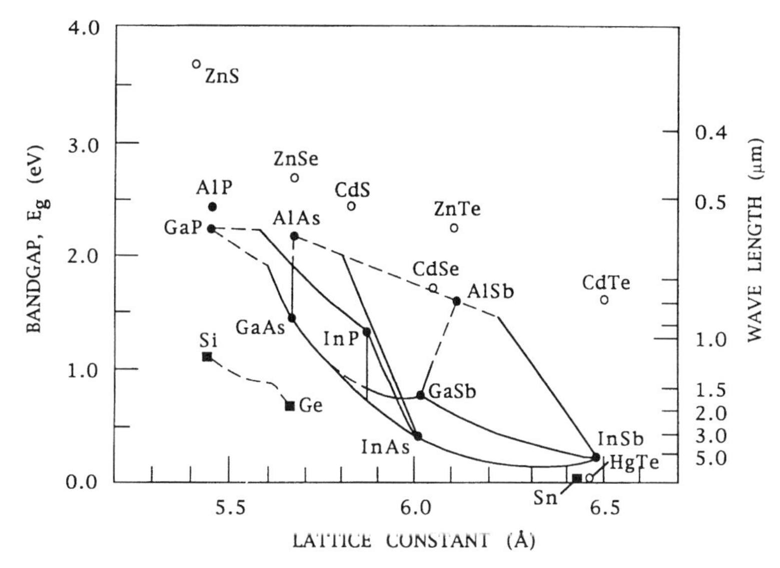

FIGURE 2.1 Energy bandgap versus lattice constant (after Jalali and Pearton, Ref. 2 © Artech House).

As the ternary compound InGaAs is varied by choice of composition on the group III sublattice from InAs to GaAs, the bandgap changes from 0.36 to 1.43 eV while the lattice constant of the crystal varies from 6.06 Å for InAs to 5.65 Å for GaAs. Clearly, we cannot grow this ternary compound over the entire composition range on a particular binary substrate. As Fig. 2.1 illustrates, it is possible to grow a specific composition of InGaAs on an InP substrate. The vertical line from InP to the InGaAs curve shows that $In_{0.53}Ga_{0.47}As$ can be grown lattice matched to a GaAs substrate. The variation of compositions on both group III and V sublattices provides additional flexibility in choosing a particular bandgap while providing lattice matching to convenient binary substrates such as GaAs and InP. Some important physical properties for Si, Ge, GaAs, InP, AlGaAs, and InGaAs are shown in Tables 2.1 and 2.2.

To achieve a broader range of alloy compositions, grown lattice matched on particular substrates, it is helpful to use quaternary alloys. Certain mixed-crystal compositions of the quaternary alloys InGaAlAs and InGaAsP are lattice matched to InP. These range from the endpoint compositions $In_{0.53}Ga_{0.47}As$ ($E_G = 0.74$ eV) to $In_{0.52}Al_{0.48}As$ ($E_G = 1.45$ eV) and $In_{0.53}Ga_{0.47}As$ to InP (1.35 eV), respectively, for the two quaternary systems. Note that the only ternary $In_xGa_{1-x}As$ composition lattice matched to InP is with $x = 0.53$. All other compositions from $x = 0$ (GaAs) to $x = 1$ (InAs) are mismatched, and the mismatch between these two endpoint binaries is 7%.

TABLE 2.1 Some Important Properties of Si, Ge, GaAs, and InP Semiconductors

	Si	Ge	GaAs	InP
ε_r	11.7	16.2	12.9	12.56
$N_C(\text{cm}^{-3})$	2.8×10^{19}	1.04×10^{19}	4.7×10^{17}	5.4×10^{17}
$N_V(\text{cm}^{-3})$	1.04×10^{19}	6.0×10^{18}	7.0×10^{18}	1.2×10^{19}
$E_G(\text{eV})$	1.12		1.424	1.34
$n_i(\text{cm}^{-3})$	1.02×10^{10}	2.33×10^{13}	2.1×10^{6}	1.2×10^{7}
μ_n at 300 K (cm/V·s)	1450	3900	8500	4600
μ_p at 300 K (cm/V·s)	500	1900	400	150
K(W/cm·°C)	1.31	0.6	0.46	

2.1.2 Velocity Overshoot

The variations of electron energy with wavenumber (momentum) in Si, Ge, and GaAs along the [100] and [111] directions in k space is shown in Fig. 2.2. The bandgap structures of Si and Ge are indirect-gap semiconductors, whereas that of GaAs represents a direct-bandgap semiconductor. For indirect semiconductors such as Si and Ge, the drift velocity is proportional to the electric field at low electric fields. The proportionality constant, called the mobility, is independent of electric field. When the fields are sufficiently large, nonlinearity in mobility and, in some cases, saturation of drift velocity are observed. For direct semiconductors such as GaAs, the velocity-field relationship is quite complicated. Consider the band structure shown in Fig. 2.2. A high-mobility valley ($\mu \approx 4000$ to $8000\ \text{cm}^2/\text{V} \cdot \text{s}$) is located at the Brillouin zone center (Γ valley), and a low-mobility satellite valley ($\mu \approx 100\ \text{cm}^2/\text{V} \cdot \text{s}$) along the [111] axes (L valley), about 0.3 eV higher in energy. The effective mass of the electrons is $0.068m_0$ in the lower valley and about $1.2m_0$ in the upper valley. Thus the density of states of the upper valley is about 70 times that of the lower valley. As the field increases, the electrons in the lower valley can be field excited to the normally

TABLE 2.2 Some Important Properties of $\text{Al}_x\text{Ga}_{1-x}\text{As}$ and $\text{In}_x\text{GaAs}_{1-x}$ Semiconductors

	$\text{Al}_x\text{Ga}_{1-x}\text{As}$	$\text{In}_x\text{GaAs}_{1-x}$
ε_r	$13.18 - 3.12x$	
$E_G(\text{eV})$	$1.424 + 1.247x\ (x < 0.45)$	$0.75\ (x = 0.53)$
	$1.9 + 0.125x + 0.143x^2\ (0.45 < x < 1.0)$	
Lattice constant (Å)	$5.6533 + 0.0078x$	$6.058 - 0.405x$
μ_n at 300 K (cm/V·s)	$8000 - 22{,}000x + 10{,}000x^2\ (x < 0.45)$	$13{,}800 (x = 0.53)$
	$-255 + 1160x - 720x^2\ (x > 0.45)$	
μ_p at 300 K (cm/V·s)	$370 - 970x + 740x^2$	

unoccupied upper valley. This results in a differential negative resistance and electron velocity overshoot in the III/V compound semiconductors. Velocity overshoot contributes to reduced transit time and capacitance.

Figure 2.3 shows velocity versus electric field for Si, GaAs, InP, and $In_{0.53}Ga_{0.47}As$ semiconductors. A higher separation between the valleys present in the band structure

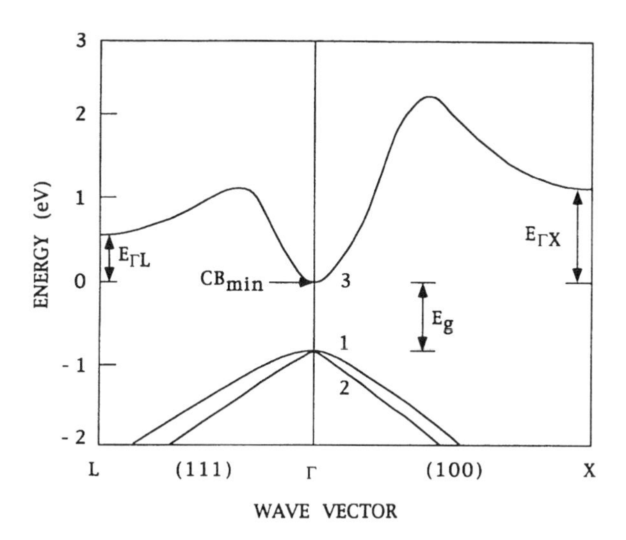

FIGURE 2.2 Electron energy with wavenumber in k space for Si, Ge, and GaAs materials.

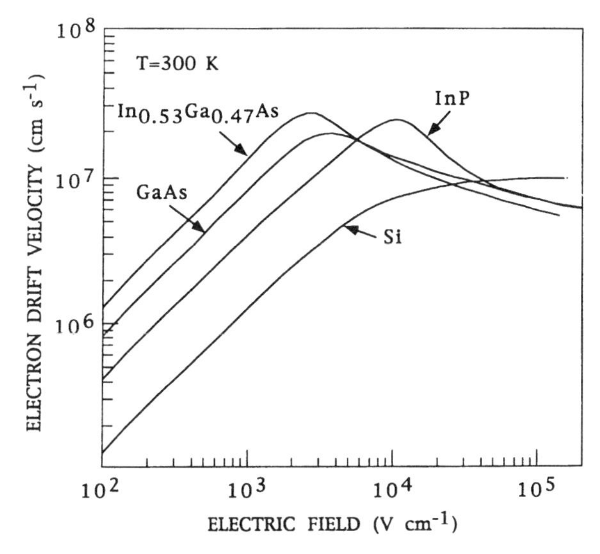

FIGURE 2.3 Velocity versus electric field for Si, GaAs, InP, and $In_{0.53}Ga_{0.47}As$ semiconductors (after Jalali and Pearton, Ref. 2 © Artech House).

of InP leads to the higher electron peak velocity. Higher overshoot velocity reduces the base–collector depletion-layer transit time of the heterojunction bipolar transistor (HBT). This increases the cutoff frequency of the HBT. The details of this are given in Chapter 4.

2.1.3 Bandgap Discontinuity

The bandgap difference ΔE_G between the emitter and the base consists of the valence-band discontinuity ΔE_V and the conduction-band discontinuity ΔE_C. In the SiGe HBT, most of the bandgap reduction results from a shift in the valence-band edge. The conduction-band discontinuity is usually a small fraction of the total bandgap difference. Gan et al. [3] measured the valence-band discontinuity of strained $Si_{1-x}Ge_x$ on unstrained Si using semiconductor–insulator–semiconductor structures with Ge compositions in the range 10 to 25%. The epitaxial heterostructures were grown by chemical vapor deposition. The experimental data indicate that the valence-band discontinuity between Si and $Si_{1-x}Ge_x$ can be approximated by $\Delta E_V = 6.4x$ meV for $0 < x < 17.5\%$. Values of ΔE_V versus Ge composition estimated by Ni et al. [4], King et al. [5], Wang et al. [6], Nauka et al. [7], Brighten et al. [8], and Gan et al. [3] are compared in Fig. 2.4.

For AlGaAs/GaAs HBTs, many data on ΔE_C and ΔE_V have been published [9–12]. Dingle et al. [9] reported that $\Delta E_C = 0.85\Delta E_G$ and $\Delta E_V = 0.15\Delta E_G$ based on the quantum-well absorption method. Miller et al. [10] showed that ΔE_V

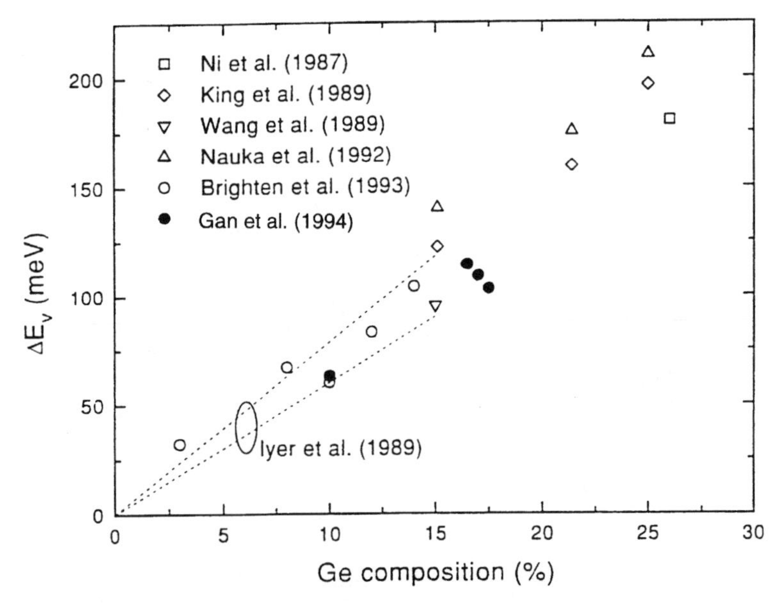

FIGURE 2.4 Valence-band discontinuity versus Ge composition (after Gan et al., Ref. 3 © IEEE).

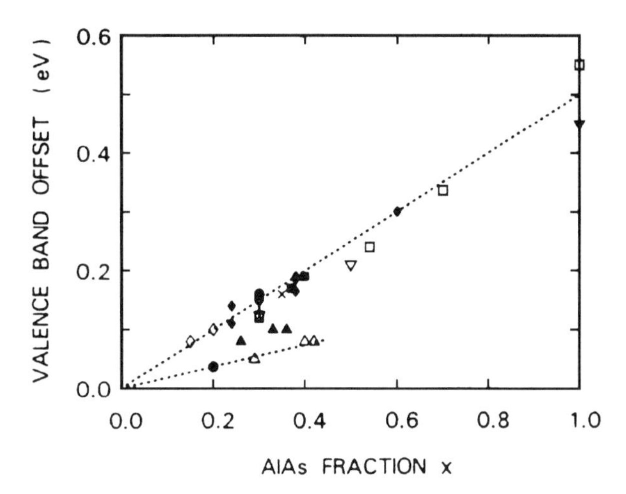

FIGURE 2.5 Valence-band discontinuity as a function of Al fraction (after Wang et al., Ref. 12 © American Institute of Physics).

is considerably larger than that proposed in [9]. Arnold et al. [11] suggest that $\Delta E_C/\Delta E_V = 0.6/0.4$ (i.e., $\Delta E_V = 0.22$ eV and $\Delta E_C = 0.15$ eV for a $Al_{0.3}Ga_{0.7}As/GaAs$ heterojunction). The valence-band discontinuity as a function of Al fraction obtained from different techniques, such as the quantum-well absorption method [8], capacitance-voltage measurement [11], and charge transfer method [12], is shown in Fig. 2.5. Most of the values correspond to 33 to 43% of the bandgap difference between $Al_xGa_{1-x}As$ and GaAs.

2.1.4 Bandgap Narrowing

The minority-carrier transport in the vertical dimension is crucial for accurate determination of collector current. Bandgap narrowing due to a heavy doping effect changes the physical constants used in the minority-carrier transport equation. It is well known that the collector current density of an HBT with flat Ge and doping profiles in the base without conduction-band spikes at the emitter–base heterojunction is given by

$$J_{C0} = \frac{qD_nN_CN_V}{G_B}e^{-E_{G,\text{eff}}} \tag{2.1}$$

where D_n is the electron diffusion constant, N_C and N_V are conduction- and valence-band densities of states, G_B is the base Gummel number, and $E_{G,\text{eff}}$, is the effective bandgap for minority-carrier concentration in the $Si_{1-x}Ge_x$ base.

Equation (2.1) can be rewritten as

$$J_{C0} = \frac{qD_{n,\text{SiGe}}}{G_{B,\text{SiGe}}}\frac{(N_CN_V)_{\text{SiGe}}}{(N_CN_V)_{\text{Si}}}n_{io,\text{Si}}^2e^{\Delta E_{G,\text{eff}}/kT} \tag{2.2}$$

where $\Delta E_{G,\text{eff}}$, is the effective bandgap reduction with respect to intrinsic Si. The ratio $(N_C N_V)_{\text{SiGe}}/(N_C N_V)_{\text{Si}}$ represents the reduction in effective densities of states in lightly doped $Si_{1-x}Ge_x$ due to strain-induced splitting of the bands. All heavy doping effects are included in $\Delta E_{G,\text{eff}}$.

There are two ways in which $\Delta E_{G,\text{eff}}$ may be found. The first is to compare J_{C0} of the HBT to that of a similar bipolar junction transistor (BJT) as a function of temperature [13,14]. If one assumes a similar temperature dependence of mobility, densities of states, and bandgap in the $Si_{1-x}Ge_x$ as in Si, one can extract $\Delta E_{G,\text{eff}}$ from the temperature dependence of the ratio of J_{C0} in the two devices. The other way is to make measurements of G_B and to make reasonable assumptions for D_n and the densities of states ratio, so that $\Delta E_{G,\text{eff}}$ can be extracted. Using the second method, the effective bandgap narrowing extracted as a function of Ge concentration for different doping levels is as shown in Fig. 2.6. For devices with similar dopings, the linear dependence on Ge concentration is obvious. Fitting the data at the same doping level gives an effective bandgap reduction with respect to Si of about 7 meV for 1% Ge. Assuming that the linear dependence on Ge concentration is independent of doping, the effects of bandgap reduction due to Ge concentration and bandgap narrowing due to heavy doping are separated [15]:

$$\begin{aligned}\Delta E_{G,\text{eff}} &= \Delta E_{G,\text{dop}} + \Delta E_{G,\text{Ge}} \\ &= 28.6 + 27.4 \log_{10} \frac{N_A}{10^{18}} + 688x \qquad \text{meV}\end{aligned} \tag{2.3}$$

Note that (2.3) is valid only for doping levels over 10^{18} cm^{-3}. The first two terms represent bandgap narrowing due to heavy doping, and the last term is the bandgap

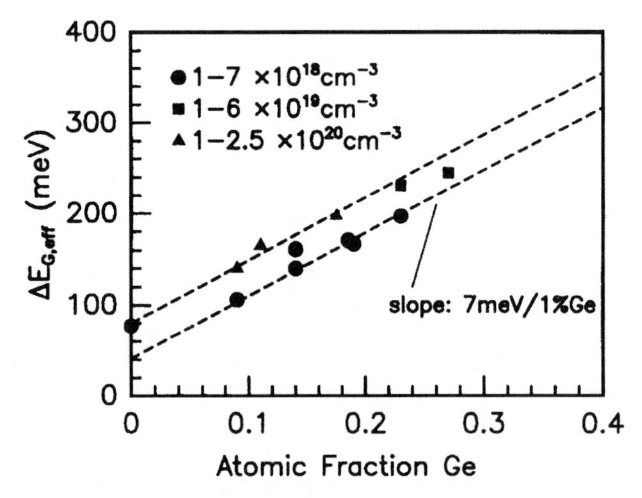

FIGURE 2.6 Effective bandgap narrowing versus Ge concentration (after Matutinovic-Krystelj et al., Ref. 15 © IEEE).

reduction due to Ge contribution. $\Delta E_{G,\mathrm{eff}}$ is not a measure of the actual bandgap reduction but the effective (apparent) bandgap reduction relevant for minority-carrier concentration and thus electron transport across the $Si_{1-x}Ge_x$ base. The apparent bandgap is larger than the true bandgap due to valence-band filling in the degenerately doped semiconductor and hence the effect of Fermi–Dirac statistics. The effective bandgap reduction is a useful parameter to model the collector current of the $Si_{1-x}Ge_x$ bipolar transistor.

Jain and Roulston [16] presented general closed-form equations for bandgap narrowing for n- and p-type Si, Ge, GaAs, and $Si_{1-x}Ge_x$ strained layers. The equations are derived by identifying the exchange energy shift of the majority band edge, correlation energy shift of the minority band edge, and impurity interaction shifts of the two band edges. The bandgap narrowing equation for $N > 10^{18}\ \mathrm{cm}^{-3}$ is given by

$$\Delta E_G = A_1 \left(\frac{N}{10^{18}}\right)^{1/3} + A_2 \left(\frac{N}{10^{18}}\right)^{1/4} + A_3 \left(\frac{N}{10^{18}}\right)^{1/2} \tag{2.4}$$

where A_1, A_2, and A_3 for n- and p-type Si, Ge, and GaAs are given in Table 2.3. For p-type $Si_{1-x}Ge_x$ alloy with a Ge content of less than 0.3, the bandgap narrowing equation is

$$\Delta E_G = 11.07(1 - 0.35x)\left(\frac{N}{10^{18}}\right)^{1/3} + 15.17(1 - 0.54x)\left(\frac{N}{10^{18}}\right)^{1/4} + 5.07(1 + 0.18x)\left(\frac{N}{10^{18}}\right)^{1/2} \tag{2.5}$$

López-González and Prat [17] used the Jain–Roulston model [16] to analyze the distribution of the bandgap narrowing between the conduction band $\Delta E_C^{\mathrm{BGN}}$ and the valence band $\Delta E_V^{\mathrm{BGN}}$. The bandgap narrowing versus doping for n-Si, p-$Si_{0.25}Ge_{0.75}$, and p-$In_{0.53}Ga_{0.47}As$ is shown in Fig. 1.8*a*, 1.8*b*, 1.8*c*, respectively. The shift of E_C

TABLE 2.3 Coefficients for Bandgap Narrowing Equation

	A_1	A_2	A_3
n-Si	10.23	13.12	2.93
p-Si	11.07	15.17	5.07
n-Ge	8.67	8.14	4.31
p-Ge	8.21	9.18	5.77
n-GaAs	16.30	7.47	90.65
p-GaAs	9.71	12.19	3.88

and E_V levels can be written as

$$\Delta E_C^{\mathrm{BGN}} = C_1 \left(\frac{N}{10^{18}} \right)^{1/a_1} + C_2 \left(\frac{N}{10^{18}} \right)^{1/2} \tag{2.6}$$

$$\Delta E_V^{\mathrm{BGN}} = C_3 \left(\frac{N}{10^{18}} \right)^{1/a_2} + C_4 \left(\frac{N}{10^{18}} \right)^{1/2} \tag{2.7}$$

Parameters C_1, C_2, C_3, C_4, a_1, and a_2 can be calculated from [17].

2.1.5 Strained Layer and Critical Thickness

In addition to the widespread use of lattice-matched epitaxial layers, the advanced epitaxial growth techniques allow for the growth of very thick layers of lattice-mismatched crystals. If the mismatch is only a few percent and the layer is thin, the epitaxial layer grows with a lattice constant in compliance with that of the seed crystal. The resulting layer is in compression or tension along the surface plane as its lattice constant adapts to the seed crystal. Such a layer is called *pseudomorphic* because it is not lattice matched to the substrate without strain. Figure 2.7 shows an accommodation of epitaxial layer with that of the substrate. Using thin alternating layers of slightly mismatched crystal layers, it is possible to grow a strain-layer superlattice in which alternate layers are in tension and compression. An unstrained SiGe layer is compared with a strained SiGe layer on the Si substrate shown in Fig. 2.8.

The thickness of the $Si_{1-x}Ge_x$ layer is an important device design consideration. The maximum thickness for pseudomorphic growth (*the critical thickness*) of $Si_{1-x}Ge_x$ alloys is an important property of the system. Van der Merwe [18] introduced the concept of critical thickness based on equilibrium theory. He defined critical thickness as the film thickness below which it was energetically favorable to contain the misfit by elastic energy stored in the distorted crystal and above which it was favorable to store part of the energy in misfit dislocations at the heteroepitaxial interface. Matthews and Blakeslee [19,20] defined critical thickness in terms of the mechanical equilibrium of a preexisting threading dislocation. Here they balanced the line tension force, which is the force of the dislocation segment residing at the heteroepitaxial interface with that component of the force per unit length acting on the threading component of the dislocation in the plane of the film. The thickness at which these two forces are equal is defined as the critical thickness: At greater film thicknesses, the dislocation assumes a misfit component at the interface that will also relieve the strain. Both of the two theories described above do not account for kinetic limitations on dislocation generation.

The experimental observation of the onset of strain relaxation is limited by the experimental technique used. Conventional x-ray diffraction and transmission electron microscopy (TEM) can detect relaxation only when it exceeds 0.1%. Refinements of these techniques, such as triple-crystal x-ray diffraction, may enable resolution of strain below 0.01%. Electron-induced-currents (EBICs) can be used to image dislocations because of the increased carrier recombination in the vicinity of the dislocation. Other electrical evaluation methods for the quality of heterojunctions include mea-

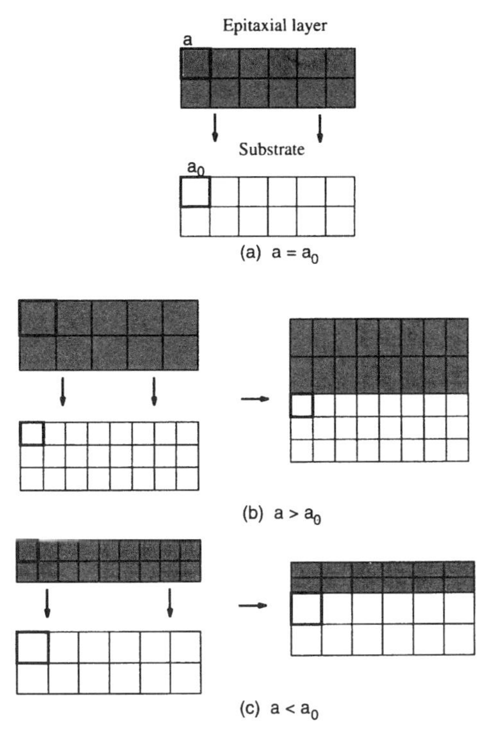

FIGURE 2.7 Accommodation of epitaxial layer with the substrate for (*a*) lattice-matched growth ($a = a_0$); (*b*) biaxial compressive strain ($a > a_0$), and (*c*) biaxial tensile strain ($a < a_0$).

surement of the reverse leakage current across a p-n junction, the ideality factor of the forward injection, and measurement of band discontinuities.

Nevertheless, critical thickness as defined either by theory or by experimental resolution does provide a frame of reference. Note that different research groups have used different substrate preparation, growth temperatures, growth rates, and measurement criteria to determine critical thickness. Figure 2.9 shows critical thickness versus Ge content. This plot shows the empirical curve of People and Bean [21] and their experimental data, EBIC-determined points of Kohama et al. [22], and interpreted data from bandgap measurements by King et al. [5]. Numbers represent the data from Iyer et al. [23] using molecular beam epitaxy (MBE) (1 and 2), chemical vapor deposition (CVD) (3), and limited reaction processing (LRP) (4). Results from thermodynamic equilibrium theory by Matthews–Blakeslee [19] and mechanical equilibrium theory by ver der Merwe [18] are also shown. For Si–Ge alloy, the onset of relaxation is gradual; film thicknesses must be greater than the critical thickness before significant relaxation occurs. In other words, if the thickness of epitaxial layers is kept below a critical thickness, the mismatch between the alloy and the Si substrate is accommodated elastically and no misfit dislocations form (pseudomorphic growth). The resulting SiGe pseudomorphic layers are considerably strained.

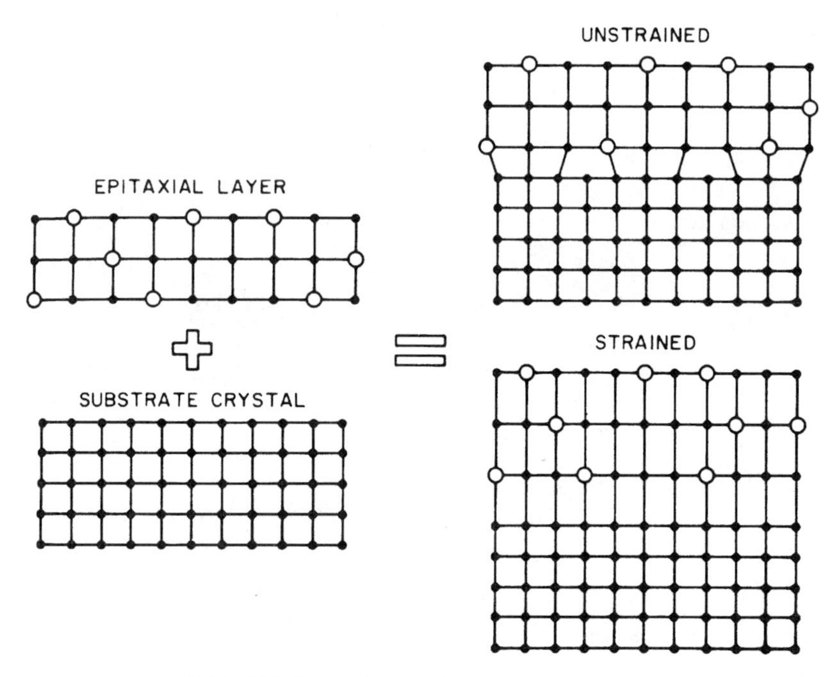

FIGURE 2.8 Strained and unstrained SiGe layer.

2.1.6 Electron Mobility

Drift mobility depends on the carrier scattering embodied in the phenomenological carrier relaxation time. To calculate the mobility, one needs to determine this relaxation time, including its functional dependence on energy. The average value of this relaxation time is then obtained by averaging over the energy states that the carrier occupies. The time constant for mobility can be calculated knowing what scattering process is dominant. If more than one is important, they all have to be evaluated and the resultant determined as a geometric mean. The time constant τ_μ is related to the inverse of the scattering rates as follows:

$$\frac{1}{\tau_\mu} = \frac{1}{\tau_1} + \frac{1}{\tau_2} + \cdots \tag{2.8}$$

The drift mobility is related to this time constant by

$$\mu_d = \frac{q\tau_\mu}{m^*} = \frac{q}{m^*}\frac{\langle v^2\tau\rangle}{\langle v^2\rangle} \tag{2.9}$$

If one associates $\mu_1, \mu_2, \ldots$ as the mobilities with the various scattering processes corresponding to $\tau_1, \tau_2, \ldots$, then

$$\frac{1}{\mu} = \frac{1}{\mu_1} + \frac{1}{\mu_2} + \cdots \tag{2.10}$$

Electron mobility in n-type lightly doped GaAs as a function of temperature together with contributions from the various sources of scattering [24] is shown in Fig. 2.10. At room temperature, ionized impurity scattering and polar phonon scattering dominate the scattering processes in doped samples. As the temperature is lowered,

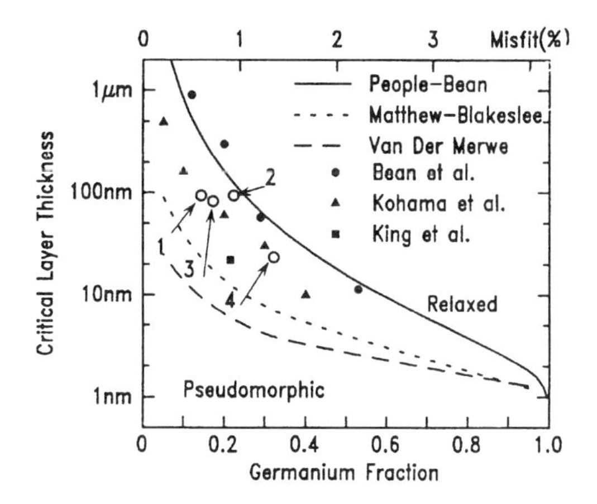

FIGURE 2.9 Critical thickness versus Ge content (after Iyer et al., Ref. 23 © IEEE).

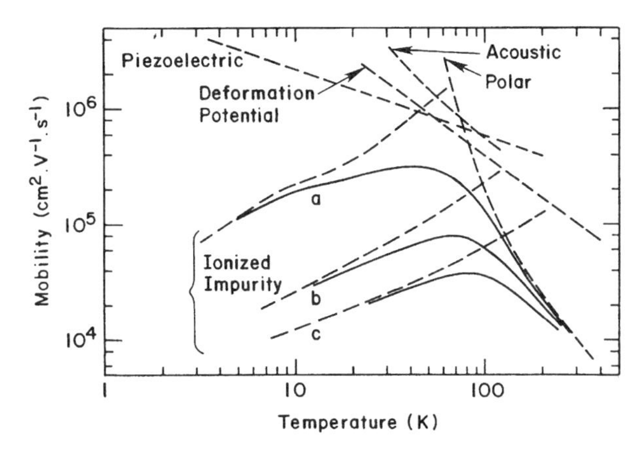

FIGURE 2.10 Electron mobility versus temperature in n-type GaAs: curve a, mid-10^3 cm^{-3} doping; curve b, $\sim 10^{15}$ cm^{-3} doping; curve c, mid -10^{15} cm^{-3} doping. (After Blakermore, Ref. 24 © American Institute of Physics).

phonon scattering becomes less efficient and the mobility is limited more by ionized impurity and deformation potential acoustic scattering. As temperature is decreased further, the ionized impurity scattering and phonon scattering are important over a broad temperature range. The carrier mobility as a function of impurity density in GaAs, for 300 K and 77 K, is shown in Fig. 2.11.

Until now, we have discussed only majority carriers. The electron, as a majority carrier, suffers coulombic scattering from the donors, other electrons, and any other residual impurities. Usually, the latter two are small. Only at high carrier densities do carrier–carrier interactions also become strong. The electron, as a minority carrier, suffers coulombic scattering from the acceptors, the holes, other electrons, and any other residual impurities. The last two, as in the case of the electron as a majority carrier, are usually small.

The behavior of the electron as a minority carrier depends on the semiconductor and on any residual field in the semiconductor device. It depends on the interaction between electron and hole. The scattering of acceptors by holes (i.e., reduction of coulombic perturbation by holes in the vicinity of the ionized acceptor) may allow for an increase in the carrier mobility. If they are considerably heavier than the electron, holes may cause scattering very similar to that due to ionized acceptors. A slight residual field in the structure causes motion of carriers in the opposite direction, which may result in a drag effect on each carrier by the other. The magnitude of the effect on each carrier is a function of their effective mass and the disparity in the effective mass. Scattering is a strong function of temperature; hence different considerations may apply to different temperatures.

Figure 2.12*a* and *b* show minority- and majority-carrier mobilities for Si and GaAs as a function of donor and acceptor density at 300 K, respectively. Silicon has a comparable mobility at low impurity concentration. As the heavy doping effect becomes important, the minority-carrier mobility is about twice the majority-carrier mobility. In GaAs the minority-carrier mobility is smaller than the majority-carrier

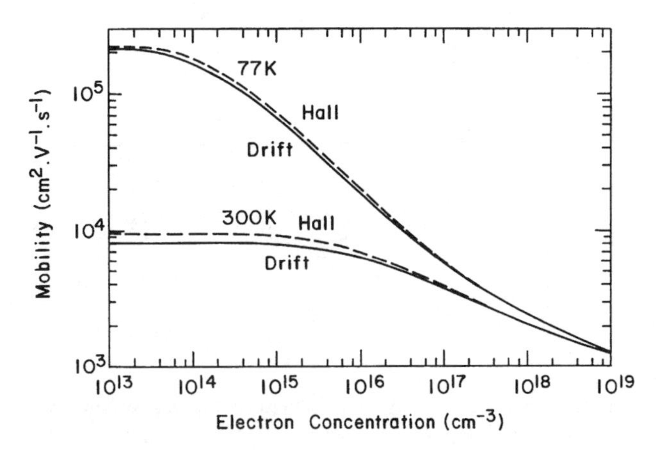

FIGURE 2.11 Electron drift mobility versus concentration.

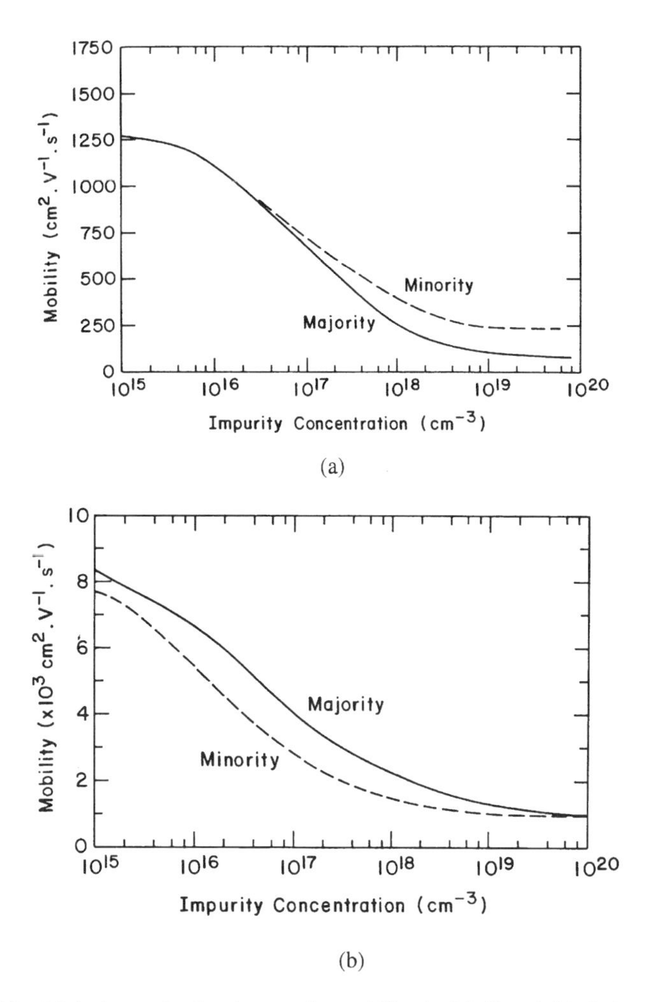

FIGURE 2.12 Majority- and minority- carrier mobility in (*a*) Si and (*b*) GaAs as a function of impurity concentration (after Tiwari and Wright, Ref. 25 © American Institute of Physics).

mobility over a wide range of impurity concentration [25]. In GaAs, electrons and heavy holes have widely differing masses. Heavy holes can be considered static for scattering purposes and lead to coulombic scattering of a magnitude similar to the ionized acceptor.

Ohmic minority and majority electron mobilities for unstrained and strained $Si_{1-x}Ge_x$ alloys up to 30% of Ge content were reported [26]. Figure 2.13 shows the low-field minority and majority drift mobilities for four different doping levels at 300 K. The solid lines in Fig. 2.13 represent the unstrained SiGe; the dashed lines present the mobility parallel to the Si/SiGe interface (μ_{xx}); the dot-dashed lines represent the mobility perpendicular to the Si/SiGe interface (μ_{zz}). As a general trend, the

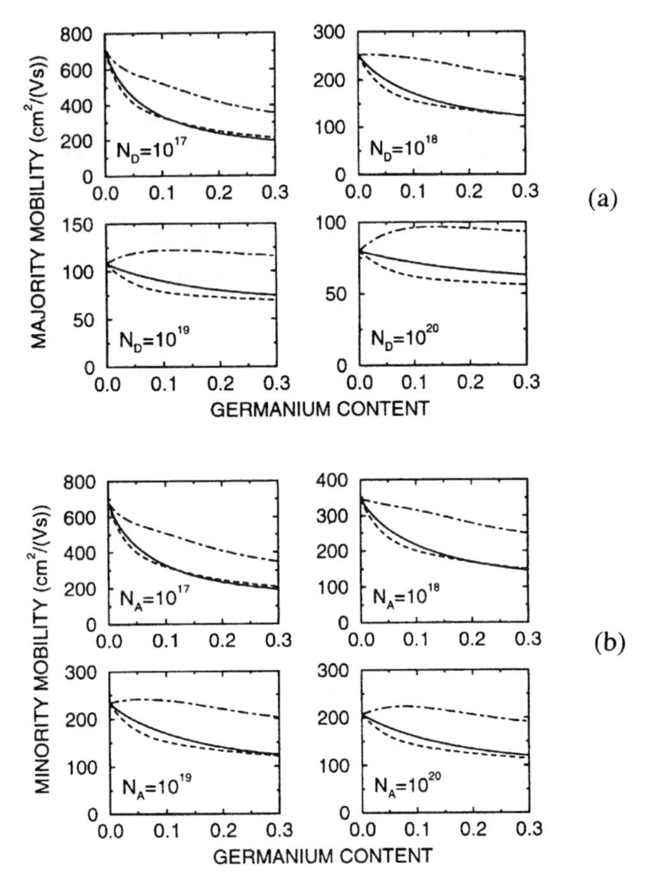

FIGURE 2.13 (*a*) Majority and (*b*) minority electron mobilities as a function of Ge content for four different impurity concentrations (after Bufler et al., Ref. 26 © IEEE).

mobility is reduced with increasing Ge content at low doping concentrations by alloy scattering, which dominates over impurity scattering in this regime. At higher doping, the relative fraction of alloy scattering in the total scattering rate is reduced. Together with the strain-induced reduction of the average effective mass in the z-direction, this increases μ_{zz} in strained SiGe. On the other hand, the reduction of μ_{xx} in strained SiGe and the mobility in unstrained SiGe is more or less equally strong.

2.1.7 Hole Drift Mobility

Because of mismatch between the pseudomorphically grown SiGe layer and the underlying silicon substrate, strained SiGe alloy changes carrier mobility due to the strain-induced energy shifts and distortions in the energy-band spectrum. In an n-p-n SiGe HBT, hole mobility determines the base resistance while electron mobility affects the emitter efficiency and unit gain frequency.

The effective drift mobility of holes in a strained MOS gate SiGe/Si heterostructure inversion layer shows a 50% enhancement in mobility over that of silicon at room temperature. Results calculated for the lattice drift mobility of strained SiGe also show an increase in the mobility relative to that of silicon [27,28]. Manku et al. [27] presented the calculation of the drift hole mobility of doped SiGe. The calculations take into account nonpolar optical scattering, acoustic scattering, alloy scattering, and ionized-impurity scattering. The hole mobility of SiGe at 300 K as a function of Ge content for four doping levels is shown in Fig. 2.14. In these plots the squares represent the mobility components parallel to the growth plane, the circles represent the mobility components transverse to the growth plane, and the triangles represent unstrained mobility. The two sets of plots represent two different alloy interaction potentials. The in-plane component is larger than the transverse component because of a larger effective mass in the in-plane component. Both in-plane and transverse components, however, are larger than the unstrained counterpart, which behaves in a manner that is very Si-like. The valence band compresses due to the in-plane compressive stress in the SiGe layer, which results in a decrease in the effective mass. Consequently, the mobility is increased relative to the unstrained counterpart.

The hole mobility calculated for SiGe (normalized to that of 0% Ge) as a function of Ge content for four different doping levels using an interaction potential of 0.23 eV is shown in Fig. 2.15. The enhancement in the mobility here is more pronounced than that using a higher interaction potential such as 0.27 eV. Furthermore, the mobilities increase almost linearly with Ge fraction. For a doping level of 10^{18} cm^{-3}, the result calculated agrees very well with the measurement of 211 cm^2/V $\cdot$ s (or 1.62 normalized) for a Ge fraction of 16.3% [29]. The enhancement in mobility can be attributed to the reduction in hole effective mass, as shown in the simple expression

$$\frac{\mu_{\text{SiGe}}(x)}{\mu_{\text{Si}}} \approx \frac{m^*_{\text{Si}}}{m^*_{\text{SiGe}}(x)} \tag{2.11}$$

The experimental evidence of the Ge effect on hole mobility is, however, not unequivocal. Matutinovic-Krstelj et al. [15] measured hole drift mobilities in a wide range of dopings and Ge concentrations. In their experimental data, no clear Ge dependence was found. The hole drift mobility as a function of the doping for different Ge contents is shown in Fig. 2.16. The drift mobility clearly decreases with increasing doping. For similar doping levels, a slight trend toward higher drift mobility with more Ge was seen. However, this trend is smaller than the error bars (25%). Thus no definite trend of hole mobility as a function of Ge content was found.

Carns et al. [28] obtained the apparent drift mobility of the compressively strained $Si_{1-x}Ge_x$ samples by measuring both the boron concentration using secondary ion mass spectrometry (SIMS) and the resistivity using van der Pauw's method. The apparent drift mobility, however, is increased by 50% over the Ge content range studied. The apparent drift mobility was determined from $\mu_d = (q\rho N_a)^{-1}$. μ_d is referred to as the *apparent drift mobility* because SIMS ensures the total boron concentration, which does not equal the free carrier concentration if some of the acceptors are not ionized. Without complete ionization, the doping concentration obtained using SIMS is greater than the free carrier concentration. The apparent drift mobility is thus lower

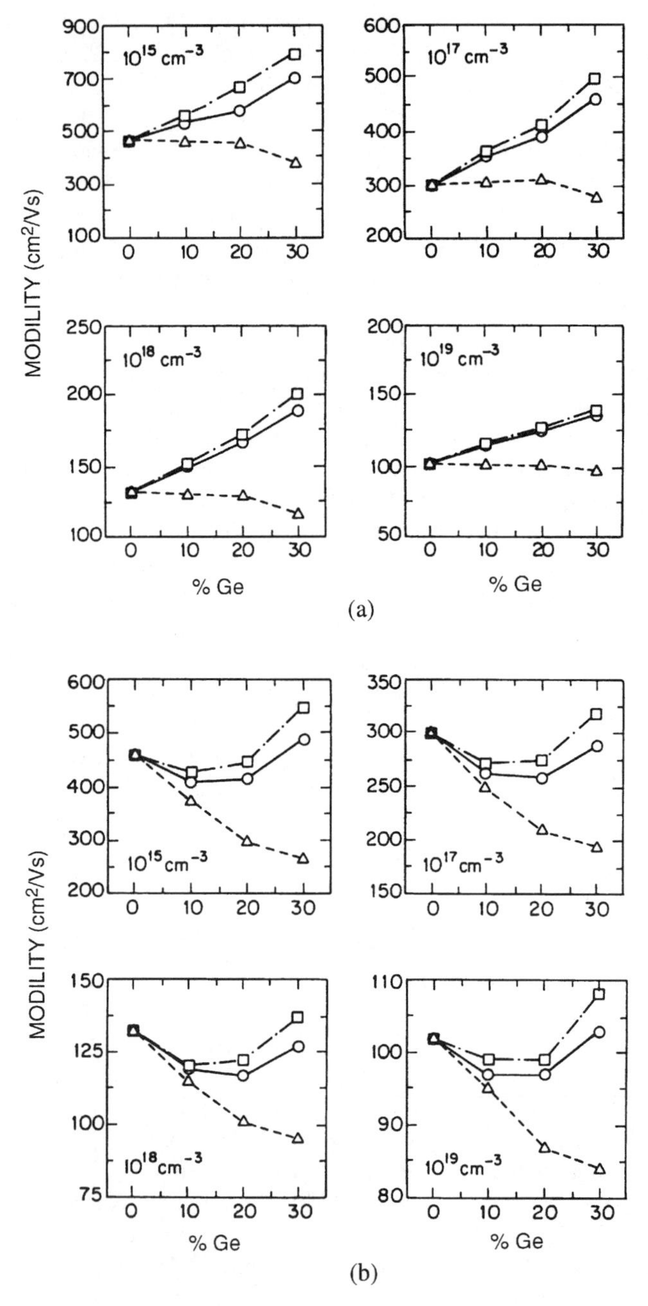

FIGURE 2.14 Hole mobility versus Ge content for (*a*) $U = 0.27$ eV and (*b*) 0.35 eV (after Manku et al., Ref. 27 © IEEE).

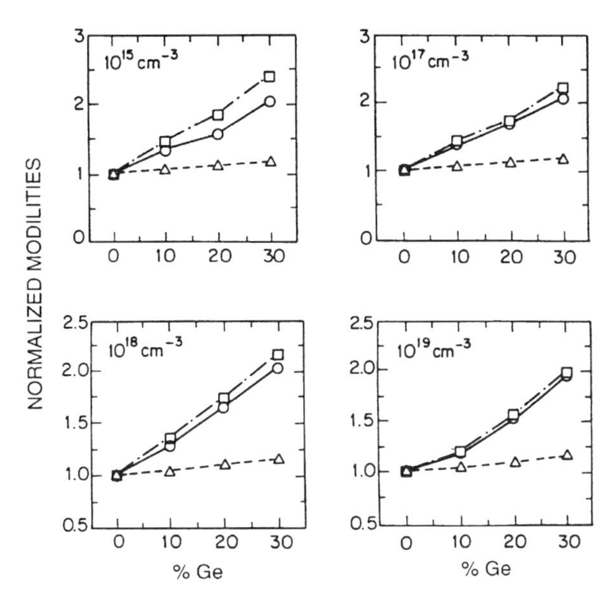

FIGURE 2.15 Normalized hole mobility versus Ge content for $U = 0.23$ eV (after Manku et al., Ref. 27 © IEEE).

than the actual drift mobility. Figure 2.17 shows the average apparent drift mobilities versus the Ge content. The error bars are used to reflect the total uncertainty of the measurements from finite contact sizes (±5%), contact resistance (±1%), SIMS (±10%), ionization (< 10%), Rutherford backscattering spectrometry (RBS) measurements of thickness (±5%) and Ge content (±5%), and changes in thickness due to surface and junction depletion (±1%). The results in Fig. 2.17 do provide an indication that increasing Ge content may enhance the drift mobility in $Si_{1-x}Ge_x$ compared to bulk Si. This is in agreement with the mobility measurements in stained SiGe and strained

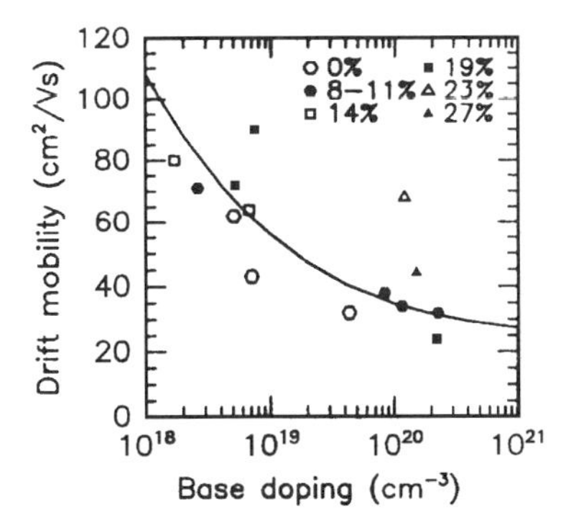

FIGURE 2.16 Drift mobility versus base doping for different Ge contents (after Matutinovic-Krystelj et al., Ref. 15 © IEEE).

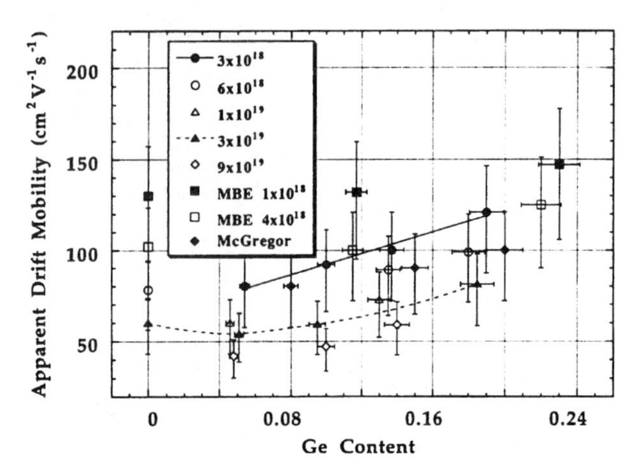

FIGURE 2.17 Average drift mobility versus Ge content (after Carns et al., Ref. 28 © IEEE).

Si channel MOSFETs published by Nayak et al. [30]. For device modeling purposes, the data for the $Si_{1-x}Ge_x$ HBT, shown in Fig. 2.17, can be fitted to the following empirical equation:

$$\mu_p = \mu_{p,\min} + \frac{\mu_{p0}}{1 + [N/(2.35 \times 10^{17})]^{0.9}} \tag{2.12}$$

where

$$\mu_{p,\min} = 44 - 20x + 850x^2$$

$$\mu_{p0} = 400 + 29x + 4737x^2$$

2.2 HETEROJUNCTION TECHNOLOGIES

Vapor-phase epitaxy and molecular beam epitaxy technologies are capable of growing epitaxial layers with high crystalline perfection and purity. Highly controlled doping levels up to 10^{19} cm^{-3} or more can be achieved, and highly controlled changes in the doping level are possible during growth with a minor adjustment in growth parameters. The doping may be changed either gradually or abruptly. Because of the relatively low growth temperatures, diffusion effects during growth are weak. With certain dopants, much more abrupt doping steps can be achieved than with any other technique. Most important in the context of heterostructures, it is possible in both techniques to change from one group III/V compound semiconductor to a different lattice-matched group III/V semiconductor with greater ease than with any of the other techniques.

Epitaxy is the regular oriented growth of a single crystal with controlled thickness and doping, over a similar single crystal or the substrate. Epitaxial techniques are used to realize junctions, ultrathin layers, and multiquantum wells with unprecedented control over purity, doping, and thickness profiles. There are many forms of epitaxy, but the most common ones are liquid-phase epitaxy, vapor-phase epitaxy, and molecular beam epitaxy. Liquid-phase epitaxy grows crystals of semiconductors from a liquid solution at temperatures well below their melting point. Since a mixture of the semiconductor with a second element may melt at a lower temperature than the semiconductor itself; it is often an advantage to grow the crystal from solution at the temperature of the mixture. Liquid-phase epitaxy was the first epitaxial process for the growth of compound semiconductors. In this technique, crystal growth on a parent substrate results from the precipitation of a crystalline phase from a saturated solution of the constituents. The crystallographic orientation of the grown layer is determined by the substrate. This technique of growth is intrinsically capable of growing very pure films but has the limitation of being unable to realize ultrathin layers. In vapor-phase epitaxy and molecular beam epitaxy techniques, the growth rates and hence the layer thicknesses can be controlled precisely. Because the growth rates themselves are low, extremely thin layers can be achieved, to the point that effects due to the finite quantum-mechanical wavelengths of the electrons can readily be generated. With both vapor-phase epitaxy and molecular beam epitaxy techniques, GaAs/AlGaAs structures with over 100 epitaxial layers have been constructed [31], and essentially arbitrary layers appear possible. With MOCVD, layer thicknesses below 50 Å have been achieved. With MBE, thicknesses below 10 Å have been reached. In either case, the capability far exceeds anything needed in the foreseeable future for transistorlike devices. Vapor-phase epitaxy and molecular beam epitaxy techniques are explained in more detail in the following sections.

2.2.1 Vapor-Phase Epitaxy

In this technique, epitaxial growth crystalline deposition of a semiconductor layer from a gaseous ambient results from a chemical reaction or decomposition. Vapor-phase epitaxy, also called chemical vapor decomposition (CVD), is one of the most widely used techniques of crystal growth. CVD refers to the formation of a condensed phase from a gaseous medium of different chemical composition. It is distinguished from physical vapor deposition processes such as sputtering and molecular beam epitaxy, where condensation occurs in the absence of a chemical change. In vapor-phase epitaxy, a mixture of gases streams through a reactor and interact on a heated substrate to grow an epitaxial layer. For example, epitaxial layers are grown on Si substrates by controlled deposition of Si atoms onto the surface from a chemical vapor containing Si. A gas of silicon tetrachlroide reacts with hydrogen gas to give Si and anhydrous HCl. If this reaction occurs at the surface of a heated crystal, the Si atoms released in the reaction can be deposited as an epitaxial layer. The HCl remains gaseous at the reaction temperature and does not disturb the growing crystal. Hydrogen gas is passed through a heated vessel in which $SiCl_4$ is evaporated. The two gases are then introduced into the reactor over the substrate crystal, along with other gases containing the doping impurities desired. The Si slice is placed on a graphite susceptor that can be heated to the reaction temperature with an RF heating coil. This method can be adapted to grow epitaxial layers of closely controlled impurity concentration on many Si slices simultaneously.

The metal-organic chemical vapor deposition (MOCVD) growth technique has emerged as a technologically important one for the production of single layers, heterojunctions, and quantum well structures, with excellent control over layer thickness and doping. In this technique organmetallic compounds such as metal alklys and hydrids as vapors are transported to the growth chamber using a carrier gas such as hydrogen. A substrate wafer is mounted on a graphic holder or susceptor that is held at the growth temperature by RF heating. A schematic of a MOCVD growth system [32] is shown in Fig. 2.18. For the growth of GaAs, the basic reaction is

$$Ga(CH_3)_3(l) + AsH_3(g) + H_2(g) \rightarrow GaAs(s) + 3CH_4(g) + H_2(g) \qquad (2.13)$$

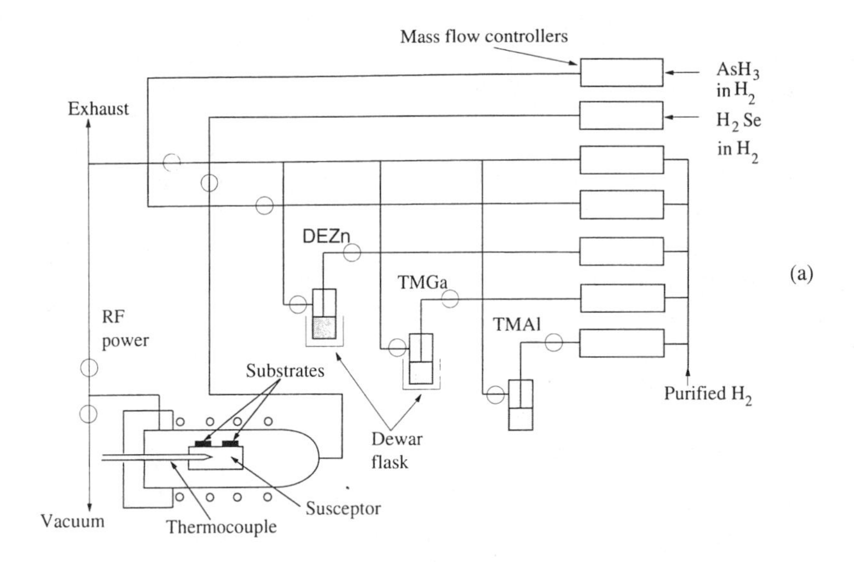

FIGURE 2.18 MOCVD (*a*) schematic and (*b*) facility (after Bhattacharya, Ref. 32, courtesy of E. Woelk, Aixtron, Germany).

Since there is no source material with which equilibrium has to be reached, compounds such as GaAs can be grown with different As/Ga ratios simply by varying the relative vapor pressures of AsH_3 and trimethyl gallium. All the reactants and dopants can be in the vapor phase and only the substrate is heated. The equipment is simple, and shorter turnaround times are possible. Proper design of the gas flow system, including precision mass flow controllers and evacuation systems for residual in-line gases, makes it possible to grow ultrathin layers and abrupt junctions with precise doping control. As source purification quality continues to increase, compounds of very high purity can be realized by the MOCVD process. The process has also demonstrated great potential for growth on large-area substrates, selective-area growth, and patterned growth. This shows great promise as a manufacturing process.

2.2.2 Molecular Beam Epitaxy

In contrast with vapor-phase epitaxy growth of semiconductor crystals under quasiequilibrium conditions, growth by molecular beam epitaxy is accomplished under nonequilibrium conditions and is governed principally by surface kinetic processes. In molecular beam epitaxy, the substrate is held in a high vacuum while molecular or atomic beams of the constituents impinge on its surface. In other words, it is a controlled thermal evaporation process under ultrahigh vacuum conditions. A schematic of the molecular beam epitaxy process [33] is shown in Fig. 2.19. In the growth of AlGaAs layers on GaAs substrates, the Al, Ga, and As components, along with dopants, are heated in separated cylindrical cells. Collimated beams of these constituents escape into the vacuum and are directed onto the surface of the substrate. The flux density of a beam incident on the substrate surface is controlled by the temperature of the effusion cell provided that the cell aperture is smaller than the mean free path of the vaporized effusing species within the cell. The individual cells are provided with externally controlled mechanical shutters whose movement times are less than the time taken to grow a monolayer. In this growth procedure, the sample is held at a relatively low temperature (about 600°C for GaAs). Abrupt changes in doping can be obtained by controlling shutters in front of the individual beam. Using very slow growth rates ($\leq 1\mu$m/h), it is possible to control the shutters to make composition changes on the scale of the lattice constant. Interfaces that are one-monolayer abrupt can be obtained fairly easily.

In the context of molecular beam epitaxy, each semiconductor has its own optimized growth temperature. For example, comparison of the congruent temperature, melting point, and heat of formation of the binaries reveals that the ratio of the magnitudes of these parameters is approximately 0.80 : 1.00 : 1.25 for InAs, GaAs, and AlAs, which are expected to be the ratio of their relative bond strengths.

2.2.3 Gas-Source MBE and Metal-Organic MBE

The use of group V hydrides such as AsH_3 and PH_3 in conjunction with elemental sources of conventional MBE has come to be known as gas-source MBE (GSMBE). Because of the difficulties associated with the use of solid phosphorus as a source for

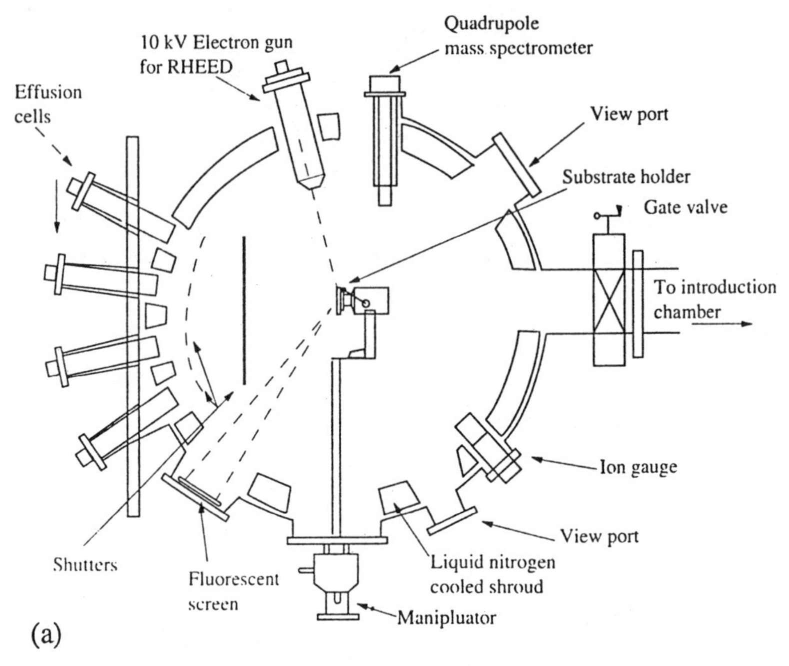

FIGURE 2.19 MBE (*a*) schematic and (*b*) facility (after Bhattacharya, Ref. 32, courtesy of Riber SA).

MBE, GSMBE is at present the preferred method for growth of InP from elemental In [34]. This method allows the advantages of MBE to be retained, while the capability of growing P-containing materials is added. In a further development, the remaining elemental sources were also replaced with gaseous precursors, primarily metal-organic compounds, in an attempt to combine the advantages of MOCVD with those of MBE. This technique is commonly referred to as metal-organic MBE (MOMBE). Very high purity InP and InGaAs have been demonstrated by MOMBE.

For both GSMBE and MOMBE, the growth apparatus shown in Fig. 2.20 is based on conventional MBE equipment. This system consists of a UHV growth chamber containing liquid-nitrogen-cooled cryopanels [2] around the substrate heater assembly and the source flange. This chamber is usually isolated from the load lock by a buffer chamber to minimize contamination of the growth environment and to allow for outgassing of wafers prior to growth. In conventional MBE, these chambers are

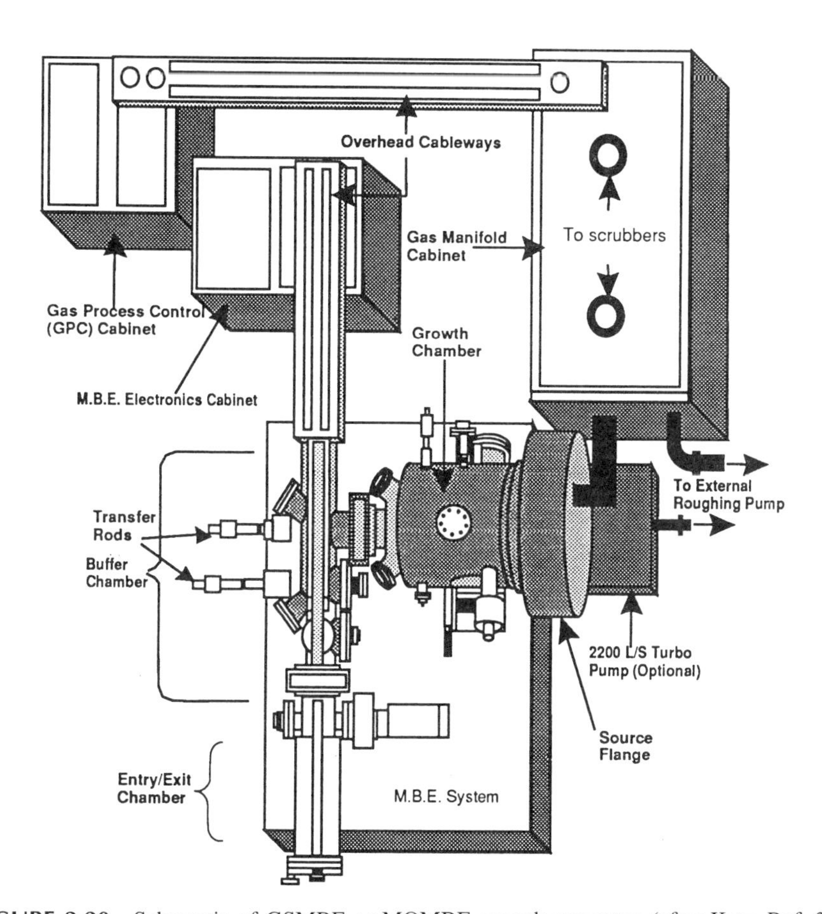

FIGURE 2.20 Schematic of GSMBE or MOMBE growth apparatus (after Katz, Ref. 34, courtesy of Intevac).

pumped with ion pumps or cryopumps. In GSMBE, large quantities of hydrogen are generated from decomposition of the hydrides; therefore, ion pumps are not generally used on the growth chamber.

As in conventional MBE, effusion cells that contain elemental charges are used to supply the various group III and dopant sources. These cells, shown in Fig. 2.21, consist of pryolytic boron nitride or carbon crucibles housed inside a restively heated cylinder. Thermocouples placed near the base of the crucible are used to monitor the temperature of the cell. The flux of the group III source is related directly to its vapor pressure. Therefore, the flux may be varied by altering the temperature of the cell. To obtain abrupt interfaces between layers, it is necessary to switch rapidly from one group V source to another during growth. This is usually accomplished by establishing the hydride flow into a vent line before introduction to the chamber. This vent line may be evacuated by any pump suitable for use with corrosive gases. Switching of the gas is accomplished through simultaneous opening of the valve on the growth chamber and closing of the valve on the vent line.

As in MBE, the incorporation kinetics of group III sources in GSMBE are roughly independent of one another and of the group V flux. The group III incorporation rate is therefore only a function of the group III flux and the sticking coefficients of the various group III elements. Below about 540°C, the sticking coefficient of In is essentially unity, and the In incorporation rate is independent of temperature. Above this temperature, the growth rate becomes strongly dependent on temperature

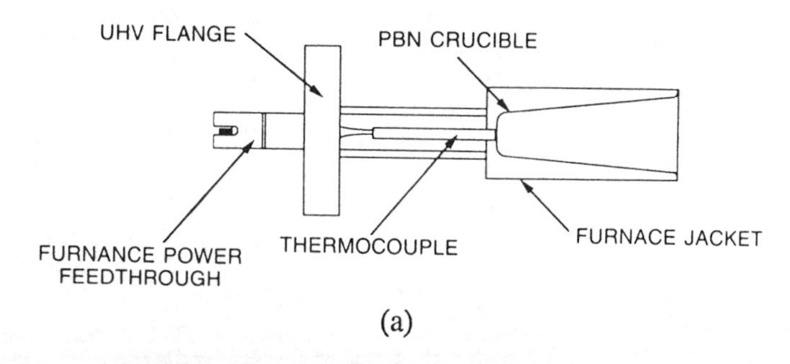

(a)

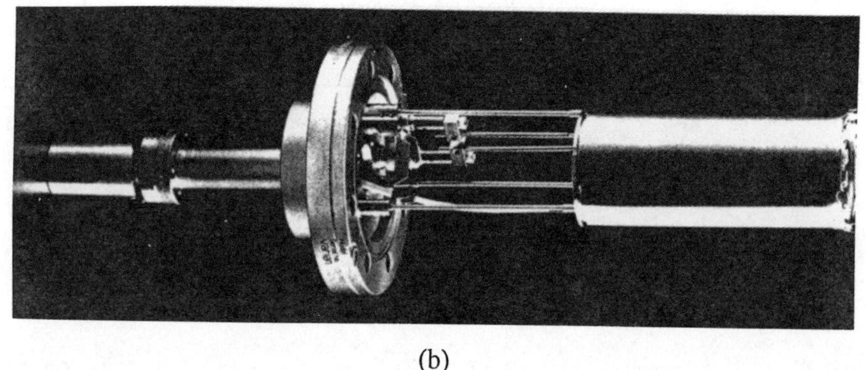

(b)

FIGURE 2.21 Group III effusion cell (after Katz, Ref. 34, courtesy of Intevac).

because of In reevaporation. Ga evaporation does not occur until about 710°C and is also unity below this temperature. Unlike the group III species, group V sources show an incorporation efficiency strongly influenced by the group III surface convergence. Generally, group V species are incorporated in a one-to-one ratio with group III elements, with the excess group V material evaporating from the surface. For this reason, growth in GSMBE is usually conducted with the minimum group V flux needed either to maintain a stable surface or to match the group III incorporation rate.

For materials containing only one group V element, the incorporation behavior is straightforward. However, for materials that contain both As and P (i.e., $In_xGa_{1-x}As_{1-y}P_y$), the incorporation of the group V species is a little more complicated. Figure 2.22 shows variation of phosphorus content, y, in $In_xGa_{1-x}As_{1-y}P_y$ versus PH_3 fraction of a hydride beam. For alloys rich in As, $y < 0.2$, the relative amounts of As and P incorporated in the grown layer are equivalent to the As/P ratio in the beam, regardless of growth temperature. For higher P compositions, the As/P ratio in the grown layer is actually higher than the As/P ratio in the beam. This disparity becomes even more pronounced for growth temperatures below 528°C. This effect has been attributed to a greater surface residence time for As than for P and to the need for thermal decomposition of P_x species on the surface, where x may be 2 or higher.

Elemental source dopants behave similarly in GSMBE and MOMBE. The most common n-type dopants are Si and Sn, due to their small activation energies and to their high sticking coefficients. Si is particularly attractive because it shows almost no tendency to diffuse or segregate at normal growth temperatures. The most common p-type dopant is Be because of its near-unity sticking coefficient. However, dramatic redistribution of Be during growth occurs for concentrations above 2×10^{18} cm^{-3}, even for growth temperatures as low as about 450°C. The cause of this dramatic

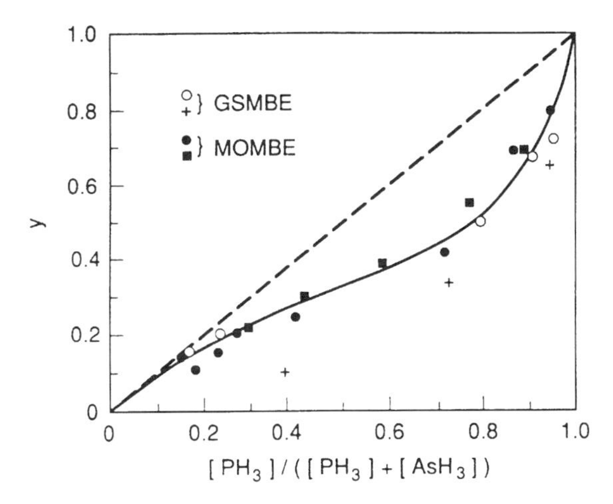

FIGURE 2.22 Phosphorus content versus PH_3 fraction of hydride beam (after Benchoimol et al., Ref. 35).

change in diffusion behavior with Be concentration is believed to be the formation of Be interstitials [35]. As the Be concentration increases, the growing layer becomes saturated with substitutional Be acceptors, leading to both rejection of Be at the growth front and to incorporation of Be interstitials. The Be interstitials diffuse away from the doped layer toward the substrate until the Be interstitial population reaches equilibrium with substitutional Be. This causes a broadening of the profile in the direction of the substrate. In additional shallow dopants, impurities that produce deep levels can also be introduced via elemental sources. In InP, Fe behaves as a deep acceptor and can be used to produce semi-insulating buffer or blocking layers.

2.3 DEVICE FABRICATION

2.3.1 SiGe HBTs

The evolution of SiGe technology has been rapid. It has gone from a mere laboratory curiosity about a decade ago to a commercial reality today. Currently, aggressively designed SiGe transistors have cutoff frequencies in excess of 100 GHz, an achievement that puts them in the same league with III/V compound semiconductor devices and ahead of the best silicon transistors by a factor of more than 2. The SiGe HBT utilizes only column IV, silicon and germanium, in the heterostructure. The processing of these semiconductors is compatible with existing silicon technology. Recent work suggests that SiGe could offer low-cost, high-speed RF and microwave integrated circuits (ICs) to the booming wireless communications market [36,37].

A Si–Ge layer may be grown by a variety of low-temperature growth techniques, including MBE and CVD. Bean et al. [38] investigated the growth-temperature range and concluded that good-quality layer-by-layer growth occurs in the neighborhood of 550°C for moderate Ge concentrations (up to 15%). At lower Ge concentrations ($< 10\%$), optimal growth temperatures can be somewhat higher, increasing with decreasing Ge content.

A heterojunction bipolar transistor was first demonstrated in the Si/SiGe system using MBE [39–41]. Because the entire device structure can be deposited using this technique, exposure to high-temperature processing steps can be avoided. In [39] the collector, base, and emitter layers were deposited epitaxially without breaking vacuum. The Ge concentration in these devices was uniform across the 100-nm base and graded in the depletion regions, as shown in Fig. 2.23*a*, with values in the base ranging from 0 to 12%. Such conditions are well below the critical thickness limits. To prevent relaxation of the SiGe layer, only low-temperature processes ($< 600°C$) were used. A schematic view of the mesa-defined SiGe HBT is shown in Fig. 2.23*b*. Dry etching to define the mesa structure, low-energy gallium implantation to improve the base contact resistance, and a low-temperature oxide deposition for junction passivation were also used.

King et al. [42] and Gibbons et al. [43] used the CVD technique known as limited reaction processing (LRP) to obtained ideal junction characteristics for SiGe/Si heterostructures. The LPR technique relies on rapid changes in substrate temperature

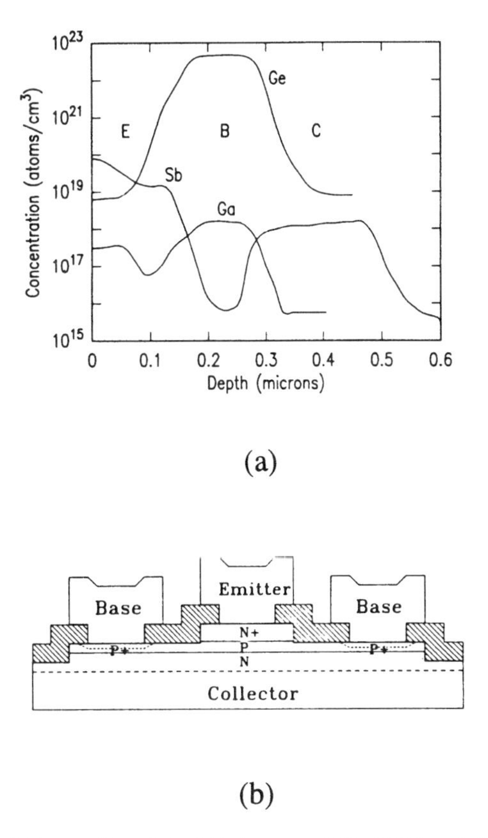

FIGURE 2.23 (*a*) SIMS profile of the MBE-grown SiGe HBT and (*b*) schematic view of the mesa-defined transistor structure (after Patton et al., Ref. 40 © IEEE).

to achieve abrupt doping and composition profiles. Epitaxial growth of the collector, base, and emitter layers took place within a quartz reactor without removing the wafer between individual layer depositions. The starting substrates were Si(100) doped with Sb to a resistivity between 0.007 and 0.02Ω·cm. Following a 1200°C, 30-s H_2 cleaning, a 2-μm-thick arsenic-doped collector was grown at 1000°C, doped to a level of 3×10^{16} cm^{-3}. The $Si_{0.69}Ge_{0.31}$ base was grown at 640°C using GeH_4, $SiCl_2H_2$, and B_2H_6 to obtain a thin layer doped to 7×10^{18} cm^{-3}. The thickness of the SiGe base layer is about 20 nm determined from cross-section transmission electron microscope (XTEM) and from grazing exit-angle Rutherford backscattering spectrometry. The emitter was then grown at 850°C for 5.5 min to a thickness of 0.41 μm with an arsenic doping level of roughly 1.4×10^{17} cm^{-3}. Layer thicknesses and doping levels were determined by spreading resistance profiling (SRP), RBS, XTEM, and SIMS. Mesa etching the Si and $Si_{1-x}Ge_x$ alloy layers with an SF_6 and CF_3Br plasma isolated individual devices. SiO_2 deposited at 400°C provided passivation for the sidewalls of the structure as well as a mask for subsequent base and emitter contact ion implants.

The base contact implant consisted of four successive implants of boron and BF_2. The boron concentration arising from these implants is about 5×10^{19} cm^{-3} from the surface to a point 0.45 μm below the surface. The emitter contact implant was made with a single arsenic implant of energy 30 keV and dose 1×10^{15} cm^{-2}. The implants were activated with a rapid thermal anneal at 850°C for 10 s in an argon ambient. Ohmic metal contacts to the devices were made by sputtering 200-nm Ti and $1 - \mu$m Al/1% Si, followed by forming a gas anneal. The device structure is shown in the inset. The SIMS plot of the doping, germanium, and oxygen profiles is shown in Fig. 2.24.

In [42,44], Si and graded-SiGe-base bipolar transistors were fabricated in a standard polyemitter bipolar process using the low-temperature (550°C) epitaxial silicon deposition process. In this technique wafers are HF passivated and then loaded into the loadlock of the ultrahigh-vacuum chemical vapor deposition (UHV/CVD) apparatus. After pumpdown below 10^{-6} torr, wafers are transferred under flowing hydrogen into the UHV section of the apparatus, and growth is begun immediately. The gaseous sources employed are silane, germane, diborane, and phosphine. Film growth rates may be varied from 0.01 to 10 nm/min as a function of temperature and file germanium content, with typical rates of 0.4 to 4 nm/min being employed. These limits are used to ensure precise dimensional control, on the order of one to two atomic layers in this instance. This level of precision is required if one is to compete effectively with the control of ion implantation, the benchmark for doping control in the processing of silicon. Furthermore, it has been found that use of compositionally graded germanium profiles within the base of these bipolar devices enables one to obtain high

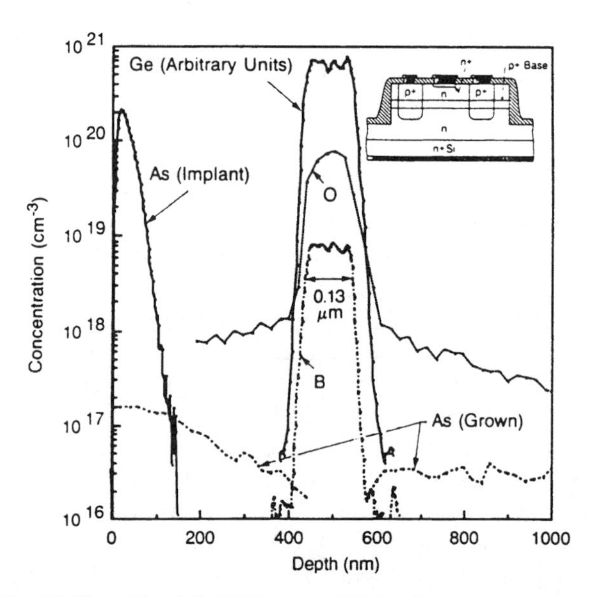

FIGURE 2.24 SIMS profile of the LPR-grown HBT (after King et al., Ref. 5 © IEEE).

levels of device performance for relatively low average germanium content. Figure 2.25*a* shows the SIMS impurity profile for the trapezoid Ge HBT developed at IBM [44,45]. A schematic device cross section of the self-aligned epitaxial-base bipolar transistor is displayed in Fig. 2.25*b*.

For a growth technique to be integrable into a technology, it must simultaneously meet all demands of the technology. To be feasible, a growth technique must employ a low thermal budget; handle patterned wafers well; and be reproducible, uniform, reliable, commercially available, and have extremely tight control of both dopant and SiGe alloy film growth. The SiGe epitaxial-base transistor technology at IBM has made the transition from the research phase to commercialization. The enabling technological enhancement has been the ability to grow device-quality films for wireless RF circuits at 5-GHz operation.

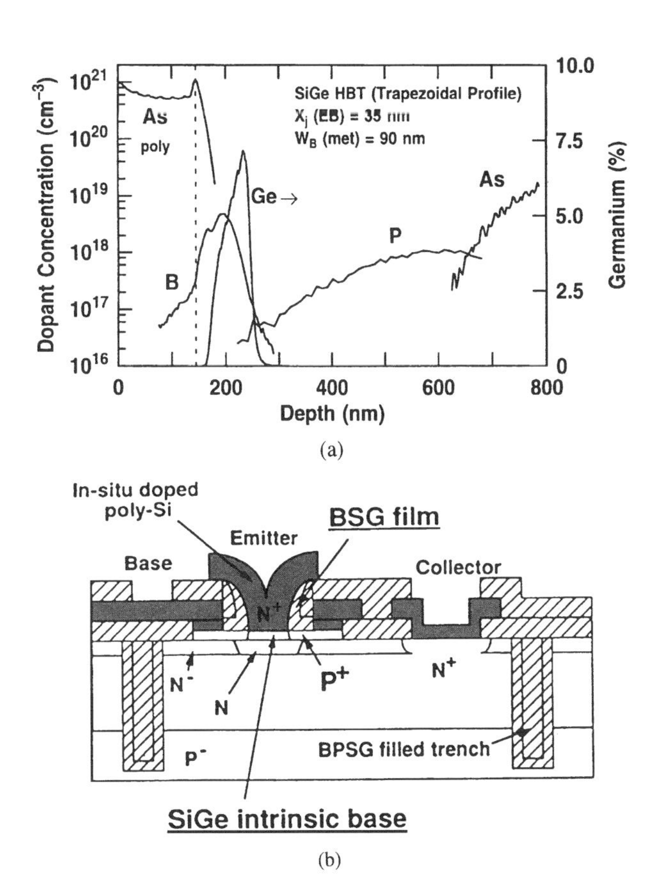

FIGURE 2.25 (*a*) SIMS profile of the epitaxial-base SiGe HBT (after Harame et al., Ref. 44) and (*b*) cross section of super self-aligned selective grown SiGe base device (after Harame et al., Ref 45 © IEEE).

2.3.2 AlGaAs/GaAs HBTs

The standard processing steps applied to a segment of the device consist of an emitter, two base contacts, and two collector contacts. The fabrication sequence basically consists of etching to reveal layers in the structure and fabricating electrical contacts to each layer. Devices are isolated and interconnections are made within each device as well as between devices. Figure 2.26 shows a typical HBT device processing sequence after each epitaxial layer has been grown on the semi-insulating substrate. Key process steps of a non-self-aligned HBT are highlighted as follows.

- *Epitaxial layer growth.* The subcollector, collector, base, and emitter layers are grown on the semi-insulating GaAs substrate using either MBE or MOCVD. A

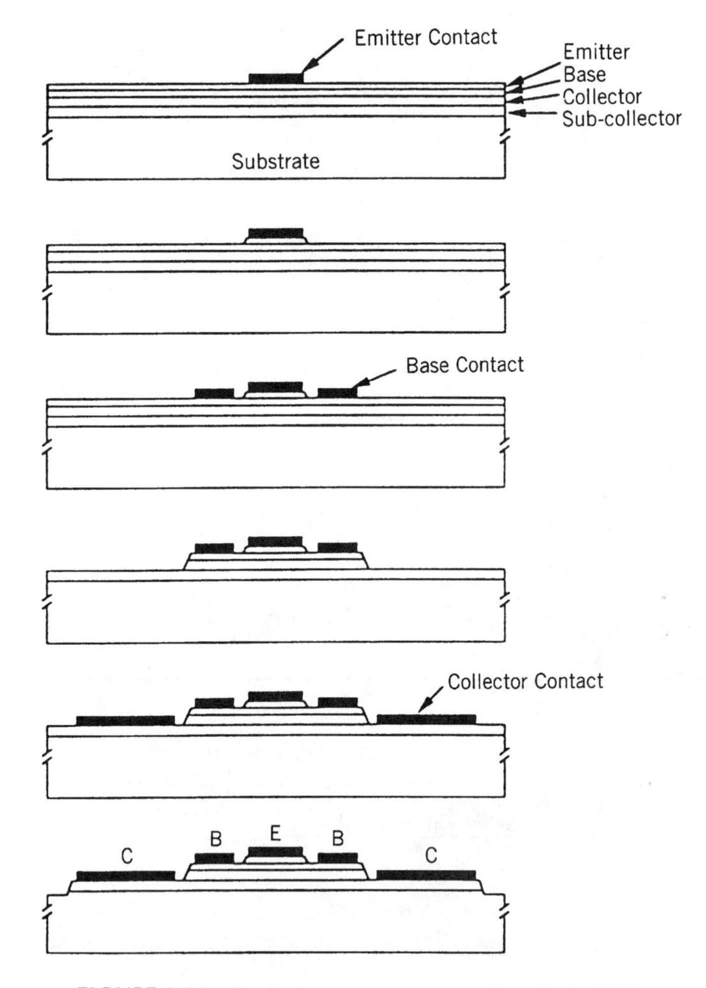

FIGURE 2.26 Typical process sequence of the GaAs HBT.

complex combination of growth temperature, arsenic/gallium ratio, base doping, base thickness, emitter doping, emitter thickness, and so on, are optimized for a given MBE processing. A key trade-off is the crystalline quality for p-type dopant diffusion. A higher growth temperature leads to better crystalline quality and low recombination currents but promotes p-type impurity dopant outdiffusion from the base to the emitter, especially if Be is incorporated interstitially. To optimize the crystalline quality and to reduce the out-diffusion effect, a two-step growth temperature technique for the base (lower temperature) and emitter (higher temperature) is used. Another solution is to use carbon instead of Be because carbon is less susceptible to diffusion for the p-type base dopant.

- *Emitter contact.* Ohmic contacts are made to the emitter contact layer by a lift-off technique. Typically, ohmic contacts consist of AuGe/Ni/Au layers sequentially evaporated and then alloyed to form an ohmic contact.
- *Base etch.* Emitter contact is masked with photoresist and the emitter layer is removed by wet-chemical or dry etching such as reactive ion etching to reveal the base layer.
- *Base contacts.* Contacts are made to the base layer by a lift-off technique. Usually, two base contacts are fabricated. The base contacts are placed as close to the emitter finger as possible to minimize the base resistance. AuBe, AuZn, or AuMg alloys are the most common ohmic contact metals to p-type GaAs for the fabrication of small-signal devices. Because the base layer of microwave HBTs is very thin ($\leq 0.1\ \mu$m), alloyed ohmic contacts can cause spikes through the base layer for a short-circuit collector–base junction. Nonalloyed contacts such as Ti/Pt/Au can be used to prevent this short circuiting at the expense of increased base-contact resistance.
- *Collector etch.* Similar to those applied for base etching, collector etching is accomplished by wet or dry etches. After collector etching, the wafer surface is brought down to the subcollector level, except in emitter and base contact areas.
- *Collector contacts.* Contacts are made to the subcollector layer by using lift-off techniques and metallization schemes similar to those employed for the emitter. At this point, the processing of individual sections of the device is completed.
- *Isolation.* Devices are isolated from one another by removing the subcollector layer between them. Note that this isolation is not applied to every device segment because each is connected to another by the subcollector layer. Ion-implantation damage can also be used instead of etching for device isolation.
- *Interconnection.* All emitter, base, and collector contacts of the device are connected together. Depending on the layout, this connection can be accomplished simply by making individual contacts to a common contact pad on the semi-insulating substrate surface. Due to the vertical nature of the HBT, the use of dielectric crossover or air-bridge connections is necessary to prevent shorting of the device around the device perimeter. Individual devices can also be

connected to each other as required by the circuit. Monolithic circuits need additional steps to fabricate passive circuit components and back-side wafer processing.

To reduce base resistance and improve device performance, it is desirable to place the emitter and base contacts as close to each other as possible. This requires device self-alignment. In a self-aligned fabrication technique, the base contact is produced in such a way that its separation from the emitter is ensured by a controlled amount because of the emitter geometry. The same technique applied to the emitter–base contact can also be used for fabrication of collector contacts self-aligned to the base. Such devices are said to be *double self-aligned.*

A self-aligned AlGaAs/GaAs HBT with InGaAs emitter cap layer has been fabricated [46]. The layers were grown by MBE on a semi-insulating GaAs substrate. The emitter cap layer consisted of three layers: an n^+-GaAs layer, an n^+-$In_xGa_{1-x}As$ ($x = 0$ to 0.5) compositionally graded layer, and an n^+-$In_{0.5}Ga_{0.5}As$ layer. The graded layer was inserted to smooth out conduction-band discontinuity. Regarding ohmic contacts to n^+-$In_xGa_{1-x}As$ layers, it has been shown that the specific contact resistance decreases as the InAs mole fraction increases. The device has a top InGaAs layer with an InAs fraction of $x = 0.5$ and doped with Si to 2×10^{19} cm^{-3}. The graded p^+-$Al_yGa_{1-y}As$ ($y = 0.12$ to 0) layer was utilized as a base layer to reduce emitter size effect and to shorten base transit time. This thin base layer (1000 Å) was heavily doped with Be (4×10^{19} cm^{-3}) to reduce base resistance. The collector layer consists of two n-GaAs layers with doping concentrations of 5×10^{16} cm^{-3}, 2000 Å and 2×10^{17} cm^{-3}, 3000 Å.

The emitter and base electrodes were formed simultaneously as shown in Fig. 2.27. After proton implantation and emitter mesa etching by reactive ion beam etching (RIBE) with BCl_3, a SiO_2 sidewall was formed as in conventional HBTs. Then the Si_3N_4 emitter mesa mask was removed by reactive ion etching (RIE) with SF_6, and Ti/Pt/Au was deposited onto the entire surface. Finally, the metal on the SiO_2 sidewall was removed by angled Ar ion milling. The collector mesa was formed by sequential etchings of the base electrode, base layer, and collector layer. Then the collector electrode was formed using an AuGe/Ni/Pt/Au-alloyed ohmic contact on an n^+-GaAs buffer layer. Alloying for AuGe/Ni/Ti/Pt/Au and sintering for Ti/Pt/Au were carried out simultaneously at 350°C for 30 s in an ambient of N_2.

Another HBT process that represents an attractive compromise between performance and producibility is a partially self-aligned base ohmic metal (SABM) HBT device developed at TRW [47]. Figure 2.28 shows the scanning electron micrograph of a cleaved cross section of a 3×10 μm^2 emitter SABM transistor after ohmic metal lift-off and the fabricated HBT. For the HBT, the self-aligned base ohmic metal essentially eliminates the parasitic external base resistance by minimizing the ohmic metal to emitter spacing. A double-photoresist lift-off technique is used for selectively patterning the HBT base ohmic metal to within 0.2 μm of the emitter edge. The active device layers are accessed by a combination of selective and nonselective wet chemical etches. In this technique a wet selective etch is used to delineate the emitter mesa. A thin, depleted, residual layer of AlGaAs associated with the selective etching

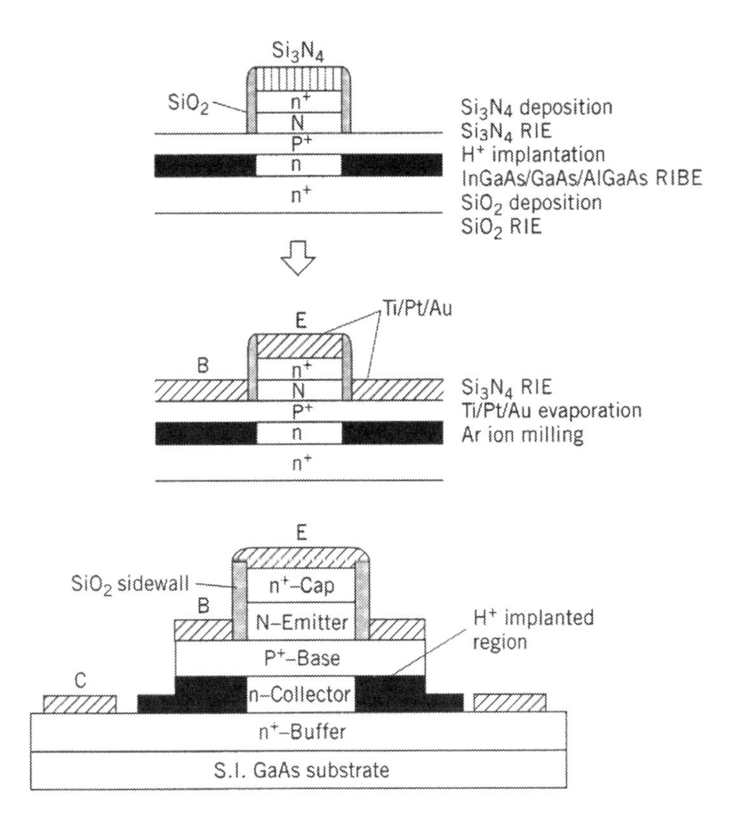

FIGURE 2.27 GaAs HBT schematic cross section (after Nagata et al., Ref. 46 © IEEE).

technique provides a wide bandgap passivation effect to reduce the external emitter–base recombination. Ohmic contacts are formed by AuBe/Pd/Au and AuGe/Ni/Ti/Au for p- and n-type ohmic contacts, respectively. A multiple boron damage implant is used for device isolation. Plasma-enhanced chemical vapor deposition of silicon nitride is used to protect the GaAs surface and serves as a dielectric insulator for metal–insulator–metal (MIM) capacitors and double-level metal interconnection.

Figure 2.29 shows a representative AlGaAs/GaAs HBT layer structure and doping profile. Emitter layers consist of $Al_xGa_{1-x}As$, with AlAs mole fraction x chosen to be on the order of 0.3. Base layers are typically made with thicknesses of 0.05 to 0.1μm, with values of doping from 5×10^{18} to 10^{20} cm^{-3}. The emitter doping concentration can be much lower than that of the base. This arrangement allows the designer to achieve a very low base resistance while maintaining a relatively low emitter–base junction capacitance. The thin base layer allows one to achieve a low base transit time. Collector layers are made with thicknesses on the order of 0.5 μm, with values of doping from 2×10^{16} to 5×10^{17} cm^{-3}. By selecting an appropriate collector doping, the collector depletion region transit time can be optimized, consistent with the need to support a high collector current density, a low collector–base

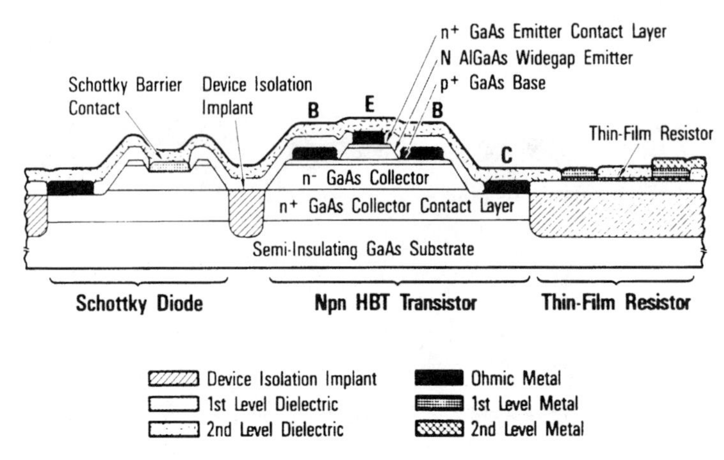

FIGURE 2.28 Schematic cross section of self-aligned HBT.

junction capacitance, and the desired level of collector breakdown voltage. With the advancement in III/V epitaxial layer growth and HBT processing technologies, the promises of physics-based device concepts actually come to fruition. By carrying out a systematic experimental investigation of the properties of AlGaAs/GaAs HBTs, Tiwari et al. [48] demonstrated that by using the existing HBT technology it is possible to achieve a high base layer doping, low ohmic contact resistances for the emitter and the base, near-unity emitter injection efficiency, and low surface recombination and bulk generation–recombination effects, particularly at high current densities ($\approx 10^5$ A/cm). Reduction of base layer thickness and collector depletion layer design optimization with electron velocity overshoot effect have led to an AlGaAs/GaAs HBT cutoff frequency between 80 and 150 GHz.

Typical emitter-up AlGaAs/GaAs HBTs are fabricated with mesa etching techniques that create exposed extrinsic base surface and emitter sidewalls, as indicated by the bold lines in Fig. 2.30*a*. Because free GaAs and AlGaAs surfaces are characterized by a high surface recombination velocity, recombination current taking place at these surfaces can be a major component to the overall base current, particularly if the emitter area is small. To increase the current gain and to reduce low-frequency noise, passivation to these areas by a thin depleted AlGaAs layer (ledge), as shown in Fig. 2.30*b*, is often used [49]. The thickness of the ledge layer is the same as that of the AlGaAs emitter, but the passivation ledge distance (between the emitter and base contacts) should be long enough (e.g., 3000 Å) to suppress the base contact recombination.

2.3.3 InP HBTs

The earliest problems faced in the growth of GaInAs and AlInAs by MBE were the achievement of lattice matching and the stabilization of the InP substrate before initiation of growth. To achieve unstained materials the lattice constant must be controlled precisely. Under normal growth conditions the arrival rate of the group III elements at

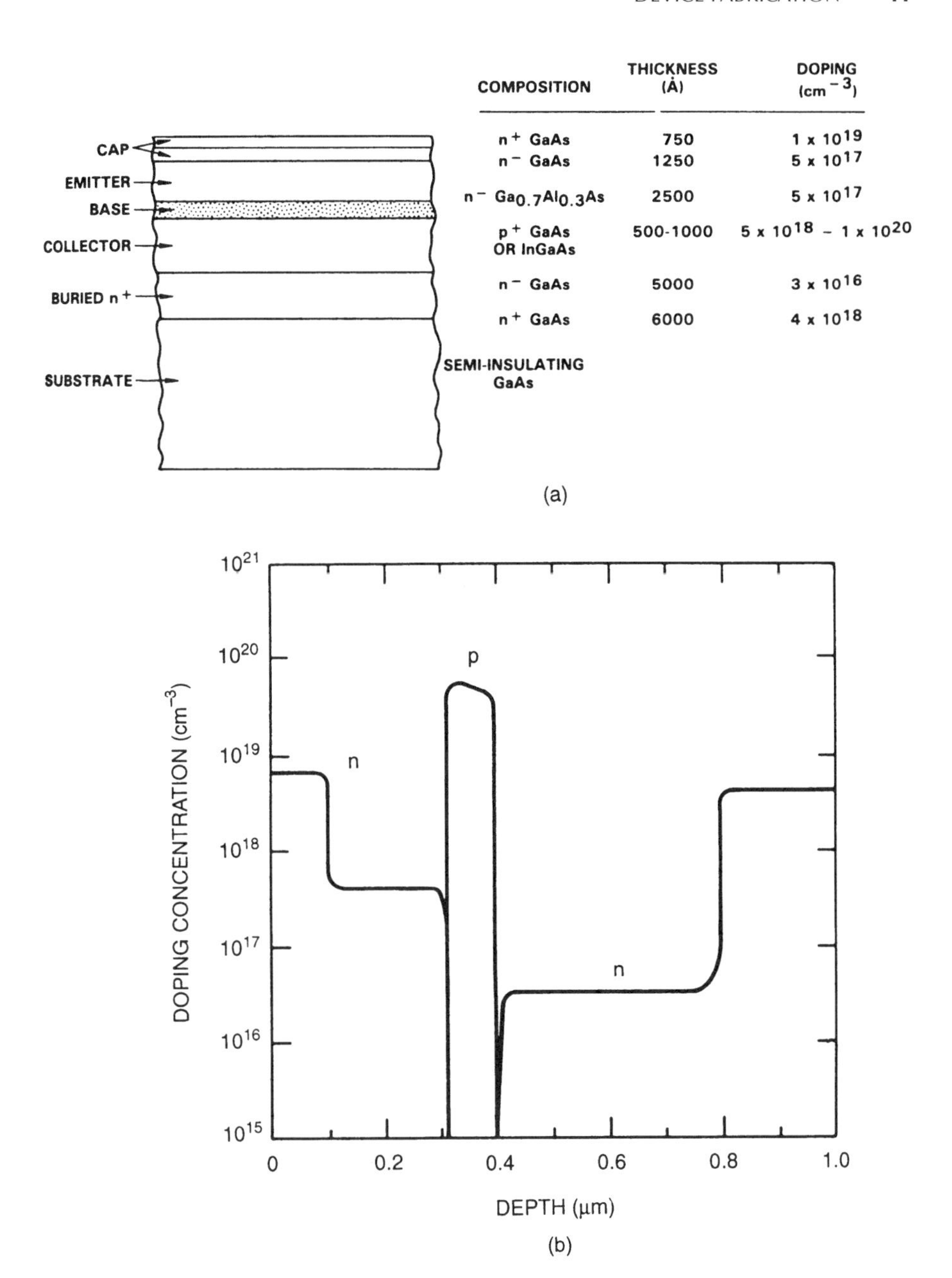

FIGURE 2.29 (*a*) Typical AlGaAs HBT layer structure and (*b*) doping profile (after Sze, Ref. 47 © John Wiley & Sons).

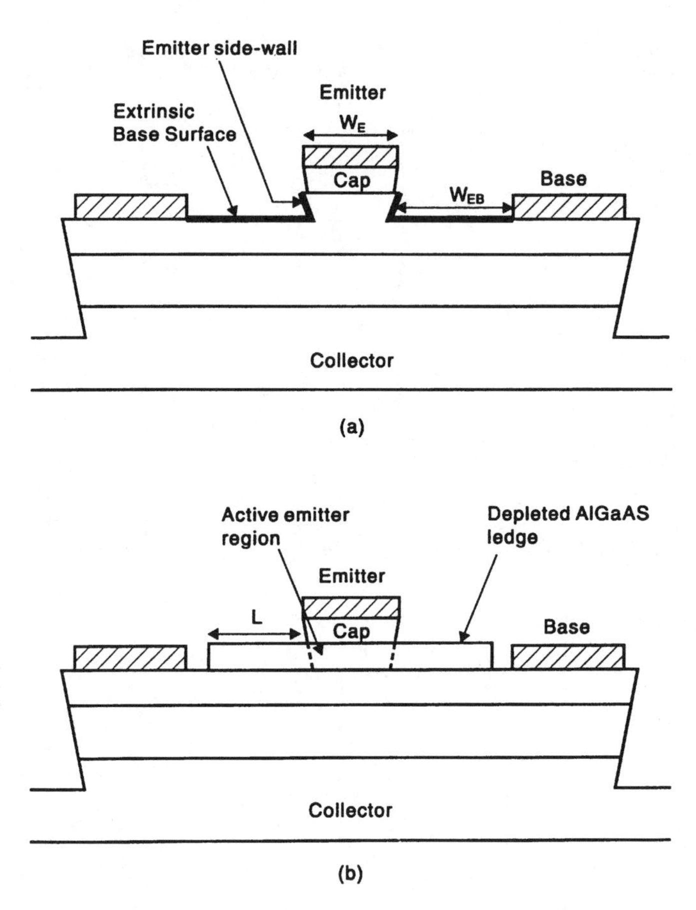

FIGURE 2.30 (*a*) Mesa etched and (*b*) passivated AlGaAs/GaAs HBT (after Hayama and Honjo, Ref. 49 © IEEE).

the growing surface determines the growth rate and the composition. The ability to determine the beam fluxes before growth has been improved dramatically due to the use of reflection high energy electron diffraction (RHEED) oscillators, which allow for direct measurement of the growth rate. This has greatly improved the reproducibility and control needed for the growth of high-quality ternaries.

Over the past decade, great progress has been made in the performance of InP-based HBTs. Much of this progress can be attributed to improvement in the quality of the materials from which these devices are fabricated. Currently, MOCVD is the major method used for deposition of InP-based materials, particularly for laser structures. Deposition of compound semiconductors in ultrahigh-vacuum (UHV) environments has become increasingly important, due to the inherent advantages of uniformity and interfacial abruptness. Conventional MBE, which employs only elemental group III and V sources, is the most common UHV deposition method for growth of GaAs structures. This technique is, however, limited by the difficulty of using elements that have low sticking coefficients or unstable flux characteristics. The allotropic nature of phosphor makes the attainment of a stable and reproducible flux from elemental P quite difficult.

Due to the problems associated with the use of a solid phosphorus source, devices containing InP have not typically been grown by conventional MBE. Instead, AlInAs has been substituted as the widegap emitter in single-heterojunction HBTs and also as the collector in double-heterojunction HBTs. The incorporation efficiency of the group III species in MBE or GSMBE is determined primarily by the respective sticking coefficients. For AlInAs, the limit species is In because this element begins desorbing at a temperature of around 540°C. Figure 2.31 shows the In desorption rate of AlInAs and InGaAs as a function of substrate temperature. The absorption rate for AlInAs and InGaAs is quite similar. The typical growth temperature for this material system is 490°C to 530°C.

The growth of AlInAs by MOMBE is similar to the growth of AlGaAs. The use of trimethylamine saline has allowed for the deposition of smooth layers with carbon levels in the low doping (10^{16} cm^{-3}) range and oxygen backgrounds in the mid-10^{17} cm^{-3} range. Unlike the case of InGaAs, there does not appear to be a strong dependence of composition on growth temperature. This is probably a reflection of the lower thermal stability of the termethylamine relative to triethylgallium (TEG). Even though similar precursors are used in MOCVD, except for the aluminum source, which is normally trimethylaluminum, much higher growth temperatures are used. This is due to the need to decompose the AsH_3 and to minimize the uptake of impurities. For atmospheric pressure MOCVD, growth temperatures typically range from 640 to 650°C. As the chamber pressure is reduced, the temperature needed to produce material of similar quality rises to 680 to 710°C. Like AlGaAs, AlInAs often shows a discrepancy between electron concentrations measured by Hall and those obtained from CV analysis. For AlInAs this disparity is believed to be due to oxygen

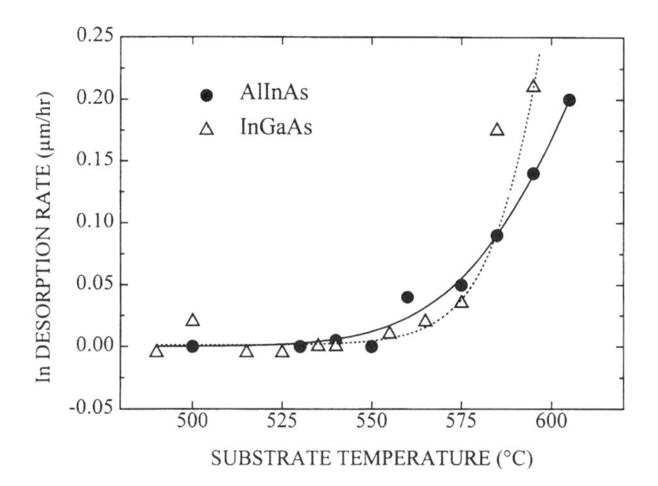

FIGURE 2.31 In desorption as a function of substrate temperature (after Jalali and Pearton, Ref. 2 © Artech House).

incorporation, where it is believed to be caused by deep levels associated with the formation of DX centers for AlGaAs [50]. The measured concentration difference for AlInAs may be reduced by growing at either higher temperatures or higher V/III ratio conditions, which are known to minimize oxygen incorporation.

The small bandgap of InGaAs makes it an ideal contacting layer for both the emitter and collector layers and an excellent base material. The use of InGaAs in the collector, however, results in lower collector–base breakdown voltage. As with AlInAs grown by MBE, growth temperatures used for InGaAs are generally kept below the In desorption temperature of 540°C so that neither growth temperature nor V/III ratio affect the composition. Unlike AlInAs and GaAs, the quality of InGaAs peaks at a temperature between 460 and 490°C.

InP/InGaAs and AlInAs/InGaAs are two dominate heterojunctions employed for fabrication of InP-based HBTs. An InP-based HBT epitaxial structure with an abrupt emitter–base heterojunction is shown in Fig. 2.32. The key features of this profile include a GaInAs base thickness of approximately 50 nm doped at 5×10^{19} cm^{-3} with Be and a GaInAs collector doped at 1×10^{16} cm^{-3} with Si at a thickness of 300 nm. The AlInAs emitter is 120 nm thick and doped at 8×10^{17} cm^{-3}. A 15-nm GaInAs spacer layer is inserted between the base and the emitter. The 15-nm spacer layer was grown at low temperature (300°C) to inhibit Be diffusion during MBE growth. Because of the abrupt emitter–base junction, a conduction-band potential spike of 0.2 to 0.3 eV is formed at the emitter–base heterojunction. This potential barrier increases the turn-on voltage of the HBT and results in a nonideal collector current with respect to base–emitter voltage.

100 nm	**GaInAs Contact**	**n = 1x10^{19} cm^{-3}**
70 nm	**AlInAs Emitter Contact**	**n = 1x10^{19}**
120 nm	**AlInAs Emitter**	**n = 8x10^{17}**
15 nm	**Low-Temp (300°C) GaInAs Spacer**	**p = 2x10^{18}**
50 nm	**GaInAs Base**	**p = 5x10^{19}**
300 nm	**GaInAs Collector**	**n = 1x10^{16}**
700 nm	**GaInAs Subcollector**	**n = 1x10^{19}**
10 nm	**GaInAs Buffer**	**Undoped**
	InP Substrate	

FIGURE 2.32 InP HBT layer structure.

For the development of an InP HBT, a dry-etch technique is required to enable a reproducible high-yield submicron process technology, thus allowing freedom from process-dependent epitaxial structures, device layout, and optimization of terminal contacts. The layer structure was grown on semi-insulating InP substrates by MBE. The growth temperature was 540°C and RHEED oscillations were used to achieve the lattice-matched ternary compositions ($Al_{0.48}In_{0.52}As$ and $In_{0.53}Ga_{0.47}As$). High-purity elemental sources of Si and Be were employed for n- and p-type doping, respectively.

Device fabrication began with the electron beam evaporation of an AuGe-based emitter contact. Emitter metal provided the mask for the formation of the emitter mesa. A 3-cm-diameter Ar ion source was used prior to evaporation to remove surface oxides and ensure good metal adhesion. The acceleration voltage of the Ar^+ ions was kept to a minimum to prevent the exposed InGaAs cap from being damaged by excessive ion bombardment. The AlInAs emitter was etched by chemical means selectively with respect to the p^+ InGaAs base. A slight undercut of AlInAs, under the InGaAs cap, was formed to allow for a break in the self-aligned base metal. The wet-etch process was developed to keep undercut of this structure to $\leq 0.2\mu$m. The self-aligned TiAu base metal was deposited by electron beam evaporation. Using a thick metal overlayer as a mask, the base mesa was formed by electron cyclotron resonance (ECR) etching. Only a narrowly defined portion of the base mesa around the emitter mesa is left electrically active to permit contact with the base. This self-aligned base mesa eliminates the photolithography step and minimizes the base–collector capacitance. Upon completion of the base-mesa etch, the masking metal overlayer was removed by wet chemical means. A via-type collector is defined by photoresist, reactive ion etching, and metal lift-off. Photoresist followed by wet etching defines the mesa isolation for the integrated devices. After PECVD dielectric passivation deposition, a new etch-back planarization process is fully scalable to contact submicron emitters. Vias to collector and base metal are opened in photoresist and etched through the dielectric. AuGe ohmic metal contacts were deposited to the n^+ subcollector after a light wet chemical etch to achieve good metal adhesion. A schematic cross section of the completed self-aligned AlInAs/InGaAs HBT is shown in Fig. 2.33.

InP/InGaAs heterojunction bipolar transistors, lattic matched to InP, with a carbon doped base were reported for the first time in 1992 [51]. The HBT structure was grown using GSMBE. AsH_3 and PH_3 were used as the sources for the group V elements. The epitaxial layer structure grown consists of an $n^+(1 \times 10^{19}$ cm$^{-3})$ 5000 Å InGaAs subcollector, an n^- $(2 \times 10^{16}$ cm$^{-3})$ 3000-Å InGaAs collector, an $p^+(6 \times 10^{19}$ cm^{-3} nominal concentration of carbon atoms) 500-Å InGaAs base, an n $(5 \times 10^{17}$ cm$^{-3})$ 800-Å InP emitter, an n $(2 \times 10^{18}$ cm$^{-3})$ 500-Å InP emitter, and an n^+ $(1 \times 10^{19}$ cm$^{-3})$ 300-Å InGaAs cap layer. For the InGaAs base, the growth temperature was 420°C and the carbon concentration from the CCl_4 flux was 6×10^{10} cm^{-3}. The structure was grown on Fe-doped semi-insulating InP substrates.

HBTs with large-area emitters were fabricated with a self-aligned emitter process. The emitter metal was used as an etch mask to etch down to the base layer. Access to the base layer was obtained by using selective wet etching, removing the InGaAs cap with $H_2SO_4/H_2O_2/H_2O$ (1:1:20) and the InP emitter with H_3PO_4/HCl (1:1). Base contacts were made by evaporating AuZn. The collector contacts were made by etching down

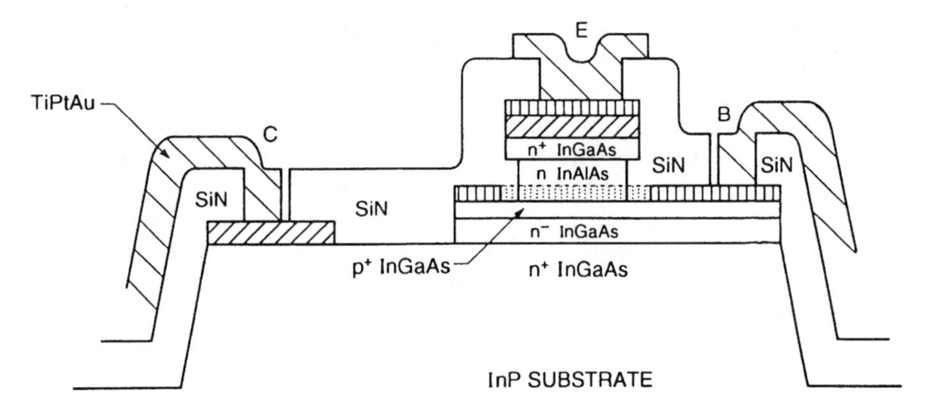

FIGURE 2.33 Schematic cross section of an AlInAs/InGaAs fabricated by dry etching (after Jalali and Pearton, Ref. 2 © Artech House).

to the subcollector using an InGaAs selective wet etch and evaporating AuGe. All contacts to the various layers were not annealed. The devices were isolated by mesa etching to the subcollector using the selective InGaAs etch.

InP/InGaAs HBTs with carbon-doped base grown by low-pressure MOCVD were also reported [52]. TMGa, TMIn, AsH_3, and PH_3 were used as dopant sources. The undoped collector (ca. 10^{15} cm^{-3}) and Si-doped emitter (ca. 1×10^{18} cm^{-3}) were grown at about 600°C, while the 1000-Å-thick C-doped $In_{0.53}Ga_{0.47}As$ base was grown at about 500°C. The temperature changes were accomplished during an 8-minute growth pause at the collector–base junction and a 4-minute growth pause at the base–emitter junction. The 8-minute pause before growth of the base is necessary to stabilize the substrate temperature, since the alloy composition of CCl_4-doped InGaAs is temperature dependent. This is due to etching of the surface by Cl-containing compounds and reduced growth efficiency of TMGa at about 500°C. The base was grown under conditions that result in a carbon doping concentration of 1.2×10^{19} cm^{-3}. However, partial hydrogen passivation during growth reduces the hole concentration in the base to 7×10^{18} cm^{-3}. Large-area devices were fabricated using a mesa etch process. Selective wet-chemical etching was employed for mesa definition.

High-speed InGaP/GaAs heterojunction bipolar transistors with a small emitter area have been fabricated [53]. The epitaxial layers were grown on a semi-insulating GaAs (100) substrate by GSMBE using Si and C for n- and p-type dopants, respectively. The subcollector layer is 5000-Å-thick n-GaAs doped to 8×10^{18} cm^{-3}. The GaAs collector layer is 2000 Å thick and undoped. The p-GaAs base layer is relatively doped to 1×10^{20} cm^{-3} to achieve low contact resistance with the WSi. The base layer thickness is 300 Å, to obtain appropriate current gain at the high base doping. InGaP is used in the emitter due to its lower surface recombination velocity. Since the InGaP emitter without surface passivation is comparable to that of AlGaAs/GaAs HBTs with an AlGaAs surface passivation layer on the extrinsic base region, the non-alloyed refractory metal of WSi can be used for the base electrode. The emitter layer is

100 Å thick, doped to 5×10^{17} cm^{-3}. Highly doped n-InGaP (500 Å, 8×10^{18} cm^{-3}), n-GaAs (100 Å, 8×10^{18} cm^{-3}), and n-InGaAs (500 Å, 4×10^{19} cm^{-3}) emitter-cap layers were grown on the emitter layer to reduce emitter resistance.

Figure 2.34 shows the main fabrication steps for the HBT with buried SiO_2 structure. First, W is deposited on the InGaAs emitter-cap layer by RF sputtering and formed into nonalloyed emitter electrode by RIE using CHF_3 and SF_6. Then the InGaAs and GaAs emitter-cap layers are etched by Cl_2/O_2 ECR plasma, using the emitter electrode as an etching mask (see Fig. 2.34*a*). The self-aligned base and collector mesas are formed by ECR plasma etching using a thick SiO_2 sidewall as an etching mask. Each device is then isolated with F^+ ion implantation (Fig. 2.34*b*). The outside of the base–collector junction is buried by SiO_2 using planarization and etching of InGaP, which is stopped when the thickness of leaving InGaP is at least 500 Å, so that the plasma does not damage the base surface. Then a thin sidewall of SiO_2 is formed to separate the emitter and the base (Fig. 2.34*c*). The remaining InGaP layer is removed by selective wet-chemical etching using a solution of dilute HCl to expose the base surface without damage. At this etching, the side facing the emitter layer is barely etched, due to the orientation dependence of the etching rate. Subsequently, WSi is deposited by RF sputtering and etched by RIE to define a base electrode, using photoresist as a mask (Fig. 2.34*d*). Annealing for 30 min at 400°C in N_2 is done to reduce the damage to the GaAs surface that occurs during RF sputtering, and therefore to reduce the contact resistance. After AuGe/W/Ni/Au/Mo is evaporated as a collector electrode on the subcollector and alloyed at 350°C for 30 min, the WSi on the emitter electrode is selectively etched by RIE with CF_4, using a photoresist mask (Fig. 2.34*e*). Finally, the metallization process follows and the device structure is completed. The final device cross section is shown in Fig. 2.34*f*.

The high-frequency characteristics of the small-signal current gain and unilateral gain were evaluated using on-wafer S-parameter measurements. The measurements were carried out at a collector–emitter voltage of 1.6 V and a collector current of 8.0 mA. The f_T and $f_{\max}$ as estimated from -20-dB/dec extrapolations were 105 GHz and 120 GHz, respectively. The estimated f_T value of a conventional HBT without using buried SiO_2 with the same dimensions was 88 GHz. The higher f_T value of our newly developed HBT is due to the reduction in base–collector capacitance, which was reduced from 33.1 fF for a conventional HBT to 16.7 fF for the buried SiO_2 HBT. The InGaP/GaAs HBT with buried SiO_2 has great potential for high-speed and low-power circuit applications.

Another fully self-aligned HBT process has been developed to fabricate submicron emitter geometries for applications requiring ultralow power consumption and very high speed performance [54]. In this process, the emitter, base, and collector ohmic contacts are all self-aligned to the emitter mesa. All three ohmic contacts are deposited in a single metallization step, thereby simplifying the fabrication process. Emitter geometries as small as 0.3 μm^2 with RF performance of over 130 GHz have been demonstrated.

Schematic fabrication steps for the submicron fully self-aligned HBT process are shown in Fig. 2.35. After sidewall formation, a light wet-chemical etching was used to remove a GaInAs thickness of approximately 0.12 μm. This critic-based etch

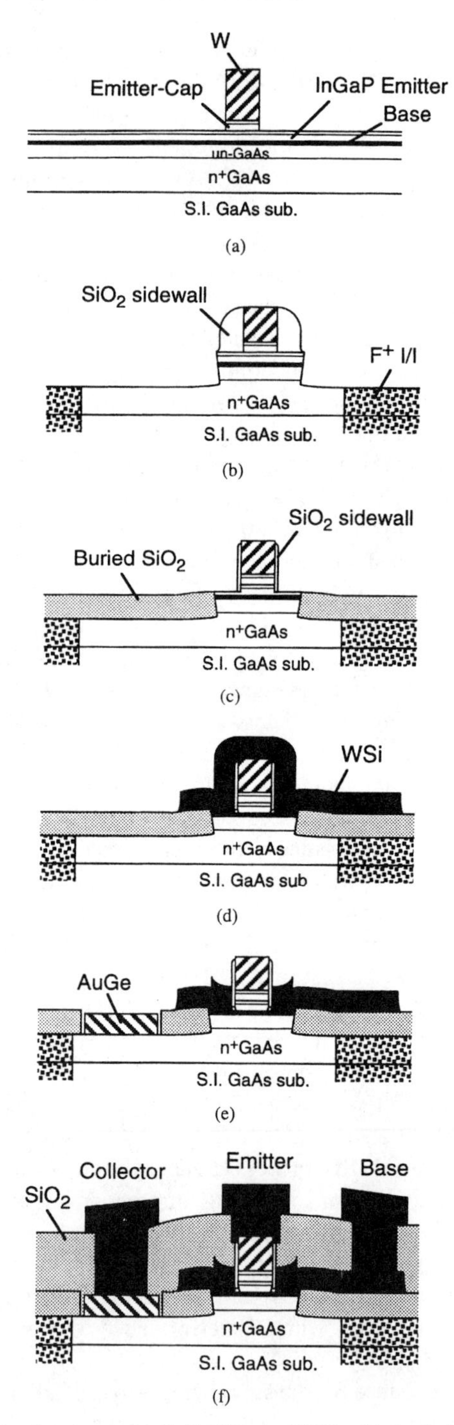

FIGURE 2.34 Fabrication steps for InGaP/GaAs HBT with buried SiO_2 (after Oka et al., Ref. 53 © IEEE).

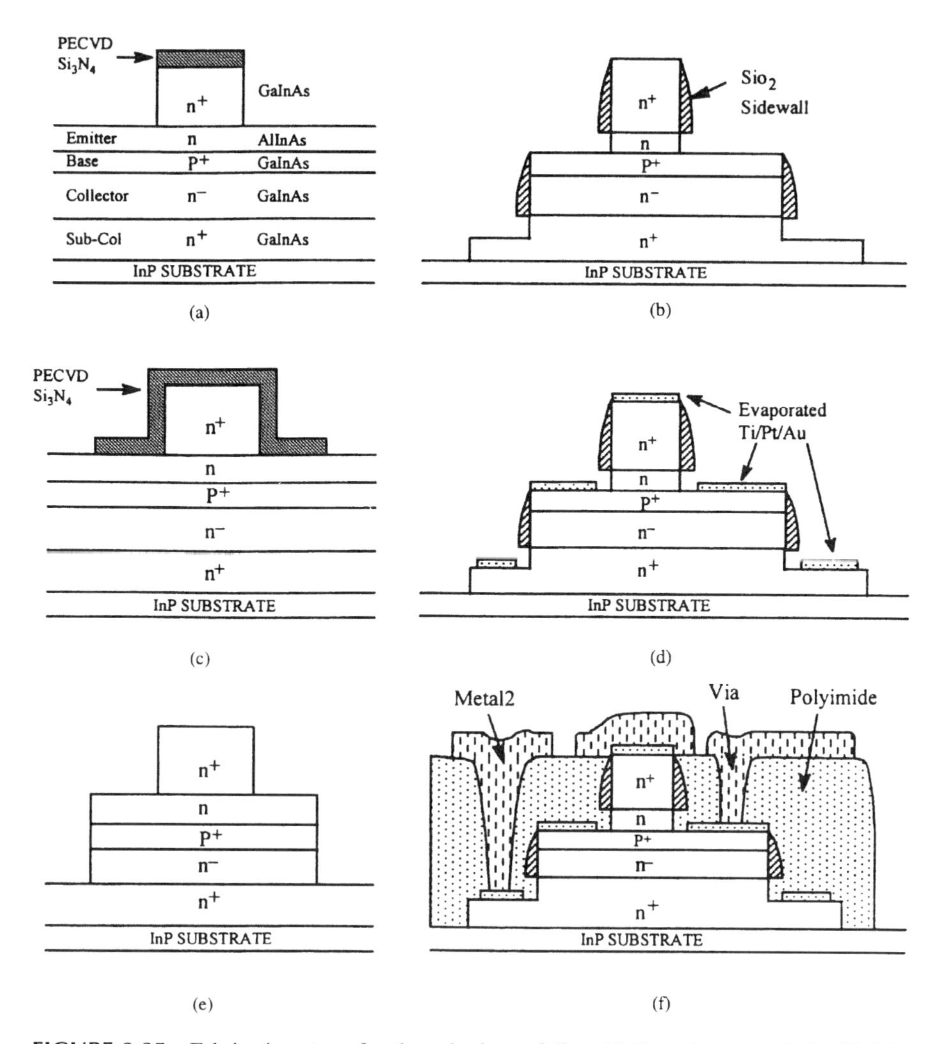

FIGURE 2.35 Fabrication steps for the submicron fully self-aligned process (after Hafizi, Ref. 54 © IEEE).

removes the emitter and causes an undercut of the SiO_2 sidewall, as shown in Fig. 2.35*d*. At the same time, the wet etching removes the subcollector and leaves an undercut of the collector sidewall as well. The overhang caused by the undercut of the sidewall facilitates fully self-aligned metallization of emitter, base, and collector contacts in a single evaporation step. The nonalloyed Ti/Pt/Au ohmic metallization had a thickness of approximately 0.045 μm. The metal on the base and subcollector break at the sidewall overhang results in a separation of approximately 0.1 μm from the mesa edges. Next, the HBT mesa structure was planarized by polymimide and etched back by RIE to expose the emitter tops. After the planarization, via holes were etched into the polymide to reach the base and collector contacts (see Fig. 2.35*f*).

REFERENCES

1. B. G. Streetman, *Solid State Electronic Devices*, 4th ed., Prentice Hall, Upper Saddle River, NJ (1995).
2. B. Jalali and S. J. Pearton, eds., *InP HBTs*: *Growth*, *Processing*, *and Applications*, Artech House, Norwood, MA (1995).
3. C. H. Gan, J. A. Del Alamo, B. R. Bennett, V. S. Meyerson, F. Crabbé, C. G. Sodini, and L. R. Reif, "Si/$Si_{1-x}G_x$ valence band discontinuity measurements using a semiconductor-insulator-semiconductor (SIS) heterostructure," *IEEE Trans. Electron Devices*, **ED-41**, 2430 (1994).
4. W.-X. Ni, J. Knall, and G. V. Hansoon, "New method to study band offsets applied to strained Si/$Si_{1-x}Ge_x$ (100) heterojunction interfaces," *Phys. Rev. B*, **36**, 7744 (1987).
5. C. A. King, J. L. Hoyt, and J. F. Gibbons, "Bandgap and transport properties of $Si_{1-x}Ge_x$ by analysis of nearly ideal Si/$Si_{1-x}Ge_x$/Si heterojunction bipolar transistors," *IEEE Trans. Electron Devices*, **ED-36**, 2093 (1989).
6. P. J. Wang, F. F. Fang, B. S. Meyerson, J. Nocera, and B. Parker, "Two-dimensional hole gas in Si/$Si_{0.85}Ge_{0.15}$/Si modulation-doped double heterostructures," *Appl. Phys. Lett.*, **54**, 2701 (1989).
7. K. Nauka, T. I. Kamins, J. E. Tuner, C. A. King, J. L. Hoyt, and J. F. Gibbons, "Admittance spectroscopy measurements of band offsets in Si/$Si_{1-x}Ge_x$ heterostructures studied with core-level x-ray photoelectron spectroscopy," *Phys. Rev. B*, **42**, 3030 (1990).
8. J. C. Brighten, I. D. Hawkins, A. R. Peaker, E. H. C. Parker, and T. E. Whall, "The determination of valence band discontinuities in Si/$Si_{1-x}Ge_x$/Si heterostructures, by capacitance-voltage techniques," *J. Appl. Phys.*, **74**, 1894 (1993).
9. R. Dingle, W. Wiegmann, and C. H. Henry, "Quantum states of confined carriers in very thin $Al_xGa_{1-x}As$-GaAs-$Al_xGa_{1-x}As$ heterostructures," *Phys. Rev. Lett.*, **33**, 827 (1974).
10. R. C. Miller, D. A. Kleinman, and A. C. Gossard, "Energy-gap discontinuities and effective massed for GaAs-$Al_xGa_{1-x}As$ quantum wells," *Phys. Rev. B*, **29**, 7085 (1984).
11. D. Arnold, A. Ketterson, T. Henderson, J. Klem, and H. Morko, "Determination of the valence-band discontinuity between GaAs and (Al,Ga)As by the use of p^+-GaAs-(Al,Ga)As-P^{1-1}-GaAs capacitors," *Appl. Phys. Lett.*, **45**, 1237 (1984).
12. W. I. Wang, E. E. Mendez, and F. Stern, "High mobility hole gas and valence-band offset in modulation-doped P-AlGaAs/GaAs heterojunctions," *Appl. Phys. Lett.*, **45**, 639 (1984).
13. J. C. Sturm, E. J. Prinz, P. M. Garone, and P. V. Schwartz, "Band-gap shifts in silicon germanium heterojunction bipolar transistors," *Appl. Phys. Lett.*, **54**, 2707 (1989).
14. P. Ashburn, H. Boussetta, M. D. R. Hashim, A. Chantre, M. Mouis, G. J. Parker, and G. Vincent, "Electrical determination of bandgap narrowing in bipolar transistors with epitaxial Si, epitaxial $Si_{1-x}Ge_x$, and ion implanted bases," *IEEE Trans. Electron Devices*, **ED-43**, 774 (1996).
15. Z. Matutinovic-Krstelj, V. Venkararaman, E. J. Prinz, J. C. Sturm, and C. W. Magee, "Base resistance and effective bandgap reduction in n-p-n Si/$Si_{1-x}Ge_x$/Si HBTs with heavy base doping," *IEEE Trans. Electron Devices*, **ED-43**, 457 (1996).

16. S. C. Jain and D. J. Roulston, “A simple expression for bandgap narrowing (BGN) in heavily-doped Si, Ge, and Ge_xSi_{1-x} strained layers,” *Solid-State Electron.*, **34**, 453 (1991).

17. J. M. López-González and Lluís Prat, “The importance of bandgap narrowing distribution between the conduction and valence bands in abrupt HBTs,” *IEEE Trans. Electron Devices*, **ED-44**, 1046 (1997).

18. J. H. van der Merwe, “Crystal interfaces: I. Semi-infinite crystals,” *J. Appl. Phys.*, **34**, 117 (1963).

19. J. M. Matthews and A. E. Blakeslee, “Defects in epitaxial multilayers: I. Misfit dislocations in layers,” *J. Cryst. Growth*, **27**, 118 (1974).

20. J. M. Matthews and A. E. Blakeslee, “Defects in epitaxial multilayers: II. Dislocation pile-ups, threading dislocations, slip lines and cracks,” *J. Cryst. Growth*, **32**, 265 (1975).

21. R. People and J. C. Bean, “Calculation of critical thickness versus lattice mismatch for Ge_xSi_{1-x}/Si strained layer semiconductor structures via plastic flow,” *Appl. Phys. Lett.*, **51**, 1325 (1987).

22. Y. Kohama, Y. Fukuda, and M. Seki, “Determination of critical layer thickness of $Si_{1-x}Ge_x$/Si heterostructures by direct observation of misfit dislocations,” *Appl. Phys. Lett.*, **52**, 380 (1988).

23. S. S. Iyer, G. L. Patton, J. M. C. Stork, B. S. Meyerson, and D. L. Harame, “Heterojunction bipolar transistors using Si–Ge alloys,” *IEEE Trans. Electron Devices*, **ED-36**, 2043 (1989).

24. J. S. Blakermore, “Semiconducting and other major properties of gallium arsenide,” *J. Appl. Phys.*, R123 (1982).

25. S. Tiwari and S. L. Wright, “Material properties of p-type GaAs at large dopings,” *Appl. Phys. Lett.*, **56**, 563 (1990).

26. F. M. Bufler, P. Graf, B. Meinerzhagen, B. Adeline, M. M. Rieger, H. Kibbel, and G. Fischer, “Low- and high-field electron-transport parameters for unstrained and strained $Si_{1-x}Ge_x$,” *IEEE Electron. Device Lett.*, **EDL-18**, 264 (1997).

27. T. Manku, J. M. McGregor, A. Nathan, D. J. Roulston, J.-P. Noel, and D. C. Houghton, “Drift hole mobility in strained and unstrained doped $Si_{1-x}Ge_x$ alloys,” *IEEE Trans. Electron Devices*, **ED-40**, 1990 (1993).

28. T. K. Carns, S. K. Chun, M. O. Tanner, K. L. Wang, T. I. Kamins, J. E. Tuner, D. Y. C. Lie, M.-A. Nicolet, and R. G. Wilson, “Hole mobility measurements in heavily doped $Si_{1-x}Ge_x$ strained layers,” *IEEE Trans. Electron Devices*, **ED-41**, 1273 (1994).

29. J. M. McGregor, “Theoretical and experimental studies related to $Si_{1-x}Ge_x$ base bipolar transistor,” Ph.D. dissertation, University of Waterloo (1992).

30. D. K. Nayak, J. C. S. Woo, J. S. Park, K. L. Wang, and K. P. MacWilliams, “High mobility of electrons in strained silicon,” *Solid State Devices and Materials Meeting*, *Extended Abstracts*, 943 (1993).

31. R. D. Dupuis, L. A. Moudym and P. D. Dapkus, “Preparation and properties of $Ga_{1-x}Al_xAs$-GaAs heterojunction grown by metal-organic chemical vapor deposition,” *First Physics Conference on Gallium-Arsenide and Related Compounds*, **45**, 1 (1979).

32. P. Bhattacharya, *Semiconductor Optoelectronic Devices*, Prentice Hall, Upper Saddle River, NJ (1994).

33. A. Y. Cho and J. R. Arthur, “Molecular beam epitaxy,” *Proc. Solid-State Chem.*, **10**, 157 (1975).

34. A. Katz, ed., *Indium Phosphide and Related Materials*: *Processing*, *Technology*, *and Devices*, Artech House, Norwood, MA (1992).

35. J. L. Benchimol, G. Le Roux, H. Thibierge, C. Daguet, F. Alexandre, and F. Brillouet, "Incorporation of growth III and V elements in chemical beam epitaxy of GaInAs p alloys," *J. Cryst. Growth*, **107**, 978 (1991).

36. D. L. Harame, J. H. Comfort, J. D. Cresller, E. F. Crabbé, J. Y.-C. Sun, B. S. Meyerson, and T. Tice, "Si/SiGe epitaxial-base transistors: II. Process integration and analog applications," *IEEE Trans. Electron Devices*, **ED-42**, 469 (1995).

37. R. Götzfried, F. Beisswanger, and S. Gerlach, "Design of RF integrated circuits using SiGe bipolar technology," *BCTM Tech. Dig.*, 51, 1997.

38. J. C. Bean, L. C. Feldman, A. T. Fiory, S. Nakahara, and I. K. Robinson, "Ge_xSi_{1-x}/Si strained-layer superlattice grown by molecular beam epitaxy," *J. Vac. Sci. Technol.*, **A2**, 436 (1984).

39. T. Tatsumi, H. Hirayame, and N. Aizaki, "Si/$Ge_{0.3}Si_{0.7}$/Si heterojunction bipolar transistor made with Si molecular beam epitaxy," *Appl. Phys. Lett.*, **52**, 2239 (1988).

40. G. L. Patton, S. S. Iyer, S. L. Delage, S. Tiwari, and J. M. C. Stork, "Silicon–germanium-base heterojunction bipolar transistors by molecular beam epitaxy," *IEEE Electron Device Lett.*, **9**, 165 (1988).

41. H. Temkin, J. C. Bean, A. Antreasyan, and R. Leibenguth, "Ge_xSi_{1-x} strained-layer heterostructure bipolar transistors," *Appl. Phys. Lett.*, **52**, 1089 (1988).

42. C. A. King, J. L. Hoyt, C. M. Cronet, J. F. Gibbons, M. P. Scott, and J. Turner, "Si/$Si_{1-x}Ge_x$ heterojunction bipolar transistors produced by limited reaction processing," *IEEE Electron Device Lett.*, **EDL-10**, 52 (1989).

43. J. F. Gibbons et al., "SiGe-base, poly-emitter heterojunction bipolar transistors fabricated by limited reaction processing," *IEDM Tech. Dig.*, 566 (1988).

44. D. L. Harame, J. H. Comfort, J. D. Cressler, E. F. Crabbé, J. Y.-C. Sun, B. S. Meyerson, and T. Tice, "Si/SiGe epitaxial-base transistors: I. Materials, physics, and circuits," *IEEE Trans. Electron Devices*, **ED-42**, 455 (1995).

45. D. L. Harame, J. H. Comfort, J. D. Cressler, E. F. Crabbé, J. Y.-C. Sun, B. S. Meyerson, and T. Tice, "Si/SiGe epitaxial-base transistors: II. Process integration and analog applications," *IEEE Trans. Electron Devices*, **ED-42**, 469 (1995).

46. K. Nagata, O. Nakajima, Y. Yamauchi, T. Nittono, H. Ito, and T. Ishibashi, "Self-aligned AlGaAs/GaAs HBT with low emitter resistance utilizing InGaAs cap layer," *IEEE Trans. Electron Devices*, **ED-35**, 2 (1988).

47. S. M. Sze, *High-Speed Semiconductor Devices*, Wiley, New York (1990).

48. S. Tiwari, S. L. Wright, and A. W. Kleinsasser, "Transport and related properties of GaAlAs/GaAs double heterostructure bipolar junction transistors," *IEEE Trans. Electron Devices*, **ED-34**, 185 (1987).

49. N. Hayama and K. Honjo, "Emitter size effect on current gain in fully aligned Al-GaAs/GaAs HBTs with AlGaAs surface passivation layer," *IEEE Electron Device Lett.*, **EDL-11**, 388 (1990).

50. R. Bhat, M. A. Koza, K. Kash, S. J. Allen, W. F. Hong, S. A. Schwarz, G. K. Chang, and P. Lin, "Growth of high quality AllnAs by low pressure organometallic chemical vapor deposition for high speed and optoelecctronic device applications," *J. Cryst., Growth*, **108**, 441 (1991).

51. R. C. Gee, T.-P, Chin, C. W. Tu, P. N. Asbeck, C. L. Lin, P. D. Kirchner, and J. M. Woodall, "InP/InGaAs heterojunction bipolar transistors grown by gas-source molecular beam epitaxy with carbon-doped base," *IEEE Electron Device Lett.*, **EDL-13**, 247 (1992).

52. A. W. Hanson, S. A. Stockman, and G. E. Stillman, "InP/$In_{0.53}Ga_{0.47}As$ heterojunction bipolar transistors with a carbon-doped base grown by MOCVD," *IEEE Electron Device Lett.*, **EDL-13**, 504 (1992).

53. T. Oka, K. Ouchi, H. Uchiyama, T. Taniguchi, K. Mochizuki, and T. Nakamura, "High-speed InGaP/GaAs heterojunction bipolar transistors with buried SiO_2 using WSi as the base electrode," *IEEE Electron Device Lett.*, **EDL-18**, 154 (1997).

54. M. Hafizi, "Submicron, fully self-aligned HBT with an emitter geometry of 0.3 μm," *IEEE Trans. Electron Devices*, **ED-18**, 358 (1997).

PROBLEMS

2.1. Discuss qualitatively how bandgap narrowing affects the collector current density and current gain of the HBT.

2.2. Use [16] to find C_1, C_2, C_3, C_4, a_1, and a_2 in Eqs. (2.6) and (2.7).

2.3. Use the hole mobility in Eq.(2.12) to calculate the base sheet resistance of an n-p-n HBT as a function of Ge content.

CHAPTER THREE

DC Performance

3.1 GENERAL STRUCTURES AND STEADY-STATE BEHAVIOR

3.1.1 Electron and Hole Currents

The central design principle of heterostructure devices utilizes energy gap variations in addition to electric fields as forces acting on electrons and holes to control their distribution and flow [1]. By a careful combination of energy gap variations and electric fields, it becomes possible, within wide limits, to control the forces acting on electrons and holes, separately. A design freedom not achievable in homostructures is obtained. The resulting greater design freedom permits a reoptimization of the doping levels and geometries of the HBT, leading to higher-speed devices.

In HBT a wide energy bandgap emitter is used. The basic theory behind a wide-gap emitter is explained as follows. Consider the energy band structure of an N-p-n bipolar transistor (the capital N represents the wide bandgap emitter), as shown in Fig. 3.1. The emitter–base band edge is sufficiently graded to eliminate any band-edge discontinuities in the conduction band. The dc currents flowing in such a transistor include a current I_n of electrons injected from the emitter into the base, a current I_p of holes injected from the base into the emitter, and a current I_s due to electron–hole recombination within the emitter–base space-charge layer. In addition, a small part I_r of the electron injection current I_n is lost due to bulk recombination. Neglecting recombination effects, the maximum current gain $\beta_{\max}$ is $\approx J_n/J_p$. The electron and hole injection current densities are of the form

$$J_n = N_E v_{nb} e^{-qv_n/kT} \tag{3.1}$$

$$J_p = N_B v_{pe} e^{-qv_p/kT} \tag{3.2}$$

where v_{nb} and v_{pe} are the mean speeds, N_E and N_B are emitter and base doping levels, and qV_n and qV_p are the heights of the potential energy barriers for electrons and holes between the emitter and base. Since the energy gap of the emitter is larger than

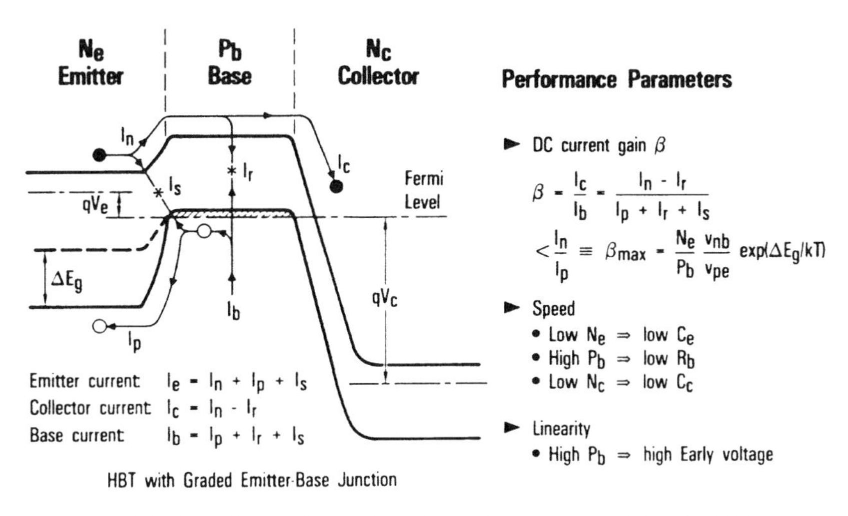

FIGURE 3.1 Current flow of the HBT (after Kroemer, Ref. 1 © IEEE).

that of the base, we obtain

$$\frac{J_n}{J_p} = \frac{N_E v_{nb} e^{\Delta E_g/kT}}{N_B v_{pe}} \tag{3.3}$$

As a result of a large energy gap difference between emitter and base, very high values of I_n/I_p can be achieved regardless of the doping ratio.

3.1.2 Abrupt and Graded Heterojunctions

For an abrupt N-p heterojunction, the conduction energy spike necessitates thermionic emission and tunneling for electron transport from the N-to-p quasineutral regions and the valence-band discontinuity suppresses hole injection from the p-to-N regions. The energy-band diagram of a heterojunction is shown in Fig. 3.2. Because of the valence-band discontinuity that suppresses hole injection into the emitter, the current gain of the heterojunction bipolar transistor has been increased significantly compared to that of the silicon bipolar transistor. The base doping of the HBT is thus increased to maintain a reasonable current gain. The increase in base doping decreases the base resistance.

The dominant transport mechanisms across an abrupt heterojunction are thermionic emission and tunneling. Thermionic emission theory relates barrier height to an injection current in the absence of tunneling. An expression for thermionic-field emitted current derived for Schottky barrier diodes can be written as

$$J_C = J_S e^{\Phi/E_0} \tag{3.4}$$

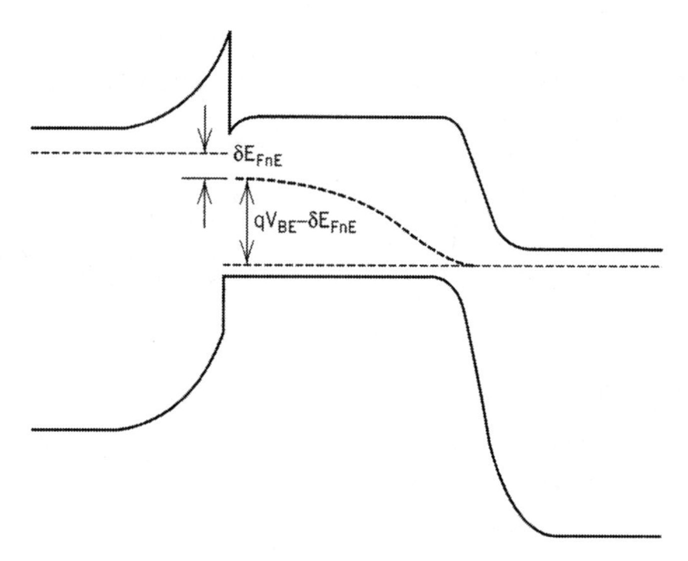

FIGURE 3.1 Energy-band diagram of an abrupt HBT.

where $E_0 = E_{00}\coth(E_{00}/kT)$, $E_{00} = (qh/4\pi)(N_d\varepsilon qm^*)^{0.5}$, Φ is the barrier height (from conduction band in bulk to top of spike), and J_S is the saturation current.

Compared to pure thermionic emission, tunneling through the conduction-band spike lowers the base–emitter bias needed for a given collector current. Reduced forward bias, in turn, implies larger energy barriers for hole injection into both the bulk and the depletion region of the emitter, which leads to improved injection efficiency. Tunneling through the spike generally serves to raise the ideality factor of the collector current. At low bias, the top of the spike is relatively narrow, and most electrons tunnel relatively far below the nominal peak. With increased forward bias, the top of the spike becomes thicker, and most electrons tunnel at higher energy levels than they do at low bias. Thus the effective barrier height increases with bias, creating an ideality factor greater than 1.

Figure 3.3 shows the measured terminal currents of abrupt AlGaAs/GaAs HBTs in both the forward and reverse active modes from many different wafers fabricated using MOCVD. In this figure the solid lines represent the emitter currents in the reverse active mode. These lines are almost identical for all the wafers measured. This indicates that diffusion is the primary mechanism of the carrier transport across the base–collector homojunction. In contrast, the dashed lines in Fig. 3.3 show different ideality factors, depending on the sample measured. All these dashed curves are different from the solid curves. The dashed curves have ideality factors between 1.11 and 1.18. This indicates that the forward terminal current in an abrupt AlGaAs/GaAs HBT is limited by conduction-band barrier transport mechanisms such as thermionic emission and tunneling.

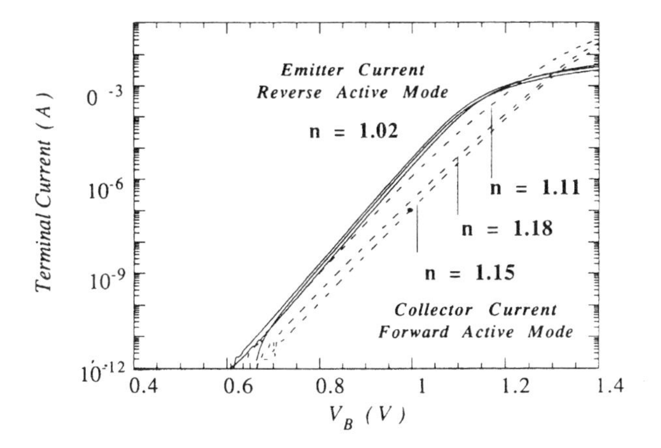

FIGURE 3.3 Measured collector current for forward and reverse active modes of AlGaAs/GaAs HBTs (after Liu et al, Ref. 2 © IEEE).

Current transport mechanisms in GaInP/GaAs HBTs are investigated. To determine the dominant current-transport mechanism in the GaInP/GaAs HBTs, the Gummel plot is displayed. Figure 3.4 illustrates measured terminal currents of the GaInP/GaAs HBT in both the forward active and reverse active modes. The device has an emitter area of $100 \times 100\ \mu m^2$, and its base–collector junction area is about $110 \times 200\ \mu m^2$. When the device is in the forward active mode, conduction carriers are emitted from the emitter–base junction, which is a GaInP/GaAs heterojunction. When the HBT is in the reverse active mode, the conduction carriers are emitted from the base–collector

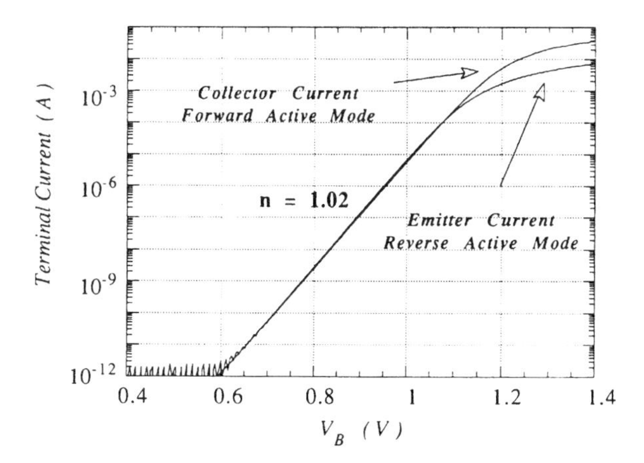

FIGURE 3.4 Measured collector current for forward and reverse active modes of GaInP/GaAs HBTs (after Liu et al, Ref. 2 © IEEE).

junction, which is a GaAs homojunction. Despite the operation differences in the forward active and reverse active modes, this figure demonstrates that the two terminal currents are nearly identical, with an ideality factor of 1.02. This indicates that carriers injected from both the GaInP/GaAs heterojunction and the GaAs homojunction are limited by the transport across the base layer, not by the conduction barrier. Note that the conduction-band discontinuity of the GaInP/GaAs HBT observed in [2] is much smaller than that of the AlGaAs/GaAs HBT.

The carrier injection efficiency can be improved if a thin region adjacent to the heterojunction is graded, especially for GaAs HBTs. Usually, a grading distance on the order of 100 to 300 Å, related to the Debye length, is sufficient to remove the discontinuity. The Debye length is related to the screening length and hence to the electrostatic field. The heterojunction grading effectively removes the spike and thus makes the thermionic and tunneling mechanisms less important. A linearly graded heterojunction with complete removal of the energy-band spike and discontinuity, shown in Fig. 3.5, has transport behavior similar to that of a homojunction.

For an abrupt heterojunction, the electron current density across the heterointerface is the difference between two opposing fluxes [3]:

$$J_n(X_J) = -q\,\frac{v_n}{4}\left[n(X_J^-) - n(X_J^+)e^{-\Delta E_C/kT}\right] \tag{3.5}$$

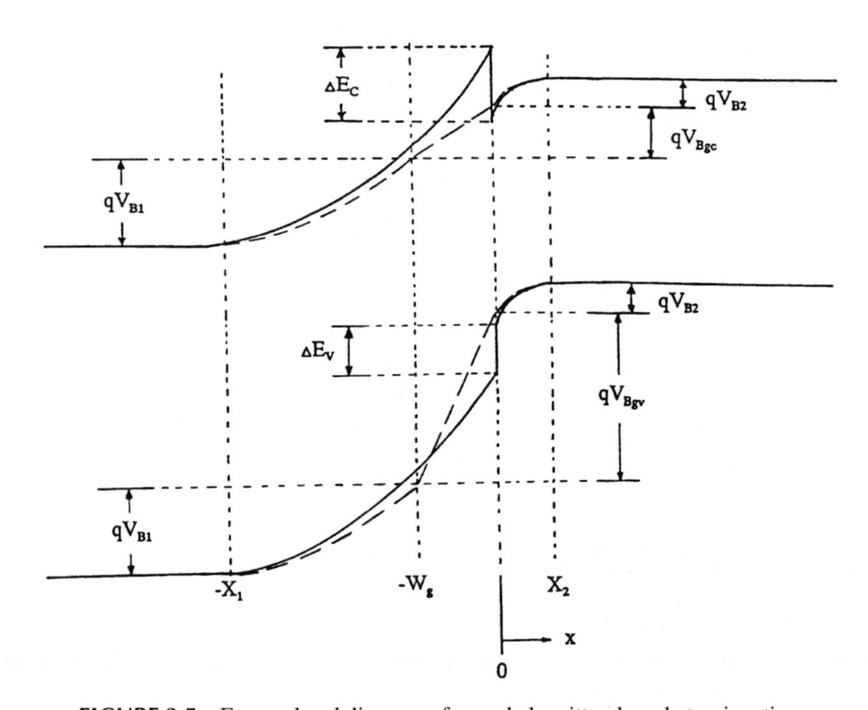

FIGURE 3.5 Energy-band diagram of a graded emitter–base heterojunction.

where $n(X_J^-)$ and $n(X_J^+)$ are the electron concentrations at each side of the heterointerface and v_n is the mean electron thermal velocity. These concentrations may be related to the electron concentrations at the boundaries of the space-charge region:

$$n(X_J^-) = n(X_1)e^{-qV_{B1}/kT} \tag{3.6}$$

$$n(X_J^+) = n(X_2)e^{qV_{B2}/kT} \tag{3.7}$$

and

$$V_{B1} = \frac{N_E\varepsilon_E(V_{bi} - V_{BE})}{N_E\varepsilon_E + N_B\varepsilon_B}$$

$$V_{B2} = \frac{N_B\varepsilon_B(V_{bi} - V_{BE})}{N_E\varepsilon_E + N_B\varepsilon_B}$$

where V_{BE} is the base–emitter voltage, ε_E the permittivity in the emitter, and ε_B the permittivity in the base.

The electron current density given by (3.5) should be equal to the electron current density due to diffusion at the boundary of the space-charge region in the base:

$$J_n(X_2) \approx -\frac{qD_n}{L_n}\frac{n(X_2)\cosh(W_B/L_n)}{\sinh(W_B/L_n)} \tag{3.8}$$

where D_n is the electron diffusion constant, L_n the electron diffusion, and W_B the base thickness. Equating (3.5) and (3.8), one can solve for $n(X_1)$ and then $J_n(X_1)$.

Figure 3.6 shows the collector current density versus the base–emitter voltage of the AlGaAs/GaAs HBT for various graded layer thicknesses ($W_g = 0$ for an abrupt heterojunction). The collector current density increases considerably when the graded layer thickness is increased from 0 to 150 Å. The collector current density increases only slightly when W_g is increased from 150 Å to 300 Å. Note that the roll-off of J_C at high V_{BE} for both $W_g = 150$ Å and 300 Å is due to the presence of the "alloy" barrier in the graded layer.

3.1.3 Undoped Setback Layer

Another approach to improve the injection efficiency is to insert a thin intrinsic spacer or setback layer at the heterojunction. The setback layer reduces the impurity out-diffusion from the heavily doped region and alters the potential barrier of the heterojunction for improving the injection efficiency. Figure 3.7 shows the energy-band diagram of an abrupt AlGaAs/GaAs HBT with an undoped setback layer. The effective potential barrier for electrons injected from emitter to base is reduced significantly. The reduction of the electron potential barrier, in turn, increases the electron injection into the base. Note that the addition of the spacer layer increases not only the injected current across the heterojunction, but also the recombination at the heterojunction space-charge region.

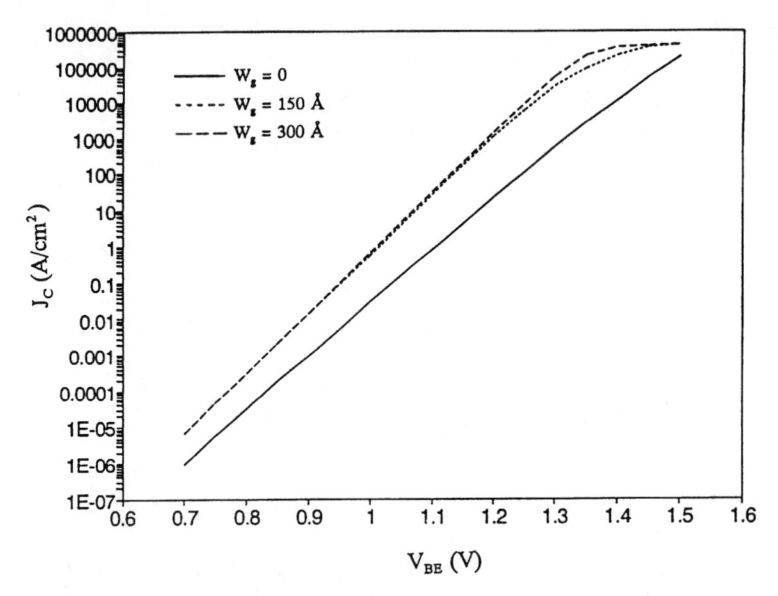

FIGURE 3.6 J_C–V_{BE} characteristics (after Liou, Ref. 4 © Artech House).

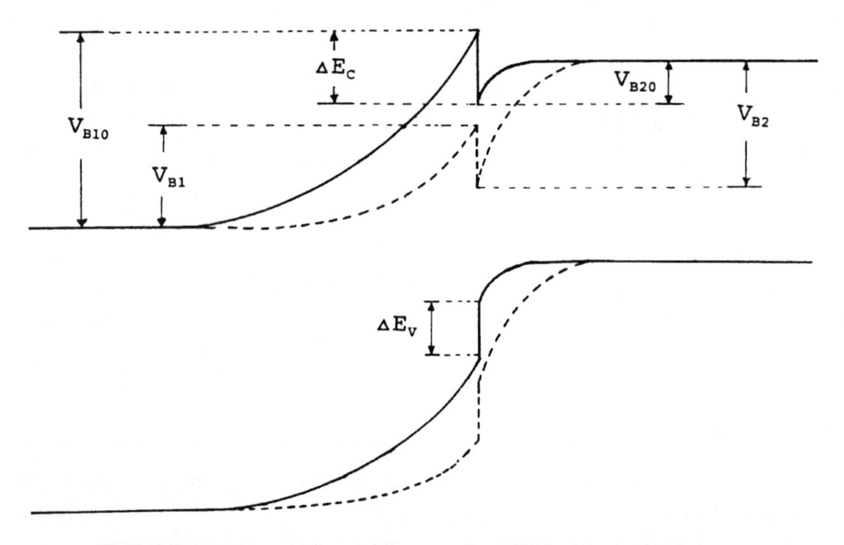

FIGURE 3.7 Energy-band diagram of an HBT with a setback layer.

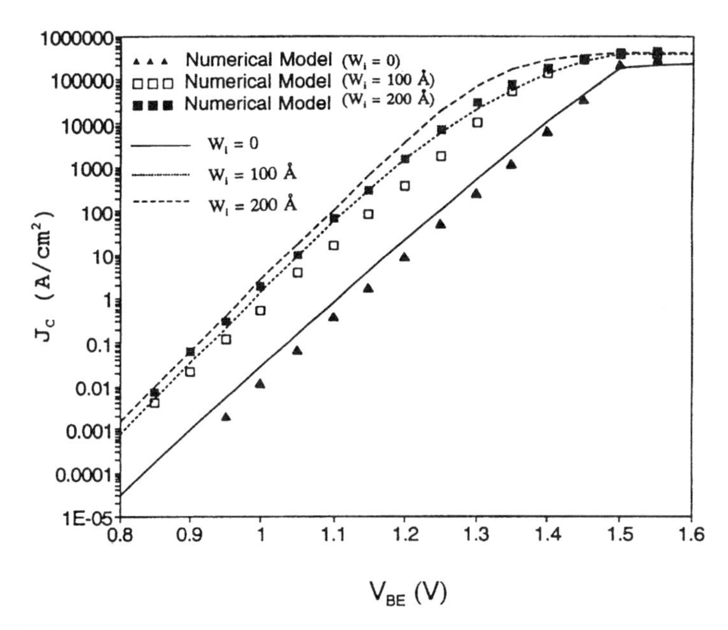

FIGURE 3.8 Effect of setback layer thickness on the collector current of an HBT (after Liou et al., Ref. 5 © Elsevier Science).

The effect of setback layer thickness (W_i) on the collector current density of an AlGaAs/GaAs HBT is shown in Fig. 3.8. In the J_C–V_{BE} characteristics, the lines represent the analytical predictions and the symbols represent the numerical simulation results. Clearly, the setback layer improves the emitter injection efficiency, particularly when the base–emitter voltage is relatively low. Also note that there is a smaller increase in collector current density when the setback layer thickness is increased from 100 Å to 200 Å than from 0 to 100 Å.

3.1.4 Graded-Base HBTs

With the ability to control semiconductor bandgap, it is also worthwhile to establish a gradual change in bandage across the base layer from E_{G0} near the emitter to $E_{G0} - \Delta E_G$ near the collector. In such a situation, the high hole conductivity ensures that the valence band is effectively flat, and the bandgap shifts establish a conduction-band energy gradient equal to $\Delta E_G / W_B$. This energy gradient constitutes a quasi-electric field that drives electrons across the base by drift as well as diffusion.

Figure 3.9 shows the energy-band diagram of a graded-base HBT. An aiding built-in field E_B is formed in the quasi-neutral base which results in a reduction in the minority-carrier transit time and an increase in the common-emitter current gain. E_B can be obtained by varying the Al or Ge mole fraction in the base, which varies the electron affinity χ and thus the conduction-band edge in the base. The magnitude of the aiding field is determined by the difference of the electron affinity at the collector–

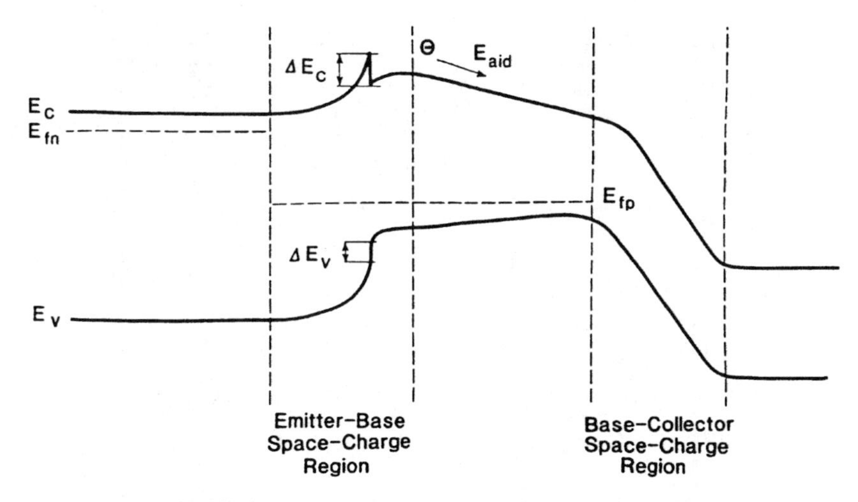

FIGURE 3.9 Energy-band diagram of a graded-base HBT.

base space-charge layer edge and at the emitter–base space-charge-layer edge in the base divided by the base width.

A change of bandgap using a linear grading at small mole fractions gives

$$E_G = E_{G0} - qE_B x \tag{3.9}$$

Since the intrinsic carrier concentration varies exponentially with the bandgap, the position-dependent intrinsic carrier concentration is written as

$$n_i^2(x) = n_{i0}^2 e^{qE_B x/kT} \tag{3.10}$$

where n_{i0} is the intrinsic carrier concentration associated with the bandgap E_{G0} at the emitter edge of the quasi-neutral base. The corresponding quantities at the collector edge are n_i^2 and E_{G1}. The current density of a graded heterojunction bipolar transistor with a linearly graded, uniformly doped base is given by [6]

$$\begin{aligned} J_n &= -\frac{qD_n}{N_B} n_{i0}^2 \frac{qE_B}{kT} \frac{e^{qV_{BE}/kT}}{e^{-qE_B W_B/kT} - 1} \\ &= \frac{q^2 E_B D_n}{N_B kT} \frac{n_{i0}^2 n_{i1}^2}{n_{i1}^2 - n_{i0}^2} e^{qV_{BE}/kT} \end{aligned} \tag{3.11}$$

3.1.5 Double Heterojunctions

Another possibility offered by bandgap engineering is to increase the bandgap of the collector. These types of devices are referred to as double heterojunction bipolar transistors (DHBTs). A wide bandgap collector suppresses the injection of holes from the base into the collector when the base–collector junction becomes forward biased. Increasing the bandgap of the collector greatly diminishes the charge storage in saturation and speeds up device turn-off from the saturation region. With double-heterojunction devices, symmetrical operation in the upward and downward directions can be established, which leads to circuit flexibility. Additional advantages of wide-bandgap collector devices are the increase in breakdown voltage and the reduction in leakage current.

The energy-band diagram of a double-heterojunction bipolar transistor is shown in Fig. 3.10. Note that the influence of the spike in the emitter–base heterojunction highlights a larger ideality factor on the corresponding current component, whereas the influence of the collector–base junction spike is characterized only by a saturation current. In an HBT with a heterojunction collector, an alloy barrier effect [7] also appears. The alloy barrier effect in the collector is essentially the same as in the emitter. At low currents, the alloy potential difference appears in the valence band, resulting in the usual hole-retarding barrier and field, while at higher currents it appears partially in the conduction band as a barrier to the flow of minority carriers. The consequences of this barrier are excess charge storage in the base and a decrease of current gain resulting from recombination.

Figure 3.11 shows two perspective plots of the conduction-band edge energy in the base of a double-heterojunction bipolar transistor at $V_{BE} = 1.3$ and 1.5 V, with zero base–collector bias $V_{BC} = 0$ V. The region plotted extends 500 Å into both the emitter and the collector. The 1.3-V surface is shifted down for clarity. The emitter junction is located in the lower right, and the collector is the broader region in the background. The base is the constant-energy region. Both heterojunctions are parabolically graded. At low bias, the potential profile is similar to that of a homojunction bipolar transistor. At higher voltages, however, a barrier due to both the alloy potential and electron space charge appears at the collector junction [8,9]. This barrier is largest

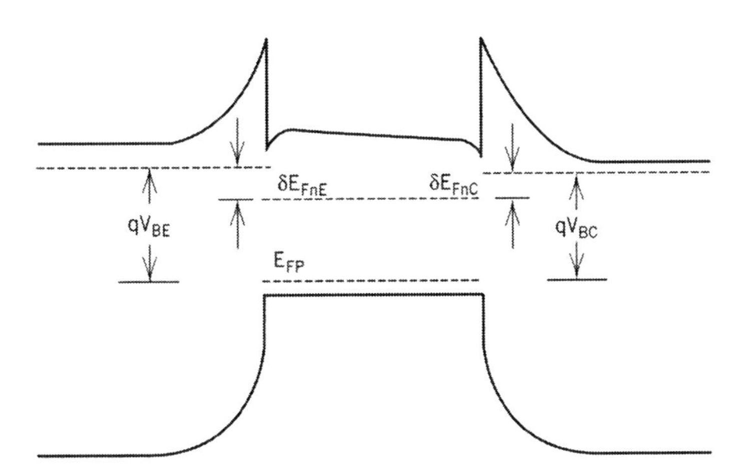

FIGURE 3.10 Energy-band diagram of a double-heterojunction bipolar transistor.

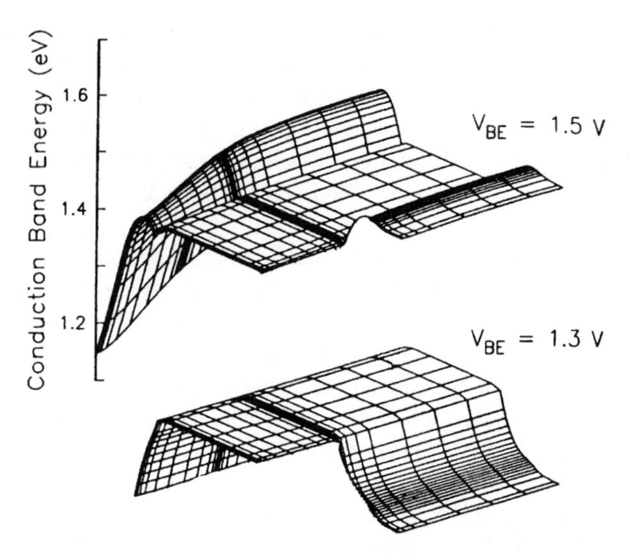

FIGURE 3.11 Conduction-band diagram at $V_{BE} = 1.3$ and 1.5 V (after Tiwari et al., Ref. 8 © IEEE).

where the current density is largest (i.e., opposite the emitter injection). The barrier is lower in the extrinsic part of the HBT because the current density is lower there. Significant spreading of the electron current goes toward the extrinsic base because the alloy barrier varies proportionally with current density.

Current gain behavior for single-heterojunction, graded-base, and double-heterojunction bipolar transistors in the absence of surface recombination ($s = 0$ cm/s) [8] is shown in Fig. 3.12. Similar behavior at a surface velocity of 2×10^5 and 2×10^6 cm/s is also observed. The graded base has larger current gains than those of single- and double-heterojunction bipolar transistors over a wide range of collector

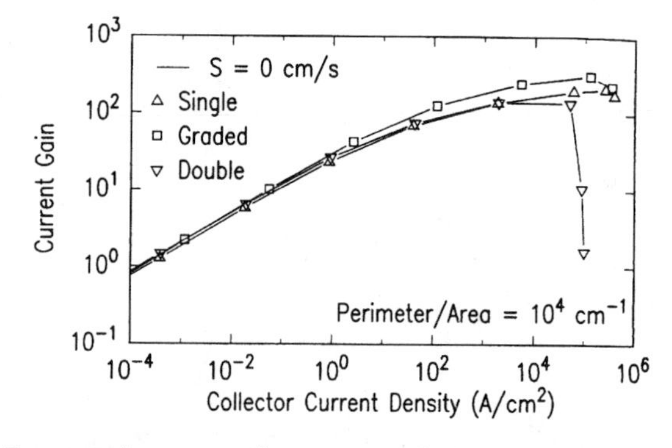

FIGURE 3.12 Current gain versus collector current in the absence of surface recombination (after Tiwari and Frank, Ref. 9 © IEEE).

current densities. The larger current gain in the graded-base HBT is due to the aiding quasi-electric field, which reduces the storage of electrons in the base and hence the neutral base recombination. The double-heterojunction HBT shows a rapid drop at high collector current densities due to the base–collector barrier effect. Although the graded-base HBT has a lower barrier-to-hole injection into the emitter, its gain actually continues to be comparable to that of the other devices at low current densities.

3.1.6 Electron Quasi-Fermi Level Splitting

In the abrupt heterojunction, a forward-biased junction reduces the barrier to electron flow into the base, yet increases the barrier to electron flow in the opposite direction. This leads to a situation where the net electron flow in the heterojunction space-charge region may not be small compared to each of the counter-direct flows. The flow of electrons thus cannot be considered as a perturbation from equilibrium. It was first proposed by Perlman and Feucht [10] that the electron quasi-Fermi level be discontinuous at the metallurgical boundary of an abrupt heterojunction. Lundstrom quantified the electron quasi-Fermi level splitting by using boundary conditions for the electron concentration on the p-side of the junction. These were arrived at by equating the thermionic emission current across the junction to the diffusion current away from the junction. The degree of electron quasi-Fermi level splitting was presented by Pulfrey and Searles [11].

The electron quasi-Fermi level splitting is given by

$$\Delta E_{Fn} = E_{Fnn} - E_{Fnp} = qV_{BE} - kT \ln \frac{n(0) + n_{b0}}{n_{b0}} \tag{3.12}$$

where $n(0)$ is the excess electron concentration at the base edge of the emitter–base space-charge region and n_{b0} is the electron concentration in the base at equilibrium. Note that ΔE_C and ΔE_V result from the formation of a heterojunction. ΔE_{Fn} is obtained due to electron current flow across the forward-biased heterojunction. ΔE_C and ΔE_V are physical constants, and ΔE_{Fn} is a function of bias.

Figure 3.13 shows the electron quasi-Fermi level splitting versus base–emitter voltage. For the abrupt heterojunction with tunneling included, the electron quasi-Fermi level splitting rises steadily with base–emitter voltage. This is because ΔE_{Fn} increases with forward bias, which leads to a reduction in the back injection of electrons and in increasing departure from quasiequilibrium. Suppression of the conduction-band spike by junction grading gives the barrier a more homojunction-like character. Thus electron quasi-Fermi level splitting is greatly reduced. By neglecting tunneling, the current at the junction is underestimated. This exaggerates the restrictive role of the junction as regards charge transport and leads to an overestimation of the electron quasi-Fermi level splitting.

The electron Fermi level splitting for both single- and double-heterojunction bipolar transistors with uniform and graded bases is shown in Fig. 3.14. For low base–emitter voltages, the DHBT virtually eliminates collector–emitter offset voltage. At $V_{BE} = 1.5$ V the electron quasi-Fermi level splitting is reduced to less than

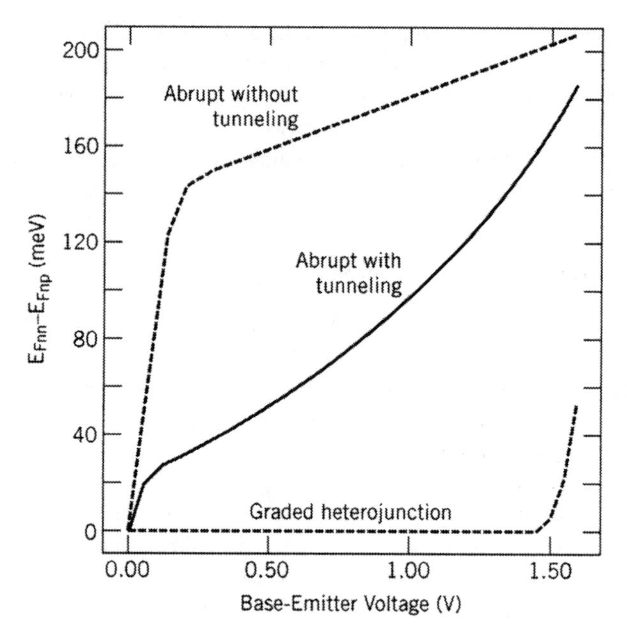

FIGURE 3.13 Electron quasi-Fermi level separation (after Pulfrey and Searles, Ref. 11 © IEEE).

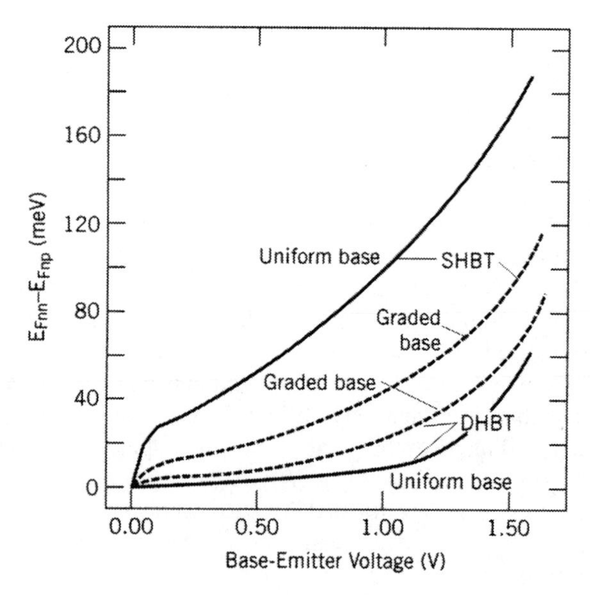

FIGURE 3.14 Electron quasi-Fermi level separation (after Pulfrey and Searles, Ref. 11 © IEEE).

50 meV for the DHBT with a uniform base, while it is about 170 meV for the SHBT. Grading the base reduces the amount of electron quasi-Fermi level splitting in an SHBT but increases it in a DHBT, although the degree of splitting is inherently less in a DHBT than in a SHBT. In the SHBT, the most important manifestation of base grading is the reduction in the conduction-band spike. This makes the junction more homojunction-like, so the electron quasi-Fermi level splitting is reduced. Interestingly, although this effect is also present in the DHBT case, it is swamped by the effect of the built-in base field, which sweeps carriers away from the emitter–base junction. This action opposes that of the electron barrier at the base–collector junction. Thus $n(0)$ is not augmented to the same degree as it is in the uniform-base case, the departure from quasiequilibrium increases and therefore the electron quasi-Fermi level splitting increases.

3.1.7 Collector–Emitter Offset Voltage

AlGaAs/GaAs single-heterojunction bipolar transistors exhibit significant collector–emitter offset voltage ΔV_{CE}, as evidenced by experimental data [12] in Fig. 3.15. This offset voltage reduces logic swing in digital circuits and increases power dissipation in analog amplifiers. The offset voltage arises because the base–emitter and base–collector junctions are very dissimilar in their current–voltage dependence. For example, the built-in voltage for electrons in the base–emitter heterojunction is larger than that in the base–collector homojunction. It may also arise because devices have a large combination current that is dependent on area and lifetime. Widely different areas for the base–emitter and the base–collector junctions can therefore give rise to large offsets because of asymmetric junction areas.

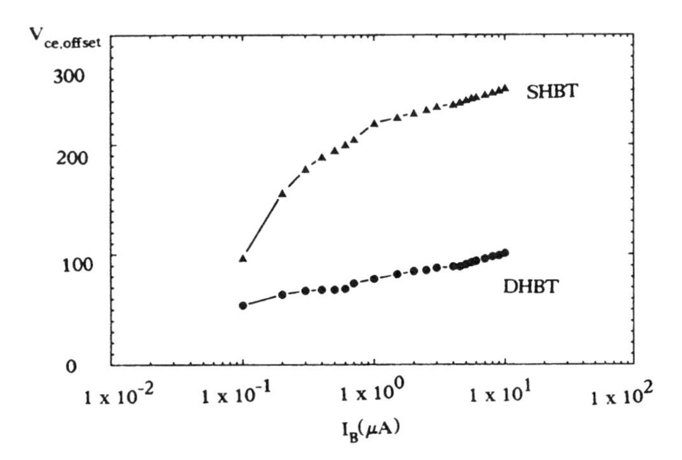

FIGURE 3.15 Offset voltage versus base current (after Won et al., Ref. 12 © IEEE).

Examine the collector current using the Ebers–Moll equation as follows:

$$I_C = \alpha_N I_{E0}\left(e^{qV_{BE}/kT} - 1\right) - I_{C0}\left(e^{qV_{BC}/kT} - 1\right) \tag{3.13}$$

where α_N is the normal common-base current gain, $\alpha_N I_{E0} = \alpha_1 I_{C0}$, where α_1 is the inverse common-base current gain. The offset voltage can be deduced by setting $I_C = 0$ in (3.13):

$$\Delta V_{CE} = \frac{kT}{q}\ln\frac{I_{C0}}{\alpha_N I_{E0}} = \frac{kT}{q}\ln\frac{A_C}{A_E} + \frac{kT}{q}\ln\frac{1}{\alpha_I} \tag{3.14}$$

Although the offset voltage of InAlAs/InGaAs HBTs is slightly less than that of AlGaAs/GaAs HBTs, the offset voltages were found to originate from the asymmetry of emitter and collector junctions [12,13]. Figure 3.16 shows variation of the offset voltage with the collector–emitter ratio (A_E/A_C) at a base current of 1 μA. The offset voltage is sensitive to A_C/A_E. The offset voltage can be lowered by increasing the injection of electrons from the collector in the offset region and decreasing A_C/A_E. The closer the inverse current transfer factor α_I is to unity, the smaller is the offset voltage. In fact, negligible offset voltage is observed when the transistor is operated in the inverted mode as shown in Fig. 3.17. In this case the appropriate offset voltage is the emitter-to-collector offset voltage, and the contribution of the first and second terms in (3.13) is negative. Thus one of the advantages of employing the collector-up structure can be the negligible offset voltage in the common-emitter configuration. For a single AlGaAs/GaAs heterojunction bipolar transistor, the collector-to-emitter offset voltage can be as large as 200 to 300 mV. This offset voltage is reduced significantly

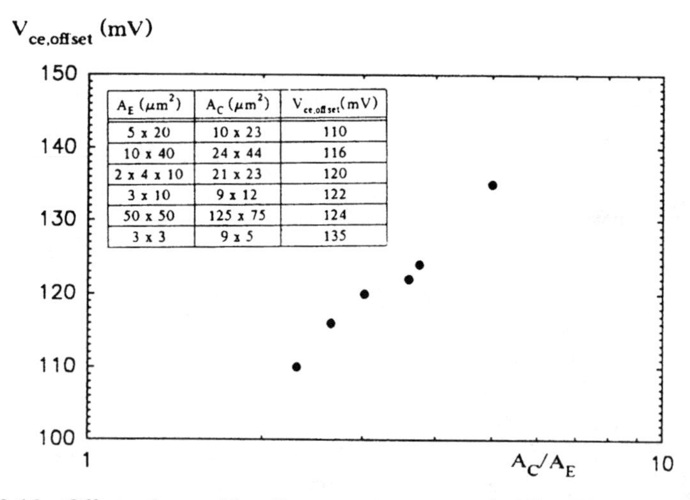

A_E (μm^2)	A_C (μm^2)	$V_{ce,offset}$ (mV)
5 x 20	10 x 23	110
10 x 40	24 x 44	116
2 x 4 x 10	21 x 23	120
3 x 10	9 x 12	122
50 x 50	125 x 75	124
3 x 3	9 x 5	135

FIGURE 3.16 Offset voltage with collector–emitter area ratio (after Won et al., Ref. 12 © IEEE).

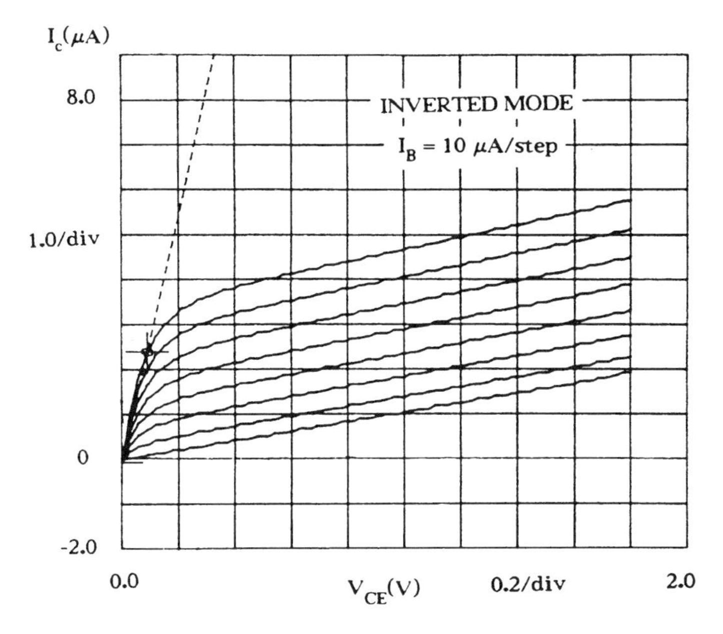

FIGURE 3.17 Collector current at the inverse active mode (after Won et al., Ref. 12 © IEEE).

when double heterojunctions are used. Grading of the base–emitter heterojunction also reduces the offset voltage [14].

The offset voltage is also a function of temperature. Figure 3.18 shows the collector current versus the collector–emitter voltage of an AlGaAs/GaAs HBT grown by chemical vapor deposition at 300, 423, 573, and 623 K. The HBT shows a low collector–emitter offset voltage of 0.1 V at 300 K. The offset voltage is increased to 0.4 V at 573 K. At higher temperatures, significant collector–base leakage current starts to degrade transistor performance. The ideality factors obtained from the Gummel plot are shown in Fig. 3.19. At 300 K the collector current exhibits an ideality factor of $n_C = 1.04$, indicating diffusion-dominated current transport. The ideal behavior is maintained until it starts to increase at 523 K. The ideality factor n_B of the base current at 300 K is 1.97, which is close to the ideal value of 2 for a pure recombination current. At temperatures higher than 523 K, the ideality factor of n_B increases to 3.

3.1.8 Early Voltage

The Early voltage (V_A) is an important parameter for the bipolar transistor. It describes the base-width modulation effect of the BJT in the forward-active mode. Since the discovery of the Early effect in 1952 [16], numerous papers on Early voltage have been

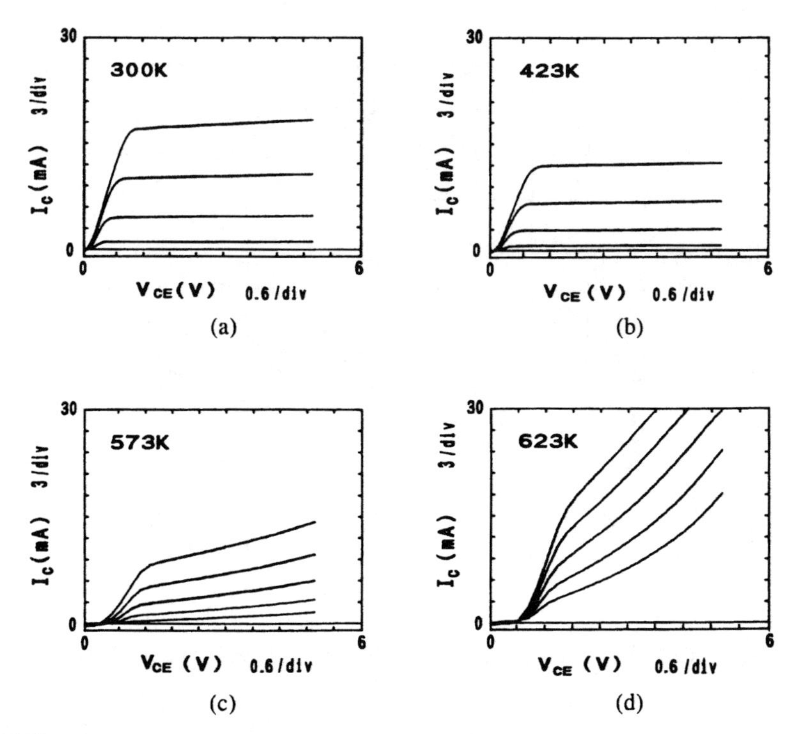

FIGURE 3.18 Collector current versus collector–emitter voltage at (*a*) 300 K, (*b*) 423 K, (*c*) 573 K, and (*d*) 623 K (after Fricke et al., Ref. 15 © IEEE).

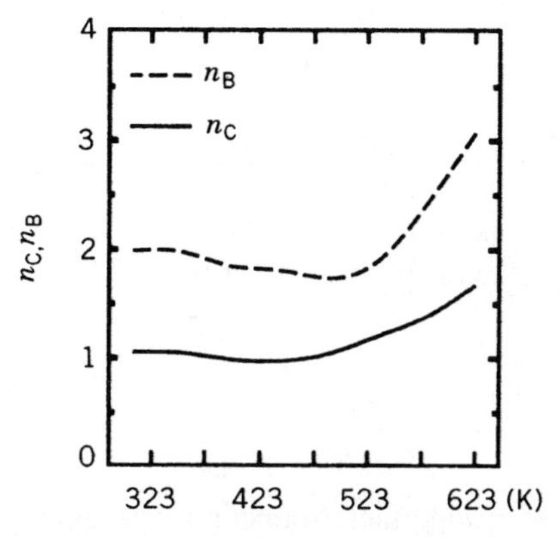

FIGURE 3.19 Ideality factors for base and collector currents versus temperature (after Fricke et al., Ref. 15 © IEEE).

published [17–20]. Roulston [17] studied mathematically the Early voltage of narrow-base bipolar transistors, including the effect of velocity saturation in the collector–base junction. Velocity saturation increases the effective Early voltage of the bipolar transistor. A more self-consistent equation was derived by Liou [18]. Yuan and Liou [19] reported that the Early voltage is influenced by both the base–emitter voltage and the collector–base voltage of the BJT. This effect is significant for modern bipolar transistors with very narrow base width W_B. Ugajin et al. [20] investigated the inverse base width modulation effect when the bipolar transistor is operated at high current densities before base pushout occurs. Unlike the Early voltage effect, the inverse base width modulation increases base width and hence decreases the collector current.

The definition of the Early voltage is given by

$$V_A \equiv J_C \frac{\partial V_{CE}}{\partial J_C} - V_{CE} \approx J_C \frac{dV_{CB}}{dJ_C} \tag{3.15}$$

where V_{CE} is the collector–emitter bias. For an HBT with high base doping, the Early voltage is relatively large. Grading in the base of the HBT also increases the Early voltage.

3.1.9 Bias-Dependent Base Resistance

The base resistance plays a significant role in the switching speed of bipolar logic and the frequency response of small-signal amplifiers. The base resistance can be divided by the intrinsic base resistance R_{bi} and the extrinsic base resistance R_{bx}. The intrinsic base resistance for a double-base contact bipolar transistor is given by

$$R_{bi} = \frac{L_E}{12q \int_0^{X_E} \int_0^{W_B} \mu_p(x, y) p(x, y)\, dx\, dy} \tag{3.16}$$

where X_E is the emitter width in the horizontal dimension and W_B is base thickness in the vertical dimension. W_B is subject to base width modulation, base conductivity modulation, emitter current crowding, and base pushout effects. This makes the base resistance current dependent [21].

The base width is modulated by the moving edge of the emitter–base space-charge region when V_{BE} changes. As V_{BE} increases, the emitter–base space-charge region contracts, and W_B expands. This reduces R_{bi} because the base cross-sectional area $W_B L_E$ increases, resulting in a larger base charge and a larger effective Gummel number.

When an n-p-n bipolar transistor is in high current operation, the hole concentration (including the excess carrier concentration) in the base will exceed the acceptor (dopant) concentration to maintain charge neutrality. As a result, the base sheet resistance under the emitter decreases as the hole injection level increases. The intrinsic base resistance is then decreased as the level of injection increases.

The nonuniform emitter current density distribution makes the effective emitter width smaller. Hence emitter current crowding reduces the effective intrinsic base resistance. For a bipolar transistor with a large emitter width at high base–emitter voltage, R_{bi} is more susceptible to emitter crowding effect.

For bipolar transistors operating at high currents, base pushout occurs. The base pushout increases the effective base width, which decreases the intrinsic base resistance. For a bipolar transistor with low epi-layer collector doping operating at high collector current density and low V_{CE}, the intrinsic base resistance is more vulnerable to the base pushout effect.

Figure 3.20 shows the base resistance versus the base current of a heterojunction bipolar transistor. The base resistance was obtained from the Gummel plot measurement. It is clear from Fig. 3.20 that the base resistance decreases with increase in base current. It is anticipated that the base current dependence will be enhanced if a lower base doping is used.

3.1.10 High Injection Barrier Effect

Heterojunction bipolar transistors rely on a difference in bandgap to obtain a selective suppression of carriers from the base into the emitter. Double-heterojunction bipolar transistors such as SiGe HBTs exhibit a unique effect at high current densities. As the collector current is larger than the onset current for base pushout, holes tend to move into the lightly doped collector when the electric field at the collector–base junction changes from negative to positive. The valence-band discontinuity at the base–collector heterojunction, however, prevents holes from moving into the collector. Thus electrons in the collector are not compensated with holes, and an electric field is established that creates a potential barrier [23]. This barrier opposes the electron flow into the collector, and an increased density at the base–collector heterojunction is required to support the collector current density.

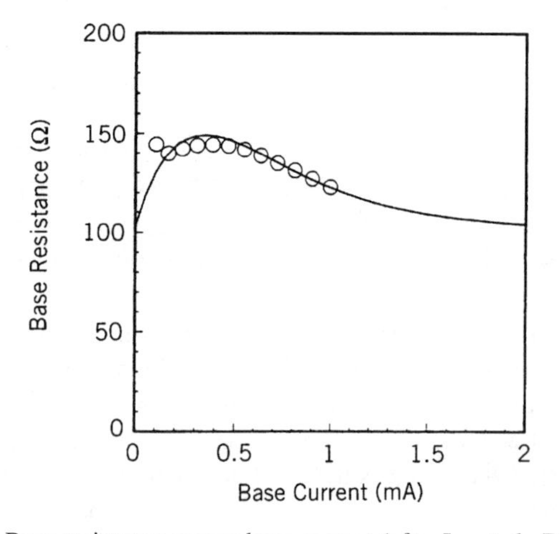

FIGURE 3.20 Base resistance versus base current (after Lu et al., Ref. 22 © IEEE).

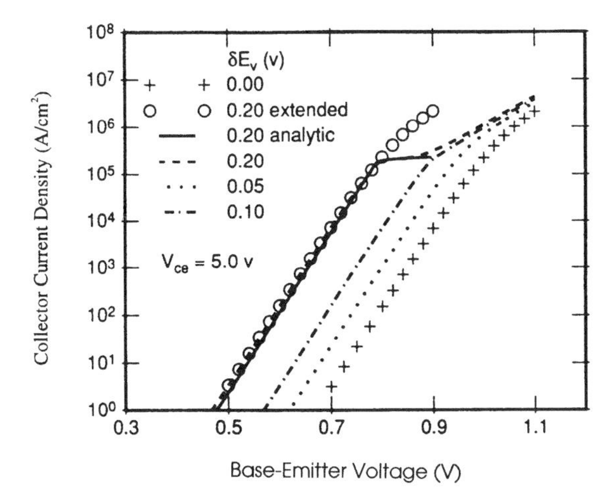

FIGURE 3.21 High current I_C versus V_{BE} (after Cottrell and Yu., Ref. 23 © IEEE).

Figure 3.21 shows the collector current density versus base–emitter voltage. The n-p-n SiGe HBT with uniform base and collector doping and varying valence band offset magnitude in the base. The base has a constant valence-band offset except for the case labeled "extended," where the narrow bandgap region is extended through the collector to the buried layer. The base width is 0.25 μm and the collector doping is 1×10^{16} cm^{-3}. The transistor structure was chosen to emphasize and isolate this effect rather than to represent a "real" design. The solid line in Fig. 3.21 represents results obtained with the analytic model and the remainder with SEDAN [24]. It is clear from Fig. 3.21 that the collector current density decreases significantly at high base–emitter voltages due to the high-injection barrier effect.

The high current barrier effect has also been shown in the AlGaAs/GaAs/AlGaAs HBT [8]. Figure 3.22 illustrates the Gummel plot of this double-heterojunction HBT for the emitter area $4 \times 12 \mu m^2$. The HBT has a base doping of 5×10^{18}cm^{-3} and thickness of 100 Å, an emitter doping of 4×10^{17}cm^{-3}, a collector doping of 1×10^{17}cm^{-3}, and an aluminum mole fraction of 0.3. In Fig. 3.22 the base current shows a rapid rise at high current densities, with the point of rapid increase dependent on the applied base–collector voltage. The onset of this effect is pushed toward higher current densities or base–emitter voltages with increased reverse biasing of the base–collector junction. The barrier decreases the electron injection into the collector and leads to large electron accumulation or hole recombination in the base. This is why the collector current is decreased and the base current is increased at high base–emitter voltages.

Consider a base–collector heterojunction with a parabolically graded GaAs/Al-GaAs junction. Under low injection conditions, the alloy grading results in a retarding chemical field for holes from the base and none for electrons because most of the alloy grading potential appears in the valence band. As injection increases, the decreasing net charge density in the depletion region, and hence reduced electrostatic

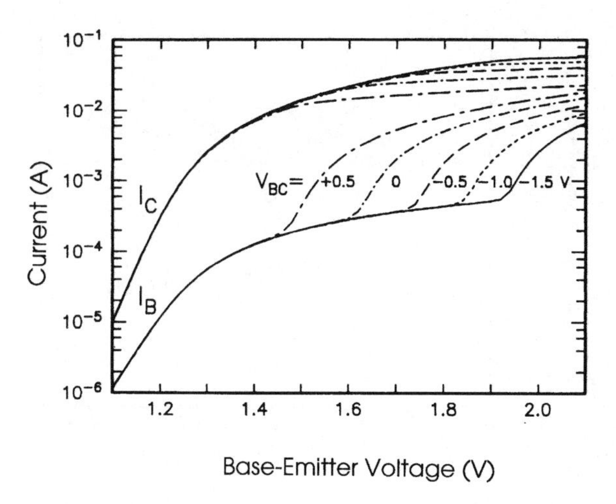

FIGURE 3.22 High current I_C versus V_{BE} (after Tiwari, Ref. 7 © IEEE).

field, leads to the appearance of some of this potential as an electron chemical barrier that retards the flow of electrons also. An electron barrier just begins to appear when the electrostatic field is the same as the electron chemical field resulting from alloy grading.

Figure 3.23 shows the conduction-band edge and quasi-Fermi level for electrons at the base–collector junction for base–emitter voltages of 1.3, 1.4, and 1.5 V and $V_{BC} = 0$ V. At high collector current densities, the mobile charge concentration becomes sufficiently large, resulting in the appearance of the bandgap grading-related

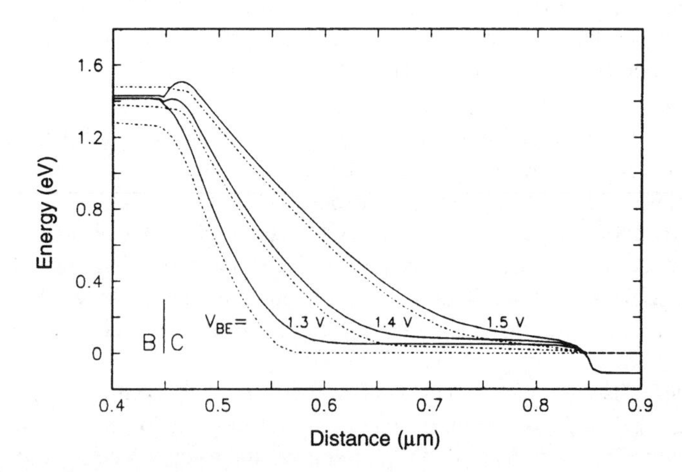

FIGURE 3.23 Conduction-band edge and electron quasi-Fermi level at $V_{BE} = 1.3$, 1.4, and 15 V (after Tiwari, Ref. 7 © IEEE).

electron chemical potential barrier. This barrier is about 0.07 eV at a forward voltage of 1.5 V at the base–emitter junction, corresponding to an emitter current density of 8×10^4 A/cm^2. The electron barrier is seen to form after the forward bias of 1.3 V and increases with forward biasing of the emitter–base junction. When a reverse bias is applied at the base–collector junction, the barrier is reduced because of an increase in the compensating electrostatic field, thus reducing the magnitude of this effect.

The degradation of collector current due to the conduction-band barrier effect may be related to the following degradation factor [23]:

$$f_c = \frac{D_n}{W_B v_s} \exp\left(\frac{-q\Phi_{\min}}{kT}\right) \tag{3.17}$$

where

$$\Phi_{\min} = -\frac{\varepsilon_C E_M^2}{2(J_C/v_s - qN_C)}$$

E_M is the electric field at the base–collector metallurgical junction, ε_C the permittivity in the collector, N_C the collector doping, and v_s the saturation velocity. The agreement between the analytic and device simulation in [23] seems to justify the simplifying assumptions and lends credibility to the physical explanation of this high-injection barrier phenomenon.

Joseph et al. [25] used a one-dimensional device simulator to examine the barrier effect of the SiGe HBT with a triangle Ge profile in the base. Electric field as a function of position in the base and collector at $J_C = 0.1$ and 4.0 mA/μm^2 is shown in Fig. 3.24. The collector–base voltage is equal to 0 V. The ambient temperature is at 200 K. As seen in Fig. 3.24, the electric field at $J_C = 4$ mA/μm^2 shows a bump at the base–collector heterojunction ($\approx$ 260 nm from emitter contact). The conduction-band

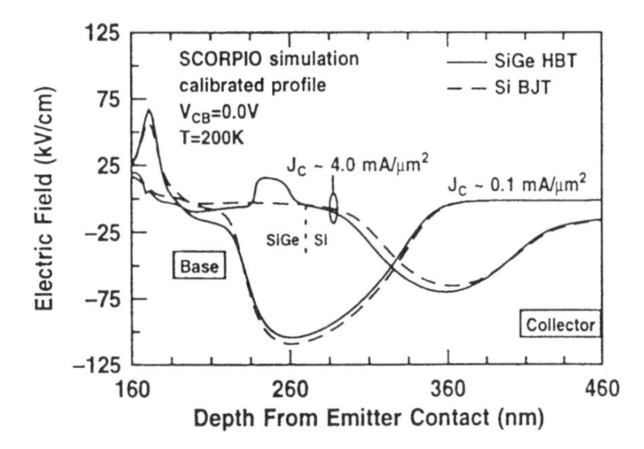

FIGURE 3.24 Electric field versus position to demonstrate the collector–base heterojunction barrier effect (after Joseph et al., Ref. 25 © IEEE).

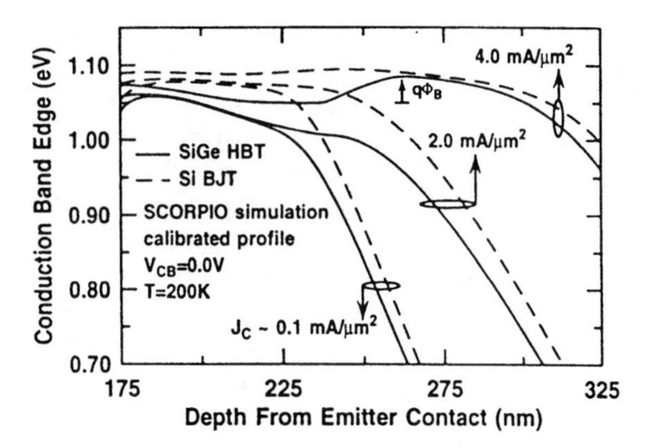

FIGURE 3.25 Simulated conduction-band edge of an SiGe HBT (after Joseph et al., Ref. 25 © IEEE).

edge in the base and collector at $J_C = 0.1, 2.0$, and 4.0 mA/μm^2 is displayed in Fig. 3.25. It is clear from this figure that a conduction barrier has been formed at a current density of 4 mA/μm^2.

3.2 SiGe HETEROJUNCTION BIPOLAR TRANSISTORS

The SiGe bipolar transistor in the BiCMOS process is attractive for low-cost, mixed-signal wireless applications, due to its high performance and process compatibility with Si technology. In this section, some unique features of the SiGe HBT not discussed in Section 3.1 are presented.

3.2.1 Current Gain and Early Voltage Product

The germanium profiles with the same total germanium content and graded-base width include uniform, trapezoid, and triangle profiles, as shown in Fig. 3.26. The grading of the Ge across the quasi-neutral base induces a drift field in the base which accelerates the electrons injected from the emitter to the collector, thereby decreasing the base transit time compared to a Si BJT. Figure 3.27 shows the SiGe/Si ratio for the three parameters of current gain, Early voltage, and gain–Early voltage product [26]. The x-axis represents the total Ge grading across the base. At $\Delta E_{g,\mathrm{Ge}}(\text{grade}) = 0$, a pure Ge box profile of 8.4% Ge is obtained, while at $\Delta E_{g,\mathrm{Ge}}(\text{grade}) = 125$ meV, a purely triangular profile of 0 to 18.6% Ge is obtained. The intermediate Ge trapezoid is located between these two extremes. The triangular profile has the largest Early voltage and gain–Early voltage product. The box profile, however, has the highest current gain since the enhancement of current gain depends exponentially on the boundary value in bandgap reduction at the emitter–base junction.

The advantages of the SiGe HBT can also be analyzed by the equations as follows. The collector current density of an n-p-n HBT where the bandgap can vary across the

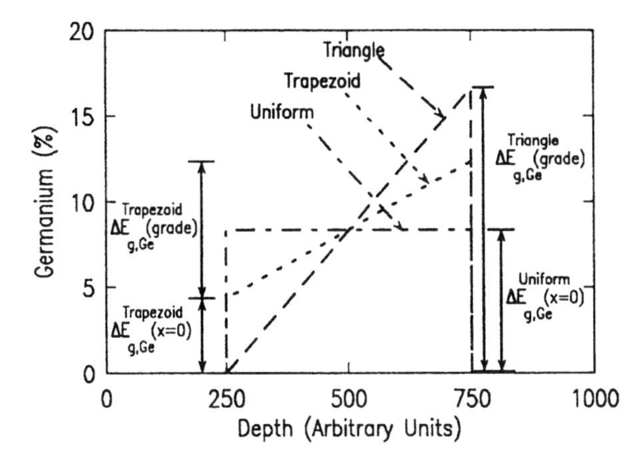

FIGURE 3.26 Uniform, triangle, and trapezoid Ge profiles in the base of an SiGe HBT.

base is given by

$$J_C = q\left[\int_0^{W_B} \frac{N_B(x)}{n_i^2(x)D_n(x)}\,dx\right]^{-1} e^{qV_{BE}/kT} \tag{3.18}$$

For the SiGe bipolar transistor with a trapezoid Ge profile, the intrinsic carrier concentration in the base is given by

$$n_i^2(x) = \begin{cases} n_i^2(0)e^{(\Delta E_{g,\mathrm{Ge}}(0)+\Delta E_g^{\mathrm{app}})/kT}\, e^{\Delta E_{g,\mathrm{Ge}}(\mathrm{grade})x/kTX_T} & \text{for } 0 \le x \le X_T \\ n_i^2(0)e^{(\Delta E_{g,\mathrm{Ge}}(W_B)+\Delta E_g^{\mathrm{app}})/kT} & \text{for } X_T \le x \le W_B \end{cases} \tag{3.19}$$

where $\Delta E_{g,\mathrm{Ge}}(0)$ is the bandgap reduction due to Ge concentration at $x = 0$, $\Delta E_{g,\mathrm{Ge}}(W_B)$ the bandgap reduction due to Ge concentration at $x = W_B$, $\Delta E_g^{\mathrm{app}}$ the apparent bandgap narrowing due to heavy doping effect in the base, and X_T the transition point between the linear and flat Ge concentrations in the base. $\Delta E_{g,\mathrm{Ge}}(\mathrm{grade}) = \Delta E_{g,\mathrm{Ge}}(W_B) - \Delta E_{g,\mathrm{Ge}}(0)$. For $X_T = 0$ the Ge base profile becomes uniform; for $X_T = W_B$, the Ge profile is linear.

Equation (3.18) is valid for the HBT without a conduction-band spike at the emitter–base heterojunction and when the base is at low level of injection. If the base current is caused only by hole injection from base into emitter, $J_B = J_{B0}\exp(qV_{BE}/kT)$, the maximum current gain at a moderate base–emitter bias is equal to

$$\beta = \frac{q}{J_{B0}}\left[\int_0^{W_B} \frac{p(x)}{n_i^2(x)D_n(x)}\,dx\right]^{-1} \tag{3.20}$$

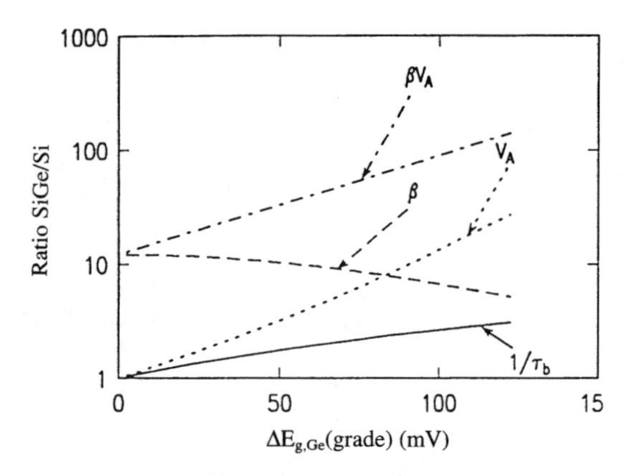

FIGURE 3.27 Early voltage and gain–Early voltage products (after Harame et al., Ref. 26 © IEEE).

where J_{B0} is the hole diffusion current density into the emitter at zero base–emitter bias. Examining Eq. (3.20), it is clear that the current gain of the SiGe HBT is larger than that of the Si BJT because of a smaller bandgap and a larger intrinsic carrier concentration in the SiGe base. Rewriting the Early voltage in (3.15) gives

$$V_A = J_C \left(\frac{\partial J_C}{\partial W_B} \frac{\partial W_B}{\partial V_{CB}} \right) \tag{3.21}$$

The change of base width with respect to collector–base voltage is

$$\frac{\partial W_B}{\partial V_{CB}} = -\frac{C_{JC}}{qN_B(W_B)} \tag{3.22}$$

where C_{JC} is the collector–base junction capacitance.

The change of collector current density with respect to base width is derived as

$$\frac{\partial J_C}{\partial W_B} = -J_C \frac{(N_B/n_i^2 D_n)_{W_B}}{\int_0^{W_B} (N_B/n_i^2 D_n)\, dx} \tag{3.23}$$

Inserting Eqs. (3.22) and (3.23) into (3.21) gives

$$V_A = \frac{qn_i^2(W_B)D_n(W_B)}{C_{JC}} \left(\int_0^{W_B} \frac{N_B}{n_i^2 D_n} dx \right) \tag{3.24}$$

From Eqs. (3.20) and (3.24) the current gain and Early voltage product is given simply by

$$\beta V_A = \frac{q^2}{J_{B0} C_{JC}} n_i^2(W_B) D_n(W_B) \tag{3.25}$$

It is clear from (3.25) that the βV_A product is a strong function of Ge concentration at the edge of the collector–base space-charge layer in the quasi-neutral base form $n_i^2(W_B)$. It is also clear that the current gain–Early voltage product of the SiGe HBT is larger than that of the Si BJT, due to a larger $n_i^2(W_B)$ value in the SiGe base.

Table 3.1 shows measured and calculated values of βV_A, J_{B0}, and C_{JC} for four devices with different Ge concentrations in the base. All devices have identical emitter and collector layers. The base consists of two 200-Å-thick SiGe layers with constant Ge profile. On both sides of the base, 40-Å-thick intrinsic SiGe spacer layers were inserted to remove spike-and-notch or conduction-band discontinuity. Using no adjustable parameters, the calculated results using (3.25) are in good agreement with the experimental data. The Early voltage is increased dramatically when the bandgap at the collector edge of the base is lower than the maximum bandgap in the base. A βV_A product of over 100,000 for device 3 with 14% Ge at the emitter side and 25% at the collector side in the base has been observed.

3.2.2 Temperature-Dependent Current Gain

Silicon bipolar transistors are generally not considered for operation at liquid-nitrogen temperatures (LNT) because of insufficient current gain [28]. The reduction in current gain is a result of the smaller bandgap in the emitter than in the base. The base current is typically limited by hole injection into the emitter, and its temperature dependence is determined largely by the effective bandgap of the emitter, which is

TABLE 3.1 Measured β, V_A, J_{B0}, and R_B with Various Ge Concentrations

	Device			
	1	2	3	4
% Ge at emitter	14	25	14	25
% Ge at collector	14	14	25	25
R_B	13.2	9.7	7.1	6.0
J_{B0}	3.7	3.9	4.9	10
Forward current gain β	750	1800	1400	1750
Early voltage V_A	18	6	120	44
Calculated βV_A	135,000	10,800	168,000	77,000
Measured βV_A	10,700	8,980	190,000	84,000

Source: After Prinz and Sturn, Ref. 27 © IEEE.

reduced because of a combination of heavy doping effects. The collector current, on the other hand, consists of electrons traversing the base, and its temperature dependence is therefore determined by the bandgap in the base. The difference in bandgap between the emitter and the base causes the current gain to decrease exponentially with temperature by two to three orders of magnitude between room temperature (RT) and liquid-nitrogen temperature. In addition to the drop in current gain, the base resistivity increases sharply because of carrier freeze-out. This effective reduction in base doping provides relatively more collector current but also increases current crowding, which is expected to degrade switching performance.

Recent advances in high-quality low-temperature epitaxial processes offer great potential for achieving the very thin, heavily doped, and abrupt base profiles required for continued vertical scaling of bipolar transistors. Because of the use of smaller bandgap and graded germanium content in the base region, the current gain of the SiGe heterojunction bipolar transistor is superior to that of Si bipolar transistors.

Proper bipolar transistor scaling and design reduce the current gain sensitivity to temperature. Figure 3.28 shows the transistor current gain as a function of reciprocal temperature for bipolar technologies of varying age. In the legend, I/I stands for ion implantation, DD stands for double diffusion, and LTE stands for low-temperature epitaxy. The SiGe technology produces a current gain at LNT which is essentially unchanged from its RT value. The high current gain at low temperature is a result of both the heavily doped base and the presence of SiGe, which offsets the bandgap narrowing in the emitter. These results suggest that silicon–germanium bipolar technology is suitable for very high speed applications at cryogenic temperatures.

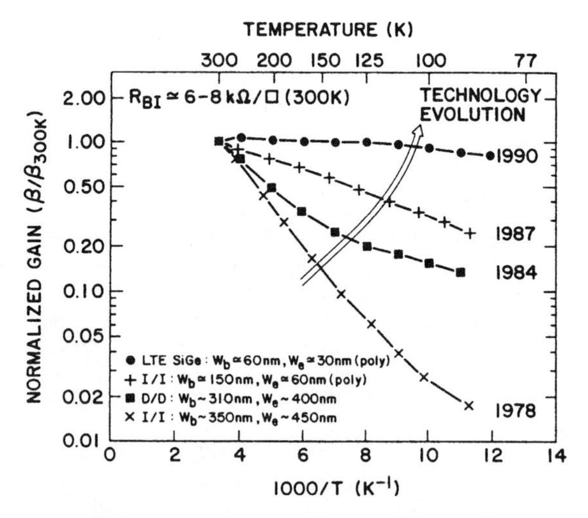

FIGURE 3.28 Current gain versus temperature for different process technologies (after Cressler et al., Ref. 29 © IEEE).

3.2.3 Current Gain Roll-off in Graded SiGe Base

For Si homojunction bipolar transistors, the current gain at a moderate injection level is relatively independent of base–emitter bias and collector current [30]. For SiGe heterojunction bipolar transistors with graded bases, however, the current gain exhibits a significant emitter–base bias dependence at moderate V_{BE} [31]. The nonideal collector current is caused by the interaction of the bias dependence of the emitter–base space-charge region width and the exponential dependence of the collector current on the Ge concentration at the edge of the emitter–base space-charge region. For example, the saturation collector current density in a graded-base HBT with a uniformly-doped base is

$$J_{C0} \approx \frac{qD_n n_i^2}{N_B W_B} \frac{\Delta E_{g,\mathrm{Ge}}(\mathrm{grade})}{kT} e^{\Delta E_{g,\mathrm{Ge}}(0)/kT} \tag{3.26}$$

The saturation current is an exponential function of $\Delta E_{g,\mathrm{Ge}}(0)$. The location of $x = 0$, however, is moving with respect to the emitter–base bias. Base-width modulation effect at the emitter–base junction changes the location of the emitter–base space-charge-layer edge in the base region. For increasing V_{BE}, the emitter–base space-charge layer shrinks. This results in a smaller $\Delta E_{g,\mathrm{Ge}}(0)$ value for a linearly graded Ge base, as indicated in Fig. 3.29. The exponential dependence of the collector current

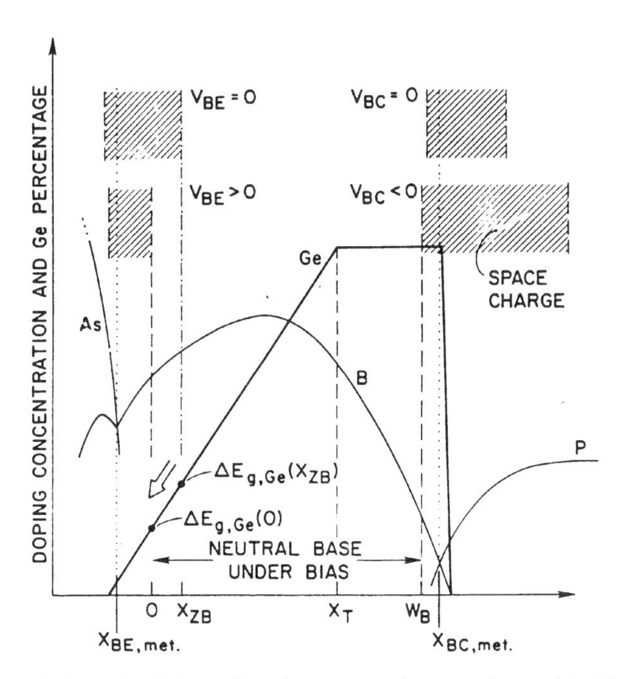

FIGURE 3.29 Schematic of the emitter–base space-charge region and its bias dependence (after Crabbé et al., Ref. 31 © IEEE).

on this bandgap reduction leads to a significant decrease in the collector current density, even for a minor variation of the width of the emitter–base space-charge region. The normalized current gain versus collector current density at 300 and 85 K is displayed in Fig. 3.30. The normalized current gain decreases with J_C due to the base-width modulation effect at the emitter–base junction. The decrease in current gain is faster when Ge grading is increased. This effect is much more pronounced at low temperatures because J_{C0} is exponentially dependent on the ratio of $\Delta E_{g,\mathrm{Ge}}(0)$ to kT.

3.2.4 Early Voltage, Including Recombination in the SiGe Base

The Early voltage predicted in (3.24) does not account for the effect of neutral-base recombination (NBR). For the heterojunction bipolar transistor with a heavily doped base, the neutral-base recombination is significant. Analytical equations to examine the output conductance of silicon bipolar transistors with large neutral-base recombination current were derived by McGregor et al. [32]. Neutral-base recombination increases the output conductance at a given collector current. A similar analytical study was published by Mohammandi and Selvakumar [33].

We now present the Early voltage of SiGe bipolar transistors with neutral-base recombination in the base. Using (3.21), the derivative of J_C with respect to W_B is given by

$$\frac{dJ_C}{dW_B} = J_C \left[\frac{q}{kT} \frac{dV_{BE}}{dW_B} - \frac{(N_B/D_n n_i^2)_{W_B}}{\int_0^{W_B} (N_B/D_n n_i^2)\, dx} \right] \tag{3.27}$$

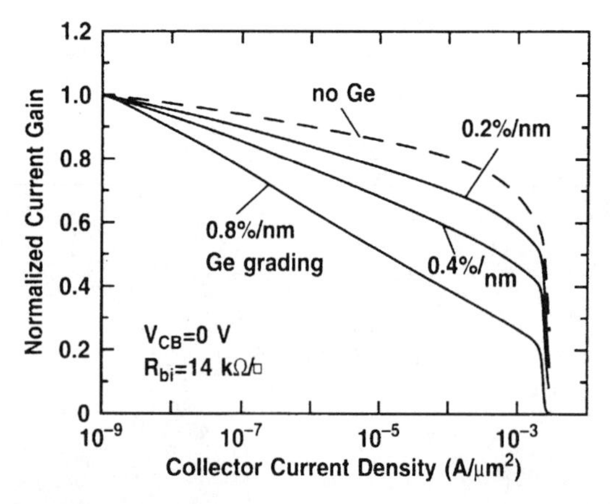

FIGURE 3.30 Normalized current gain versus collector current density (after Crabbé et al., Ref. 31 © IEEE).

Neglecting emitter–base space-charge-layer recombination, the total base current density is written

$$J_B = J_{B0} e^{qV_{BE}/kT} + q \int \frac{n(x, W_B)}{\tau(x)}\, dx \tag{3.28}$$

where J_{B0} is the preexponential base current density for hole injection into the emitter and τ is the position-dependent electron lifetime in the base. The first term in Eq. (3.28) represents the recombination current in the emitter and the second term represents the recombination in the base. At constant J_B, dV_{BE}/dJ_B could be set to zero. The derivative of (3.28) gives

$$J_{B0} e^{qV_{BE}/kT} \frac{q}{kT} \frac{dV_{BE}}{dW_B} + q \frac{d}{dW_B} \int_0^{W_B} \frac{n(x, W_B)}{\tau(x)}\, dx = 0 \tag{3.29}$$

Using Leibniz's rule for the differentiation of integrals yields

$$\frac{d}{dW_B} \int_0^{W_B} \frac{n(x, W_B)}{\tau(x)}\, dx = \int_0^{W_B} \frac{1}{\tau(x)} \frac{\partial n(x, W_B)}{\partial W_B}\, dx + \frac{n(W_B, W_B)}{\tau(W_B)} \tag{3.30}$$

Since $n(W_B)$ is negligible at the forward active mode, (3.30) can be rewritten

$$J_{B0} \frac{e^{qV_{BE}/kT}}{kT/q} \frac{dV_{BE}}{dW_B} = -q \int_0^{W_B} \frac{1}{\tau} \frac{\partial n}{\partial W_B}\, dx \tag{3.31}$$

Using

$$n(x) = \frac{e^{qV_{BE}/kT}}{\int_0^{W_B} (N_B/D_n n_i^2)\, dx} \frac{n_i^2}{N_A} \int_x^{W_B} \frac{N_B}{D_n n_i^2}\, dy \tag{3.32}$$

and solving (3.31) gives

$$\frac{q}{kT} \frac{\partial V_{BE}}{\partial W_B} = \frac{\left(\dfrac{N_A}{D_n n_i^2}\right)_{W_B} \left[\left(\dfrac{J_{C0}}{q}\right) \int_0^{W_B} \left(\dfrac{n_i^2}{\tau N_B}\right) \int_x^{W_B} \left(\dfrac{N_B}{D_n n_i^2}\right) dy\, dx - \int_0^{W_B} \left(\dfrac{N_B}{D_n n_i^2}\right) dx\right]}{\left(\dfrac{J_{B0}}{J_{C0}}\right) + \int_0^{W_B} \left(\dfrac{n_i^2}{\tau N_B}\right) \int_x^{W_B} \left(\dfrac{N_B}{D_n n_i^2}\right) dy\, dx} \tag{3.33}$$

where

$$J_{C0} = \frac{q}{\int_0^{W_B} (N_B/D_n n_i^2)\, dx} \tag{3.34}$$

Substituting Eq. (3.33) into (3.27) leads to

$$\frac{\partial J_C}{\partial W_B} = -J_C \left(\frac{N_B}{D_n n_i^2} \right)_{W_B} \frac{(J_{B0}/q) + \int_0^{W_B} (n_i^2/\tau N_B)\, dx}{(J_{B0}/J_{C0}) + \int_0^{W_B} (n_i^2/\tau N_B) \int_x^{W_B} (N_B/D_n n_i^2)\, dy\, dx} \tag{3.35}$$

Inserting (3.22) and (3.35) into (3.21) gives [34]

$$V_A = \frac{q D_n(W_B) n_i^2(W_B)}{C_{BC}} \times \frac{(J_{B0}/J_{C0}) + \int_0^{W_B} \left[n_i^2(x)/\tau(x) N_B(x)\right] \int_x^{W_B} \left[N_B(y)/D_n(y) n_i^2(y)\right] dy\, dx}{(J_{B0}/q) + \int_0^{W_B} \left[n_i^2(x)/\tau_n(x) N_B(x)\right] dx}. \tag{3.36}$$

The effect of neutral-base recombination on the Early voltage of SiGe bipolar transistors can be evaluated by the recombination factor defined as follows:

$$f_R \equiv \frac{V_A(\tau = \text{finite})}{V_A(\tau \to \infty)} = \frac{J_{B0} + J_{C0} \int_0^{W_B} \left[n_i^2(x)/\tau(x) N_B(x)\right] \int_x^{W_B} \left[N_B(y)/D_n(y) n_i^2(y)\right] dy\, dx}{J_{B0} + q \int_0^{W_B} \left[n_i^2(x)/\tau(x) N_B(x)\right] dx} \tag{3.37}$$

If the recombination current in the emitter is much larger than that of the base, $\tau \to \infty$ and $f_R \to 1$. If the recombination current in the base is much larger than that of the emitter, $J_{B0} \approx 0$. For the bipolar transistor with a trapezoid Ge concentration (including linear profile also) in the uniformly doped base, (3.37) reduces to

$$f_R\big|_{J_{B0}\to 0} = \frac{A + e^{-A} - 1 - A(1 - e^{-A})(1 - W_B/X_T) + 0.5A^2(1 - W_B/X_T)^2}{e^A + e^{-A} - 2 - A(e^A - e^{-A})(1 - W_B/X_T) + A^2(1 - W_B/X_T)^2} \tag{3.38}$$

where $A = \Delta E_{g,\text{Ge}}(\text{grade})/kT$.

Figure 3.31 shows the effect of X_T/W_B on the Early voltage of a SiGe bipolar transistor normalized to that of the Si bipolar transistor. In this plot the solid lines represent the analytical model in (3.36) and the empty squares represent the predictions using [35]. Good agreement between the analytical predictions using (3.36) and in [35] has been obtained. Note that the results in [35] only account for linear Ge grading in the base (i.e., $X_T/W_B = 1$), without neutral-base recombination. It is clear from Fig. 3.31 that increasing X_T/W_B enhances the normalized Early voltage. The normalized Early voltage is plotted against $\Delta E_{g,\text{Ge}}(\text{grade})/kT$ in Fig. 3.32. The empty squares represent the linear Ge concentration in the base. As shown in Fig. 3.32, the larger the Ge grading, the larger the normalized Early voltage. The increasing rate is nearly exponential. The normalized Early voltage for a flat Ge profile ($X_T/W_B = 0$) is independent of $\Delta E_{g,\text{Ge}}(\text{grade})/kT$, which is consistent with device physics.

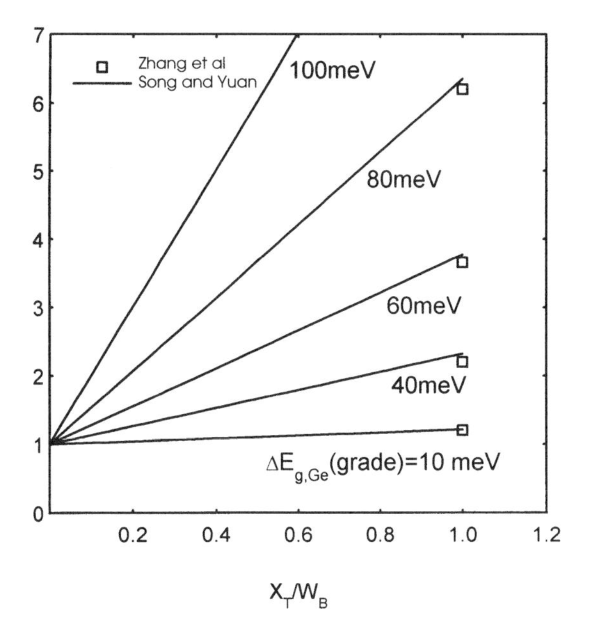

FIGURE 3.31 Normalized Early voltage versus X_T/W_B.

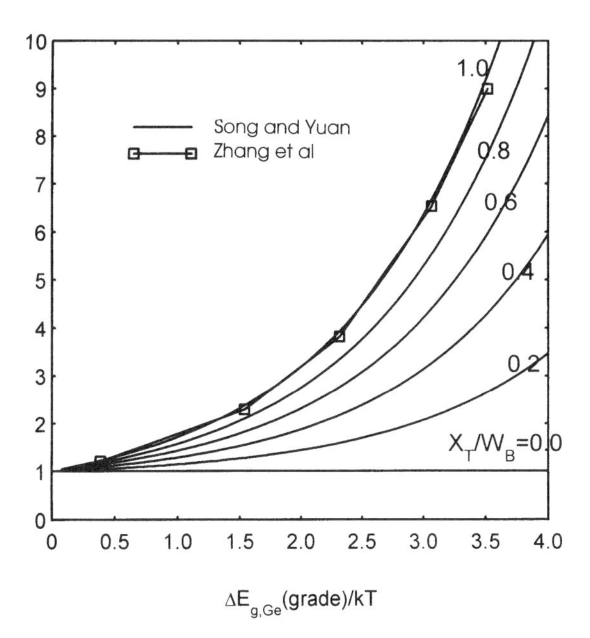

FIGURE 3.32 Normalized Early voltage versus $\Delta E_{g,\mathrm{Ge}}(\mathrm{grade})/kT$.

As indicated in [32,33], neutral-base recombination current changes the Early voltage and output conductance of bipolar transistors. The normalized Early voltage versus $\Delta E_{g,\mathrm{Ge}}(\mathrm{grade})/kT$ between two extreme conditions (0 and 100% neutral-base recombination) is shown in Fig. 3.33. In this figure the dashed line represents the HBT without neutral-base recombination, the dotted line represents the device with 100% neutral-base recombination (i.e., no recombination in the emitter), the solid line represents the model predictions with both neutral-base recombination and back injection into the emitter, and the solid circles represent the experimental data [27]. In this plot, the SiGe HBTs have two stepped flat Ge profiles with positive and negative $\Delta E_{g,\mathrm{Ge}}(\mathrm{grade})/kT$. The agreement between the analytical predictions using (3.36) and experimental data is very good. The analytical predictions without taking into account neutral-base recombination are overestimated. The predictions taking into account 100% neutral-base recombination (or no back injection into the emitter) give the normalized Early voltage of 0.5, independent of the Ge grading.

The recombination factor f_R in (3.38) for the Early voltage versus X_T/W_B is displayed in Fig. 3.34. At $X_T/W_B = 0$, the recombination factor is 0.5. This is identical to that published in [32] for a silicon bipolar transistor without Ge content in the base. When both Ge grading [i.e., $\Delta E_{g,\mathrm{Ge}}(\mathrm{grade})/kT$] and X_T/W_B increase, the recombination factor decreases significantly. This indicates that the neutral-base recombination current must be considered in evaluation of the Early voltage when Ge grading is increased. This plot also demonstrates that a fast degradation occurs for X_T/W_B from 0 to about 0.4. The degradation is more significant for larger Ge grading.

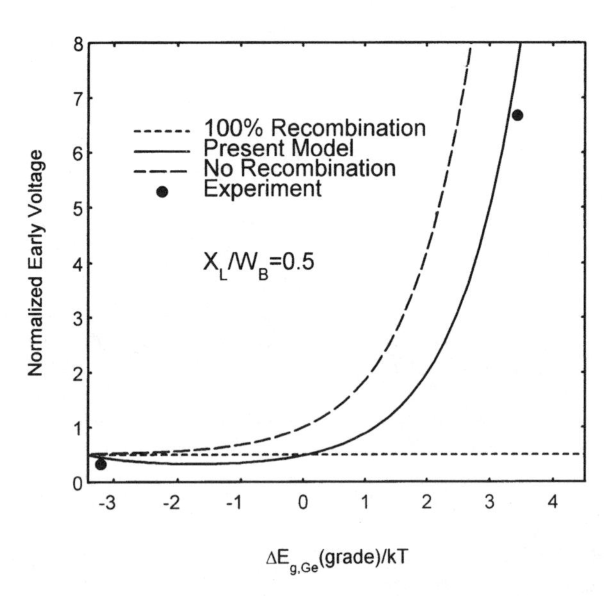

FIGURE 3.33 Normalized Early voltage versus $\Delta E_{g,\mathrm{Ge}}(\mathrm{grade})/kT$ (with and without neutral-base recombination).

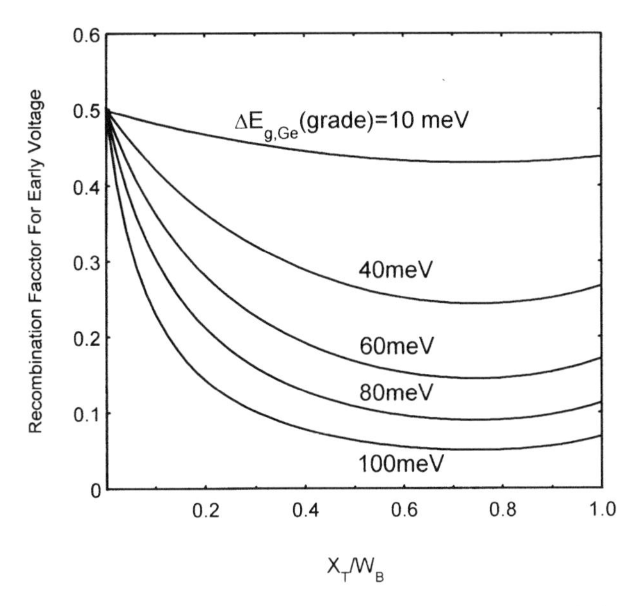

FIGURE 3.34 Recombination factor for the Early voltage versus X_T/W_B for various $\Delta E_{g,\mathrm{Ge}}(\mathrm{grade})/kT$ values.

It is clear that base bandgap and temperature significantly affect the βV_A product of transistors in the presence of neutral-base recombination. Figures 3.35 and 3.36 show the typical common-emitter output characteristics ($I_C - V_{CE}$) for a SiGe HBT obtained using forced I_B and forced V_{BE} at 358 and 200 K, respectively. In a forced-I_B situation, V_{BE} is allowed to change in such a way as to maintain constant I_B. Due to the fact that I_B decreases with increasing V_{CB} in the presence of neutral-

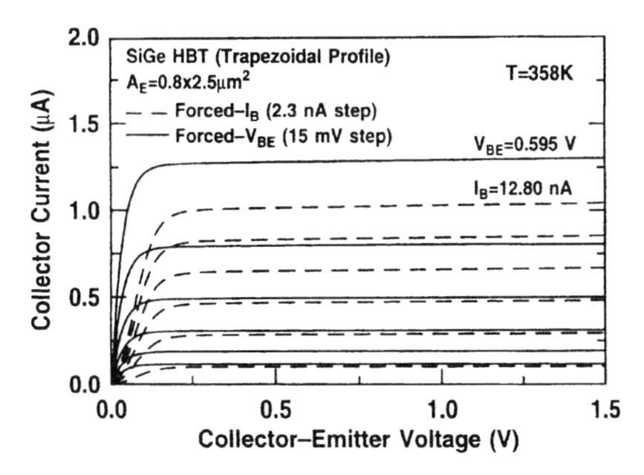

FIGURE 3.35 Collector current versus collector–emitter voltage under forced V_{BE} and I_B conditions at 358 K (after Joseph et al., Ref. 36 © IEEE).

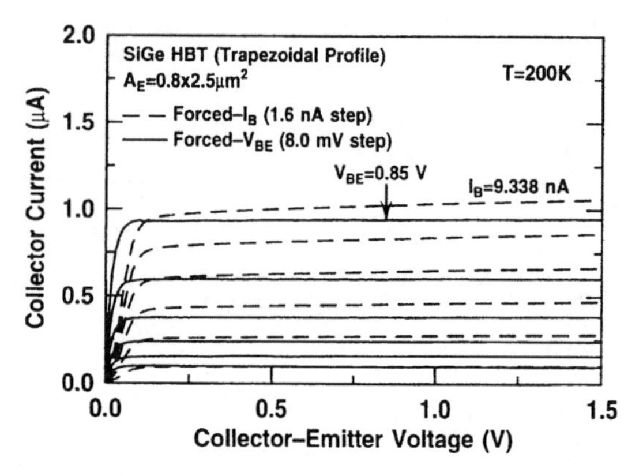

FIGURE 3.36 Collector current versus collector–emitter voltage under forced V_{BE} and I_B conditions at 200 K (after Joseph et al., Ref. 36 © IEEE).

base recombination, V_{BE} is forced to increase so as to maintain a constant I_B. The analytical equation of V_A under forced-V_{BE} conditions is given by

$$V_A(\text{forced} - V_{BE}) = J_C(0)\left(\left.\frac{\partial I_C}{\partial V_{CB}}\right|_{V_{BE}}\right)^{-1} \approx \left\{\frac{(\partial/\partial V_{CB})\left[\int_0^{W_B}(dx/n_{ib}^2)\right]}{\int_0^{W_B}(dx/n_{ib}^2)}\right\}^{-1} \tag{3.39}$$

In a forced-V_{BE} situation, V_{BE} is kept constant. The increase of I_C is due to the decrease in base width for an increase of V_{CB}. The Early voltage under forced-I_B conditions is approximated as

$$V_A(\text{forced} - I_B) = J_C(0)\left(\left.\frac{\partial I_C}{\partial V_{CB}}\right|_{I_B}\right)^{-1} \approx \left\{\frac{q}{kT}\left(\frac{\partial V_{BE}}{\partial V_{CB}}\right)_{I_B} - \frac{(\partial/\partial V_{CB})\left[\int_0^{W_B}(dx/n_{ib}^2)\right]}{\int_0^{W_B}(dx/n_{ib}^2)}\right\}^{-1} \tag{3.40}$$

It is apparent from Figs. 3.35 and 3.36 that with cooling, the slope of I_C with respect to V_{CE} increases in forced-I_B measurement, as opposed to a decrease in the same quantity for a forced-V_{BE} measurement. Figure 3.37 displays V_A obtained for both Si and SiGe transistors using forced-I_B and forced-V_{BE} conditions as a function of reciprocal temperature. The exponential degradation of V_A measured using a forced-

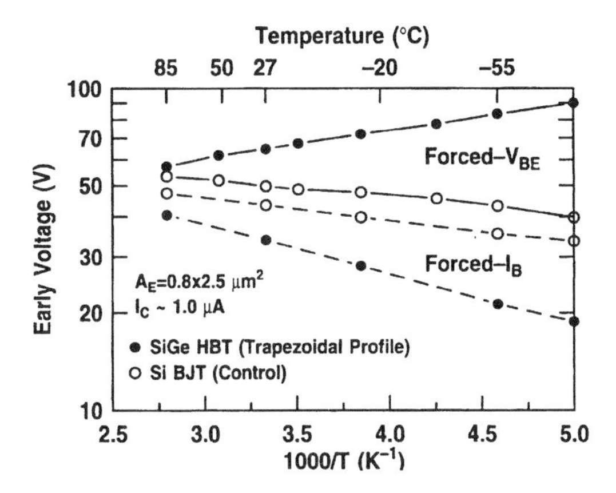

FIGURE 3.37 Early voltage versus inverse temperature (after Joseph et al., Ref. 36 © IEEE).

I_B technique in the trapezoidal-base SiGe HBT is due to the thermally activated nature of the neutral-base recombination of I_B through $\Delta E_{g,\mathrm{Ge}}(x = 0)$.

Precision current sources (CSs), which are used extensively in many analog circuits, rely on the Early voltage of a transistor to maintain constant current output for large output voltage swings, and thus require large output resistance. Due to the significant differences between the V_A of a SiGe HBT, depending on how the device is biased at the input, it is expected that the presence of neutral-base recombination will have a strong impact on the temperature characteristics of such SiGe HBT current sources. Figure 3.38 shows the cascode and Wilson current sources used in high-precision analog circuits. The impact of NBR on the temperature characteristics of these circuits has been examined using SPICE. The temperature dependence of NBR was modeled by introducing a collector-to-base resistance whose values were determined experimentally from the inverse slope of the I_B versus V_{CB} characteristics. Figure 3.39 shows the close agreement between the modified SPICE model predictions and the measured common-emitter output characteristics for the SiGe HBT at 358 and 200 K, respectively. The output resistance of cascode and Wilson current sources versus temperature predicted by SPICE simulation with and without NBR is shown in Fig. 3.40. Under the ideal situation (i.e., without NBR), both current sources show an increase in the output resistance with cooling. In the presence of NBR in the transistors, however, the output resistance of both current sources is not only smaller compared to the ideal case but also has the opposite temperature dependence.

3.2.5 Inverse Base Width Modulation Effect

Base pushout is the usual mechanism underlying the high current degradation of bipolar transistors. At high current densities, mobile carriers modulate the collector–base junction space charge and push it into the epitaxial collector. The effective quasi-neutral base widens. For SiGe bipolar transistors with high collector doping concentration, base pushout effect is reduced. Two physical mechanisms, inverse base

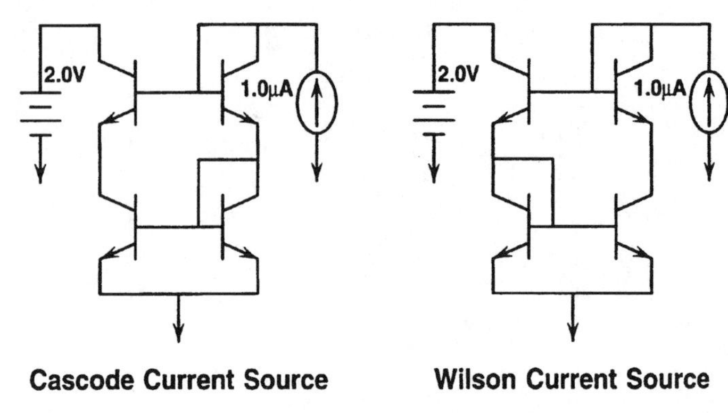

FIGURE 3.38 Cascode and Wilson current sources.

width modulation and collector space-charge-region widening, become important before base pushout. Inverse base width modulation is a widening of the quasi-neutral base width. The collector–base space-charge-region layer on the base side shrinks as the collector current density increases. Mobile carriers modulate the space charge and diminish the electric field at the junction. The inverse base width modulation decreases the current gain and increases the base transit time of the HBT [20].

The base widening due to inverse base width modulation and collector space-charge-layer widening is illustrated in Fig 3.41. The magnitude of the peak electric field in the collector–base space-charge region diminishes with increasing collector current density because of current-induced perturbation and ohmic drops in the collector region. This causes the collector–base space-charge-layer thickness on the base side to decrease and hence the quasi-neutral base width to increase. The inverse base width modulation begins at a collector current density below the onset current density for base pushout.

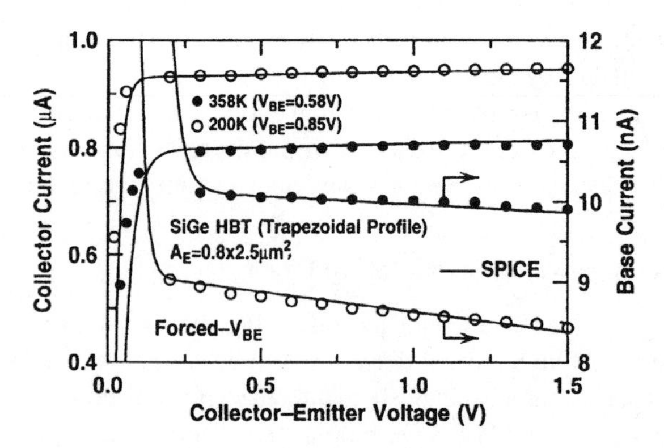

FIGURE 3.39 Collector current versus collector–emitter voltage (SPICE prediction versus experiment) (after Joseph et al., Ref. 36 © IEEE).

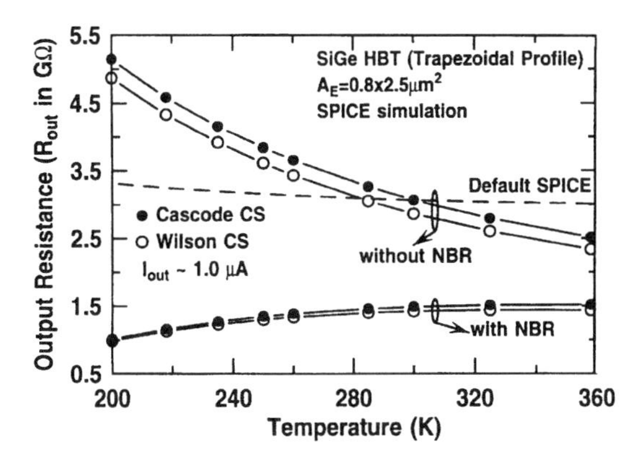

FIGURE 3.40 Output resistance versus temperature (after Joseph et al., Ref. 36 © IEEE).

Assuming that the base doping profile is exponential, the thicknesses of W_{SC} and W_{SCC} are given by [20]

$$W_{SC} \approx \frac{W_{BM}}{\eta} \ln\left(\frac{\varepsilon \eta E_M}{q W_{BM} N_C} + 1\right) \tag{3.41}$$

$$W_{SCC} = \sqrt{\frac{2\varepsilon_C A v_s (\phi_C - V'_{BC} - I_C R_{EPI})}{I_0 - I_C}} \tag{3.42}$$

where $N_C = N_{A0} \exp(-\eta)$, R_{EPI} is the epitaxial collector resistance, $I_0 = qAv_s N_{EPI}$, and $V'_{BC} = V_{BC} + I_C R_C - I_B R_B$.

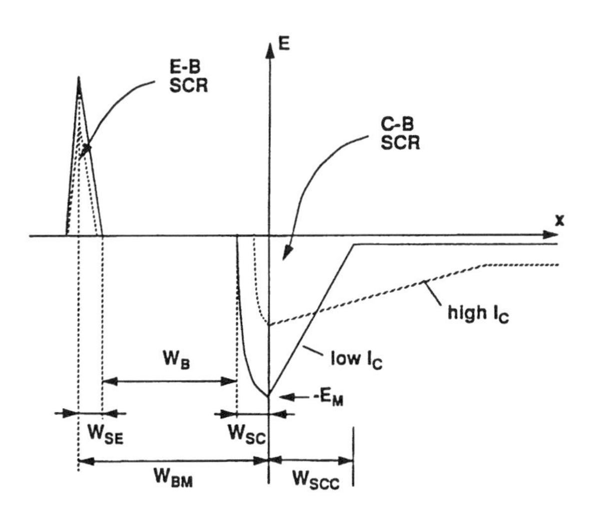

FIGURE 3.41 Bipolar doping profile for demonstrating inverse base width modulation effect (after Ugain et al., Ref. 20 © IEEE).

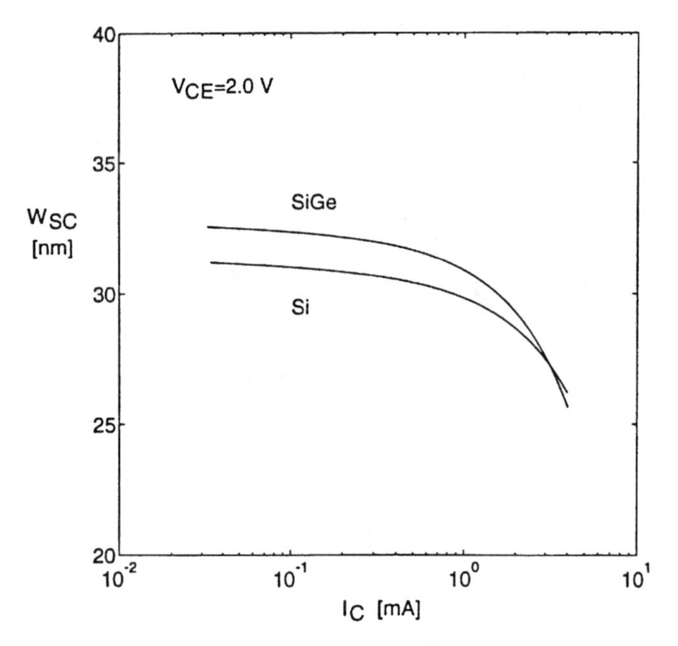

FIGURE 3.42 W_{SC} versus I_C (after Ugajin et al., Ref. 20 © IEEE).

Figure 3.42 shows the dependence of W_{SC} on the collector current with and without Ge incorporation. The two transistors have an effective emitter area of $0.2 \times 3.0 \mu m^2$. Although the low-I_C values of W_{SC} differ between the two devices (because of slight differences in base doping near the collector junction), both devices show significant decrease in W_{SC} at $I_C \approx 2$ mA. This result supports the significance of inverse base width modulation at high collector currents in highly scaled transistors.

3.3 III/V COMPOUND HETEROJUNCTION BIPOLAR TRANSISTORS

In this section we address some unique dc characteristics of III/V compound HBTs not presented in Section 3.1.

3.3.1 Self-Heating Effect

Heterojunction bipolar transistors using GaAs semi-insulating substrates exhibit self-heating effects due to low thermal conductivity of the GaAs semi-insulating substrate (about one-third of Si thermal conductivity) [37]. The collector current of the heterojunction bipolar transistor shows a negative differential resistance at high power dissipation levels, as seen in Fig. 3.43. Unlike the silicon bipolar transistor, which has a positive temperature coefficient for current gain, the heterojunction bipolar transistor exhibits a decrease in current gain with increasing collector–base voltage

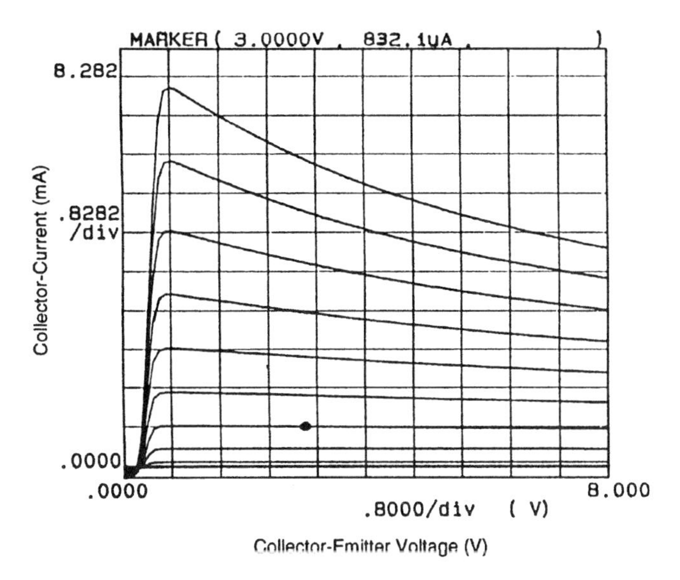

FIGURE 3.43 Collector current versus collector–emitter voltage to demonstrate self-heating effect.

due to self-heating. The negative temperature coefficient of the current gain results from bandgap narrowing in the base of the HBT. For example, self-heating causes the increase in base and collector currents. The increase in base current is larger than that of the collector current. Thus, for a constant base current in the I_C–V_{CE} plot, the collector current decreases with increasing collector–emitter voltage. The negative resistance effect becomes significant when the power dissipation is large. Therefore, reducing the self-heating effect is very important for power heterojunction bipolar transistor circuit design.

Self-heating could change analog circuit response significantly. For example, consider the current mirror in Fig. 3.44, with two discrete heterojunction bipolar transistors Q_1 and Q_2 and two external resistors R_1 and R_2. Since the collector–emitter voltage of Q_1 equals the base–emitter voltage V_{BE1}, the bipolar transistor Q_1 usually does not experience self-heating. However, the collector–emitter voltage of the bipolar transistor Q_2 is $V_{CC} - I_{C2}R_2$ and the bipolar transistor Q_2 can have significant self-heating if the collector current I_{C2} and the collector–emitter voltage of Q_2 are high. Self-heating of Q_2 can thus result in a totally different output current, which is much higher than the designed current at the isothermal condition. For a fixed resistor R_1, for instance, the collector current I_{C1} of the reference transistor Q_1 is constant. The collector current I_{C2} of transistor Q_2 follows the reference current I_{C1} when $R_1 = R_2$. If the resistance of R_2 is decreased, the collector current I_{C2} will increase due to the increase of collector–emitter voltage at Q_2. The output current I_{C2} will then be much larger than the reference current.

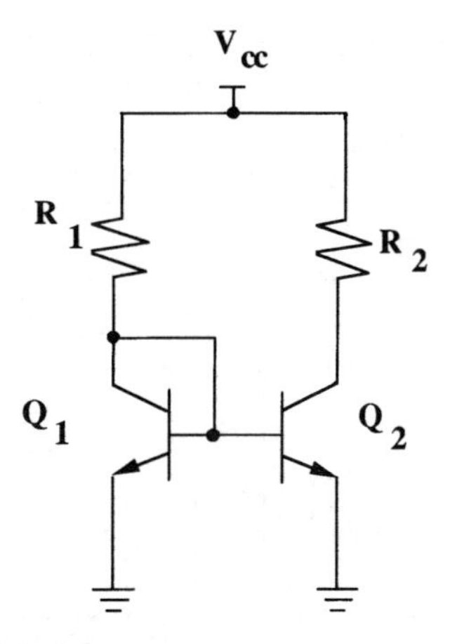

FIGURE 3.44 Schematic of a current mirror.

Figure 3.45 shows the sensitivity of the collector currents for reference transistor Q_1 and mirror transistor Q_2 as a function of $1/R_2$ at $V_{CC} = 15$ V. The resistor R_1 is selected at 2 kΩ in order to make the base–emitter voltage V_{BE} equal to 1.4 V. The mirror current designed to operate at the isothermal condition should follow the reference current (solid line in Fig. 3.45). The predicted mirror current with

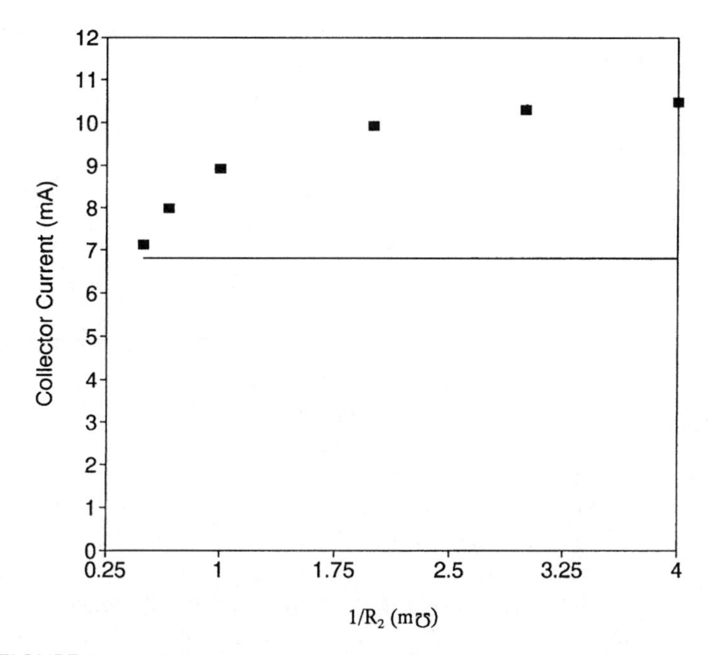

FIGURE 3.45 Collector current versus $1/R_2$ (after Yuan, Ref. 38 © IEEE).

self-heating, however, increases with decreasing R_1 due to an increase in the collector–emitter voltage for Q_2. The solid squares of Fig. 3.45 represent R_1 simulated at 2 $k\Omega$, 1.5 $k\Omega$, 1 $k\Omega$, 500 Ω, 333 Ω, and 250 Ω, respectively. The corresponding relative error $(I_{C2} - I_{C1})/I_{C1}$ is 4.85, 17.3, 31, 46, 51.3, and 54.1%, respectively.

Since the size of the intrinsic HBT (e.g., intrinsic emitter, base, and collector) is much smaller than that of the extrinsic HBT (e.g., subcollector and semi-insulating substrate), the temperature in the intrinsic HBT can be considered spatially independent. Heat generated in the intrinsic HBT is dissipated primarily through an effective heat diffusion area in the semi-insulating substrate. The heat power P_S generated in the HBT is thus

$$P_S = J_C V_{CE} A_E \tag{3.43}$$

The temperature increment due to self-heating is related to power dissipation and thermal resistance R_{th}:

$$\Delta T = T - T_0 = P_S R_{\text{th}} \tag{3.44}$$

where T is the lattice temperature and T_0 is the ambient temperature (300 K). Note that the heat is dissipated throughout the semi-insulating substrate with a lateral diffusion angle θ [39] as shown in Fig. 3.46, and the thermal conductivity of GaAs is proportional to $(T/T_0)^{-b}$. This, together with the Kirchhoff transformation [40],

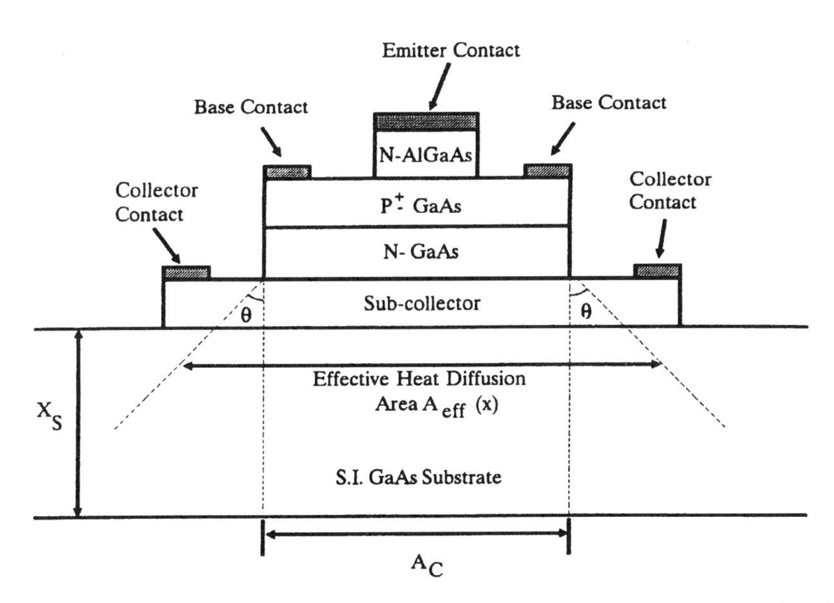

FIGURE 3.46 Heat diffusion of an HBT throughout the semi-insulating substrate (after Liou et al., Ref. 39 © IEEE)

yields

$$R_{\text{th}} = \frac{\eta - T_0}{P_S} \tag{3.45}$$

where

$$\eta = \left[\frac{1}{T_0^{b-1}} - \frac{(b-1)R_{\text{th0}}P_s}{T_0^b} \right]^{-1/b-1} \tag{3.46}$$

Here R_{th0} is the thermal resistance for the case that the thermal conductivity in the semi-insulating substrate is temperature independent and R_{th} is the temperature-dependent thermal resistance.

For microwave HBTs, the multiemitter finger structure is frequently used. Such a structure allows less power to be generated in each HBT unit cell, thus making the self-heating effect less prominent than to its single-emitter finger counterpart. The thermal effect in the multiemitter HBT is more complicated than that in the single-emitter finger HBT, because thermal coupling among the neighboring fingers is also important in the multiemitter finger HBT. The combination of self-heating and thermal coupling is examined in detail in Chapter 7. Note that InAlAs/InGaAs HBTs on InP substrates have a relatively smaller self-heating effect because of the higher substrate thermal conductivity of InP substrates (0.7 versus 0.46 W cm^{-1} K^{-1}). A list of thermal conductivities for key semiconductor materials is given in Table 3.2.

The self-heating effect in III/V HBTs could be enhanced in high-temperature operation. The dc performance of a single-emitter finger HBT operating at high ambient temperature (between 300 and 500 K) has been investigated [41]. Both large ($100 \times 100\mu\text{m}^2$)- and small ($2 \times 10\mu\text{m}^2$)-emitter-area HBTs were considered. The

TABLE 3.2 Thermal Conductivity of Elemental and Compound Semiconductors

Semiconductor	κ(W $\text{cm}^{-1}\text{K}^{-1}$)
Si	1.5
Ge	0.6
AlP	0.9
AlAs	0.8
AlSb	0.57
GaAs	0.46
GaP	0.77
GaSb	0.39
InAs	0.273
InP	0.68
InSb	0.166

self-heating effect of the large-emitter-area HBT is less significant than that of the small emitter area HBT because the large emitter area allows heat generation to be dissipated quickly.

3.3.2 Recombination Currents

The major sources of generation–recombination effects in the heterojunction bipolar transistors are the emitter–base and collector–base space-charge regions, the quasi-neutral base and emitter, the surface at the extrinsic base, and the surface at the junction depletion region edges. The emitter quasi-neutral recombination can be ignored in most cases because hole injection is suppressed efficiently by the valence-band discontinuity ΔE_V. The recombination in the collector–base space-charge region is not important because the junction is reverse biased under normal operating conditions. Figure 3.47 shows the relative recombination rates of recombination in the space-charge region and at the surface compared to quasi-neutral recombination. This figure shows that the electrons injected into the base recombine copiously at the surface of the GaAs base and at the base–emitter junction and that holes injected from the base recombine at the GaAlAs emitter surface. Of these, the base surface and the base–emitter space-charge recombination are the dominant components. The quasi-neutral base recombination is the smallest.

The space-charge recombination in the base–emitter junction has a $2kT$ exponential dependence. Surface recombination at the extrinsic base behaves similar to bulk recombination in the quasi-neutral base. Surface recombination at the junction depletion region edges behaves similarly to conventional recombination in the depletion region. Tiwari and Frank [9] have analyzed the surface recombination numerically using the single acceptor–donor model. At low currents, the surface recombination

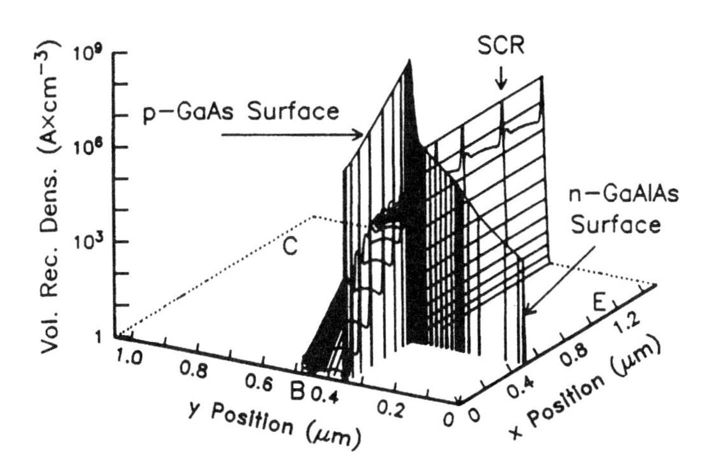

FIGURE 3.47 Relative recombination rates of a heterojunction bipolar transistor (after Tiwari and Frank, Ref. 9 © IEEE).

current density follows a $1.8kT$ behavior. At high currents and high surface recombination velocity, the ideality factor shows a $1.2kT$ dependence. The effective way of separating these "surface" recombination mechanisms from the "bulk" recombination mechanisms is to observe the device size dependence of base current and current gain.

The semiconductor surface is characterized by a surface recombination velocity and a surface-state density. Compared to InP HBTs with GaAs counterparts, InGaAs lattice matched to InP substrates have a smaller surface recombination velocity than that of GaAs (10^3 cm/s versus 10^6 cm/s). Therefore, the surface recombination current of InP HBTs should be much smaller. For AlGaAs/GaAs HBTs, however, the extrinsic base surface recombination can be a major component of the overall base current. Surface recombination can be minimized by various technologies, including the use of surface chemical passivation or the use of a wide-gap material at the surface of the extrinsic base. This could be a p-type GaAlAs obtained by converting the polarity of the injection emitter, or a depleted GaAlAs emitter region (by thinning of this material). The use of a graded-base layer may also reduce the surface recombination.

The extrinsic base surface passivated with a thin depleted AlGaAs layer virtually eliminates the surface recombination current and increases the current gain of AlGaAs/GaAs HBTs. Experimental results [42] of passivated and unpassivated AlGaAs/GaAs HBTs of various emitter areas are compared here. The HBTs have a linearly graded emitter–base heterojunction with a AlGaAs grading layer of 300 Å. In smaller devices, the emitter width W_E is varied from 4, 8, 12, to 16 μm, and the emitter length L_E is 10 μm. For bigger devices, the emitter area is 100×100 μm^2. The relation between the current gain and the emitter dimensions is given by

$$\frac{1}{\beta} = \frac{J_{BB} + J_{B\text{SCR}} + J_{BE}}{J_C} + 2\frac{K_{B\text{surf}}}{J_C}\left(\frac{1}{W_E} + \frac{1}{L_E}\right) \tag{3.47}$$

where J_{BB}, $J_{B\text{SCR}}$, and J_{BE} are the recombination current densities in the base, emitter–base space-charge region, and emitter. $K_{B\text{surf}}$ is the surface recombination current divided by the emitter periphery.

Figure 3.48 shows measured J_B $(= J_C/\beta)$ versus $(1/W_E + 1/L_E)$ for passivated and unpassivated HBTs at $J_C = 10^4$ A/cm^2. The intercept of the curves with the y-axis represents the sum of J_{BB}, $J_{B\text{SCR}}$, and J_{BE}. The slope represents $2K_{B\text{surf}}$. It is clear from this plot that the unpassivated HBTs have significant $K_{B\text{surf}}$. The passivated HBTs show virtually no variation with $(1/W_E + 1/L_E)$, demonstrating that the surface recombination current in the passivated devices is indeed negligible. Figure 3.49 shows measured Gummel plots for HBTs with emitter areas of 4×10 and 12×10 μm^2. The ideality factors of collector currents are about 1, indicating that the collector current is diffusion limited in the base layer. This is because the emitter–base heterojunction has a linearly graded AlGaAs layer. More important, Fig. 3.49 shows that the ideality factors for the base currents are 1.71 for the passivated device and 1.33 for the unpassivated device with emitter area of 4×10 μm^2. For the passivated and unpassivated devices with an emitter area of 12×10 μm^2, the ideality factors are 1.76 and 1.49, respectively. For passivated devices, the surface recombination is negligible.

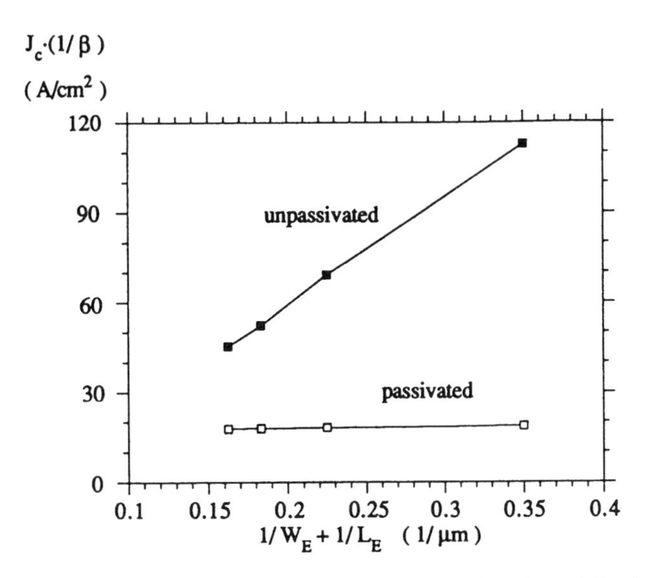

FIGURE 3.48 Measured $J_C(1/\beta)$ versus $(1/W_E + 1/L_E)$ (after Liu and Harris, Ref. 42 © IEEE).

The overall base recombination current dominates by space-charge recombination. This is why the ideality factor is close to 2. For unpassivated devices, the overall base current is the sum of the space-charge region recombination and the surface recombination. The ideality factors in this case indicate that the surface recombination current increases with the base–emitter voltage with an ideality factor of between 1 and 1.33. Figure 3.50 shows the current gain versus collector current. The HBTs in this plot have a linearly graded base–emitter heterojunction. The emitter area of the devices is $4 \times 10\ \mu\text{m}^2$. As demonstrated in Fig. 3.50, the current gain of passivated devices improves significantly.

3.3.3 Temperature-Dependent Current Gain of AlGaAs/GaAs HBTs

GaAs has a wide bandgap which suppresses the thermal generation of carriers and thus reduces leakage currents more effectively than does silicon. Microwave devices with an extended temperature range for space applications may be realized because of GaAs's high electron mobility and saturation velocity. High-temperature operational amplifiers with excellent frequency response may be realized using AlGaAs/GaAs HBTs.

Figure 3.51 displays the small-signal current gain h_{FE}, β, and I_C as a function of temperature. The current gain is defined by

$$\beta = \frac{I_{C|(I_B=350\mu\text{A})} - I_{C|(I_B=0\mu\text{A})}}{350\mu\text{A}}$$

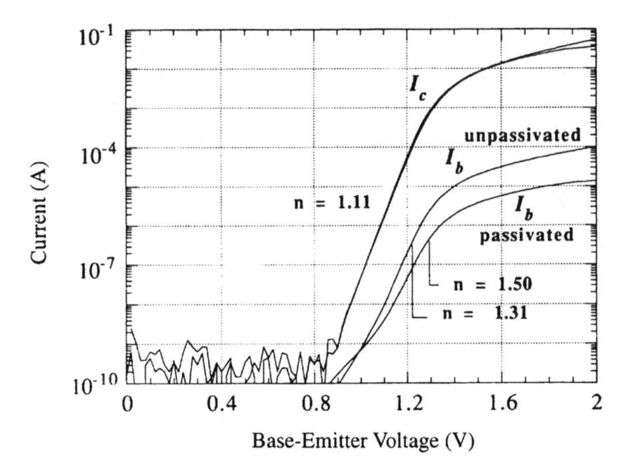

FIGURE 3.49 Gummel plot of an HBT (after Liu and Harris, Ref. 42 © IEEE).

At increased temperatures, h_{FE}, β, and I_C are reduced by the same order of magnitude. The increase in current gain at the highest temperature measured (523 K) is due to the additional hole leakage current from the collector into the emitter.

An analytical description that highlights the important physical parameters influencing the temperature dependence of the current gain in heterojunction bipolar transistors has been developed [43]. Each of the possible base current components was discussed and its relative importance to temperature dependence was assessed. For a typical abrupt AlGaAs/GaAs HBT, the current gain versus collector current at various temperatures was simulated, and the result is shown in Fig. 3.52*a*. The collector and base current were computed self-consistently using generalized Ebers–

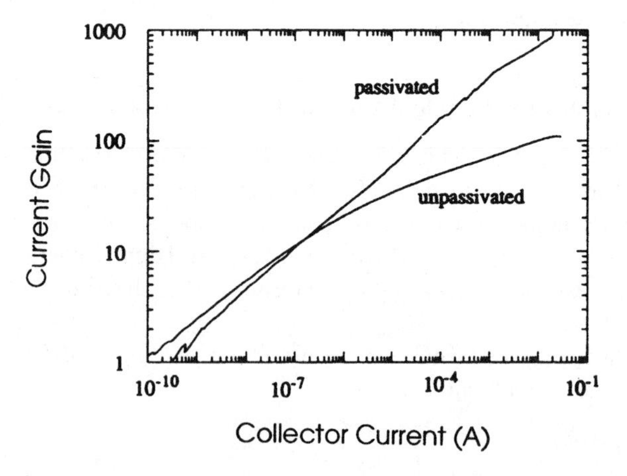

FIGURE 3.50 Current gain versus collector current (after Liu and Harris, Ref. 42 © IEEE).

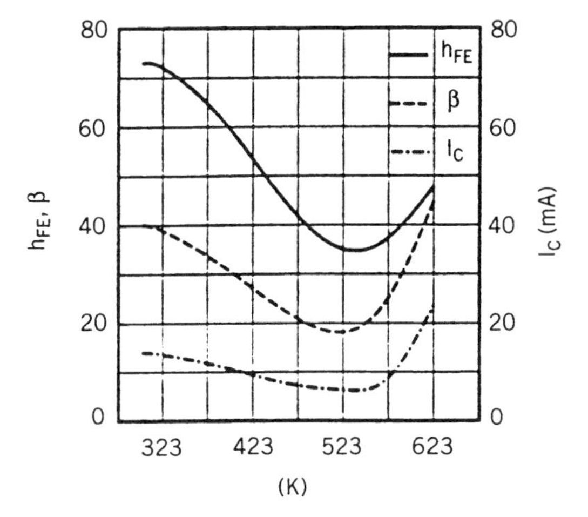

FIGURE 3.51 Dependence of h_{FE}, β, and I_C on temperature (after Fricke et al., Ref. 15 © IEEE)

Moll transport equations taking into account thermal effects and electrical effects, including tunneling through the conduction-band spike, electron quasi-Fermi level splitting, base bandgap narrowing, and series resistances in the emitter and collector. The space-charge-region recombination currents are evaluated numerically using Shockley–Read–Hall statistics.

The dominance of the large variations with temperature of SCR recombination at low currents is shown clearly in Fig. 3.52*a*. The current gain versus temperature for a 200-Å graded AlGaAs/GaAs HBT is also shown (Fig. 3.52*b*). Similar temperature dependence of current gain for the AlGaAs/GaAs HBT measured at different substrate temperatures is shown in Fig. 3.53.

In abrupt HBTs, the $\Delta E_C/\Delta E_V$ partitioning ratio determines the relative importance of reverse hole current compared to injected electron current. The conduction-band discontinuity ΔE_C suppresses the electron current, causing an increase in the relative importance of the reverse hole current. Tunneling significantly enhances the current gain at low currents, but its role at high currents is less effective because of the flattening of the conduction-band spike. At high temperatures, the thermionic emission component of the electron current becomes more important. Also, the reverse hole injection could be significant as the temperature is raised.

In general, the following temperature-dependent characteristics have been observed [43]:

- The gain of HBTs in the base–emitter bias region is governed by SCR recombination in the emitter and is subject to large reductions with increasing temperature despite the fact that the recombination lifetime is generally thought to increase with temperature.

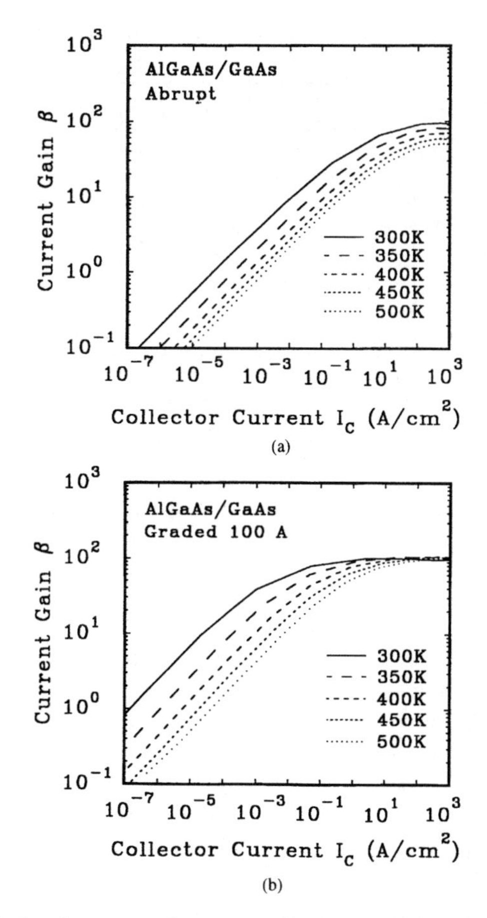

FIGURE 3.52 Calculated current gain versus collector current at different temperatures for (*a*) an abrupt AlGaAs/GaAs HBT and (*b*) a linearly graded AlGaAs/GaAs HBT (after Ng et al., Ref. 43 © IEEE).

- Extrinsic base surface recombination in GaAs is unlikely to cause significant temperature dependence of current gain even in unpassivated structures, where it can dominate the base current.
- At high current densities, reverse hole injection can contribute to limiting the gain and may be the main cause of negative differential resistance (NDR) in abrupt AlGaAs/GaAs HBTs, where the conduction-band discontinuity is large.
- At very high base doping ($> 2 \times 10^{19}$ cm^{-3}), in devices whose gain is limited by base bulk recombination, the negative temperature dependence of the dominate Auger recombination lifetime in GaAs can result in NDR.

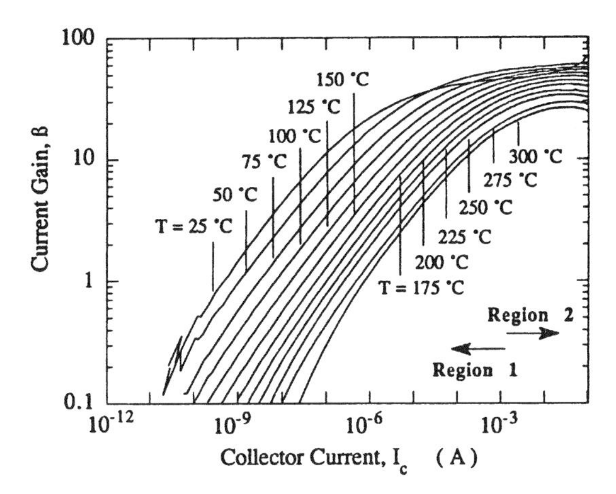

FIGURE 3.53 Measured current gains versus collector current of AlGaAs/GaAs HBTs at various temperatures (after Liu et al., Ref. 44 © IEEE).

3.3.4 Temperature-Dependent Current Gain of InP-Based HBTs

AlInAs/GaInAs heterojunction bipolar transistors grown on InP substrate have gained considerable attention for high-performance, low-power, and low-voltage applications. These devices have demonstrated excellent reliability and stable device characteristics under high-temperature stress. AlInAs/GaInAs HBT technology is also attractive for cryogenic applications because of the narrow bandgap of the GaInAs base layer, which leads to a low turn-on voltage for the device.

An AlInAs/GaInAs HBT on the InP substrate grown by MBE was studied [45]. The base–emitter turn-on voltage was measured from -196 to $250°C$. The base–emitter turn-on voltage is 0.64 V at $J_C = 5 \times 10^3$ A/cm^2. The low turn-on voltage of this HBT is due to the use of the low bandgap GaInAs base. The temperature coefficient of the turn-on voltage is approximately -0.95 mV/°C. The current gain over the temperature range from -196 to $250°C$ is relatively stable, as shown in Fig. 3.54. The dc current gain is measured from a $2 \times 5 \mu m^2$ HBT at $V_{BC} = 0$ V and $J_C = 5 \times 10^3, 5 \times 10^4$, and 2.5×10^5 A/cm^2. From room temperature up to 250°C, the current gain is decreased by 15%. From room temperature to liquid-nitrogen temperature ($-196°C$), the current gain is increased by 5% at $J_C = 2.5 \times 10^5$ A/cm^2.

Similar temperature behavior has been observed in $Al_xGa_{0.52-x}In_0.48P$/GaAs HBTs grown by MOCVD [46]. Nearly-ideal dc characteristics observed at room temperature are maintained up to at least $T = 623$ K across the entire range of aluminum composition x. The excellent quality of the emitter for direct bandgap compositions means that the current gain is limited by the base bulk recombination current, giving rise to temperature-insensitive current gain performance and a lack of negative differential resistance.

Figure 3.55 shows a plot of current gain versus collector current from 300 to 623 K for the $Al_xGa_{0.52-x}In_{0.48}P$/GaAs HBTs with two different Al composition

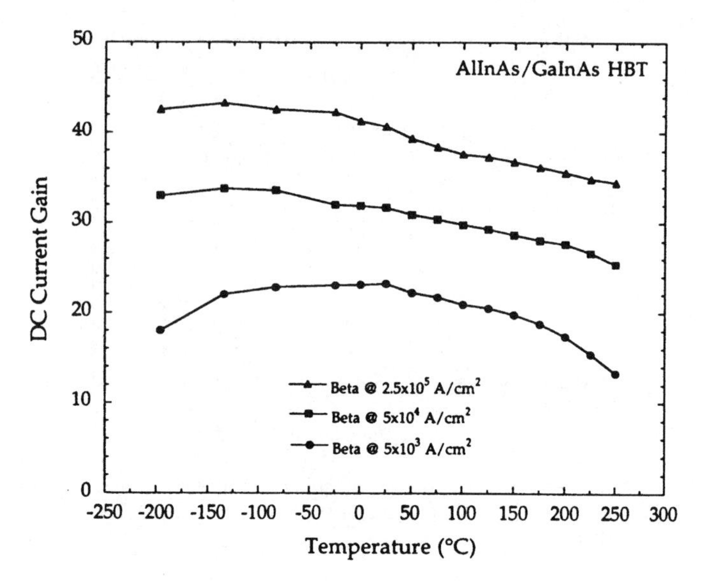

FIGURE 3.54 Current gain versus temperature of an AlInAs/GaInAs HBT (after Hafizi et al., Ref. 45 © IEEE).

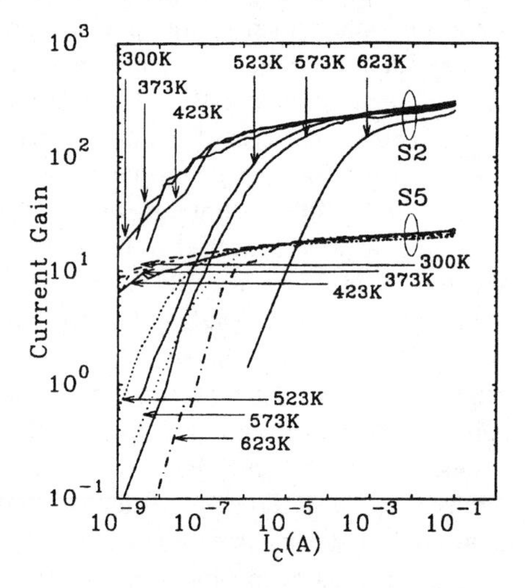

FIGURE 3.55 Temperature-dependent current gain characteristics of $Al_xGa_{0.52-x}In_{0.48}P$/GaAs HBTs with $x = 0.18$ (S2) and $x = 0.52$ (S5) (after Yow et al., Ref. 46 © IEEE).

values ($x = 0.18$ and 0.52). The current gain decreases substantially only at low currents and only after the temperature has risen to more than 500 K, while at higher currents the gain is stable with both collector current and temperature. The larger bandgap emitter device (S5) demonstrates a greater range of current gain stability at low currents.

REFERENCES

1. H. Kroemer, "Theory of wide-gap emitter for transistor," *Proc. IRE*, **45**, 1535 (1957).
2. W. Liu, S.-K. Fan, T. S. Kim, E. A. Beam III, and D. B. Davito, "Current transport mechanism in GaInP/GaAs heterojunction bipolar transistors," *IEEE Trans. Electron Devices*, **ED-40**, 1378 (1993).
3. M. Shur, *GaAs Devices and Circuits*, Plenum, New York (1987).
4. J. J. Liou, *Principles and Analysis of AlGaAs/GaAs Heterojunction Bipolar Transistors*, Artech House, Norwood, MA (1996).
5. J. J. Liou, C. S. Ho, L. L. Liou, and C. I. Huang, "An analytical model for current transport in AlGaAs/GaAs abrupt HBTs with a setback layer," *Solid-State Electron.*, **36**, 819 (1993).
6. H. Kroemer, "Heterostructure bipolar transistors and integrated circuits," *Proc. IEEE*, **70**, 13 (1982).
7. S. Tiwari, "A new effect at high currents in heterostructure bipolar transistors," *IEEE Electron Device Lett.*, **EDL-6**, 142 (1988).
8. S. Tiwari, S. L. Wright, and A. W. Kleinsasser, "Transport and related properties of (Ga,Al)As/GaAs double heterostructure bipolar junction transistors," *IEEE Trans. Electron Devices*, **ED-34**, 185 (1987).
9. S. Tiwari and D. J. Frank, "Analysis of the operation of GaAlAs/GaAs HBTs," *IEEE Trans. Electron Devices*, **ED-36**, 2105 (1989).
10. S. S. Perlman and D. L. Feucht, "P-n heterojunction," *Solid-State Electron.*, **27**, 911 (1964).
11. D. L. Pulfrey and S. Searles, "Electron quasi-Fermi level splitting at the base–emitter junction of AlGaAs/GaAs HBTs," *IEEE Trans. Electron Devices*, **ED-40**, 1183 (1993).
12. T. Won, S. Iyer, S. Agarwala, and H. Morkoç, "Collector offset voltage of heterojunction bipolar transistors grown by molecular beam epitaxy," *IEEE Electron Device Lett.*, **EDL-10**, 274 (1989).
13. S.-C. Lee, J.-N. Kau, and H.-H. Lin, "Origin of high offset voltage in an AlGaAs/GaAs heterojunction bipolar transistor," *Appl. Phys. Lett.*, **45**(10), 1114 (1984).
14. N. Chand, R. Fisher, and H. Morkoç, "Collector–emitter offset voltage in AlGaAs/GaAs heterojunction bipolar transistors," *Appl. Phys. Lett.*, **47**(3), 313 (1985).
15. K. Fricke, H. L. Hartnagel, W.-E. Lee, and J. Würfl, "AlGaAs/GaAs HBT for high-temperature applications," *IEEE Trans. Electron Devices*, **ED-39**, 1977 (1992).
16. J. M. Early, "Effects of space-charge layer widening in junction transistors," *Proc. IRE*, **40**, 1401 (1952).
17. D. J. Roulston, "Early voltage in very-narrow-base bipolar transistors," *IEEE Electron Device Lett.*, **EDL-11**, 88 (1990).
18. J. J. Liou, "Comments on 'Early voltage in very-narrow-base bipolar transistors,' " *IEEE Electron Device Lett.*, **EDL-11**, 236 (1990).

19. J. S. Yuan and J. J. Liou, "An improved Early voltage model for advanced bipolar transistors," *IEEE Trans. Electron Devices*, **ED-38**, 179 (1991).

20. M. Ugajin, G.-B. Hong, and J. G. Fossum, "Inverse base-width modulation and collector space-charge-region widening: degradation effects at high current densities in highly scaled BJTs (and HBTs)," *IEEE Trans. Electron Devices*, **ED-41**, 266 (1994).

21. J. S. Yuan, J. S. Yuan, and W. R. Eisenstadt, "Physics-based current-dependent base resistance mode for advanced bipolar transistors," *IEEE Trans. Electron Devices*, **ED-35**, 1055 (1988).

22. K. Lu, A. Perry, and T. J. Brazil, "A new large-signal AlGaAs/GaAs HBT model including self-heating effects, with corresponding parameter-extraction procedure," *IEEE Trans. Microwave Theory Tech.*, **MTT-43**, 1433 (1995).

23. P. E. Cottrell and Z. Yu, "Velocity saturation in the collector of Si/Ge_xSi_{1-x}/Si HBTs," *IEEE Electron Device Lett.*, **EDL-11**, 431 (1990).

24. *SEDAN*: *One-Dimensional Device Analysis Program*, Technology Modeling Associates, Inc. Palo Alto, CA, 1984.

25. A. J. Joseph, J. D. Cressler, D. M. Richey, and D. L. Harame, "Impact of profile scaling on high-injection barrier effects in advanced UHV/CVD SiGe HBTs," *IEDM Tech. Dig.*, 253 (1996).

26. D. L. Harame, J. H. Comfort, J. D. Cressler, E. F. Crabbé, J. Y.-C. Sun, B. S. Meyerson, and T. Tice, "Si/SiGe epitaxial-base transistors: I. Materials, physics, and circuits," *IEEE Trans. Electron Devices*, **ED-42**, 455 (1995).

27. E. J. Prinz and J. C. Sturm, "Current-gain–Early voltage products in heterojunction bipolar transistors with nonuniform base bandgaps," *IEEE Trans. Electron Devices*, **ED-12**, 661 (1991).

28. W. P. Dumke, "The effect of base doping on the performance of Si bipolar transistors at low temperatures," *IEEE Trans. Electron Devices*, **ED-28**, 494 (1981).

29. J. D. Cressler, J. H. Comfort, E. F. Crabbé, G. L. Patton, J. M. C. Stork, J. Y.-C. Sun, and B. S. Meyerson, "On the profile design and optimization of epitaxial Si- and SiGe-base bipolar technology for 77 K applications: I. Transistor dc design considerations," *IEEE Trans. Electron Devices*, **ED-40**, 525 (1993).

30. S. M. Sze, *Physics of Semiconductor Devices*, 2nd ed., Wiley, New York (1981).

31. E. F. Crabbé, J. D. Cressler, G. L. Patton, J. M. C. Stork, J. H. Comfort, and J. Y.-C. Sun, "Current gain rolloff in graded-base SiGe heterojunction bipolar transistors," *IEEE Electron Device Lett.*, **EDL-14**, 193 (1993).

32. J. M. McGregor, D. J. Roulston, J. P. Noel, and D. C. Houghton, "Output conductance of bipolar transistors with large neutral-base recombination current," *IEEE Trans. Electron Devices*, **ED-39**, 2569 (1992).

33. S. Mohammandi and C. R. Selvakumar, "Analysis of BJT's, pseudo-HBT's, and HBT's by including the effect of neutral base recombination," *IEEE Trans. Electron Devices*, **ED-41**, 1708 (1994).

34. J. Song and J. S. Yuan, unpublished data.

35. W. R. Zhang, W. G. Wu, Z. Zeng, and J. S. Luo, "Effect of Ge composition linear grading in the base on the Early voltage of Si/$Si_{1-x}Ge_x$/Si HBTs," *Solid-State Electron.*, **38**, 1842 (1995).

36. A. J. Joseph, J. D. Cressler, D. M. Richey, R. C. Jaeger, and D. L. Harame, "Neutral base recombination and its influence on the temperature dependence of Early voltage and current gain–Early voltage product in UHV/CVD SiGe heterojunction bipolar transistors," *IEEE Trans. Electron Devices*, **ED-44**, 404 (1997).

37. D. P. Maycock, "Thermal conductivity of silicon, germanium, III-V compound and III-V alloys," *Solid-State Electron.*, **10**, 161 (1967).

38. J. S. Yuan, "Thermal and reverse base current effects on heterojunction bipolar transistors and circuits," *IEEE Trans. Electron Devices*, **ED-43**, 789 (1995).

39. J. J. Liou, L. L. Liou, C. I. Huang, and B. Bayraktaroglu, "A physics-base heterojunction bipolar transistor including thermal and high-current effects," *IEEE Trans. Electron Devices*, **ED-40**, 1570 (1993).

40. W. B. Joyce, "Thermal resistance of heat sink with temperature-dependent conductivity," *Solid-State Electron.*, **18**, 321 (1975).

41. S.-F. Lin, "Modeling the AlGaAs/GaAs heterojunction bipolar transistor operating in the temperature range of 300 and 500 K," M.S. thesis, University of Central Florida (1995).

42. W. Liu and J. S. Harris, Jr., "Diode ideality factor for surface recombination current in AlGaAs/GaAs heterojunction bipolar transistors," *IEEE Trans. Electron Devices*, **ED-39**, 2726 (1992).

43. C.-M. S. Ng, P. A. Houston, and H.-K. Yow, "Analysis of the temperature dependence of current gain in heterojunction bipolar transistors," *IEEE Trans. Electron Devices*, **ED-44**, 17 (1997).

44. W. Liu, S.-K. Fan, T. Henderson, and D. Davito, "Temperature dependencies of current gains in GaInP/GaAs and AlGaAs/GaAs heterojunction bipolar transistors," *IEEE Trans. Electron Devices*, **ED-40**, 1351 (1993).

45. M. Hafizi, W. E. Stanchina, R. A. Metzger, P. A. Macdonald, and F. Williams, Jr., "Temperature dependence of dc and RF characteristics of AlInAs/GaInAs HBTs," *IEEE Trans. Electron Devices*, **ED-40**, 1583 (1993).

46. H.-K. Yow, P. A. Houston, C.-M. Ng, C. Button, and J. S. Roberts, "High-temperature dc characteristics of $Al_xGa_{0.52-x}In_{0.48}P$/GaAs heterojunction bipolar transistors grown by metal organic vapor phase epitaxy," *IEEE Trans. Electron Devices*, **ED-43**, 2 (1996).

PROBLEMS

3.1. Calculate and plot the collector current density versus V_{BE} for a uniform $Si_{0.8}Ge_{0.2}$ base, a linear $Si_{1-x}Ge_x$ base, and a trapezoidal SiGe base. For the $Si_{1-x}Ge_x$ base, $x = 0$ at the emitter–base heterojunction and $x = 0.2$ at the collector–base heterojunction. For the trapezoidal SiGe base, the transition point between the linear and flat Ge regions is at $0.5W_B$. The uniform base doping is 10^{18} cm^{-3} and the base width is 0.08 μm.

3.2. Compare the Early voltage of a linearly graded SiGe HBT with and without neutral-base recombination. The SiGe HBT has the base width $W_B = 0.08\mu$m, the base doping $N_B(x) = 10^{18}\exp(-2x/W_B)$ cm^{-3}, and the Ge content $x = 0.2$ at the collector–base junction.

3.3. Show that the recombination factor f_R in (3.37) equals 0.5 for a two-step SiGe profile.

3.4. Discuss qualitatively how bandgap narrowing affects the collector current density, current gain, and Early voltage of the HBT.

3.5. Plot the base sheet resistance of an n-p-n SiGe HBT as a function of Ge content. The base doping is 10^{18} cm^{-3} and the base thickness is 0.08 μm.

3.6. **(a)** If a SiGe heterojunction bipolar transistor has two base contacts (one on either side of an emitter stripe), find the intrinsic base resistance expression of a uniformly doped base at low injection.

(b) If the base doping is exponential, $N_B(x) = 10^{18} \exp(-2x/W_B)$, what is the intrinsic base resistance for a single-base-contact HBT at low injection? The base thickness W_B is 0.08 μm, the emitter width X_E is 1 μm, and the emitter length L_E is 5 μm.

3.7. Explain qualitatively how the base width modulation at the emitter–base and collector–base junctions together at high I_C affect the base transit time of a SiGe HBT at high collector current densities.

3.8. Design a collector profile to reduce the high-injection-induced barrier effects of a double-heterojunction bipolar transistor.

3.9. The ideality factor of an HBT is measured to be 1.2. Is it a graded or an abrupt heterojunction?

3.10. Consider an $Al_{0.3}Ga_{0.7}As$/GaAs graded junction HBT with a GaAs collector. The emitter region is doped 2×10^{18} cm^{-3}, and the grading is across 300 Å. The base is doped 1×10^{19} cm^{-3} and is 1000 Å in thickness. The collector is doped 5×10^{16} cm^{-3} and is 0.5 μm in thickness, and the subcollector is doped 5×10^{18} cm^{-3} and is 0.2 μm in thickness.

(a) For a one-dimensional structure, draw the energy-band diagram, including the conduction-band edge, the valence-band edge, and the electron and hole quasi-Fermi levels in the forward active, saturation, and inverse active mode of operation.

(b) At what current density will the Kirk effect become important? Assume a saturated velocity of 8×10^6 cm/s.

3.11. Discuss the thermal behavior of abrupt and graded AlGaAs/GaAs heterojunction bipolar transistors. Both transistors are operated at $V_{BE} = 1.4$ V and $V_{CE} = 8.0$ V. Assume both transistors have identical dopings and device sizes.

3.12. Plot various base recombination currents versus the base–emitter voltage of an AlGaAs/GaAs HBT. The emitter doping is 7×10^{18} cm^{-3}, the base doping is 5×10^{19} cm^{-3}, the collector doping is 5×10^{17} cm^{-3}; the emitter thickness is 0.1 μm; the base thickness is 0.08 μm, and the collector thickness is 0.5 μm.

CHAPTER FOUR

RF and Transient Performance

4.1 GENERAL DEVICE BEHAVIOR

4.1.1 Output Conductance

The output conductance describes the base width modulation effect on the collector current in the small-signal equivalent circuit. The common-emitter output conductance is known by

$$g_o = \frac{dJ_C}{dW_B}\frac{dW_B}{dV_{CE}} \approx \frac{dJ_C}{dW_B}\frac{dW_B}{dV_{CB}} \tag{4.1}$$

It is clear from (4.1) that the larger dW_B/dV_{CB}, the larger the output conductance. For a bipolar transistor with low base doping and narrow base width, W_B is very sensitive to the change in the collector–base voltage. This gives a large output conductance (or a small output resistance) for the bipolar transistor.

4.1.2 Transconductance

The transconductance of the bipolar device is defined as the change of the collector current with respect to the base–emitter voltage:

$$g_m \equiv \frac{\partial I_C}{\partial V_{BE}} \tag{4.2}$$

The common-emitter small-signal voltage gain is proportional to the transconductance. The transconductance depends linearly on the collector current if I_C is given by $I_S \exp(qV_{BE}/kT)$.

4.1.3 Heterojunction Junction Capacitance

Heterojunction capacitances play an important role in determining device switching speed and high-frequency response. Because of heavy doping in the quasi-neutral base and emitter regions, minority-carrier charge storage in the emitter and base is small. The junction capacitance can be much larger than the diffusion capacitance of the heterojunction bipolar transistor. Using the depletion approximation, one finds the potentials between the p- and n-type regions of a heterojunction:

$$V_{B1} = \frac{\varepsilon_1 N_1}{\varepsilon_1 N_1 + \varepsilon_2 N_2}(V_{bi} - V) \tag{4.3a}$$

$$V_{B2} = \frac{\varepsilon_2 N_2}{\varepsilon_1 N_1 + \varepsilon_2 N_2}(V_{bi} - V) \tag{4.3b}$$

The depletion widths X_1 and X_2 are given by

$$X_1 = \sqrt{\frac{2N_2\varepsilon_1\varepsilon_2(V_{bi} - V)}{qN_1(\varepsilon_1 N_1 + \varepsilon_2 N_2)}} \tag{4.4a}$$

$$X_2 = \sqrt{\frac{2N_1\varepsilon_1\varepsilon_2(V_{bi} - V)}{qN_2(\varepsilon_1 N_1 + \varepsilon_2 N_2)}} \tag{4.4b}$$

The depletion capacitance per unit area is expressed as

$$\frac{C_{\text{depl}}}{A} = \frac{\varepsilon_1\varepsilon_2}{\varepsilon_1 X_2 + \varepsilon_2 X_1} \tag{4.5}$$

Although it serves adequately for reverse bias, the depletion model is questionable for junctions under forward bias because of free carriers in the space-charge region. For physical understanding [1] the junction capacitance is the sum of the depletion capacitance and the free carrier capacitance due to modulation of free carrier charge in that region. Thus

$$C_J = \frac{\varepsilon_1\varepsilon_2 A}{\varepsilon_1 X_2 + \varepsilon_2 X_1} + qA\int_{-X_1}^{X_2} \frac{\partial n}{\partial V}\,dx = C_{\text{depl}} + C_F \tag{4.6}$$

In general, from (4.6), C_J can be obtained by numerical methods. To treat C_J analytically, one needs to estimate the electron or hole distribution in the space-charge region so that the integral of (4.6) can be carried out in closed form. Detailed analytical equations for the heterojunction capacitance accounting for high and very high forward biases of a SiGe heterojunction are given in [2].

The heterojunction capacitance as a function of bias is shown in Fig. 4.1. In this figure the solid line represents the physics-based model prediction, and the dashed line represents the depletion approximation. At low bias, the junction capacitance predicted by the analytical result is larger than the depletion approximation. This is because the physics-based model accounts for holes and electrons in the depletion region, which increase the junction capacitance. The heterojunction capacitance predicted by the depletion approximation increases monotonically and approaches

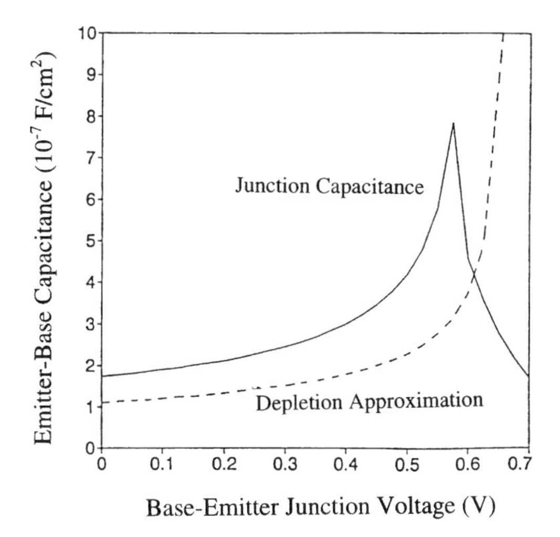

FIGURE 4.1 SiGe heterojunction capacitance versus base–emitter voltage (after Yuan, Ref. 2. © Elsevier Science.)

infinity as the base–emitter voltage approaches the built-in voltage. Note that the built-in voltage of a SiGe heterojunction is smaller than that of the Si junction because the Ge has a smaller intrinsic concentration. The depletion capacitance causes the total free-carrier charge stored in the emitter–base space-charge region to approach infinity even though the space-charge region vanishes when the base–emitter voltage approaches the built-in voltage. The analytical result, however, correctly represents the device physics of the heterojunction at high forward biases.

4.1.4 Base Transit Time

Base transit time has been studied extensively in the past. For bipolar transistors, Van Wijnen and Gardner [3] showed that a uniform base profile gives a higher cutoff frequency than a graded profile for a given base resistance and peak base concentration. Suzuki [4] showed that the exponential base profile is the optimal profile for a given base resistance and base width. Yuan [5] studied the base transit time of uniform and exponential base profiles under high levels of injection. Base transit time decreases with the level of injection for a uniform base, but increases with injection for an exponential base. For SiGe bipolar transistors, Winterton et al. [6] used an iterative procedure to find the optimal Ge profile for base transit time optimization. If the total germanium content in the base is specified, the method produces a profile having a steeper slope near the collector and a shallower slope near the emitter than does a linear ramp. If the germanium content is specified at the ends of the base, the prescribed profile approximates a linear ramp over the central portion of the base but is sharply retrograde near the collector and flat near the emitter. Winterton et al. [7] also demonstrated that the optimum doping profile in a homojunction transistor is not close to the exponential decrease from emitter to collector predicted by earlier studies in [7]. Unfortunately, though, the optimum profiles produced in [8] are difficult to realize in fabrication.

For the bipolar transistor, the well-known base transit time is given by

$$\tau_B = \int_0^{W_B} \frac{n_i^2(z)}{p_B(z)} \int_z^{W_B} \frac{p_B(y)\,dy}{D_n(y)n_i^2(y)}\,dz \tag{4.7}$$

where $p_B = N_B$ at low injection and $p_B > N_B$ at high injection. For a uniform base HBT at low injection, $\tau_B = W_B^2/2D_n$.

4.1.5 Collector–Base Space-Charge-Layer Delay

The collector space-charge-layer time constant is considered as $x_d/2v_s$, where x_d is the space-charge-layer width. This time constant can be derived using charge-control analysis [9]. In charge-control analysis it is assumed that the field in the space-charge region is sufficiently large to cause the electrons carrying the collector current to move with the scattering-limited saturation velocity over essentially the entire region. Since electrons move at a constant velocity, a mobile charge density Q_C/v_s exists in the space-charge region, adding to the negative-charge density in the p-region and subtracting from the positive-charge density in the n-region, as shown in Fig. 4.2. If the collector–base voltage is held constant, the area under the field curve must remain unchanged. Assume a one-sided collector–base junction, which is the typical case of HBTs with a heavily doped base and a lightly doped collector. Most of the voltage across the junction is dropped on the collector side. Solution of Poisson's equation in the space-charge layer yields

$$V_{CB} + \phi_C \approx \frac{qN_d x_{d1}^2}{2\varepsilon} = \frac{qN_d - Q_C}{2\varepsilon} x_{d2}^2 \tag{4.8}$$

where x_{d1} is the space-charge-layer width in the collector with no current present and x_{d2} is the width when the current flows. The relation between x_{d1} and x_{d2} is given by

$$\frac{x_{d2}}{x_{d1}} = \sqrt{\frac{1}{1 - Q_c/qN_d}} \approx 1 + \frac{Q_c}{2qN_d} \tag{4.9}$$

The extra positive charge exposed on donors in the collector space-charge-layer edge is thus

$$Q = qN_d(x_{d2} - x_{d1}) = Q_C \frac{x_{d1}}{2} \tag{4.10}$$

Equation (4.10) shows that only a half of $Q_C x_{d1}$ is neutralized by the added positive charge density on the collector side. The contribution to the time constant associated

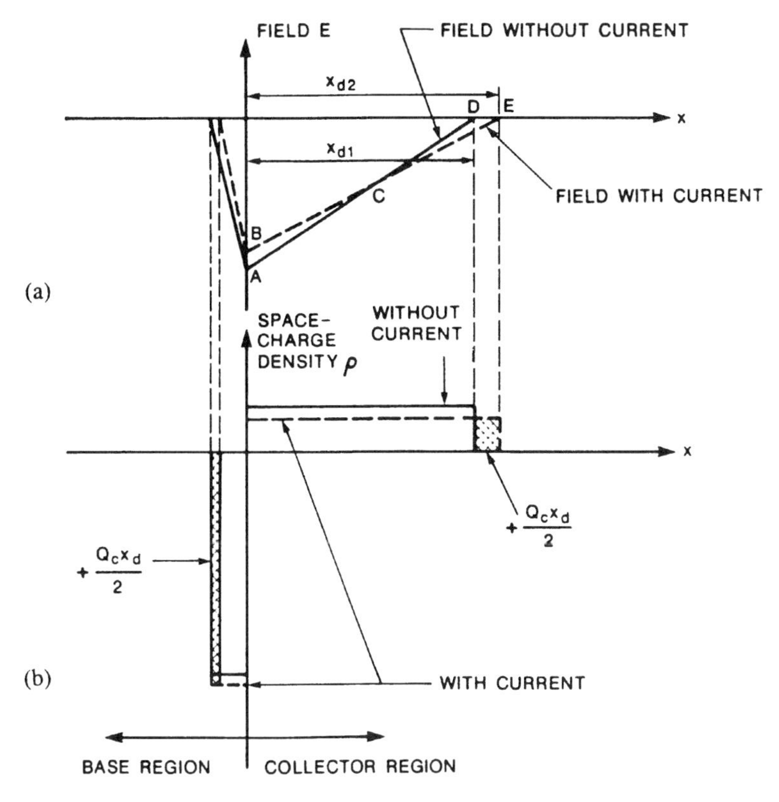

FIGURE 4.2 (*a*) Electric field and (*b*) space-charge density in the collector–base space-charge region including the effects of mobile electrons due to collector current.

with the charge traversing the collector–base space-charge region is thus $Q_C x_{d1}/2$ divided by the collector current density J_C :

$$\tau_C = \frac{Q_C x_{d1}}{2J_C} = \frac{Q_C x_{d1}}{2Q_C v_s} = \frac{x_{d1}}{2v_s} \tag{4.11}$$

The well-known expression in (4.11) equating the collector signal delay to one-half of the transit time (x_{d1}/v_s) is questionable in the presence of a nonconstant carrier velocity, as occurs for velocity overshoot. For example, for the strongly nonuniform velocity profiles calculated for AlGaAs/GaAs HBTs in [10], this expression overestimates the collector signal delay by 20 to 60%. A constant electron velocity in the collector depletion layer cannot explain the dependence of f_T on the collector–emitter voltage measured from S-parameters [11]. The difference between the simple model predictions using (4.11) and the experimental data is caused by velocity overshoot in the collector space-charge region, which is a function of the collector–emitter bias.

The velocity overshoot effect is enhanced for $Ga_{0.47}In_{0.53}$. As compared to GaAs, as shown in Fig. 2.3.

Assuming that (1) the carriers move with velocity $v(x)$ through the depletion region, as a concentration that is small compared to the local ionized doping concentration; (2) the sinusoidal conduction current in the depletion region consists of a stream of carriers that is density modulated but not velocity modulated [i.e., the local velocity is always equal to $v(x)$]; and (3) the carriers enter the depletion region with the same velocity $v(0)$ (i.e., there is no incoming velocity spread), the collector transport factor ζ is obtained by summing the contributions to the total current induced at the base ($x = 0$) by the carriers in transit through the depletion region as follows [12]:

$$\begin{aligned}\zeta &= \frac{1}{x_m}\int_0^{x_d} dx \exp\left[-j\omega\int_0^x \frac{dy}{v(y)}\right] \\ &= \frac{1}{x_d}\int_0^{\tau_C} dt\; v(x(t))e^{-j\omega t}\end{aligned} \tag{4.12}$$

The low-frequency result for ζ can be written as

$$\begin{aligned}\zeta_{LF} &= 1 - \frac{j\omega}{x_m}\int_0^{x_d} dx \left[\int_0^x \frac{dy}{v(y)}\right] \\ &= 1 - j\omega\int_0^{x_d} \frac{dx}{v(x)}\left(1 - \frac{x}{x_d}\right)\end{aligned} \tag{4.13}$$

where the second equality is obtained by an integration by parts. From (4.13) the low-frequency estimate for the collector signal delay is

$$\tau_{C,LF} = \int_0^{x_d} \frac{dx}{v(x)}\left(1 - \frac{x}{x_d}\right) \tag{4.14}$$

To minimize $\tau_{C,LF}$ the weighting function $(1 - x/x_d)$ dictates that $v(x)$ should be largest near $x = 0$, with smaller values of $v(x)$ resulting in progressively less penalty near $x = x_d$. This low-frequency result is readily derived from a charge-control perspective as well. Such a derivation clarifies that this weighting function accounts for unequal terminating charges induced at the two ends of the depletion layer by the nonuniform mobile charge density in the depletion region.

When carriers traverse a region where there is a significant gradient in the electric field, they undergo a velocity overshoot in parts of the region. Velocity overshoot is a term that implies that the local velocity at a local field becomes higher than what it would reach for a slowly varying field of the same magnitude. A local nonequilibrium occurs. It arises because the carrier energy and the electric field are no longer in equilibrium. When the electric field becomes large, higher carrier energy results. Higher carrier energy causes a larger amount of scattering, which results in lower mobility. Thus velocity, a product of mobility and electric field with mobility varying inversely with field, becomes a constant. When a large gradient of electric field exists, the carrier does not immediately achieve the energy that corresponds to the local field if the field is rapidly varying. It actually has a lower energy and hence suffers less scattering, yet is accelerated by the high field. The carrier is therefore said to have a velocity overshoot. This large velocity occurs despite the fact that it has a lower energy.

Laux and Lee [13] analyzed the collector signal delay, including velocity overshoot profiles in the collector space-charge region. Assuming that the position-dependent velocity profile is a simple piecewise constant velocity with $v(x) = v_1$ for $0 \leq x \leq x_1$ and $v(x) = v_2$ for $x_1 \leq x \leq x_d$ as shown in Fig. 4.3, the quasi-neutral base is at $x < 0$, where the collector is at $x > x_d$. The signal delay versus normalized radian frequency is displayed in Fig. 4.4. In this plot, the three cases $v_1 = 2v_2$, $v_1 = v_2$, and $v_1 = 0.5v_2$ with $x_1 = 0.5x_d$ are shown. The signal delay is normalized by the transit time $\tau_C = \int_0^{x_d} dx/v(x)$.The conventional estimate $\tau_C'/\tau_C = 0.5$ is true only in the constant velocity case ($v_1 = v_2$); a different result is obtained for the other two cases. When the larger velocity is near the base ($v_1 = 2v_2$), $\tau_C'/\tau_C < 0.5$. The collector signal delay decreases further, from 0 to 2π. When the larger velocity is near the collector ($v_1 = 0.5v_2$), $\tau_C'/\tau_C > 0.5$. The signal delay increases as the normalized radian frequency is increased from 0 to 2π.

Contours of equal signal delay are shown in a two-dimensional plane in Fig. 4.5. The x-axis gives the position of x_1 in the depletion region by the ratio x_1/x_d and the y-axis specifies the ratio of the two velocity values v_1/v_2. These contours further clarify that $\tau_C' = \tau_C/2$ only when $v_1 = v_2$.

For the evaluation of collector signal delay, a modeling approach has been reported for the steplike velocity profile [14] and the piecewise-linear velocity profile [15], although Eq. (4.14) implicitly accounts for any type of velocity profile in $v(x)$. A comparison of collector signal delay versus overshoot width for steplike and piecewise linear profiles is given in Fig. 4.6a and 4.6b, respectively. For a given overshoot width and velocity ratio (v_1/v_2), the piecewise linear profile results in an expected higher collector signal delay.

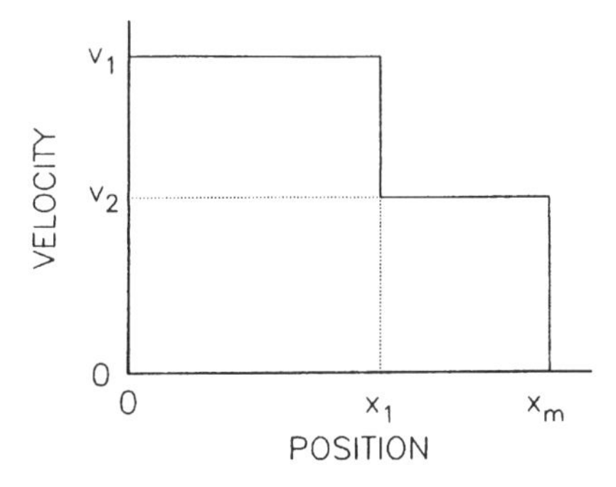

FIGURE 4.3 Ideal position-dependent velocity profile for theoretical analysis.

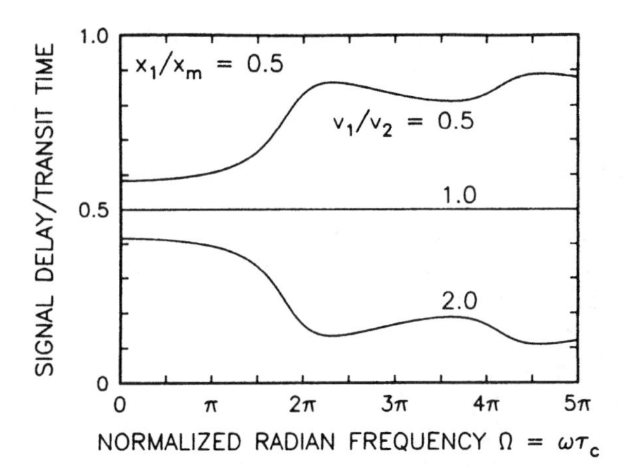

FIGURE 4.4 Normalized collector signal delay versus normalized radian frequency (after Laux and Lee, Ref. 13 © IEEE).

The HBT signal delay at high collector currents has been evaluated [16]. Simulations were performed for a one-dimensional device structure. Profiles for the electric field and the hole concentrations at different collector current densities are shown in Fig. 4.7*a* and 4.7*b*, respectively. The electric field evolves in much the same way as in silicon bipolar transistors. The field changes from a high value near the base–collector junction to a high value near the collector high–low junction. The field is almost constant at $J_C = 1.7 \times 10^5 \mathrm{A/cm}^2$. This indicates the collapse of the space-charge region at the collector–base junction. Thereafter, a nearly neutral region

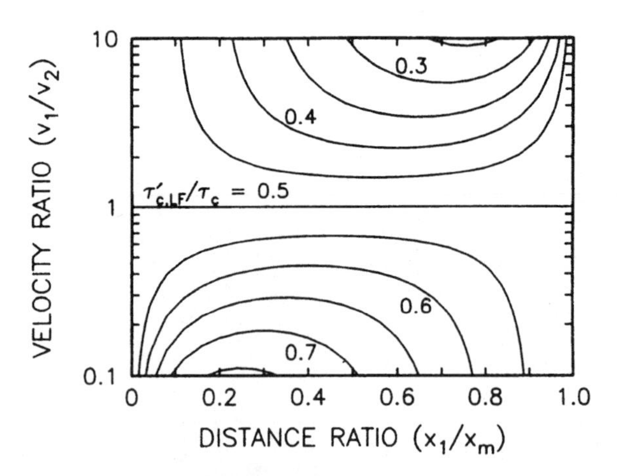

FIGURE 4.5 Contours of collector signal delay as a function of velocity ratio and distance ratio (after Laux and Lee, Ref. 13 © IEEE).

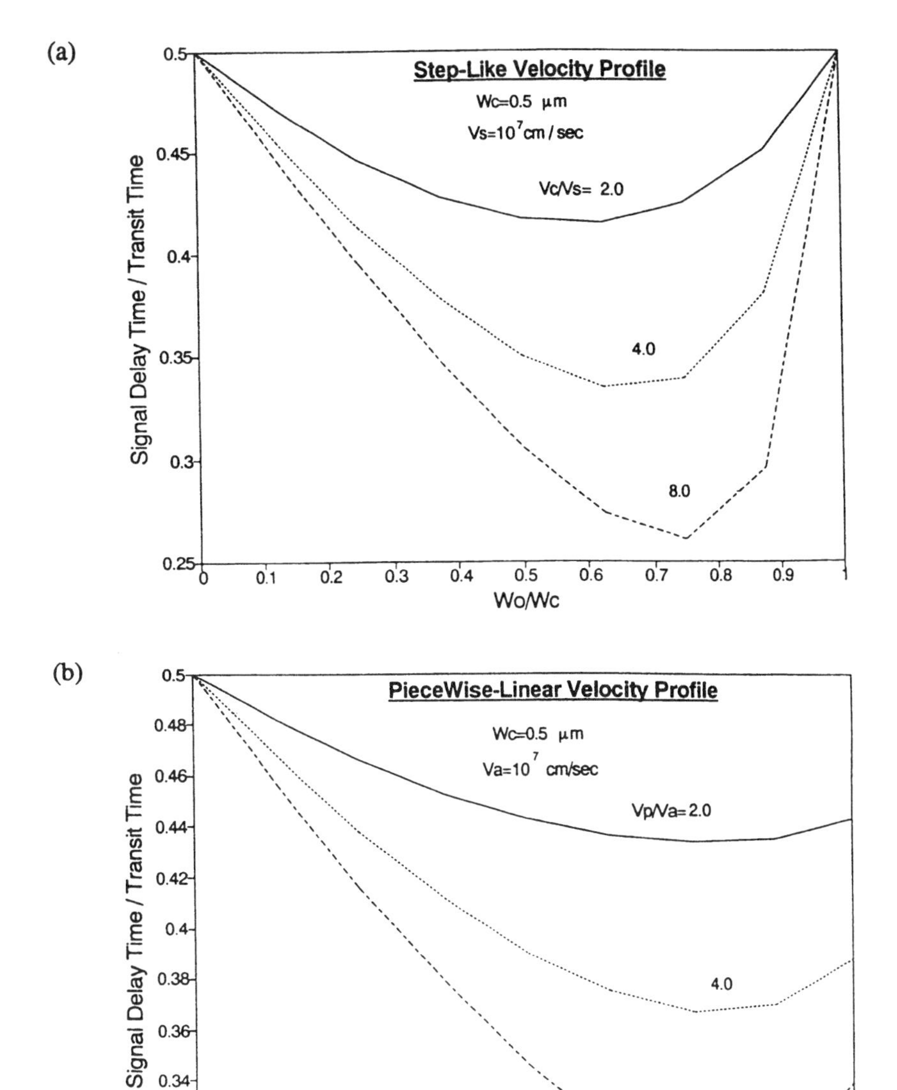

FIGURE 4.6 Normalized collector signal delay versus distance ratio for (*a*) steplike and (*b*) piecewise linear velocity profiles (after Liou and Shakouroi, Ref. 15© Elsevier Science).

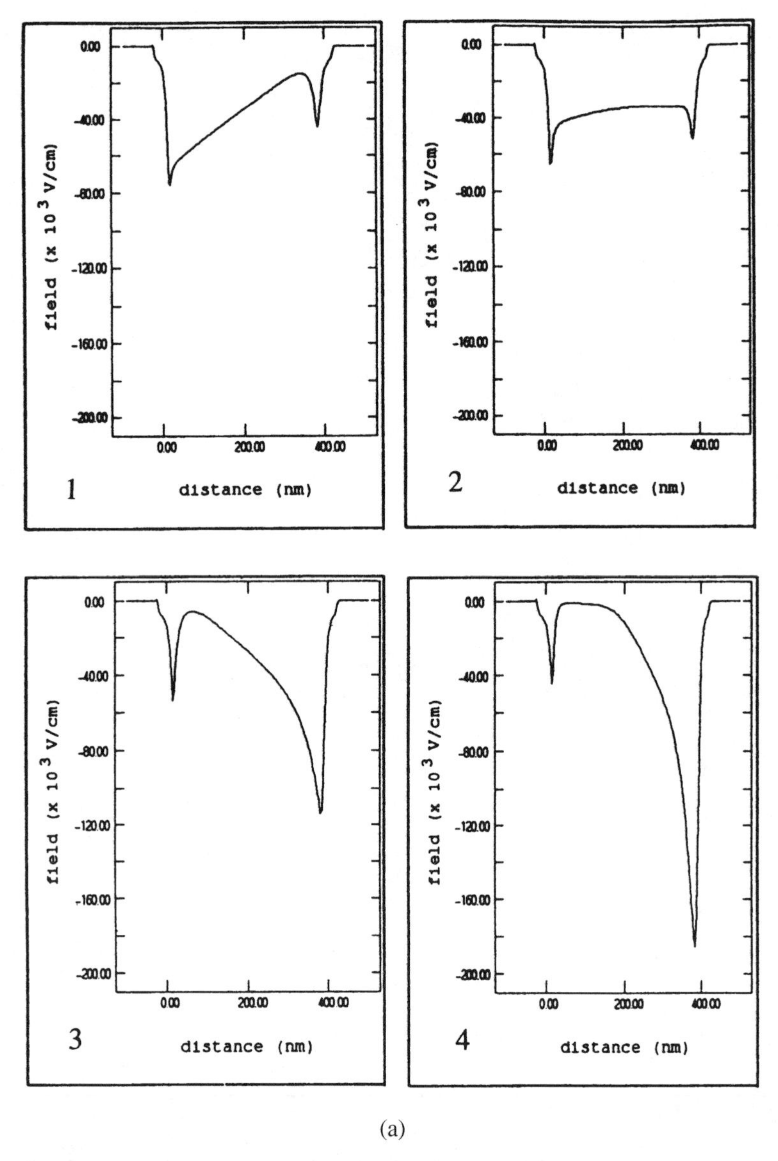

(a)

FIGURE 4.7 (a) Electric field and (b) hole concentration versus collector current density (after Zhou and Pulfrey, Ref. 16 © IEEE).

appears on the collector side of the junction. This signifies a high injection condition in the collector as evidenced by the hole concentration plot in Fig. 4.7b.

The signal delay ($\tau_B + \tau_C'$) versus the collector current density is shown in Fig. 4.8. The signal delay, at first, decreases and then increases with J_C. Decreases in the

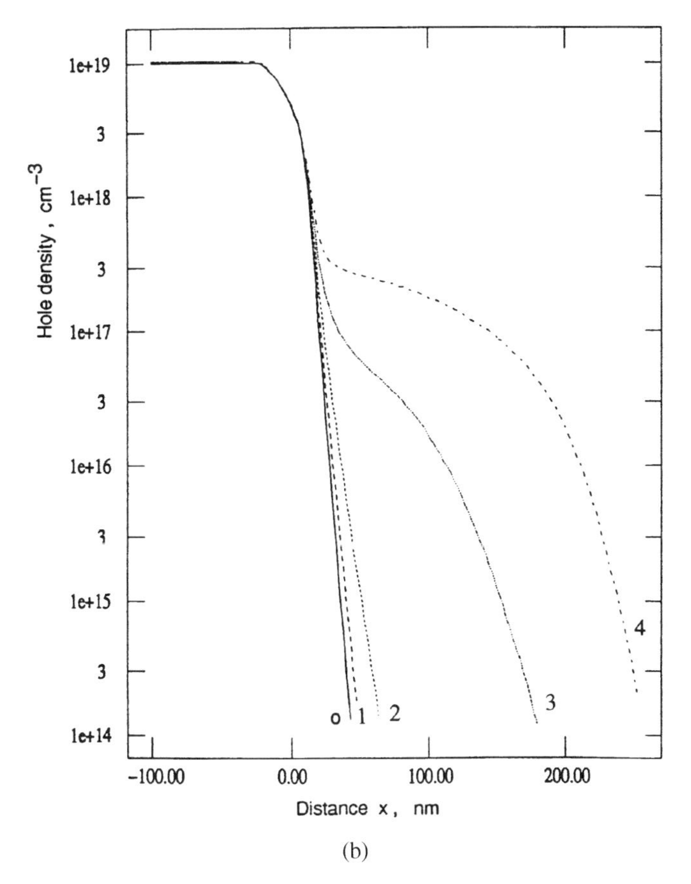

(b)

FIGURE 4.7 *(Continued)*

signal delay are attributed to the enhancement of velocity overshoot in the lowered-field region of the collector. When base pushout begins, the signal delay increases with J_C rapidly when both electrons and holes are accounted for in the simulation. The reduction in electron scattering to the upper valley in the pushout-out base turns out to be a minor effect. The overall current transport mechanism is dominated by ambipolar diffusion in a nearly neutral region. Under these conditions the effective diffusion constant of each carrier is nearly equal to that of the slower particle (i.e., the hole).

4.1.6 Cutoff Frequency

The cutoff frequency of the HBT is a figure of merit. This parameter is important for analyzing the bandwidth of small-signal amplifiers and the power gain of power

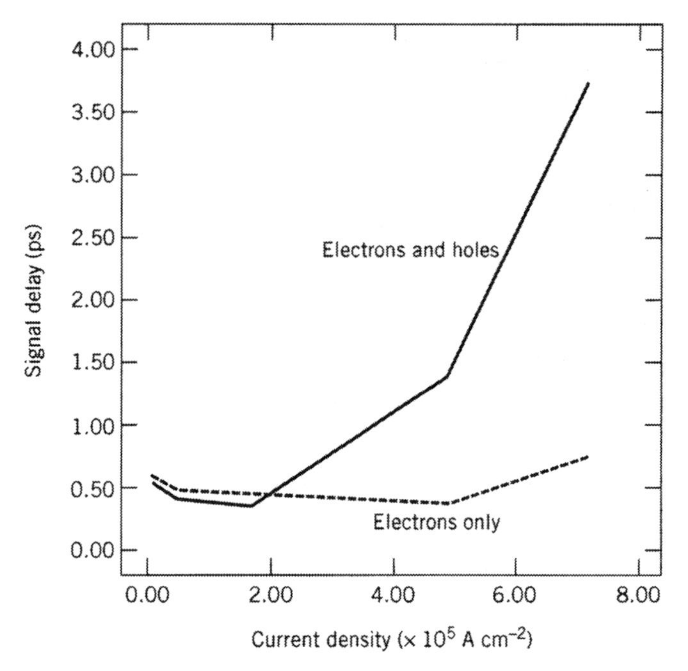

FIGURE 4.8 Signal delay versus collector current density (after Zhou and Pulfrey, Ref. 16© IEEE).

amplifiers. The cutoff frequency is given by the well-known approximation as follows:

$$f_T = \frac{1}{2\pi\tau_{EC}} \approx \frac{1}{2\pi[r_E(C_{JE} + C_{JC}) + \tau_B + \tau_C + R_C C_{JC}]} \tag{4.15}$$

where $r_E \approx kT/qI_C$ and R_C is the collector resistance. Equation (4.15) is generally valid for SiGe, GaAs, and InP heterojunction bipolar transistors. The cutoff frequency is a function of collector current. At low collector currents, the cutoff frequency increases with collector current, due to the decrease in emitter charging time. The cutoff frequency reaches the maximum at a moderate collector current, at which the peak cutoff frequency is determined by the combination of τ_B and τ_C. The cutoff frequency then falls off at high collector currents, due to base pushout effect.

4.1.7 Maximum Frequency of Oscillation

The maximum frequency of oscillation is approximated by

$$f_{\max} \approx \sqrt{\frac{f_T}{8\pi R_B C_{JC}}} \tag{4.16}$$

It is clear from (4.16) that the maximum frequency of oscillation is determined by the combined effect of f_T, τ_B, and $R_C C_{JC}$. The increase in f_T due to a reduced τ_C by higher collector doping, for example, could increase C_{JC} and may cause a decrease in $f_{\max}$, depending on the relative importance of τ_C in f_T.

Figure 4.9 shows measured f_T and $f_{\max}$ versus collector current for a SiGe HBT at $T = 309$ K. The HBT has a $0.6 \times 19.3 \mu m^2$ emitter area and a collector doping of $2 \times 10^{17} cm^{-3}$. Both f_T and $f_{\max}$ increase with I_C from low to moderate collector currents and decrease with I_C at high collector currents. Figure 4.10 shows the maximum frequency of oscillation as a function of temperature for a $2 \times 10 \mu m^2$ AlInAs/GaInAs HBT. The cutoff frequency and the maximum frequency of oscillation decrease with temperature. From room temperature to 125°C, the cutoff frequency decreases by about 10%. The cutoff frequency increases by about 20% from room temperature to liquid-nitrogen temperature. The plot of the emitter charging time τ_E and the base transit time τ_B plus collector–base space-charge-layer transit time τ_C, corresponding to peak cutoff frequency, as a function of temperature, is shown in Fig. 4.11. The linear increase in $\tau_B + \tau_C$ with temperature is the dominate factor in the decay observed in cutoff frequency with increasing temperature. The saturation velocity of the InGaAs collector decreases with increasing temperature. This causes an increase in τ_C with temperature.

4.1.8 NPN Versus PNP on RF Performance

To date, most wide-gap emitter HBTs are N-p-n heterojunction structures because the high electron mobility of III/V compound materials decreases base transit time considerably. The detrimental effect of lower hole mobility can be offset by use of a larger base width for analog applications. In contrast, for digital circuits, the gate propagation delay is not sensitive to the base transit time, but to the base resistance, load resistance, and load capacitance. High electron mobility in the base of a P-n-p

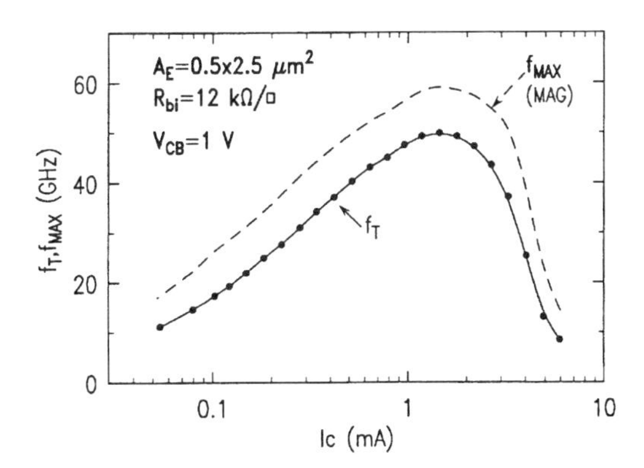

FIGURE 4.9 Measured f_T and $f_{\max}$ versus collector current for a SiGe HBT (after Harame et al., Ref. 17 © IEEE).

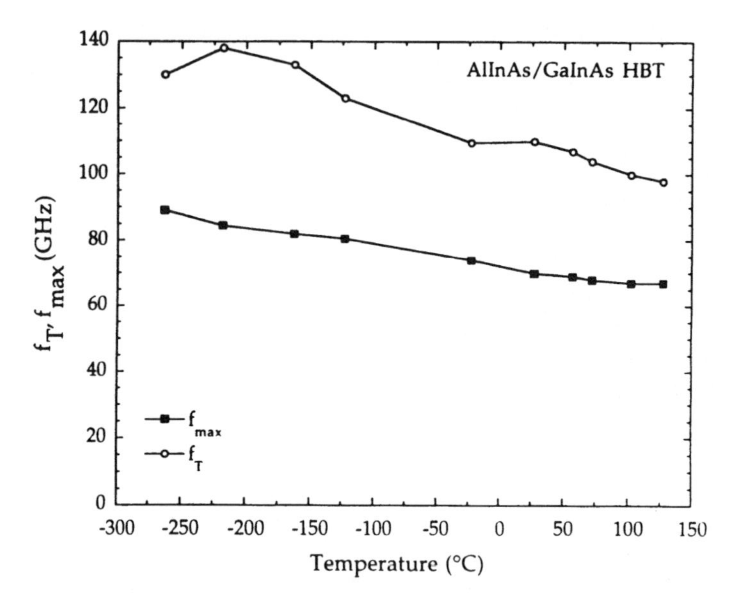

FIGURE 4.10 f_T and $f_{\max}$ versus temperature (after Jalali and Pearton, Ref. 18 © Artech House).

HBT results in low base resistance, which reduces the gate propagation delay of digital circuits. Individual optimization of heterojunction doping profiles for N-p-n and P-n-p HBTs in microwave and digital circuits is thus needed [19].

Consider the case where the doping profiles of P-n-p and N-p-n HBTs are identical. The heterojunction bipolar transistors used have an emitter doping density of $5 \times 10^{17}\text{cm}^{-3}$, a base doping of $4 \times 10^{19}\text{cm}^{-3}$, a collector epitaxial layer doping density of $5 \times 10^{16}\text{cm}^{-3}$, a base width of 0.1 μm, an epi-layer thickness of 0.2 μm, an emitter area of 10 μm^2, and a collector area of 28 μm^2. Regarding the heterojunction bipolar transistor process, the specific resistance of contacts to n-type GaAs ($\rho_{cn} \approx 1 \times 10^{-6}\Omega\cdot\text{-cm}^2$) is much lower than that of p-type GaAs ($\rho_{cp} \approx 2 \times 10^{-5}\Omega\cdot\text{-cm}^2$). Since the emitter and collector contact resistances are only small parts of the total emitter and collector resistances and the emitter and collector resistive charge times (e.g., $R_C C_{JC}$) are a small part of the total emitter–collector transit time, the emitter and collector contact resistance effect is negligible for determining the peak cutoff frequency. To compare the optimal performance of P-n-p and N-p-n heterojunction bipolar transistors, the peak cutoff frequency right before the onset of base pushout of HBTs is evaluated. The peak cutoff frequency at $I_C \approx 20$ mA ($J_C = 2 \times 10^5\text{A/cm}^2$) and $V_{CB} = 2.0$ V for the N-p-n heterojunction bipolar transistor is 74 GHz. The peak cutoff frequency at $I_C \approx 10$ mA ($J_C = 1 \times 10^5\text{A/cm}^2$) and $V_{CB} = 2.0$ V for the P-n-p heterojunction bipolar transistor is 55.5 GHz. It is obvious that the N-p-n heterojunction bipolar transistor has a higher cutoff frequency than that of the P-n-p HBT. This is due to the higher onset base pushout current I_0 ($\propto qA_e v_{\text{eff}}$) for electrons and lower hole velocity through the quasi-neutral base region and collector–base

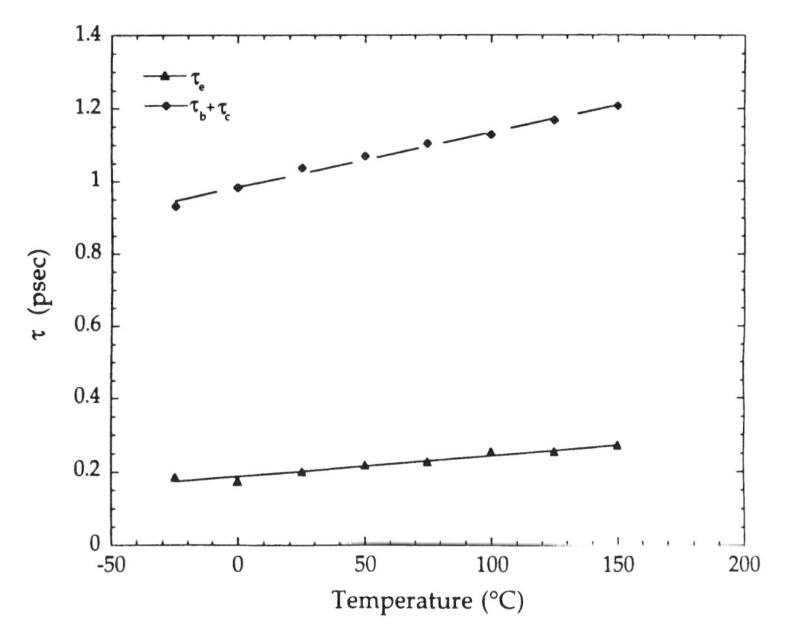

FIGURE 4.11 Transit time versus temperature (after Jalali and Pearton, Ref. 18 © Artech House).

space-charge layer, which increases the base transit time and collector–base depletion layer transit time. The unit-power gain frequency for a P-n-p and an N-p-n HBT using a device profile is also compared. The unit-power gain frequency for a P-n-p HBT (74 GHz) is more comparable with that of an N-p-n HBT (55.5 GHz). This is due to the high electron mobility in the base of a P-n-p heterojunction bipolar transistor, which decreases R_B and increases $f_{\max}$.

We now proceed to discuss the doping profile design for achieving the highest cutoff frequency, maximum frequency of oscillation, and the lowest gate propagation delay. With regard to the peak cutoff frequency, the base transit time and the collector–base space-charge layer transit time are important. To reduce base transit time, a thinner base width must be used. A limit on the base width is set by the emitter-to-collector punch-through. The punch-through is due to the fact that at high collector–base biases, the collector–base space-charge layer extends all the way across the moderately doped base and reaches the emitter–base space-charge layer. To reduce the collector–base space-charge layer width and transit time, a larger epi-layer concentration is designed. A larger epi-layer doping density with a narrower epi-layer thickness also increases the onset current for base pushout, which increases the peak cutoff frequency. As the collector doping concentration increases, however, there is a corresponding decrease in the collector breakdown voltage. Thus the collector junction breakdown voltage provides a constraint for maximum permissible collector concentration.

To design for maximum frequency of oscillation, it is not necessarily desirable to achieve the highest cutoff frequency. To optimize $f_{\max}$ typically requires using relatively small values of base resistance and collector–base junction capacitance. Lower R_B and C_{JC} can be designed by introducing higher base resistivity and lower collector concentration. The resulting low base diffusion coefficient and large collector–base depletion width tend to increase the base transit time and collector transit time, which reduce the cutoff frequency. The effect of a lower base resistance offsets a larger base transit time and a larger collector junction capacitance. As long as the decreasing rate of $R_B C_{JC}$ is larger than that of f_T, $f_{\max}$ increases with the decrease in base and collector doping.

4.1.9 Collector-Up Versus Collector-Down Influence on RF Performance

The collector-up or "inverted" HBT structure offers the advantage of eliminating the extrinsic base–collector junction area beneath the base contact. This reduction in the base–collector capacitance should result in improved microwave performance, particularly the power gain and $f_{\max}$. In the collector-up structure, however, an extrinsic emitter–base junction area is created, which reduces the emitter-injection efficiency. The parasitic current injection through the extrinsic base–emitter junction area beneath the base contact leads to a roll-off in the current gain at low bias levels. The current injection from the extrinsic emitter area can be suppressed by deep oxygen implantation to create a resistive layer beneath the base contact. This is particularly applicable to InP-based materials systems because unlike GaInAs, the wide-bandgap AlInAs can be made highly resistive using O^+ ion implantation.

An AlGaAs/GaAs collector-up HBT with a heavily carbon-doped base layer was fabricated using oxygen-ion implantation and zinc diffusion [20]. The cross-sectional view of this HBT is shown in Fig. 4.12. The microwave performance was characterized by the S-parameters, measured on wafers from 0 to 50 GHz. The cutoff frequency and maximum oscillation frequency versus collector current density are shown in Fig. 4.13. The extrapolated f_T and $f_{\max}$ values for 6-dB/octave falloff were 68 and 128 GHz, respectively. The collector-up HBT gives higher $f_{\max}$ than f_T values due to lower C_{JC} and R_B values. The collector-up structure has been demonstrated in SiGe

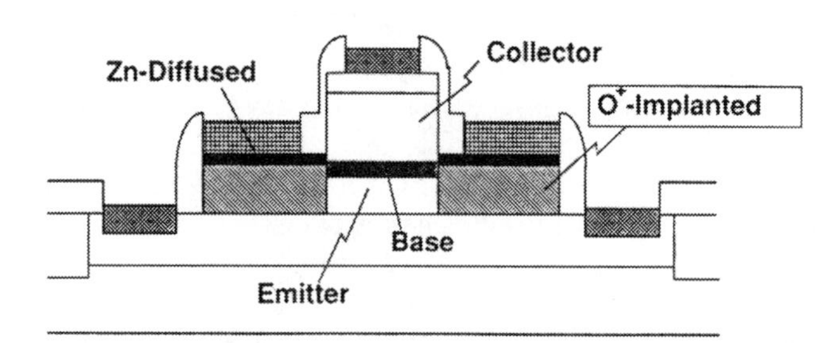

FIGURE 4.12 Cross-sectional view of a collector-up HBT fabricated using O^+ implantation and Zn diffusion (after Yamahara et al., Ref. 20 © IEEE).

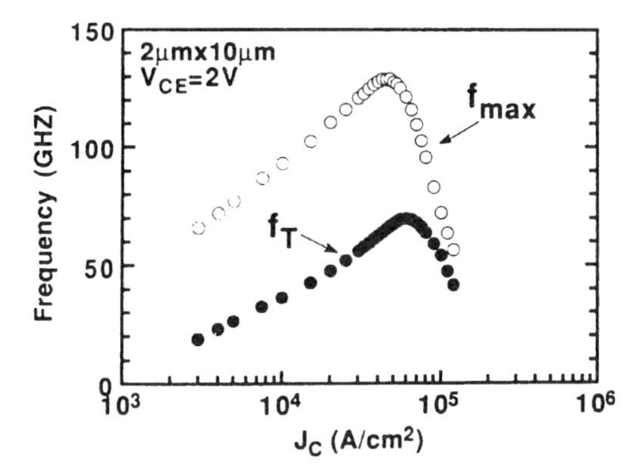

FIGURE 4.13 f_T and f_{max} versus J_C (after Yamahara et al., Ref. 20 © IEEE).

HBTs [21] and Ge/GaAs HBTs [22]. A schematic of the self-aligned collector-up (or emitter-down) SiGe HBT is shown in Fig. 4.14.

For the emitter-up structure, reduction of base–collector capacitance by undercutting the collector and subcollector of an GaInAs/InP DHBT [23] and by using a selective buried subcollector of a GaInP/GaAs HBT [24] have been proposed. To realize the undercut during the wet-chemical etching process, the edges of the base mask must be parallel to the ⟨001⟩ or ⟨010⟩ crystal directions. When the edge is parallel to the ⟨011⟩ or ⟨011⟩ directions, the etching terminates at the edge of GaInAs layer because of very little etching of the ⟨111⟩ facet. The cross section of the InP

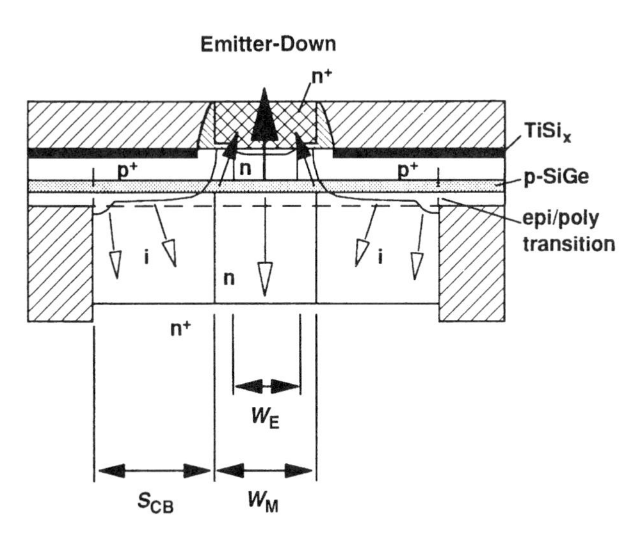

FIGURE 4.14 Cross-sectional view of a SiGe collector-up HBT (after Burghartz et al., Ref. 21 © IEEE).

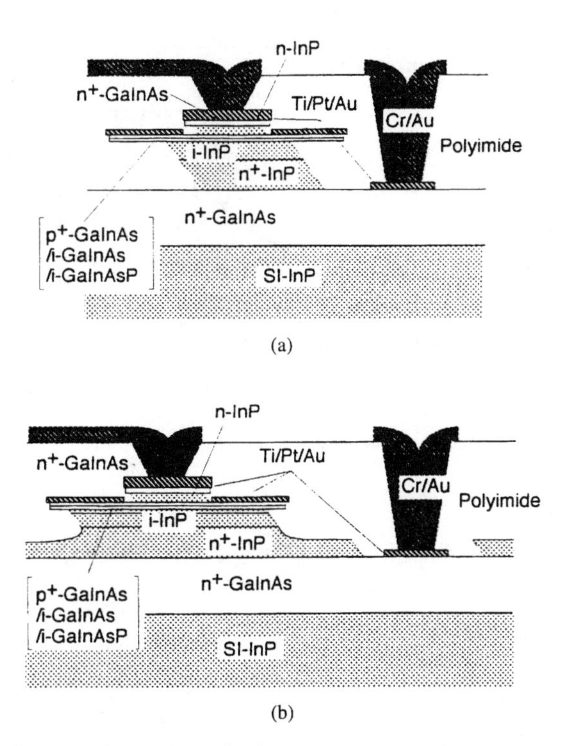

FIGURE 4.15 Cross-sectional view of a GaInAs/InP DHBT for (*a*) undercut structure and (*b*) conventional structure (after Miyamoto et al., Ref. 23 © IEEE).

pedestal which became a parallelogram is shown in Fig. 4.15*a*. The lateral extent of the undercut by wet-chemical etching was controlled by using dummy patterns on the base with various widths. Conventional processing with no undercut is shown in Fig. 4.15*b* for comparison. The comparison of f_T and $f_{\max}$ values versus collector current for the undercut device (sample A) and the conventional device (sample B) is given in Fig. 4.16. The reduction of C_{JC} in sample A compared to that of sample B was estimated to be in the region 54 to 61%. Use of the collector undercut technique to reduce the base–collector capacitance for InGaP/GaAs HBTs has also been reported [25]. An f_T value of 80 GHz and an $f_{\max}$ value of 171 GHz were achieved (see Fig. 4.17). Compared to HBTs without undercuts, the undercut HBTs show a 1.38-fold improvement in the highest achievable $f_{\max}$ value, due to the significant reduction in base-collector capacitance, as shown in Fig. 4.18.

Figure 4.19 shows the cross section and schematic layout of an HBT with selective buried subcollector. The cutoff frequency, maximum frequency of oscillation, and base–collector capacitance versus collector current density are shown in Fig. 4.20. The HBT with selective buried subcollector exhibits a higher $f_{\max}$ value because of reduced base–collector capacitance.

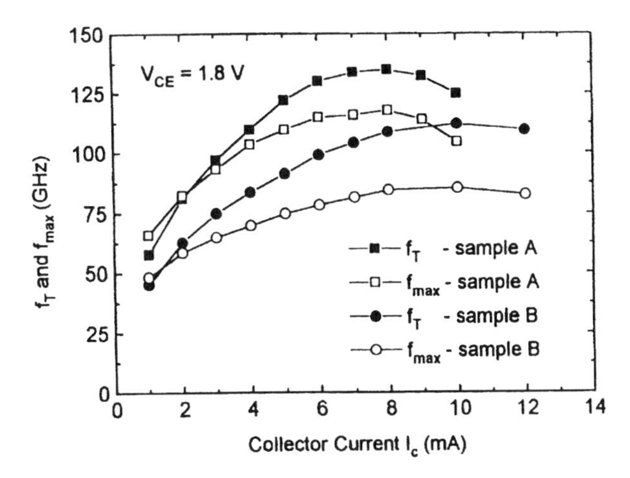

FIGURE 4.16 f_T and f_{max} versus I_C (after Miyamoto et al., Ref. 23 © IEEE).

4.1.10 Noise

The formal study of noise in electrical devices has its origin in the work of Schottky, who in 1918 showed that fluctuations in electrical current are intimately related to Brownian motion. The main goal of the early work on noise was to use noise measurements as a method for determining fundamental physical constants such as electron charge and Boltzmann's constant. In modern days, the low-frequency noise

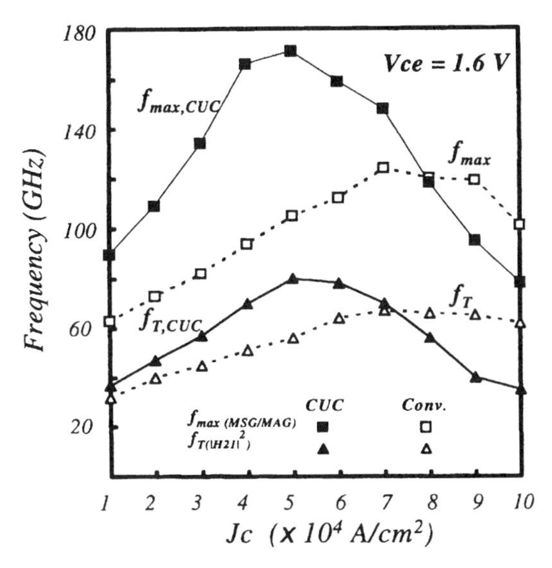

FIGURE 4.17 f_T and f_{max} versus collector current density (after Chen et al., Ref. 24 © IEEE).

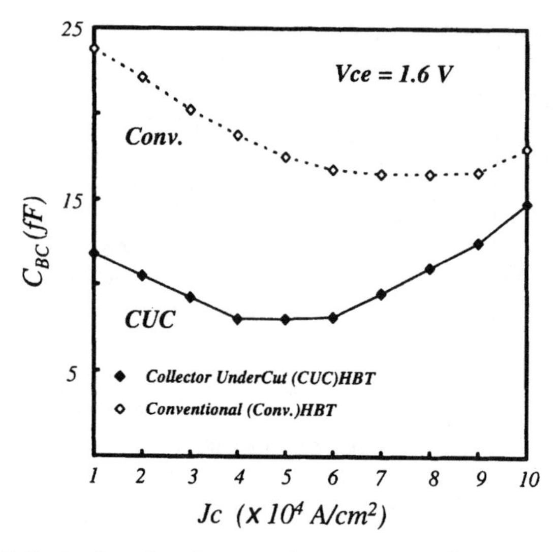

FIGURE 4.18 Collector–base junction capacitance versus collector current (after Chen et al., Ref. 24 © IEEE).

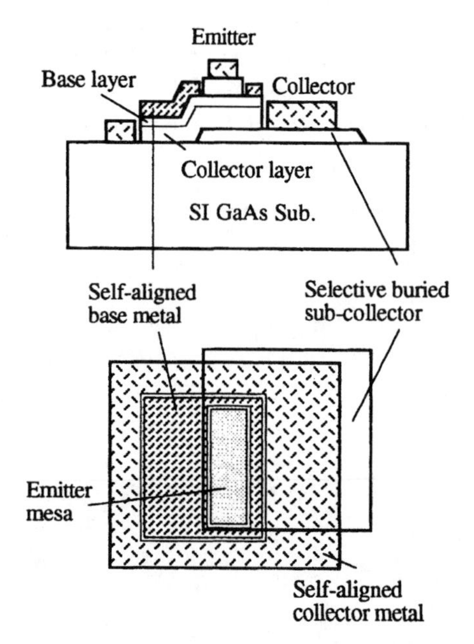

FIGURE 4.19 Cross-sectional view of a GaInP/GaAs HBT using a selective buried subcollector (after Yang et al., Ref. 25 © IEEE).

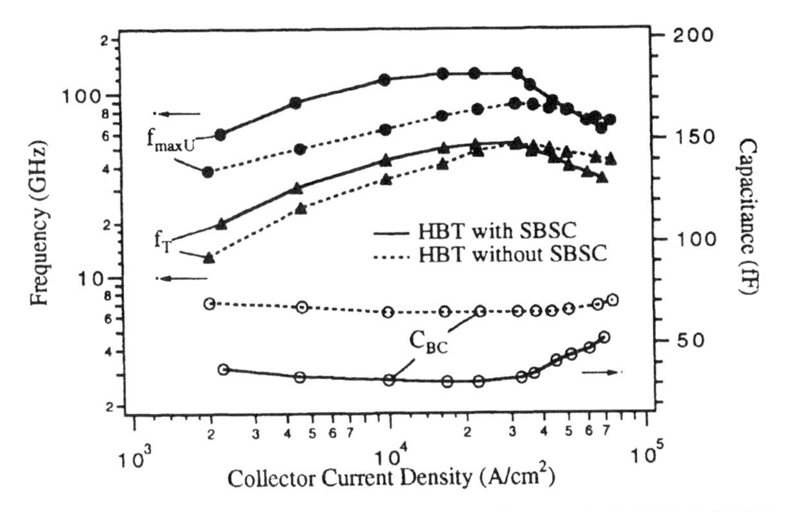

FIGURE 4.20 f_t, $f_{\max}$, and C_{BC} versus J_C (after Yang et al., Ref. 25 © IEEE).

is a key parameter of interest from the point of view of reliability, phase noise, and an understanding of the effects of trapping on device performance. As the noise of an oscillator is added to the carrier and fed back through the resonator, the resulting signal is both amplitude and phase modulated. The limiting amplifier will remove the amplitude-modulated component but does not affect the phase modulation. Phase noise dominates the near-carrier spectrum of the oscillator and often determines one endpoint of the dynamic range in a system. The maximum undistorted signal depends on system linearity, while the minimum is set by noise.

Electrical noise can be classified into three main categories: thermal noise, shot noise, and flicker ($1/f$) noise. Thermal agitation of a conductor gives rise to the random motion of electrons, resulting in thermal noise. Shot noise originates from the quantized nature of charge flow: When electrons overcome a potential barrier, the emission process occurs as independent, random events. The flicker noise originates from surface states and bulk defects and is less intuitive than the thermal and shot-noise sources. Its frequency spectrum has a near-$1/f$ characteristic. Both shot noise and thermal noise are primarily white-noise sources. The flicker noise plays a major role in the performance of broadband amplifiers, as well as oscillators in which it is up-converted to the oscillation frequency.

The low-frequency noise was suggested to be the fluctuation of the surface recombination velocity of minority carriers. Due to the presence of deep-level traps in the emitter–base forward depletion region, low-frequency noise can also arise through the generation–recombination process [26,27]. The base surface states related noise, and the emitter depletion region generation–recombination noise in a practical device usually consists of many components, due to discrete surface states and bulk generation–recombination centers with distributed energies. As a consequence of this energy distribution, the spectral response of this noise becomes

$1/f$-like for extended frequencies. Low-frequency noise measured in high-current-gain AlGaAs/GaAs DHBTs has been shown to originate from the base and is found to be interface $1/f$ and generation–recombination noise [28].

The HBT's vertical current flow through well-shielded junctions leads to small trapping effects and low $1/f$ noise. Although the measured $1/f$ noise corner frequency, on the order of tens to several hundred kilohertz, is higher than Si bipolar transistors, it is a factor of 10 to 100 times lower than GaAs FETs, which suffer significant surface states and surface trapping effects. A direct effect is observed in the phase noise performance of the HBT in an oscillator circuit. Oscillator phase noise is dependent on the overall quality factor of the resonator and flicker noise of the active devices.

For modeling purposes, $1/f$ noise is expressed as [29]

$$i_c^2 = \frac{a_f \Delta f I_C^{kf}}{f} \tag{4.17}$$

where a_f is the flicker noise coefficient, k_f the flicker noise exponent, and Δf the bandwidth of the noise spectral.

At high frequencies, the various low-frequency noise sources can be totally neglected. High-frequency noise is related to the collector and base shot and thermal noise. The shot noise, due to fluctuation of charge carriers across the emitter–base junction, and the thermal noise, due to the fluctuation of semiconductor mobilities at the collector and base, are physically uncorrected. The shot noise can be expressed as

$$\bar{i}_c^2 = 2q\Delta f I_C \tag{4.18a}$$

$$\bar{i}_b^2 = 2q\Delta f I_B \tag{4.18b}$$

and the thermal noise generators associated with the base and emitter resistance are expressed as

$$\bar{e}_b^2 = 4kT\Delta f R_B \tag{4.19a}$$

$$\bar{e}_e^2 = 4kT\Delta f R_E \tag{4.19b}$$

The measured low-frequency noise in high-performance, UHV/CVD epitaxial Si- and SiGe-base bipolar transistors has been reported [30]. Figure 4.21 shows noise power spectral density at a base current of 2.25 μA for multistripe Si and SiGe transistors with an emitter area of $0.5 \times 10.0 \times 3$ μm^2. Both types of transistors have comparable doping profiles. The inferred noise corner frequencies are 480 and 373 Hz for Si and SiGe transistors, respectively. At low frequencies, $S_{IB}(f)$ rises over the shot-noise and thermal-noise background and exhibits the expected $1/f$ dependence. For 35 different samples measured, having both Si and SiGe profiles, the $1/f^{\alpha}$ exponent varies from $0.84 < \alpha < 1.16$ over the frequency range 1 to 100 Hz. At higher frequencies, where the shot noise dominates, the spectra approach a

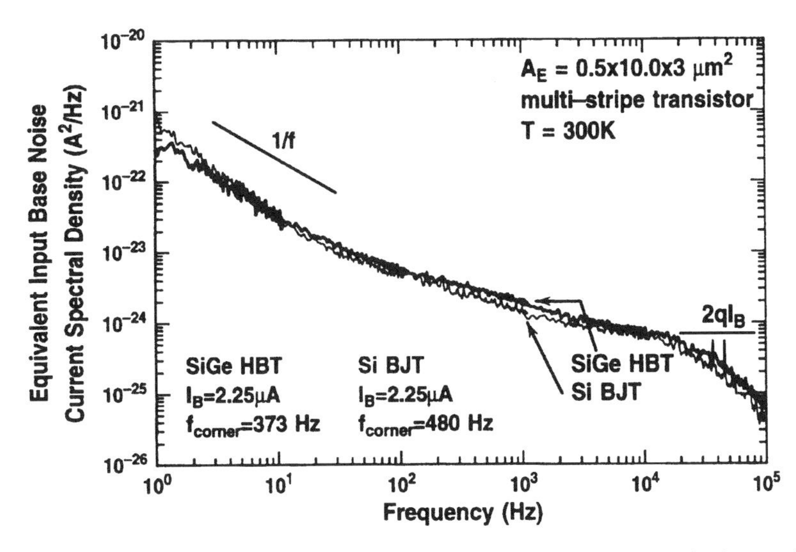

FIGURE 4.21 Equivalent input base noise versus frequency for Si and SiGe bipolar transistors (after Cressler et al., Ref. 30 © IEEE).

constant value with respect to frequency and are in good agreement with their expected shot-noise values of $2qI_B$. The roll-off of the spectra above 10 kHz is due to the Miller capacitance associated with the device and packaging [31].

GaInP/GaAs HBT low-frequency noise performance was demonstrated [32]. The $1/f$ noise spectra were measured on common-emitter devices with emitter dimensions of $3 \times 1.4\ \mu$m $\times$ $8.5\ \mu$m at $V_{CE} = 2$ V and $I_C = 15$ mA. The setup is shown in Fig. 4.22 using the HP 3048 to collect $1/f$ noise data. Base bias was supplied through a 4-$k\Omega$ resistor and collector bias through a 1.1-$k\Omega$ resistor (R_L). The output noise voltage (in V²/Hz) was measured across R_L. Dividing by R_L^2 converts this output

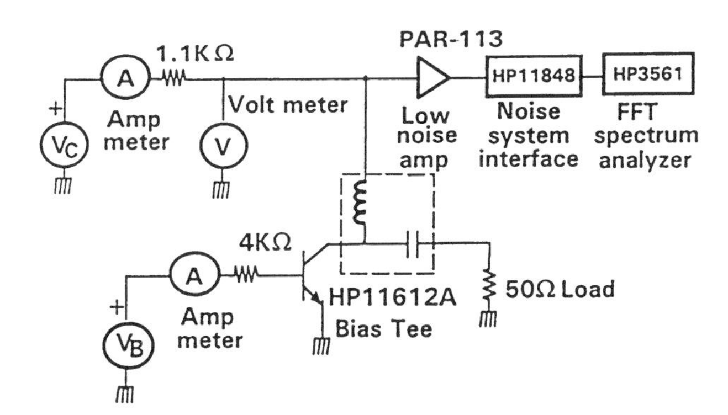

FIGURE 4.22 Noise measurement setup (after Ho et al., Ref. 32 © IEEE).

noise voltage to an equivalent output noise current. The bias tee and 50-Ω load at the device output keep the device stable at higher frequencies, 40 MHz and above.

Measured noise spectral density for the Si BJT, the GaInP HBT, and the AlGaAs HBT is shown in Fig. 4.23. The measured noise level of GaInP HBTs was -92 dB · V^2/Hz close to dc and -150 dB · V^2/Hz at 100 kHz. Comparing with AlGaAs HBTs and commonly used low-noise silicon bipolar transistors (NE219) with emitter area of $14 \times 1.5\ \mu m \times 29\ \mu m$ at the same frequency range, GaInP HBTs were 20 to 30 dB better than the AlGaAs HBT and comparable to the silicon BJT. It is also noteworthy that the noise bump often observed in the AlGaAs HBTs between 10 and 100 kHz was not visible in the GaInP HBTs. This may be due to a lack of DX centers in GaInP material [33].

Burst noise in bipolar transistors is often associated with traps or generation–recombination centers near the emitter–base space-charge region. It could be attributed to the repeated switching between two generation–recombination centers, at intervals on the order of a millisecond. Previous phenomenological studies showed that the spectral density of burst noise takes the form of a Lorenztian expression [34]:

$$S_{I\text{burst}}(f) \propto \frac{\tau I_C^m}{1 + (2\pi\tau f)^2} \tag{4.20}$$

where τ is the time constant of the trap involved and m is an empirical constant between 0.5 and 2. The time constant is thermally activated by an energy E_a:

$$\tau = \frac{\tau_0}{T^2} e^{E_a/kT} \tag{4.21}$$

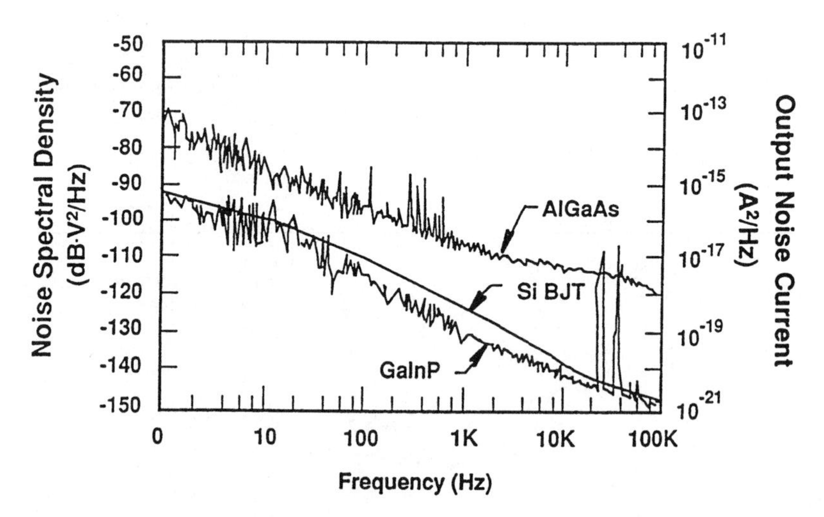

FIGURE 4.23 Noise power spectral density and output noise current versus frequency (after Ho et al., Ref. 32 © IEEE).

The Lorenztian spectrum has a large temperature dependence, as shown in Fig. 4.24, because the time constant decreases with increasing temperature.

Low-frequency noise at room temperature (298 K) and liquid-nitrogen temperature (80 K) was compared [35]. Figure 4.25*a* shows noise power versus frequency for different base currents at 296 K. The collector–emitter bias was fixed at 2.0 V. As can be seen in Fig. 4.25*a*, the noise characteristics vary considerably as the base current changes. On the other hand, no marked changed was observed in the characteristics when V_{CE} was changed. The fact that the low-frequency noise is more sensitive to changes in base current and fairly insensitive to changes in V_{CE} suggests that the noise source is located closed to the emitter–base junction. The low-frequency noise arises mainly from fluctuations in the occupancy of the traps. For $I_B = 0.5\ \mu$A and 1 μA, the noise is close to the $1/f$ type at low frequencies and is of the generation–recombination type at higher frequencies. The results suggest the presence of traps at the emitter–base heterointerface and their significant contribution to the low-frequency noise.

The noise characteristics at 80 K is shown in Fig. 4.25*b*. The results are similar to 296-K spectra in the sense that the noise magnitude increases as the base current increases. For lower base currents the spectrum is of the $1/f$ type, whereas at higher base currents the spectra exhibit generation–recombination components also. The presence of the generation–recombination components at lower temperatures suggest that different traps are active in different temperature regimes.

4.1.11 *S*-Parameters

S-parameters are used to characterize the heterojunction transistor's f_T and $f_{\max}$ as well as for the design of RF circuits. Consider a two-port network with incident (a_1

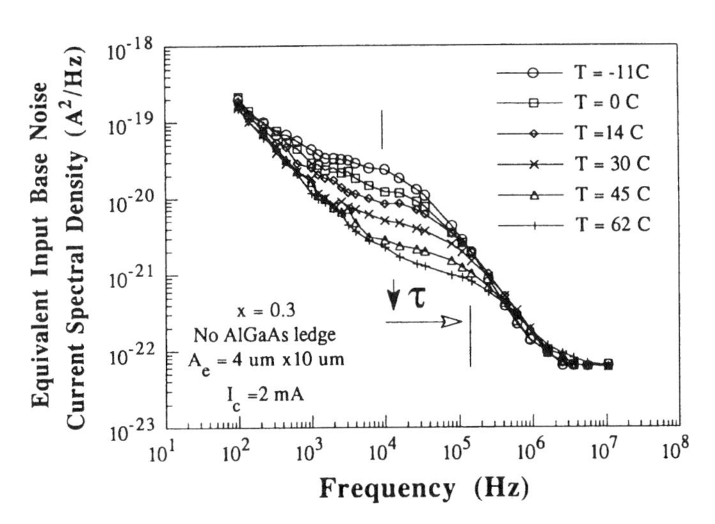

FIGURE 4.24 Equivalent input base noise versus frequency as a function of temperature (after Costa and Harris, Ref. 34 © IEEE).

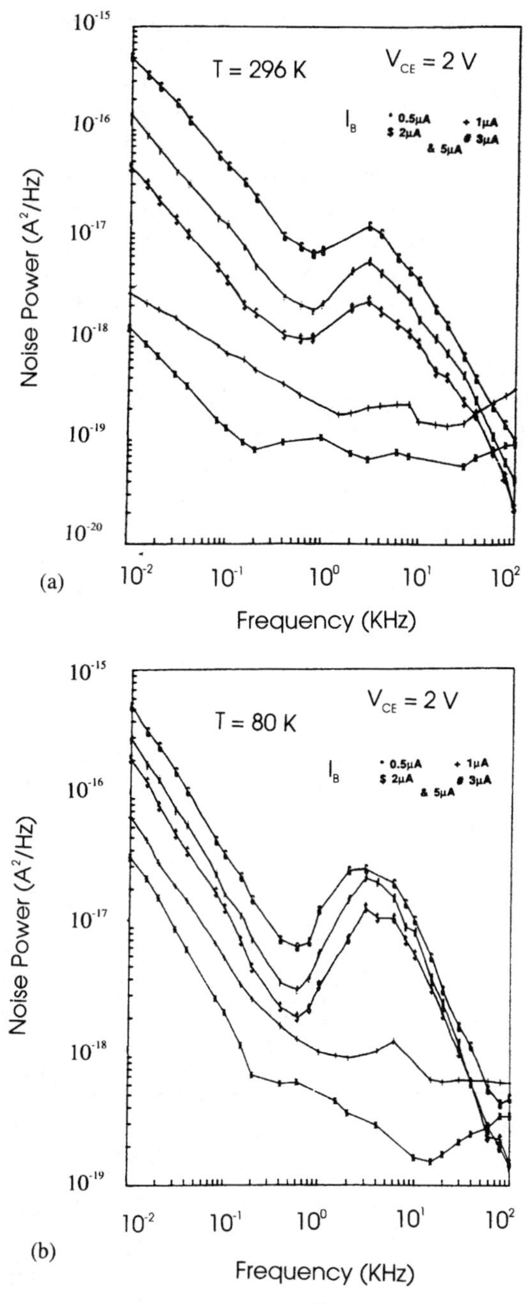

FIGURE 4.25 Noise power versus frequency for different base currents at (*a*) 296 K and (*b*) 80 K (after Raman et al., Ref. 35 © IEEE).

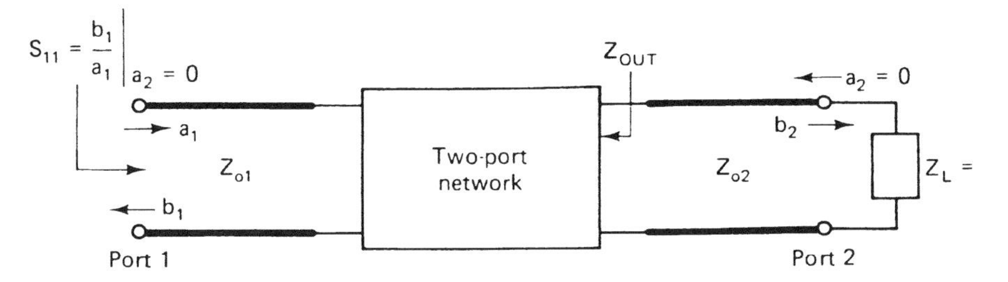

FIGURE 4.26 Two-port network for S-parameter definition.

and a_2) and reflected waves (b_1 and b_2) shown in Fig. 4.26. The scattering parameters are defined by

$$\begin{bmatrix} b_1 \\ b_2 \end{bmatrix} = \begin{bmatrix} s_{11} & s_{12} \\ s_{21} & s_{22} \end{bmatrix} \begin{bmatrix} a_1 \\ a_2 \end{bmatrix} \tag{4.22}$$

where

$$s_{11} = \left.\frac{b_1}{a_1}\right|_{a_2=0}$$

(input reflection coefficient with output terminated by a matched load)

$$s_{12} = \left.\frac{b_1}{a_2}\right|_{a_1=0}$$

(reverse transmission gain with input terminated by a matched load)

$$s_{21} = \left.\frac{b_2}{a_1}\right|_{a_2=0}$$

(forward transmission gain with output terminated in a matched load)

$$s_{22} = \left.\frac{b_2}{a_2}\right|_{a_1=0}$$

(output reflection coefficient with input terminated by a matched load)

The unilateral gain is the forward power gain in a feedback amplifier with its reverse power gain set to zero by adjusting a lossless reciprocal feedback network around the HBT. This gain is given by [36]

$$U = \frac{s_{11}s_{22}s_{12}s_{21}}{(1 - s_{11}^2)(1 - s_{22}^2)} \tag{4.23}$$

Figure 4.27 shows the scattering parameters for the Siemns HBT from 0.5 to 12 GHz. The HBT was biased at $I_C = 78.5$ mA and $V_{CE} = 3$ V. The S-parameters for the

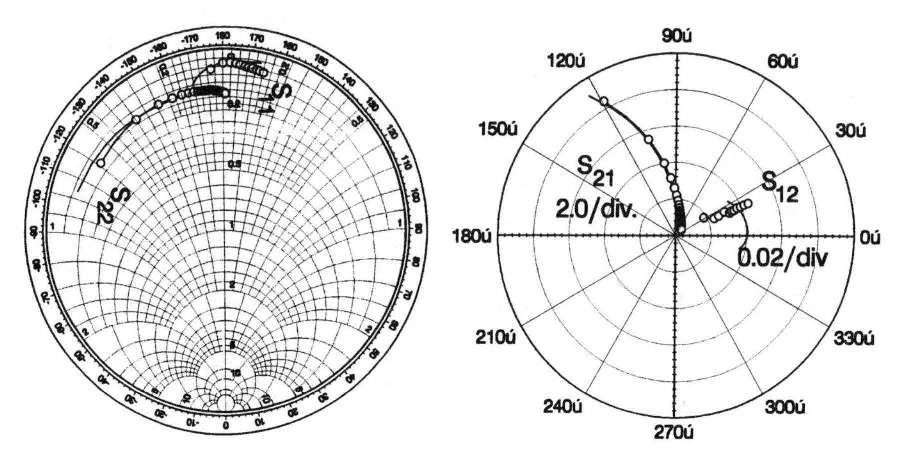

FIGURE 4.27 S-parameter experimental data of Siemens HBTs (after Lu et al., Ref. 37 © IEEE).

Rockwell HBT from 0.57 to 9.8 GHz are shown in Fig. 4.28. The device was biased at $I_C = 55$ mA and $V_{CE} = 3.0$ V. Unlike the current gain and cutoff frequency of the HBT, it is very difficult to relate S-parameters to device doping profiles and device dimensions for physical insight and device design.

Temperature-dependent forward transit time and base transit time as a function of temperature for an abrupt InP/$In_{0.53}Ga_{0.47}As$ HBT have been extracted from the measured S-parameter data [38]. S-parameters at $V_{CE} = 1.1$ V, $I_C = 13$ mA, and $T = 55$ K are displayed in Fig. 4.29. Extracted delay versus temperature is shown in Fig. 4.30. In this figure solid circles represent the experimental data, open circles

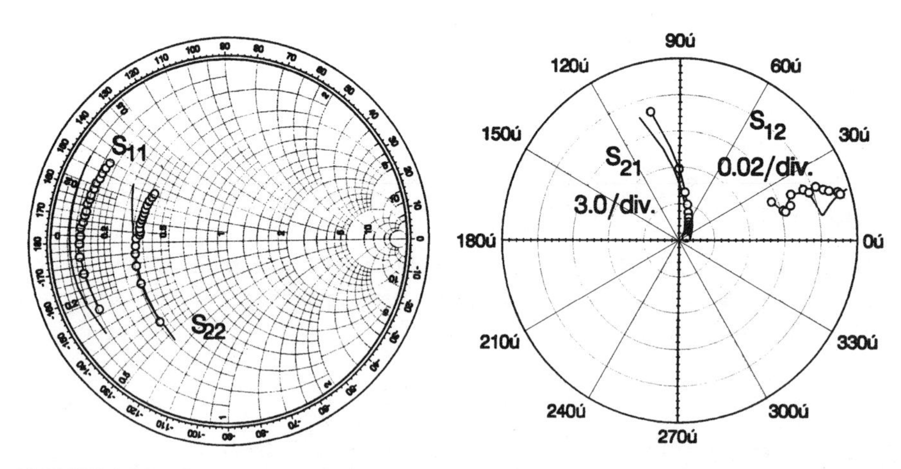

FIGURE 4.28 S-parameter experimental data of Rockwell HBTs (after Lu et al., Ref. 37 © IEEE).

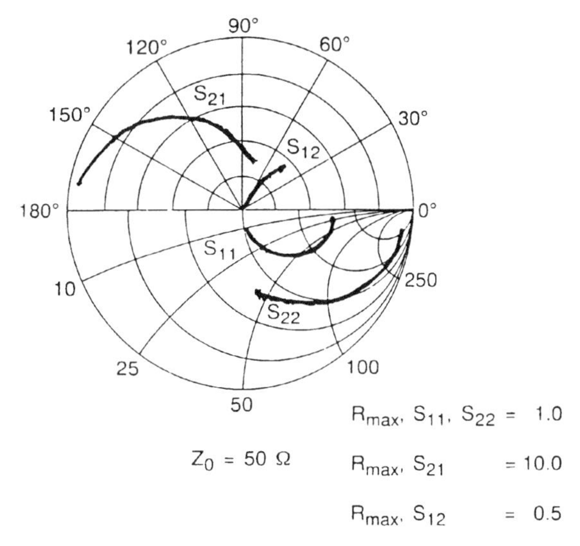

FIGURE 4.29 S-parameters at 55 K (after Laskar et al., Ref. 38 © IEEE).

represent the Monte Carlo simulation data, and the solid lines represent a least-squares fit to the simulation data. The HBT has an emitter area of $1.2 \times 10^{-7}\text{cm}^{-2}$. The n-InP emitter has a thickness of 2000 Å and a doping of $1 \times 10^{18}\text{cm}^{-3}$. The n-InGaAs emitter cap layer has a doping of $7 \times 10^{19}\text{cm}^{-3}$ with a thickness of 2000 Å. The p-InGaAs base has a thickness of 500 Å and a doping of $1 \times 10^{20}\text{cm}^{-3}$. The n-InGaAs collector has a doping of 2×10^{16} with a 3000 Å thickness. The n-InGaAs subcollector has a doping of $5 \times 10^{19}\text{cm}^{-3}$ with a thickness of 2500 Å grown on semi-insulating InP substrates. At a temperature $T = 340$ K the forward delay τ_F is 0.5 ps. With decreasing temperature, τ_F becomes smaller, saturating at a value of 0.28 ps for $T \leq 150$ K. At low temperatures ($T \leq 50$ K), the base transit delay is about one-third of the total forward delay. The HBT device is biased at $V_{CE} = 1.2$ V and $I_C = 20$ mA; $\tau_F \sim 0.30$ ps and $\tau_B \sim 0.11$ ps. The average electron velocity in the base is 4.5×10^7 cm/s, and in the collector it is 8×10^7 cm/s. At room temperature ($T = 300$ K) the base transit time $\tau_B \sim 0.19$ ps is almost half the total delay $\tau_F \sim 0.43$ ps. In this case $v_B \sim 2.6 \times 10^7$ cm/s and $v_C \sim 6 \times 10^7$ cm/s. The high average electron velocity in the base arises from nonequilibrium electron transport. The observed temperature dependence of intrinsic forward delay is due to variation in electron scattering with temperature and carrier density. This may be illustrated by considering the inelastic scattering rate for a central Γ-valley conduction band electron of initial kinetic energy $E = 150$ meV. In the neutral base, electrons may scatter elastically from statistically screened ionized p-type impurities or inelastically, changing kinetic energy by $\hbar\omega$ and changing momentum by the scattering wave vector.

The discussion of τ_F above refers to a collector–base voltage $V_{CE} = 0.3$ V. In this situation base transport is more temperature sensitive than collector transport. The relative temperature-dependent contributions of Γ-L and Γ-X scattering do not

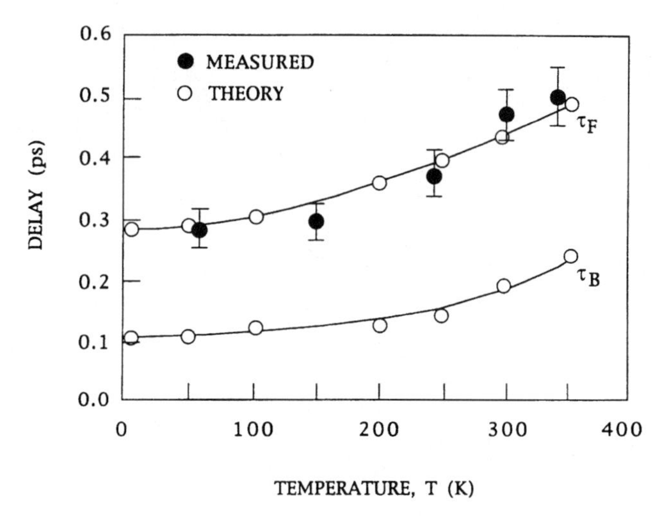

FIGURE 4.30 Forward transit time versus temperature (after Laskar et al., Ref. 38 © IEEE).

dominate. At larger reverse bias (e.g., $V_{CB} = 1$ V), more electrons scatter into the X- and L-valleys and the relative importance of the temperature dependence of phonon and electron–electron scattering in these valleys increases. Figure 4.30 shows the enhanced temperature sensitivity of τ_F with increasing temperature and increasing V_{CE}. At values of $V_{CB} \leq 0.2$ V, the temperature-sensitive collector diffusion capacitance causes the experimentally determined τ_F to increase.

4.1.12 Turn-off Transient

An AlGaAs/GaAs heterojunction bipolar transistor has a homojunction between base and collector. When the HBT is biased in saturation, significant minority carriers inject into the lightly doped collector region. These excess carriers cause significant switching delay when the HBT is moving out of saturation. Assuming that electrons are governed by diffusion once they pass the emitter–base heterojunction, the electron concentration in the uniformly doped base is determined by the continuity equation:

$$\frac{\partial n}{\partial t} = D_n \frac{\partial^2 n(x,t)}{\partial x^2} \tag{4.24}$$

The boundary conditions for the switch-off transient under constant voltage approach are

$$\Delta n(0,t) = 0 \tag{4.25a}$$

$$\Delta n(W_B,t) = 0 \tag{4.25b}$$

Since the electron diffusion length in the base is much larger than the base width, the minority-carrier concentration in the base is linear. The initial condition of the differential equation in (4.24) can be written as

$$\Delta n(x, 0) = \Delta n(0) - \left[\Delta n(0) - \Delta n(W_B)\right] \frac{x}{W_B} \tag{4.26}$$

Using Eqs. (4.25*a*–*b*) and solving (4.24), we obtain

$$\Delta n(x, t) = \sum_{s=1}^{\infty} \left[\frac{2\Delta n(0)}{s\pi} - \frac{2\Delta n(W_B)}{s\pi} (-1)^s \right] e^{-s^2\pi^2 D_n t / W_B^2} \sin \frac{s\pi x}{W_B} \tag{4.27}$$

Equation (4.27) describes the position- and time-dependent electron concentration in the base during the switch-off transient.

When the HBT is turning off, electrons in the base are removed through the emitter–base and collector–base junctions, as shown in Fig. 4.31. The stored electron charge in the base at an arbitrary time is obtained by integrating the electron concentration from $x = 0$ to $x = W_B$:

$$\begin{aligned} Q_n(t) &= -qA \int_0^{W_B} \Delta n(x, t)\, dx \\ &= qA \sum_{s=1}^{\infty} \left[\frac{2\Delta n(0)}{s\pi} - \frac{2\Delta n(W_B)}{s\pi} (-1)^s \right] \\ &\qquad \times e^{-s^2\pi^2 D_n t / W_B^2} [(-1)^s - 1] \frac{W_B}{s\pi} \end{aligned} \tag{4.28}$$

When the heterojunction bipolar transistor is in the forward active mode, minority hole injection into the collector is negligible. As the base–collector junction becomes forward biased, excess holes inject into the lightly doped collector by diffusion. Neglecting recombination, the continuity equation for holes in the collector is expressed

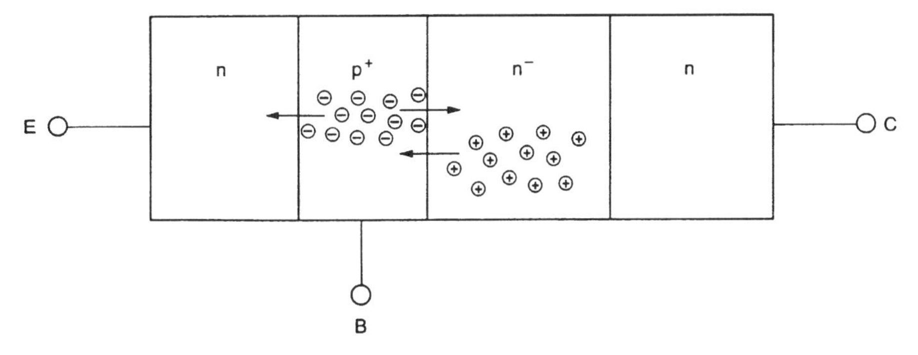

FIGURE 4.31 Carrier flow during the HBT turn-off.

as

$$\frac{\partial \Delta p}{\partial t} = D_p \frac{\partial^2 p(x,t)}{\partial x^2} \tag{4.29}$$

The boundary conditions at the collector high–low junction are

$$\Delta p(0,t) = 0 \tag{4.30a}$$

$$\Delta p(W_C,t) \approx 0 \tag{4.30b}$$

and the initial condition is

$$\Delta p(x,0) = \Delta p(0) - \frac{\cosh[(W_C - x)/L_p]}{\cosh(W_C/L_p)} \tag{4.31}$$

Note that the position $x = 0$ has been changed to the edge of the collector–base space-charge region in the collector side. Using (4.30*a–b*) and (4.31), the minority-carrier concentration in the collector is solved as

$$\Delta p(x,t) = \sum_{s=1}^{\infty} \frac{-4\Delta p(0)(s\pi/W_C^2)}{(s\pi/W_C)^2 + 1/L_p^2} e^{-s^2\pi^2 D_p t/W_C^2} \sin\frac{s\pi x}{W_C} \tag{4.32}$$

During the switch-off transient, holes are removed from the collector through the collector–base junction. Integrating Δp from the collector–base junction to $x = W_C$ yields the minority hole charge at an arbitrary time:

$$\begin{aligned} Q_p(t) &= qA \int_0^{W_C} \Delta p(x,t)\,dx \\ &= qA \sum_{s=1}^{\infty} \left[\frac{-4\Delta p(0)/W_C}{(s\pi)^2 + 1/L_p^2} \right] [(-1)^s - 1] e^{-s^2\pi^2 D_p t/W_C^2} \end{aligned} \tag{4.33}$$

Figure 4.32 shows the electron concentration in the base of a single heterojunction bipolar transistor during the switch-off transient. The heterojunction bipolar transistor has an emitter area $20 \times 10^8 \text{cm}^{-2}$, a base doping $5 \times 10^{19} \text{cm}^{-3}$, a base width 0.1 μm, a collector concentration $5 \times 10^{16} \text{cm}^{-3}$, and a collector width 0.5 μm. The HBT is driven into saturation ($V_{BE} = 1.4$ V and $V_{BC} = 1.2$ V) before switching. During the switch-off transient, electrons are removed both from the emitter–base junction and the collector–base junction. The electron concentration is plotted at $t = 10^{-13}$ s, 3×10^{-13} s, 5×10^{-13} s, 1×10^{-12} s, and 3×10^{-12} s, respectively [39]. At $t \sim 3 \times 10^{-12}$ s the electron charge in the base is removed completely. The fast charge removal is due to high electron mobility and diffusivity in the base of the AlGaAs/GaAs HBT. The percentage of minority charge removed from the base $[R(\%) = [Q_n(t = 0) - Q_n(t)]/Q_n(t = 0)]$ with respect to time is shown in Fig. 4.33. It only takes 2×10^{-13} s to remove 50% electron charge, it takes 8×10^{-13} s to remove 90% electron charge, and it takes 3×10^{-12} s to remove 100% electron charge.

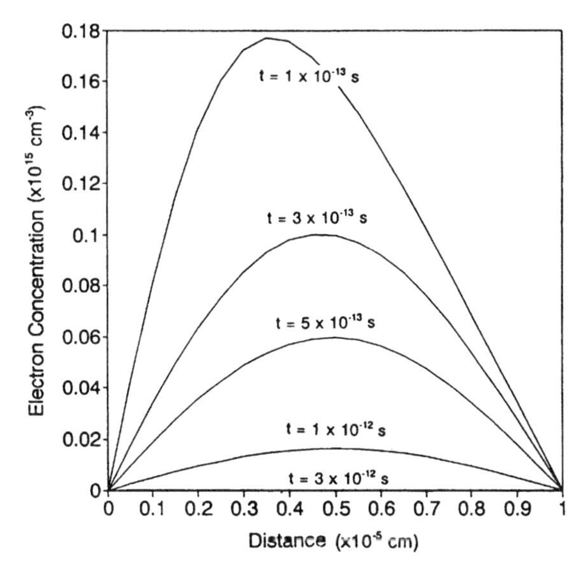

FIGURE 4.32 Electron concentration in the base.

Figure 4.34 shows the hole concentration in the collector during the switch-off transient. This figure is plotted at $t = 2 \times 10^{-13}$ s, 2×10^{-12} s, 1×10^{-11} s, 3×10^{-11} s, and 1×10^{-10} s. It is seen from Fig. 4.34 that the holes are removed from the collector–base junction first. The excess carriers are then removed from the collector high-low

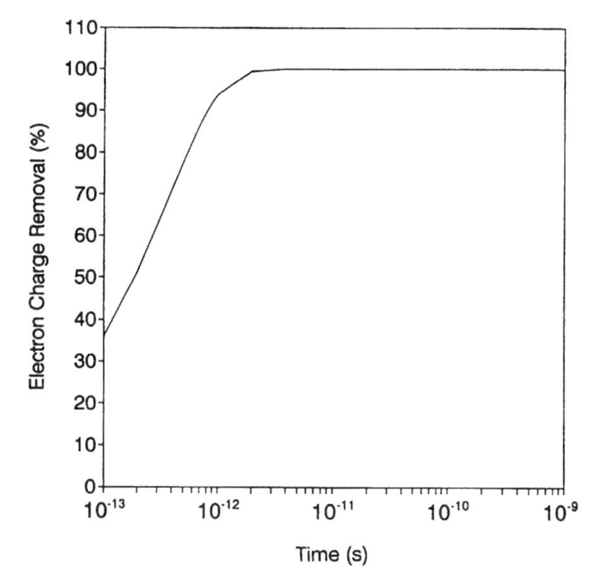

FIGURE 4.33 Electron charge removed as a function of time.

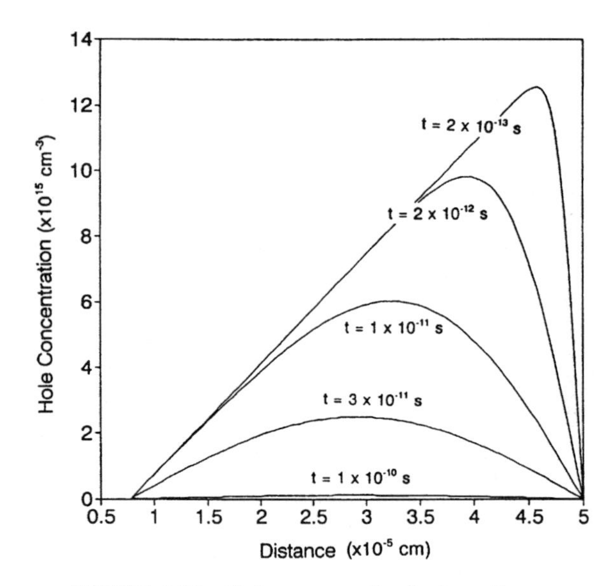

FIGURE 4.34 Hole concentration in the collector.

junction. The percentage of hole charge removed from the collector is displayed in Fig. 4.35. It takes 10 ps to removed 50% Q_p, 40 ps to removed 90% Q_p, and 200 ps to remove Q_p completely. Comparing Figs. 4.39 and 4.41, the storage time for the entire charge removed from the collector is more than 50 times larger than that of the electron charge removed from the base. This clearly indicates that the storage time of the HBT is determined by excess holes stored in the collector, not excess electrons stored in the base. Excess holes stored in the collector depend on the doping concentration and the thickness of collector layers, as well as the base–collector junction voltage. To reduce excess holes in the collector, the collector concentration should be increased and the collector thickness should be reduced. However, higher collector doping and narrower collector thickness give a lower collector breakdown voltage.

In silicon integrated circuits, base-collector p-n junctions of bipolar transistors are often used as diode elements for logic circuits. The junction shows a high-voltage blocking capability and has a convenient structure for fabricating many diodes that have a common n-type region. However, the junction diode cannot operate fast because of minority-carrier storage in the n-type collector region.

This problem is overcome by using the SiGe-HBT structure [40,41]. The collector–base heterojunction of the SiGe HBT hardly shows minority-carrier accumulation in forward conduction, so high-speed diodes for LSI or VLSI circuits can be realized using this junction. The SiGe-base HBTs also show little carrier accumulation even when operated in the saturation region, so they are capable of operating at high speed in saturated mode-switching applications. This is because the hole injection is prevented by the large potential barrier at the heterojunction interface and holes are hardly accumulated in the n-type collector region. Figure 4.36 shows the simulation

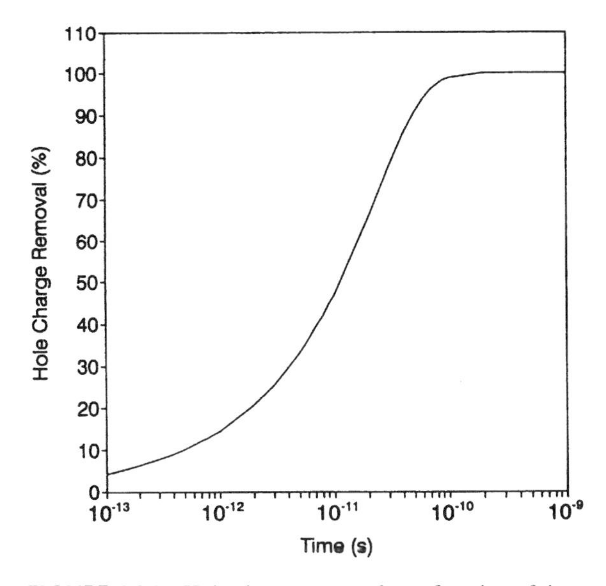

FIGURE 4.35 Hole charge removed as a function of time.

result of carrier storage in the SiGe diode and the silicon homojunction diode as a function of forward current density [40]. The SiGe bandgap decrement is 0.3 eV. Q_n shows the absolute value of electron storage in the p-type region, and Q_p shows hole storage in the n-type region. It is clear from this figure that the SiGe diode has a total hole storage two orders less than that of the silicon diode. Lower minority-carrier charge storage will result in a smaller storage time and faster turn-off switching.

4.2 SILICON–GERMANIUM HETEROJUNCTION BIPOLAR TRANSISTORS

In this section, effects of Ge profiles on g_o, τ_B, f_T, $f_{\max}$ and inverse base width modulation on τ_B and τ_C as well as transconductance clipping at high currents and low temperatures are presented.

4.2.1 Effect of Ge Profiles on g_o and τ_B

The output conductance including neutral-base recombination effect can be derived as

$$g_o = J_C \frac{C_{JC}}{qD_n(W_B)n_i^2(W_B)} \times \frac{J_{B0}/q + \int_0^{W_B} \left[n_i^2(x)/\tau(x)p_B(x)\right]\,dx}{J_{B0}/J_{C0} + \int_0^{W_B} \left[n_i^2(x)/\tau(x)p_B(x)\right] \int_x^{W_B} \left[p_B(y)/D_n(y)n_i^2(y)\right]\,dy\,dx} \quad (4.34)$$

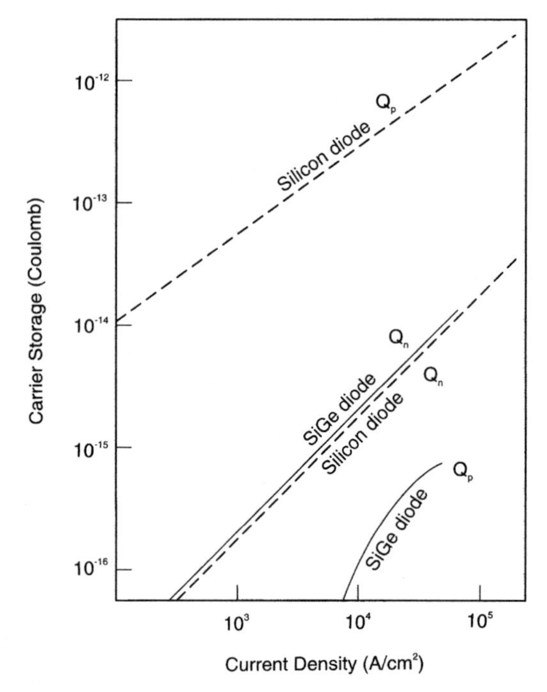

FIGURE 4.36 Charge storage of a SiGe diode (after Ugajin and Amemiya, Ref. 40 © Elsevier Science).

The output conductance versus Ge grading is shown in Fig. 4.37. The Ge grading increases the output conductance for larger X_T/W_B.

The integration of (4.7) is expanded for the examination of the Ge effect on the base transit time as follows [42]:

$$\tau_B = \int_0^{x_T} \frac{n_i^2(z)}{p_B(z)} \left[\int_z^{x_T} \frac{p_B(y)\,dy}{D_n(y)n_i^2(y)} + \int_{x_T}^{W_B} \frac{p_B(y)\,dy}{D_n(y)n_i^2(y)} \right] dz + \int_{x_T}^{W_B} \frac{n_i^2(z)}{p_B(z)} \int_z^{W_B} \frac{p_B(y)\,dy}{D_n(y)n_i^2(y)}\,dz \tag{4.35}$$

Inserting (3.19) into (4.35) and performing the resulting integration gives

$$\tau_B = \frac{X_T^2}{D_n} \frac{kT}{\Delta E_{g,\mathrm{Ge}}(\mathrm{grade})} \left\{ 1 + \left(e^{-\Delta E_{g,\mathrm{Ge}}(\mathrm{grade})/kT} - 1 \right) \times \left[1 + \frac{kT}{\Delta E_{g,\mathrm{Ge}}(\mathrm{grade})} - \frac{W_B}{X_T} \right] + \frac{\Delta E_{g,\mathrm{Ge}}(\mathrm{grade})}{2kT} \left(\frac{W_B}{X_T} - 1 \right)^2 \right\} \tag{4.36}$$

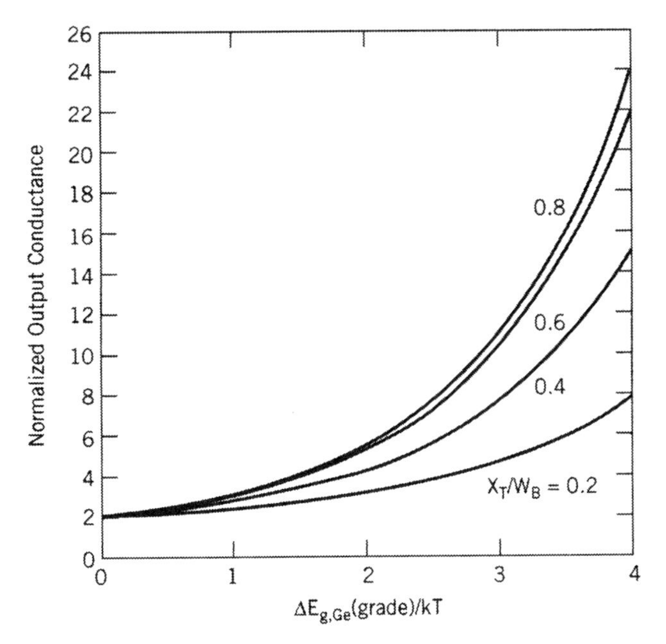

FIGURE 4.37 Output conductance versus Ge grading at different X_T/W_B.

The ratio of base transit time in the transistor with the Ge trapezoid to that of a pure Si device is

$$\Theta = \frac{\tau_{B,\text{SiGe}}}{\tau_{B,\text{Si}}} = \frac{2}{a^2 x}\left[1 - (1 - e^{-x})\left(\frac{1}{x} + 1 - a\right) + \frac{x}{2}(a-1)^2\right] \tag{4.37}$$

The asymptotic behavior of (4.37) is

$$\lim_{x \to 0} \Theta = 1$$

$$\lim_{x \to \infty} \Theta = \left(1 - \frac{1}{a}\right)^2$$

The normalized base transit time versus Ge grading using (4.37) is shown in Fig. 4.38. In this plot $\Theta = 1$ at $X_T/W_B = 0.01$ and $\Delta E_{g,\text{Ge}}(\text{grade})/kT = 1$. The present analytical predictions of base transit time are consistent with device physics. For a SiGe HBT with $X_T/W_B = 0.6$, Cressler et al. [43] measured $\Theta = 0.522$ at 309 K. Using (4.4), one obtains $\Theta = 0.543$, which is in good agreement with the experimental data.

The closed-form equation to find the optimum trapezoidal Ge profile in the base for base transit time minimization is derived using (4.37). Rewriting (4.37), the normalized base transit time of a trapezoid Ge profile in the uniformly doped base is

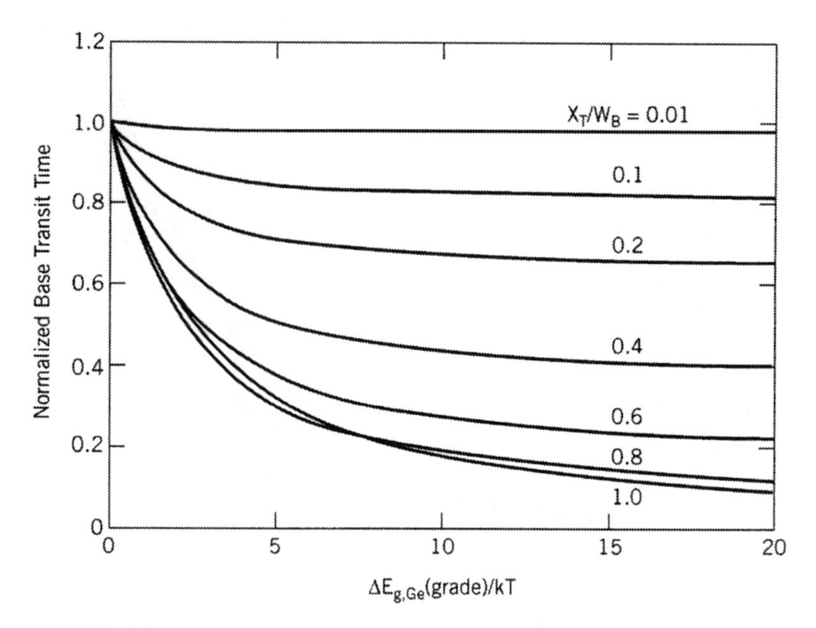

FIGURE 4.38 Normalized base transit time versus $\Delta E_{G,\mathrm{ge}}$(grade)/kT as a function of X_T/W_B (after Song and Yuan, Ref. 42 © IEEE).

given by

$$\Theta = \left(1 - \frac{X_T}{W_B}\right)^2 + \frac{2kT\left(\dfrac{X_T}{W_B}\right)^2\left[1 + \left(e^{-\Delta E_{g,\mathrm{Ge}}(\mathrm{grade})/kT} - 1\right)\left(1 + \dfrac{kT}{\Delta E_{g,\mathrm{Ge}}(\mathrm{grade})} - \dfrac{W_B}{X_T}\right)\right]}{\Delta E_{g,\mathrm{Ge}}(\mathrm{grade})} \tag{4.38}$$

The optimum base transit time is determined by setting

$$\frac{d\Theta}{d(X_T/W_B)} = 0$$

This results in [44]

$$\left.\frac{X_T}{W_B}\right|_{\mathrm{opt}} = \frac{1 + \dfrac{kT\left(e^{-\Delta E_{g,\mathrm{Ge}}(\mathrm{grade})/kT} - 1\right)}{\Delta E_{g,\mathrm{Ge}}(\mathrm{grade})}}{1 + \dfrac{2kT[1 + \left(e^{-\Delta E_{g,\mathrm{Ge}}(\mathrm{grade})/kT} - 1\right)\left(1 + kT/\Delta E_{g,\mathrm{Ge}}(\mathrm{grade})\right)]}{\Delta E_{g,\mathrm{Ge}}(\mathrm{grade})}}. \tag{4.39}$$

Figure 4.39 shows the normalized base transit time versus X_T/W_B at $T = 309$ K. At $X_T/W_B = 0$, the HBT has a uniform Ge profile in the base. In this case the normalized base transit time is equal to 1, as expected. At $X_T/W_B = 1$, the HBT has a linear Ge profile in the base. The normalized base transit time for the trapezoidal Ge profile decreases with increasing X_T/W_B, reaches a minimum, and then increases slightly with X_T/W_B. For $X_T/W_B = 0.6$ and $\Delta E_{g,\mathrm{Ge}}(\mathrm{grade}) = 60$ meV, the experimental data (solid square in Fig. 4.39) for the HBT with a trapezoidal Ge profile in the base [43] is shown. The agreement between the model prediction and experiment is fairly good. As shown in Fig. 4.39, the minimum base transit time is not at $X_T/W_B = 1$ (linear Ge profile), but between 0.75 and 0.85, depending on the value of $\Delta E_{g,\mathrm{Ge}}(\mathrm{grade})$.

Their temperature-dependent behavior is evaluated below. The normalized base transit times at 85 and 533 K are shown in Fig. 4.40*a* and *b*, respectively. Again, the minimum base transit is not given by a linear Ge profile. In Fig. 4.40 the agreement between the model prediction and the experimental data (solid square) is very good. Comparing Figs. 4.39 and 4.40*a* and *b* it is clear that θ increases with temperature for given X_T/W_B and $\Delta E_{g,\mathrm{Ge}}(\mathrm{grade})$. Optimum X_T/W_B decreases with temperature for a given germanium gradient (bandgap difference). Optimum X_T/W_B versus $\Delta E_{g,\mathrm{Ge}}(\mathrm{grade})$ for $T = 85$, 309, and 533 K is depicted in Fig. 4.41. Optimum X_T/W_B is sensitive to temperature and $\Delta E_{g,\mathrm{Ge}}(\mathrm{grade})$, as expected. For example, optimum X_T/W_B for $\Delta E_{g,\mathrm{Ge}}(\mathrm{grade}) = 60$ meV is 0.785 at $T = 533$ K, but 0.905 at $T = 85$ K.

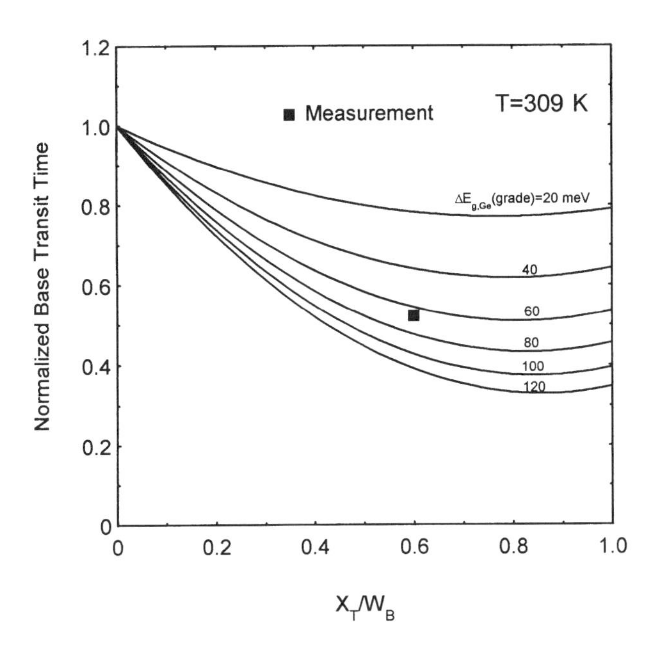

FIGURE 4.39 Normalized base transit time versus X_T/W_B (after Song and Yuan, Ref. 44 © Elsevier Science).

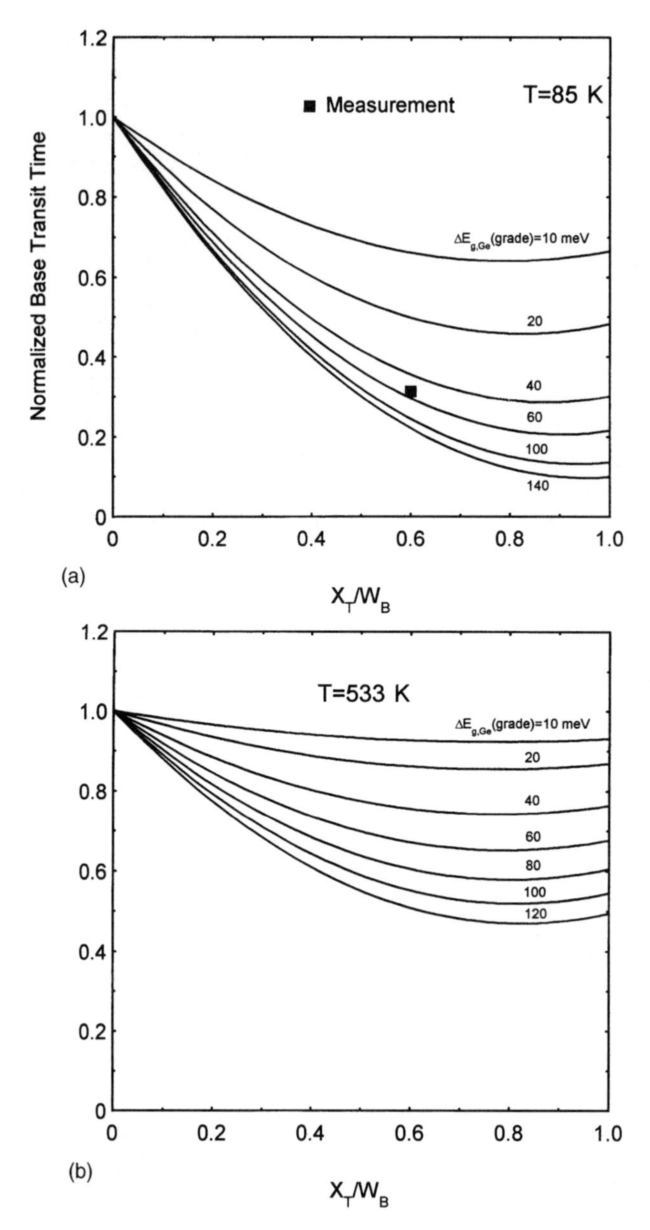

FIGURE 4.40 Normalized base transit time versus X_T / W_B for different $\Delta E_{g,\mathrm{Ge}}$(grade) at (*a*) 85 K and (*b*) 533 K (after Song and Yuan, Ref. 44 © Elsevier Science).

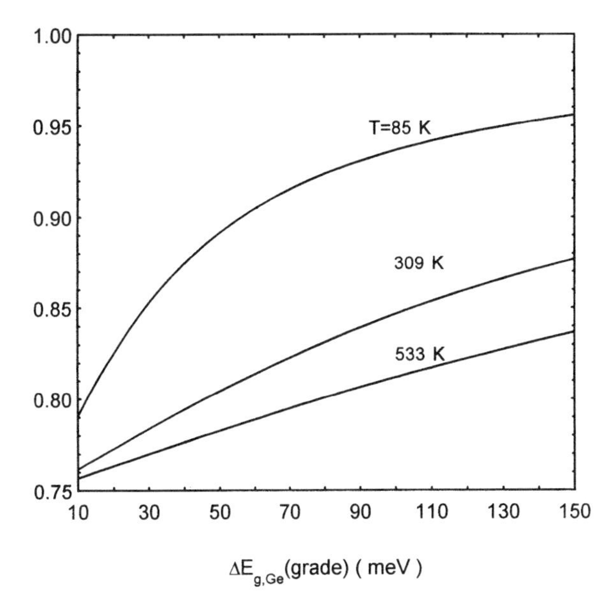

FIGURE 4.41 Optimized X_T/W_B versus Ge grading for base transit time minization (after Song and Yuan, Ref. 44 © Elsevier Science).

4.2.2 Effect of Ge Profiles on f_T and $f_{\max}$

It is known that the most important device figures of merit will vary depending on the particular circuit application. For example, the current gain–Early voltage product βV_A has great relevance to precision current sources, voltage references, analog-to-digital converters ADCs, and digital-to-analog converters DACs, but $f_{\max}$ is more important for high-frequency amplifiers, mixers, and oscillators. Therefore, the "best" Ge profile may also vary, depending on the circuit application and the trade-off associated with it.

The box (uniform Ge content) profile in the base produces the largest values of current gain, but extremely large values may also produce undesirably low values of breakdown voltage BV_{CE0}. The box Ge profile also produces the largest values of f_T in the scaled profile device. Figure 4.42 shows τ_B and τ_E for the bipolar transistors with no Ge, box, and triangle (linearly graded Ge content) profiles at each stability point. The stability points [46] are defined in Fig. 4.43 of the effective thickness versus effective strain plot. The box profile is best for optimizing τ_E, while the triangle profile produces the smallest τ_B. If τ_B becomes small compared to τ_E, the box Ge profile produces the largest enhancement in f_T. At stability point 3 in Fig. 4.43, the box Ge profile has the smallest τ_E, and $\tau_B \leq \tau_E$ for both Ge profiles.

The maximum frequency of oscillation of the SiGe HBT normalized to that of the Si BJT as a function of temperature is illustrated in Fig. 4.44. $f_{\max}$ (SiGe)/$f_{\max}$(Si) increases with decreasing temperature. For the box Ge profile, $f_{\max} = 40$ GHz at

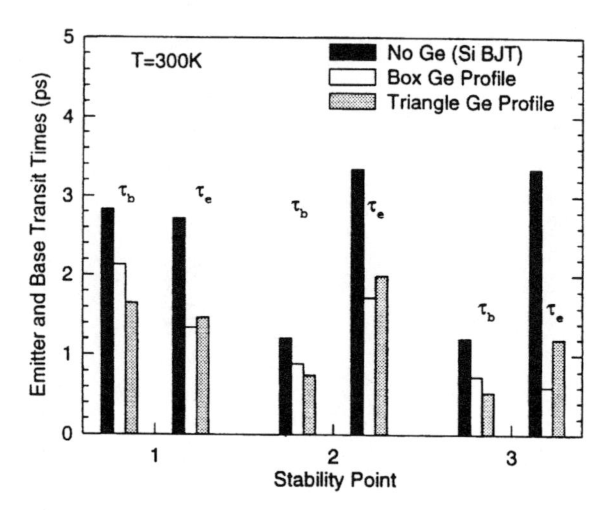

FIGURE 4.42 τ_B and τ_E versus Ge profile (after Richey et al., Ref. 45 © IEEE).

stability point 1 and 104 GHz at stability point 3, at 300 K. For the triangle Ge profile, $f_{\max} = 40$ GHz at stability point 1 and 86 GHz at stability point 3, at 300 K.

The triangle Ge profile produces the largest enhancement in Early voltage [47] and a reasonable enhancement in current gain and f_T. Figure 4.45 shows βV_A(SiGe) normalized to βV_A(Si) as a function of temperature for box and triangle Ge profiles at stability points 1 and 3. Although the triangle Ge profile produces a modest enhancement in $f_{\max}$, this profile produces the largest enhancement in the βV_A product. In addition, if temperature behavior is an important design constraint, β and f_T for

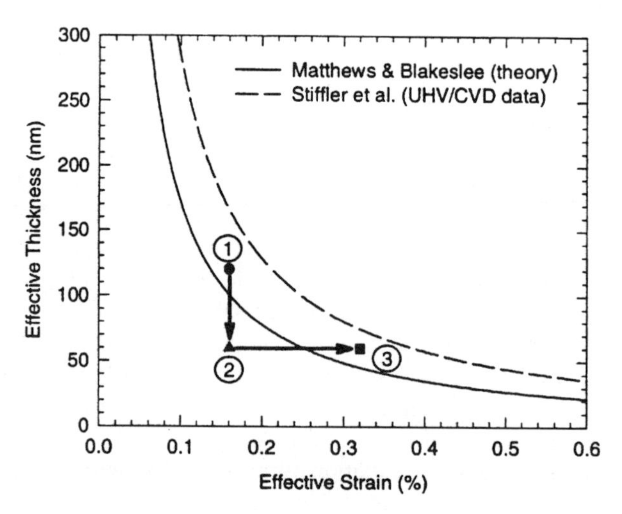

FIGURE 4.43 Definition of stability points.

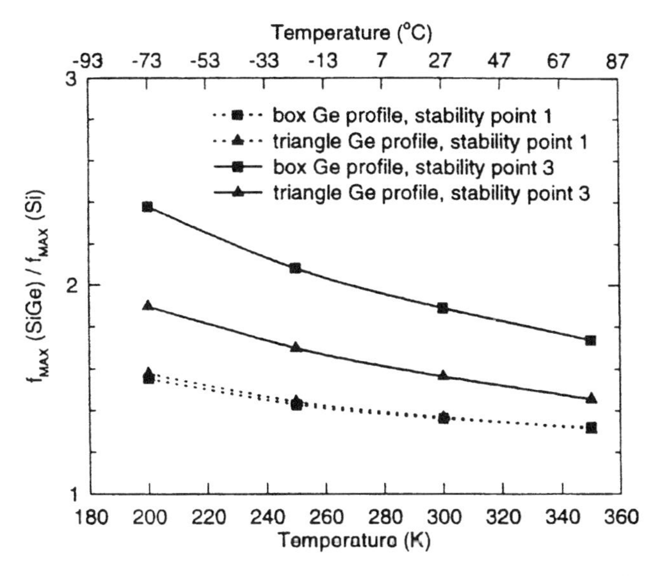

FIGURE 4.44 Normalized $f_{\max}$ versus temperature (after Richey et al., Ref. 45 © IEEE).

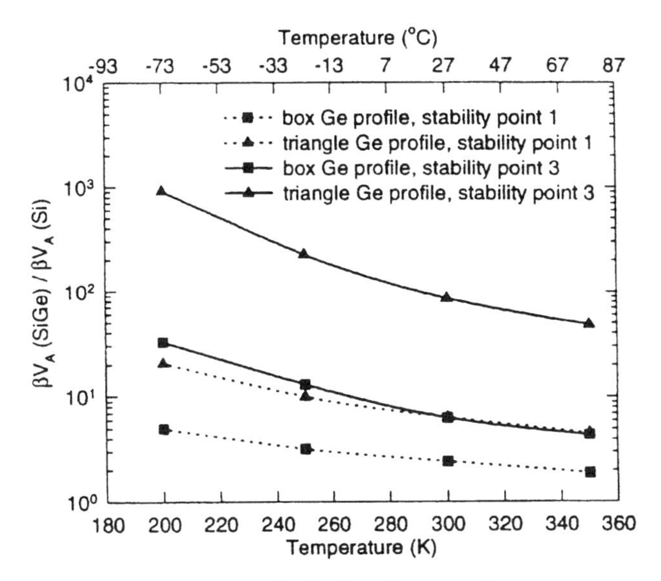

FIGURE 4.45 Normalized current gain–Early voltage product versus temperature (after Richey et al., Ref. 45 © IEEE).

the box Ge profile are more sensitive functions of temperature than for the triangle Ge profile.

4.2.3 Effect of Inverse Base Width Modulation on τ_B and τ_C

The base transit time of minority carriers is usually a dominant component of the total transit time in silicon-based bipolar transistors. In SiGe HBTs, however, the base transit time has been significantly reduced due to the aiding field in the base from Ge grading. Thus, other time constants become important in determining the cutoff frequency of the device. In particular, Ugajin and Fossum [48,49] have shown that inverse base width modulation—a widening of base width—and widening of the collector–base space-charge region become the predominant mechanisms underlying the degradation of cutoff frequency at high current densities. The falloff of cutoff frequency at high current densities could occur well below the onset current density for base pushout. The early degradation is triggered by the incremental transit time resulting from the collector–base space-charge region widening and the inverse base-width modulation effect. High collector current increases mobile carriers in the collector–base space-charge region, which compensate for the donor and acceptor concentrations in that region. The collector–base space-charge layer is effectively widening and τ_C is increased. The inverse base width modulation occurs when the collector–base junction space-charge region on the base side shrinks as the collector current increases. This effect causes an increase of base transit time.

The increase of τ_B and τ_C could be derived analytically. Consider the doping profile around the collector–base space charge of a bipolar transistor and its electric field distribution illustrated in Fig. 4.46, where W_{SC} is the collector–base space charge layer width in the base side, W_{SCC} is the collector–base space-charge-layer width in the collector side, and W_{EPI} is the epitaixal collector width. The base transit time increment due to high current effect is

$$\Delta\tau_B = \frac{d\Delta Q_B}{dI_C} = I_C \frac{d\tau_B}{dW_B}\frac{dW_B}{dI_C} \approx -I_C \frac{d\tau_B}{dW_B}\frac{dW_{\mathrm{SC}}}{dI_C} \tag{4.40}$$

Equation (4.40) implicitly assumes that the base width modulation at the base–collector junction is much larger than that at the emitter–base junction (i.e., $dW_B/dI_C \approx -dW_{\mathrm{SC}}/dI_C$). The derivative of τ_B with respect to I_C can be obtained using (4.7). For the SiGe HBT with uniform doping and a trapezoidal Ge profile in the base,

$$\frac{d\tau_B}{dW_B} \approx \frac{\Delta E_{g,\mathrm{Ge}}(\mathrm{grade})}{2kT}\left(\frac{W_B}{X_T} - 1\right) \tag{4.41}$$

The collector–base space-charge-layer transit time increment is approximated by

$$\Delta\tau_C = \frac{d\Delta Q_{BC}}{dI_C} \approx qAn\frac{d(W_{\mathrm{SC}} + W_{\mathrm{SCC}})}{dI_C} - qAN_D(W_{\mathrm{SCC}})\frac{dW_{\mathrm{SCC}}}{dI_C} \tag{4.42}$$

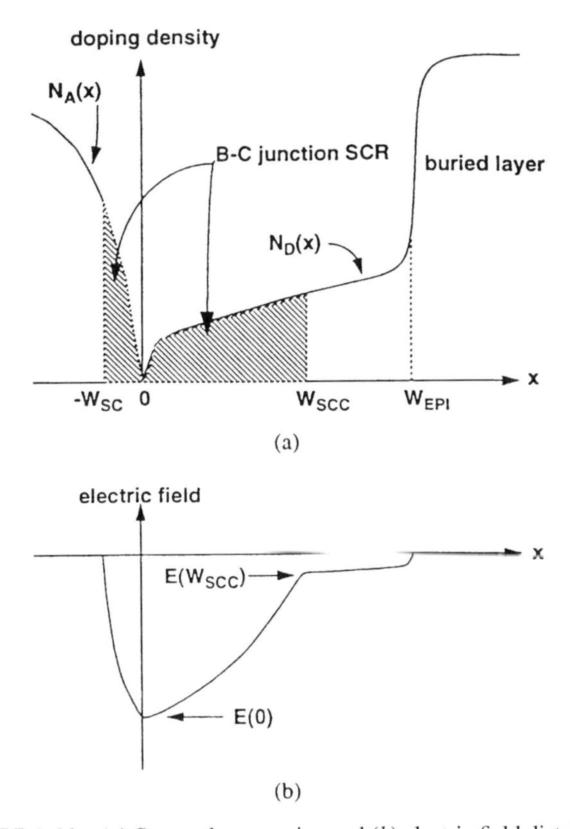

FIGURE 4.46 (*a*) Space-charge region and (*b*) electric field distribution.

where n is the electron density in the collector–base space-charge region ($n \approx I_C/qAv_s$) and $N_D(W_{\text{SCC}})$ is the epitaxial collector doping density at $x = W_{\text{SCC}}$. Assuming that the integral of the electric field across the entire epi-collector region and the collector–base space-charge region is equal to the junction built-in potential plus the reverse bias $-V'_{\text{BC}}$, one obtains [49]

$$\begin{aligned}\phi_C - V'_{BC} \approx & - E(0)W_{\text{SCC}} - \int_0^{W_{\text{SCC}}} \int_0^x \frac{q[N_D(y) - n]}{\varepsilon} \, dy \, dx \\ & + \int_{W_{\text{SCC}}}^{W_{\text{EPI}}} \frac{I_C}{qA\mu_n(x)N_D(x)} dx \\ & - E(0)W_{\text{SC}} - \int_{-W_{\text{SCC}}}^0 \int_x^0 \frac{q[N_A(x) + n]}{\varepsilon} dy dx \qquad (4.43)\end{aligned}$$

where

$$E(0) \approx -\int_{-W_{SC}}^{0} \frac{q[N_A(x)+n]}{\varepsilon}\,dx$$
$$\approx E(W_{SCC}) - \int_{0}^{W_{SCC}} \frac{q[N_D(x)-n]}{\varepsilon}\,dx \tag{4.44}$$

By differentiating Eqs. (4.18) and (4.19) with respect to V'_{BC} we have

$$\frac{\partial W_{SC}}{\partial V'_{BC}} \approx -\frac{\varepsilon}{q(W_C+W_{SCC})[N_A(-W_{SC})+n]} \tag{4.45}$$

$$\frac{\partial W_{SCC}}{\partial V'_{BC}} \approx -\frac{\varepsilon}{q(W_C+W_{SCC})[N_D(W_{SCC})-n]} \tag{4.46}$$

Differentiating (4.18) and (4.19) with respect to I_C gives

$$\frac{\partial W_{SC}}{\partial I_C} \approx \frac{1}{q[N_A(-W_{SC})+n]}\left(-\frac{W_{SC}+W_{SCC}}{2Av_s} - \frac{\varepsilon R_{EPI}}{W_{SC}+W_{SCC}}\right) \tag{4.47}$$

$$\frac{\partial W_{SC}}{\partial I_C} \approx \frac{1}{q[N_D(W_{SCC})-n]}\left[\frac{W_{SC}+W_{SCC}}{2Av_s} - \frac{\varepsilon(R_{SCR}+R_{EPI})}{W_{SC}+W_{SCC}}\right] \tag{4.48}$$

where

$$R_{EPI} = \int_{W_{SCC}}^{W_{EPI}} \frac{dx}{qA\mu_n(x)N_D(x)} \tag{4.49}$$

$$R_{SCR} = \frac{W_{SCC}+W_{SC}}{qA\mu_n(W_{SCC})N_D(W_{SCC})} \tag{4.50}$$

When the collector doping is constant [i.e., $N_D(x) = N_{EPI}$], $R_{EPI} = (W_{EPI} - W_{SCC})/qA\mu_n N_{EPI}$ and $R_{SC}R = (W_{EPI} + W_{SC})/qA\mu_n N_{EPI}$.

From (4.41)–(4.48) the current-induced variations of the two portions of the space-charge region are

$$\frac{dW_{SC}}{dI_C} = \frac{\partial W_{SC}}{\partial I_C} + \frac{\partial V'_{BC}}{\partial I_C}\frac{\partial W_{SC}}{\partial V'_{BC}}$$
$$= \frac{\partial W_{SC}}{\partial I_C} + \left(R_C + R_E + \frac{1}{g_m}\right)\frac{dW_{SC}}{dV_{BC}}$$
$$\approx \frac{1}{q[N_A(-W_{SC})+n]}$$
$$\times\left[-\frac{W_{SC}+W_{SCC}}{2Av_s} - \frac{\varepsilon(R_{EPI}+R_C+R_E+1/g_m)}{W_{SC}+W_{SCC}}\right] \tag{4.51}$$

$$
\begin{aligned}
\frac{dW_{SCC}}{dI_C} &= \frac{\partial W_{SCC}}{\partial I_C} + \frac{\partial V'_{BC}}{\partial I_C}\frac{\partial W_{SCC}}{\partial V'_{BC}} \\
&= \frac{\partial W_{SCC}}{\partial I_C} + \left(R_C + R_E + \frac{1}{g_m}\right)\frac{dW_{SCC}}{dV'_{BC}} \\
&\approx \frac{1}{q[N_D(W_{SCC}) - n]} \\
&\quad \times \left[-\frac{W_{SC} + W_{SCC}}{2Av_s} - \frac{\varepsilon(R_{SCR} + R_{EPI} + R_C + R_E + 1/g_m)}{W_{SC} + W_{SCC}}\right]
\end{aligned} \tag{4.52}
$$

Inserting (4.41) and (4.51) into (4.40), we obtain

$$
\begin{aligned}
\Delta\tau_B &\approx \frac{I_C(d\tau_B/dW_B)}{qA[N_A(-W_{SC}) + n]} \\
&\quad \times \left[\frac{W_{SC} + W_{SCC}}{2v_s} + \frac{\varepsilon A}{W_{SC} + W_{SCC}}\left(R_{EPI} + R_C + R_E + \frac{1}{g_m}\right)\right]
\end{aligned} \tag{4.53}
$$

Similarly, inserting (4.51) and (4.52) into (4.42) gives $\Delta\tau_C$.

4.2.4 Transconductance Degradation at High Current Densities and Low Temperatures

Experimental evidence of a clipping of transconductance for 84 K for the SiGe HBT is observed [50] as shown in Fig. 4.47. The abrupt clipping of the transconductance at low temperature is believed to be a signature of the collector–base heterojunction barrier effect. It is known that the bandgap shrinkage associated with the incorporation

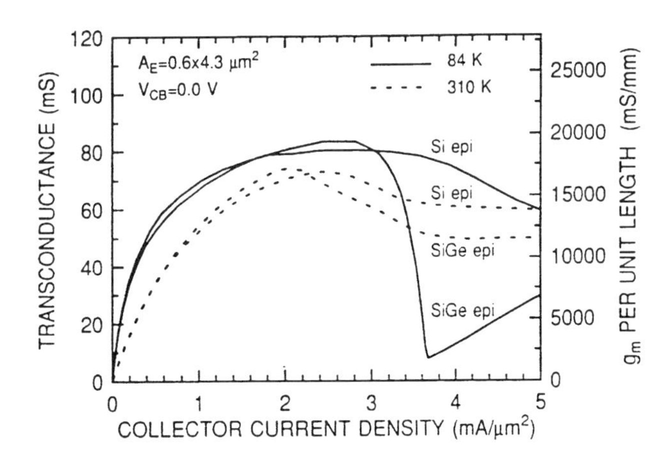

FIGURE 4.47 Transconductance clipping (after Cressler et al., Ref. 43 © IEEE).

of Ge in the base occurs in the valence band. Hence there is a bandgap discontinuity associated with the transition from SiGe to Si. Under high-level injection, the collector–base space-charge region begins to collapse, eventually exposing the valence-band barrier. Qualitatively, as the quasi-neutral base begins to push out, the valence-band barrier prevents holes from flowing into the collector. Charge neutrality cannot be maintained and a barrier to electron transport is induced in the conduction-band. The induced barrier is thermally activated. Thus even while the barrier may be so small as to be undetectable at room temperature, its effects should be increased greatly at low temperatures.

4.2.5 Ge and Collector Doping Profile Design to Improve the Clipping Effect

The conventional method to suppress base widening is simply to adopt a highly doped uniform collector profile (and/or thin epi-layer collector). Consequently, base widening is suppressed at the expense of low BV_{CE0}. One of the methods to increase BV_{CE0} effectively while suppressing base widening is to introduce a retrograde collector profile [51]. The retrograde collector increases the onset current density for pushout. An analytical model to predict the critical current density of bipolar transistors with a retrograde collector profile is given as follows [52]. A schematic representation of a retrograde collector profile is shown in Fig. 4.48. Emitter and base regions are not shown for simplicity. A Gaussian function is assumed for the n^- collector doping profile. J_0/qv_s stands for the electron concentration passing through the n^- collector region at J_0. W_x is the width over which the electric field exists. W_C is the distance from metallurgical base–collector junction (x_{jbc}) to the n^+ buried layer. The n^- collector doping profile is given by

$$N_C(x) = N_{\text{RGW}} e^{-(x-W_C)^2/\sigma^2} \tag{4.54}$$

where σ is the characteristic length of the Gaussian function and N_{RGW} is shown in Fig. 4.49.

Over the collector region, Poisson's equation is

$$\frac{dE(x)}{dx} = \frac{1}{\varepsilon}\left[qN_C(x) - \frac{J_0}{v_s}\right] \tag{4.55}$$

where hole current is ignored because of its small value. Electric field $E(x)$ can be obtained by assuming that $E(x_{jbc}) \approx 0$. Using $E(x) = -dV/dx$, we have

$$\int_{V(0)}^{V(W_c)} dV = V_{CB} + \phi_{bi} = \frac{q}{\varepsilon}\left[\frac{J_0}{qv_s}\frac{W_x^2}{2} - \int_0^{W_x}\int_0^{x} N_C(y)\,dy\,dx\right] \tag{4.56}$$

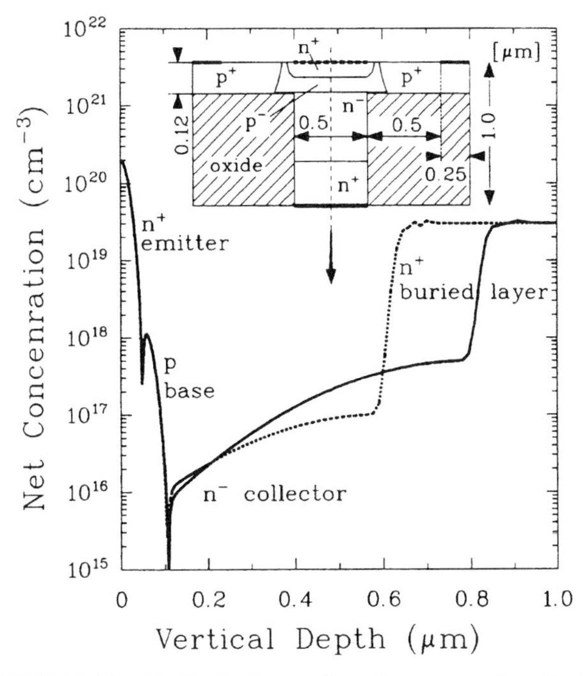

FIGURE 4.48 Bipolar doping profile with a retrograde collector.

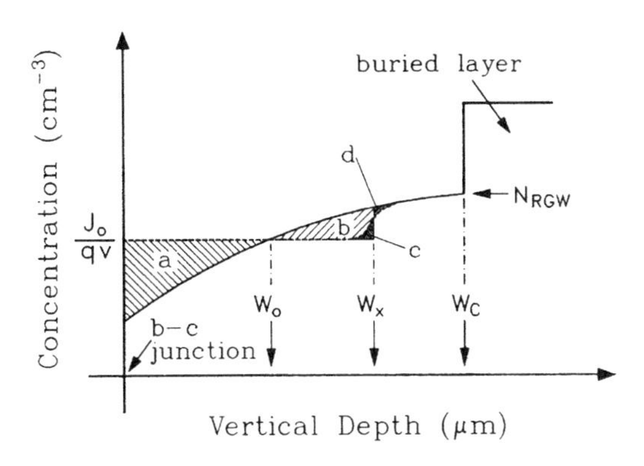

FIGURE 4.49 Schematic representation of a retrograde collector profile (after Lee et al., Ref. 52 © IEEE).

Solving (4.56) and rearranging terms for J_0 yields

$$J_0 = \frac{q v_s}{W_x} \left\{ \frac{2\varepsilon(V_{CB} + \phi_{bi})}{q W_x} + \frac{\sigma\sqrt{\pi} N_{\mathrm{RGW}}}{W_x} \left[\int_0^{W_x} \operatorname{erfc}\left(\frac{W_C - x}{\sigma}\right) dx - W_x \operatorname{erfc}\left(\frac{W_C}{\sigma}\right) \right] \right\} \quad (4.57)$$

Equation (4.57) has two unknowns, J_0 and W_x. Because of this, another equation is needed to calculate the two unknowns. Since the integration of net charge from 0 to W_x is zero, one can write

$$\int_0^{W_x} \left[\frac{J_0}{q v_s} - N_C(x) \right] dx = 0 \quad (4.58)$$

Using (4.55)–(4.57), one obtains

$$J_0 = q v_s \left\{ \frac{N_{\mathrm{RGW}}}{W_x} \frac{\sigma\sqrt{\pi}}{2} \left[\operatorname{erfc}\left(\frac{W_C - W_x}{\sigma}\right) - \operatorname{erfc}\left(\frac{W_C}{\sigma}\right) \right] \right\} \quad (4.59)$$

From (4.57) and (4.59), J_0 and W_x can be obtained by numerical calculation. Figure 4.50 shows the J_0 and W_x results from the device simulation and the derived model. In Fig. 4.50, W_x for a V_{CB} bias larger than 2.5 V is equal to W_C, and J_0 can be calculated using (4.57) only. Fairly good agreement between the device simulation and the model has been obtained. Error between the device simulation and the model is less than 5% over a typical V_{CB} bias range. The analytical model is useful for predicting the onset current density for transconductance clipping of the SiGe bipolar transistor with a retrograde collector profile.

Joseph et al. [53] investigated the high current cutoff frequency clipping of the SiGe HBT using the retrograde collector profile. Three different retrograde profile shapes were designed as shown in Fig. 4.51. The stability constraint applicable to UHV/CVD SiGe epitaxial films for these three retrograde profiles are compared with the calibrated SiGe profile as shown in Fig. 4.52. The gradually varying retrograde profile shapes (retrograde 1 and 2) provide similar f_T but larger barrier onset current density compared to the original calibrated SiGe HBT profile as shown in Fig. 4.53. Retrograde 2 and 3 profiles also show a weaker roll-off in f_T at high current density compared to the calibrated profile. The retrograde profile (2) is not optimum because only marginal improvement in $J_{C,\mathrm{barrier}}$ is achieved at the cost of a significant lowering of the film stability. One the other hand, an abrupt box-shaped Ge profile (retrograde 3) which has a stability constraint close to retrograde 1 obtains similar f_T but with a much higher barrier onset current density. However, a sharper degradation in f_T beyond the barrier onset occurs in retrograde 3 profile when compared to other retrograde designs.

Future-generation SiGe HBTs will have very narrow base width ($\leq$ 50 nm) with larger base doping and increased Ge concentration. Therefore, the impact of high-

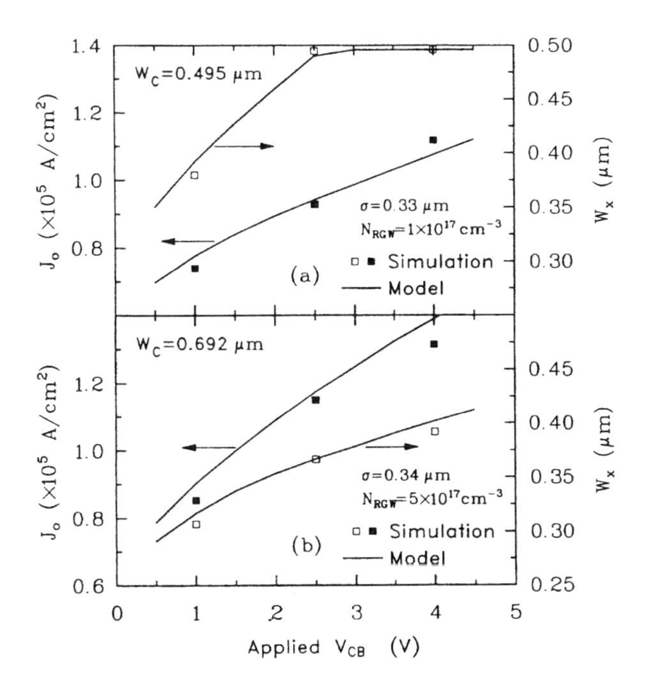

FIGURE 4.50 Critical current density J_0 and width W_x as a function of V_{CB} (after Lee et al., Ref. 52 © IEEE).

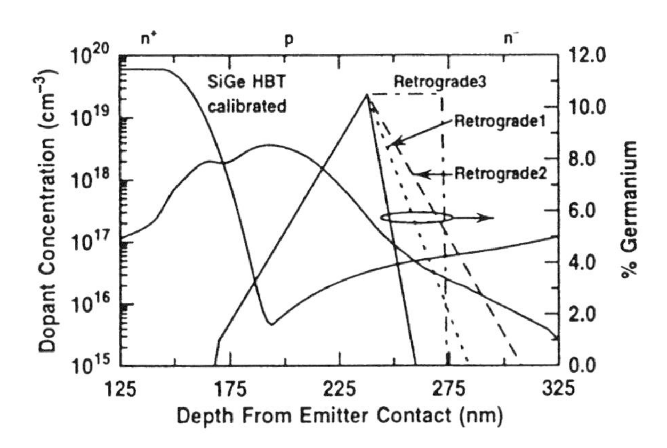

FIGURE 4.51 Ge retrograde profiles for investigating barrier effects (after Joseph et al., Ref. 53 © IEEE).

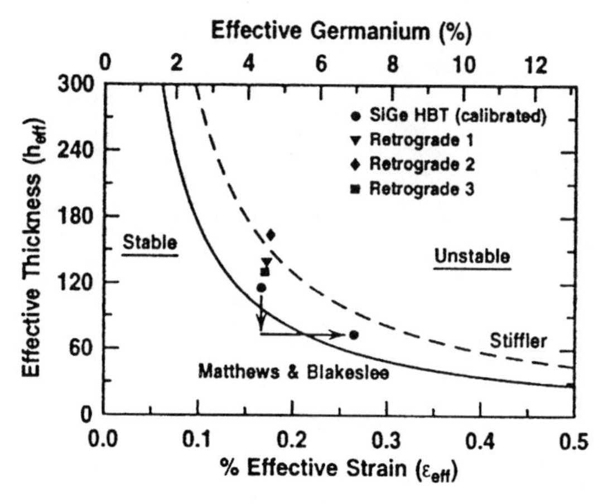

FIGURE 4.52 Effective thickness versus effective strain (after Joseph et al., Ref. 53 © IEEE).

injection barrier effects of these devices will be different from the present state-of-the-art devices. A vertically scaled UHV/CVD, trapezoidal Ge profile SiGe HBT that has emitter and collector profiles identical to the calibrated profile so as to maintain the emitter drive-in and the identical BV_{CE0} has been simulated (see Fig. 4.54). The base width is scaled by half (keeping R_{bi} constant) and the Ge content is doubled so as to maintain the same stability as in the present SiGe HBTs. A box Ge profile SiGe HBT that has the same integrated Ge content (same stability) was also constructed for this scaled profile. It is shown in Fig. 4.55 that a box Ge profile SiGe HBT shows a peak cutoff frequency of about 100 GHz at 300 K and a barrier onset current density larger than the unscaled profile.

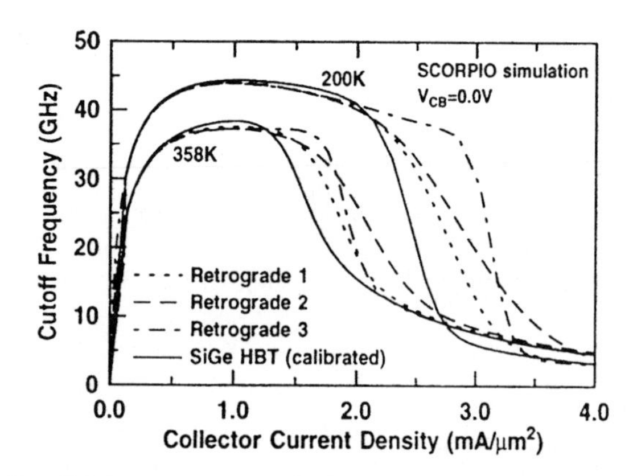

FIGURE 4.53 Cutoff frequency versus collector current density (after Joseph et al., Ref. 53 © IEEE).

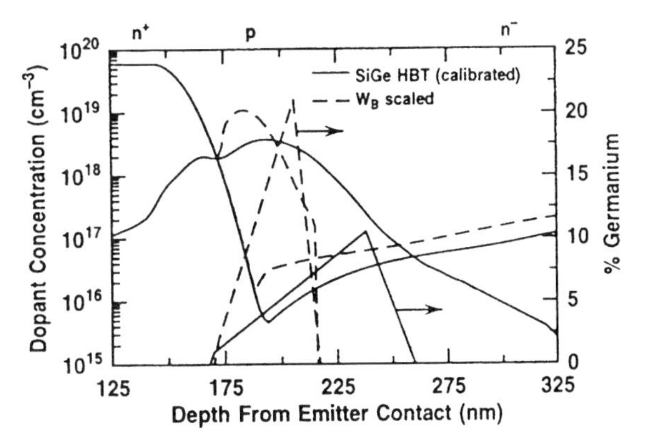

FIGURE 4.54 Doping profile for a vertically scaled SiGe HBT (after Joseph et al., Ref. 53 © IEEE).

4.3 III/V COMPOUND HETEROJUNCTION BIPOLAR TRANSISTORS

4.3.1 Emitter Delay

The emitter delay has traditionally been neglected due to its insignificance in bipolar transistors with a heavily doped emitter. The emitter delay is comparable to base transit time of III/V compound heterojunction bipolar transistors with a heavily doped base and a moderately doped emitter [54]. The emitter delay is defined by

$$\tau_E = \frac{dQ_{pE}}{dJ_n} = \frac{dQ_{pE}}{dV_{BE}} \frac{dV_{BE}}{dJ_n} \tag{4.60}$$

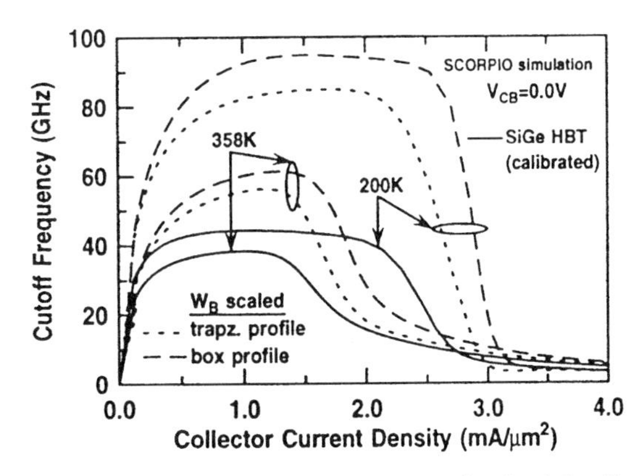

FIGURE 4.55 Cutoff frequency versus collector current density (after Joseph et al., Ref. 53 © IEEE).

where Q_{pE} is the minority hole charge in the emitter and J_p is the hole current in the emitter. The excess hole charge in the emitter is given by

$$Q_{pE} = \frac{qW_E \Delta p(0)'}{2} \tag{4.61}$$

where

$$\Delta p(0') = \frac{n_{iE}^2}{N_E(1 + S_{dp}/S_{ep} + S_{dp}/S_{ip})} n \left(e^{qV_{BE}/kT} - 1\right) \tag{4.62}$$

W_E is the emitter thickness, $x = 0'$ is the location at the edge of the emitter–base junction in the emitter side, $S_{dp} = D_{pE}/W_E$ is the diffusion velocity of holes in the emitter, S_{ep} is the drift and diffusion velocities of the holes across the space-charge region, S_{ip} is the effective interface velocity of the holes, and n_{iE} is the effective intrinsic carrier concentration, including the bandgap narrowing effect in the emitter.

The hole and electron current densities in the emitter and base can be written as

$$J_n = \frac{qD_{nB}}{W_B} \Delta n(0) \tag{4.63}$$

where

$$\Delta n(0) = \frac{n_{iB}^2}{N_B(1 + S_{dn}/S_{\text{en}} + S_{dn}/S_{\text{in}})} \left(e^{qV_{BE}/kT} - 1\right) \tag{4.64}$$

$x = 0$ is the location at the edge of the emitter–base junction in the base side, $S_{dn} = D_{nB}/W_B$ is the diffusion velocity of electrons in the base, S_{en} is the drift and diffusion velocities of the electrons across the space-charge region, S_{in} is the effective interface velocity of the electrons, and n_{iB} is the effective intrinsic carrier concentration, including the bandgap narrowing effect in the base. Using (4.61) and (4.63) for the derivatives in (4.60) gives

$$\tau_E = \frac{N_B W_E W_B}{2N_E D_{nB}} \frac{1 + S_{dn}/S_{\text{en}} + S_{dn}/S_{\text{in}}}{1 + S_{dp}/S_{ep} + S_{dp}/S_{ip}} e^{-\Delta E_G/kT} \tag{4.65}$$

where ΔE_G is the bandgap difference between the base and the emitter.

In homojunction bipolar transistors, both the hole and electron currents are controlled by their respective diffusion velocities in the quasi-neutral regions near the emitter–base junction (i.e., $S_{en} \gg S_{dn}$, $S_{ep} \gg S_{dp}$, $S_{in} \gg S_{dn}$, $S_{ip} \gg S_{dp}$). Equation (4.65) thus reduces to

$$\tau_E = \frac{N_B W_E W_B}{2N_E D_{nB}} e^{-\Delta E_G/kT} \tag{4.66}$$

The emitter delay can be neglected compared to the base transit time ($W_B^2/2D_{nB}$).

For the heterojunction bipolar transistor with an abrupt emitter–base heterojunction, the interface velocity is much larger than the diffusion velocity, the electron current is controlled by the drift and diffusion velocity of the carriers across the space-charge region barrier ($S_{en} \ll S_{dn}$), and the hole current is limited by diffusion velocity within the quasi-neutral base ($S_{ep} \gg S_{dp}$). From (4.65) we have

$$\tau_E = \frac{N_B}{N_E}\frac{S_{dn}}{S_{\text{en}}} e^{-\Delta E_G/kT} \tag{4.67}$$

Using $e^{-\Delta E_G/kT} = 5 \times 10^7$ at room temperature, $N_B/N_E = 100$, $S_{en} = 75$ cm/s [55], $S_{dn} = 2.5 \times 10^6$ cm/s ($W_B = 0.1\mu$m, $D_{nB} = 25\text{cm}^2/\text{s}$) and Al mole fraction of 0.3, $\tau_E = 1.67(W_E W_B/D_{nB})$. For high-speed HBTs, $W_B < 0.1\mu$m. If $W_E > 0.1\mu$m, the emitter delay time is more than twice the base transit time.

For a graded emitter–base heterojunction, the hole and electron currents are controlled by their respective diffusion velocities. In this case, the emitter delay time is much less than the base transit time and can therefore be neglected.

4.3.2 AlGaAs and InGaAs Graded Bases

It is well known that base bandgap grading reduces base transit time and increases current gain from the aiding field in the base [56]. Results of bipolar transistors utilizing a strained $\text{Si}_x\text{Ge}_{1-x}$ base [57] and AlGaAs graded base both display reduced base transit times and increased current gains [58]. Experimental data of $f_T = 83$ GHz and $f_{\max} = 197$ GHz of self-aligned InGaP/GaAs heterojunction bipolar transistors with a compositionally graded $\text{In}_x\text{Ga}_{1-x}\text{As}$ base have been demonstrated [59]. In addition to producing a field that enhances electron transport through the base, the $\text{In}_x\text{Ga}_{1-x}\text{As}$ grade is better than $\text{Al}_x\text{Ga}_{1-x}\text{As}$ grade. This is because the $\text{In}_x\text{Ga}_{1-x}\text{As}$ grade does not alter the emitter–base band alignment, maintaining the same large valence-band discontinuity as in an HBT with a nongraded base. The minority-carrier mobility in $\text{In}_x\text{Ga}_{1-x}\text{As}$ is also higher than GaAs and AlGaAs.

The effect of the Al mole fraction on the cutoff frequency and current gain of AlGaAs graded-base HBTs has been analyzed using transport equations [60–62]. The analyses in [60–62] were based on the HBTs with a linearly graded base. An aiding built-in field is formed in the quasi-neutral base, which results in a reduction in the minority-carrier transit time and an increase in the common-emitter current gain. The effect of different base-grading profiles on the performance of AlGaAs HBT's are evaluated [63]. These graded bases, including the linear, exponential, parabolic, Gaussian, and square root profiles, are compared. In an $\text{Al}_y\text{Ga}_{1-y}\text{As}$ (or $\text{In}_y\text{Ga}_{1-y}\text{As}$) base, the aluminum (or indium) mole fraction y is expressed as

$$y = y_0 \left(\frac{x - x_3}{x_2 - x_3}\right)^{\xi} \tag{4.68}$$

where x is the position variable, x_2 and x_3 are the edges at the emitter–base and collector–base space-charge layers in the base, y_0 is the mole fraction at x_2, and ξ is a constant. $\xi = 1$ represents the linear base grading, $\xi = 2$ represents parabolic base grading, and $\xi = \frac{1}{2}$ represents the square root base grading. The various aluminum mole fractions in the base are displayed in Fig. 4.56.

The evaluation of base transit time for different base profiles is based on the equation

$$\tau_B = \frac{q\int_{x_2}^{x_3} n(x)\,dx}{J_n} \tag{4.69}$$

where the electron current density in the base is given by

$$J_n = -qnv_d + kT\mu_n \frac{dn}{dx} \tag{4.70}$$

Since E_B is a function of position, μ_n and v_d are position dependent. Taking the derivative on both sides of (4.70) gives

$$kT\mu_n \frac{d^2n}{dx^2} + \left(kT\frac{d\mu_n}{dx} - qv_d\right)\frac{dn}{dx} - qn\frac{dv_d}{dx} = \frac{dJ_n}{dx} \tag{4.71}$$

For the AlGaAs HBT, $v_d \approx 10^7 + 200|E_B|$ and $\mu_n \approx v_d/E_B$ [64]. The built-in field in the base is $qE_B = dE_i/dx = q\,d\kappa(x)/dx = 1.1ny_0(x - x_3)^{\xi-1}/(x_2 - x_3)^{\xi}$, where E_i is the intrinsic Fermi level and κ is the electron affinity given by $4.07 - 1.1y$ for $0 \leq y \leq 0.45$ [65]. Substituting $\mu_n \approx v_d/E_B$ and $v_d \approx 10^7 + 2000|E_B|$ into

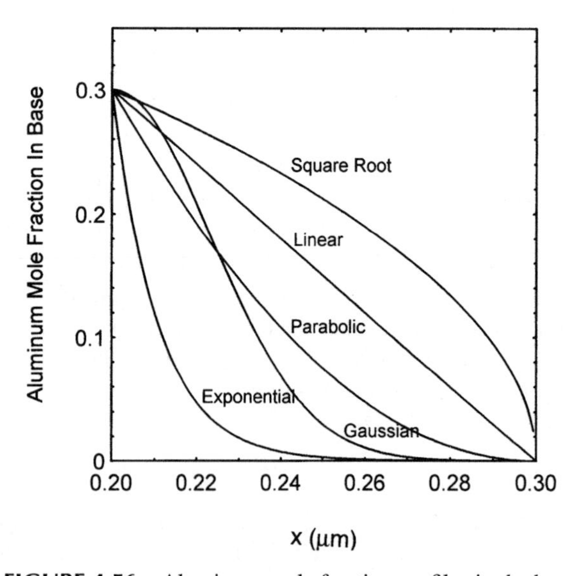

FIGURE 4.56 Aluminum mole fraction profiles in the base.

(4.71) and assuming that the base recombination is negligible (i.e., $dJ_n/dx \approx 0$), we obtain

$$\frac{d^2n}{dx^2} + f_1 \frac{dn}{dx} + f_2 n = 0 \tag{4.72}$$

where

$$f_1 = \frac{C_2(1-\xi)(x-x_3)^{-\xi} - (q/kT)[10^7 + C_1(x-x_3)^{\xi-1}]}{2\times 10^3 + C_2(x-x_3)^{\xi-1}}$$

$$f_2 = \frac{-qC_1(\xi-1)(x-x_3)^{\xi-2}}{kT[2\times 10^3 + C_2(x-x_3)^{\xi-1}]}$$

$$C_1 = 2.2\times 10^3 \frac{\xi y_0}{(x_2-x_3)^{\xi}}$$

$$C_2 = 10^7 \frac{(x_2-x_3)^{\xi}}{1.1\xi y_0}$$

Similarly, one can find the result for the InGaAs graded-base HBT using appropriate κ, μ_n, and v_d for $In_yGa_{1-y}As$. If $\xi = 1$, the base grading is linear. Equation (4.72) reduces to

$$\frac{d^2n}{dx^2} - \frac{qv_d}{kT\mu_n}\frac{dn}{dx} = 0 \tag{4.73}$$

The solution to (4.73) is

$$n(x) = A + B\exp\left(\frac{qv_d x}{\mu_n kT}\right) \tag{4.74}$$

where A and B can be obtained using the boundary conditions of $n(x_2)$ and $n(x_3)$ derived from the thermionic model [66]. The electron current density J_n at the collector terminal is then equal to $-qn(x_3)v_d + kT\mu_n\, dn(x_3)/dx$.

For the exponential base grading, the mole fraction is $C_3\exp(-x/C_4)$, where $C_3 = y_0\exp(x_2/C_4)$ and $C_4 = (x_3 - x_2)/\ln m$, and where m is the ratio of the mole fraction at $x = x_2$ to that at $x = x_3$. The differential equation for determining the electron concentration in the base is derived as

$$\frac{d^2n}{dx^2} + f_3\frac{dn}{dx} + f_4 n = 0 \tag{4.75}$$

where

$$f_3 = \frac{\alpha_3 e^{x/C_4} + \alpha_4 e^{-x/C_4} + \alpha_5}{\alpha_1 + \alpha_2 e^{x/C_4}}$$

$$f_4 = \frac{\alpha_6 e^{-x/C_4}}{\alpha_1 + \alpha_2 e^{x/C_4}}$$

and $\alpha_1 = 2 \times 10^3 kT/q$, $\alpha_2 = 10^7 C_4 kT/1.1qC_3$, $\alpha_3 = \alpha_2/C_4$, $\alpha_4 = -2.2 \times 10^3 C_3/C_4$, $\alpha_5 = -10^7$, and $\alpha_6 = 2.2 \times 10^3 C_3/C_4^2$.

For the Gaussian base grading, $y = y_0 \exp[-\eta(x - x_2)^2]$, where $\eta = \ln m/(x_3 - x_2)^2$, the electron density equation is given by

$$\frac{d^2n}{dx^2} + f_5 \frac{dn}{dx} + f_6 n = 0 \tag{4.76}$$

where

$$f_5 = \frac{V_T(d\mu_n/dx) - v_d}{V_T \mu_n}$$

$$f_6 = \frac{-d\mu_n/dx}{V_T \mu_n}$$

and $V_T(d\mu_n/dx) - v_d = -V_T \cdot 10^7[1 - 2\eta(x - x_2)^2]/\{2.2y_0\eta(x - x_2)^2 \exp[-\eta(x - x_2)^2]\} - 10^7 - 2000|E_B|$, $dv_d/dx = 4400y_0\eta[1 - 2\eta(x - x_2)^2]\exp[-\eta(x - x_2)^2]$, $V_T\mu_n = V_T(2000 + 10^7/|E_B|)$, $|E_B| = 2.2y_0\eta(x - x_2)\exp[-\eta(x - x_2)^2]$ and $V_T = kT/q$.

The base transit time for different AlGaAs graded bases has been calculated. The heterojunction bipolar transistor has an emitter doping of 5×10^{17} cm^{-3}, a base doping of 1×10^{19} cm^{-3}, a collector doping of 5×10^{17} cm^{-3}, an emitter–base junction depth X_{JE} of 0.2μm, and a collector–base junction depth X_{JC} of 0.3μm. The electron carrier concentrations in the base for various profiles are displayed in Fig. 4.57. The electron concentration distribution is very sensitive to the Al fraction in the base. The electron concentration and current density together determine the base transit time of the HBT. The base transit time as a function of the Al mole fraction is shown in Fig. 4.58. For an Al mole fraction from 0.05 to 0.3, the parabolic base grading results in the lowest base transit time followed by exponential, Gaussian, square root, and linear profiles. Note that the analysis here is based on the conventional analysis without taking into account the quantum mechanical effect.

4.3.3 Heterojunction Capacitance, Including the Composition Grading and a Setback Layer

The heterojunction capacitance in Eq. (4.5) is applicable for an abrupt heterojunction without a graded layer within the heterojunction. For III/V compound HBTs such as AlGaAs/GaAs heterojunction bipolar transistors, a graded AlGaAs layer and/or an undoped GaAs layer is used to improve the injection efficiency and RF performance [67]. The generalized heterojunction capacitance, including the effect of heterojunction grading and an undoped spacer (setback) layer, is derived [68]. Figure 4.59 shows a one-dimensional schematic of the linearly graded heterojunction with a setback

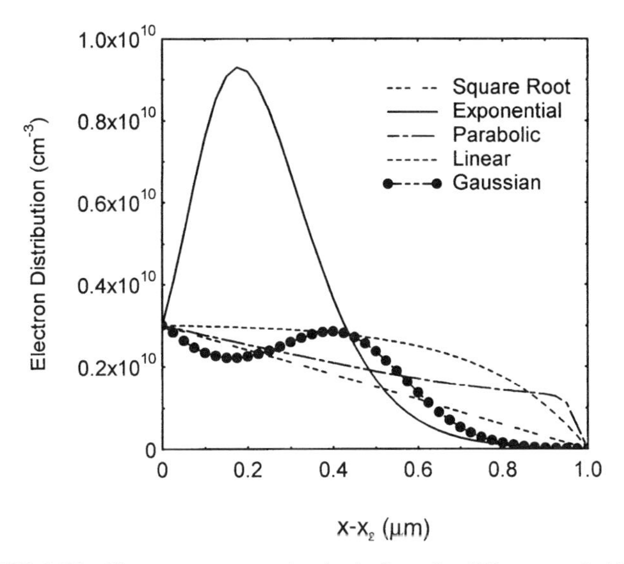

FIGURE 4.57 Electron concentration in the base for different graded bases.

layer. In Fig. 4.59, region I ($-X_1 \leq x \leq -X_p$) has a constant permittivity ε_1 next to the emitter. Region II ($-X_p \leq x \leq 0$) has a graded layer with position-dependent permittivity $\varepsilon(x) = \varepsilon_2 - (\varepsilon_1 - \varepsilon_2)x/X_0$. Region III ($0 \leq x \leq X_S$) has an undoped setback layer with a constant permittivity ε_2. Region IV ($X_S \leq x \leq X_2$) has a constant permittivity ε_2 next to the base. Assuming free carrier concentrations are

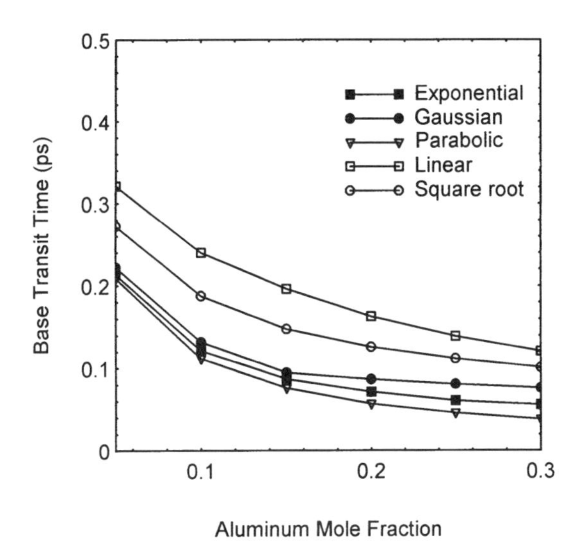

FIGURE 4.58 Base transit time versus Al mole fraction for different graded bases.

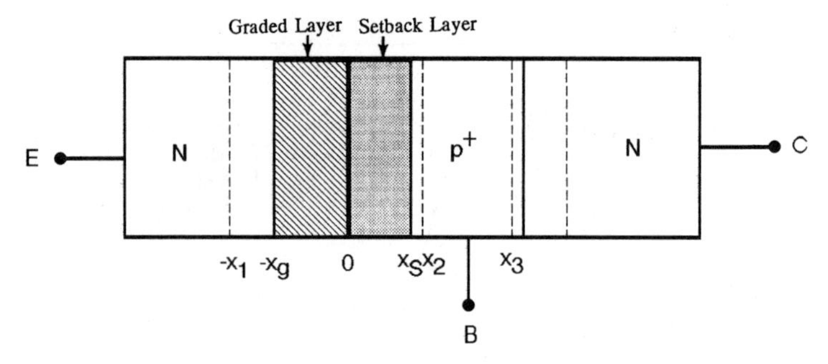

FIGURE 4.59 One-dimensional HBT schematic including graded and setback layers.

much less than donor and acceptor concentrations (i.e., the depletion approximation), one finds the electric field in each region using the equations

$$E_1(x) \approx \frac{qN_D(x + X_1)}{\varepsilon} \qquad \text{for} - X_1 \leq x \leq -X_g \tag{4.77a}$$

$$E_g(x) = \frac{\rho x}{\varepsilon(x)} + \frac{\varepsilon_2 E_g(0)}{\varepsilon(x)} \qquad \text{for} - X_g \leq x \leq 0 \tag{4.77b}$$

$$E_s(x) \approx \frac{qN_A(X_2 - X_{s)}}{\varepsilon_2} \qquad \text{for } 0 \leq x \leq X_s \tag{4.77c}$$

$$E_2(x) \approx \frac{qN_A(X_2 - x)}{\varepsilon_2} \qquad \text{for } X_s \leq x \leq X_2 \tag{4.77d}$$

The electric displacement ($D = \varepsilon E$), $D_1(-X_g) = D_g(-X_g)$, provides the boundary condition to solve for X_2:

$$X_2 = X_S + X_1 \frac{N_D}{N_A} \tag{4.78}$$

Furthermore, using $V_{bi} - V = -V_2(X_2)$ yields

$$X_2 = \frac{-b + \sqrt{b^2 - 4ac}}{2a} \tag{4.79}$$

where

$$a = \frac{qN_D}{2\varepsilon_1}\left(\frac{N_A}{N_D}\right)^2 + \frac{qN_A}{2\varepsilon_2}$$

$$b = -\frac{qN_D}{\varepsilon_1}\left(\frac{N_A}{N_D}\right)^2 X_S - \frac{qX_g N_A}{\varepsilon_1} + \frac{qN_A X_g}{\varepsilon_2 - \varepsilon_1} \ln \frac{\varepsilon_2}{\varepsilon_1}$$

$$c = \frac{qN_D}{2\varepsilon_1}\left(\frac{N_A}{N_D}X_S + X_g\right)^2 - \frac{qN_A X_S^2}{2\varepsilon_2} + \frac{qN_D X_g^2 - qN_A X_g X_S \ln(\varepsilon_2/\varepsilon_1)}{\varepsilon_2 - \varepsilon_1}$$
$$- \frac{qN_D X_g^2 \varepsilon_2}{(\varepsilon_2 - \varepsilon_1)^2}\ln\left(\frac{\varepsilon_2}{\varepsilon_1}\right) - (V_{bi} - V)$$

The modulation of space charge in the depletion region with respect to the applied bias results in a depletion capacitance. The depletion capacitance of a heterojunction is expressed as

$$\begin{aligned} C_J &= \left|\frac{dQ_n}{d(V_{bi} - V)}\right| \\ &= qN_A\left|\frac{dX_2}{d(V_{bi} - V)}\right| \end{aligned} \tag{4.80}$$

Using (4.78) in deriving $dX_2/d(V_{bi} - V)$ and inserting the resulting equation into (4.80) gives the depletion capacitance,

$$C_J = \frac{qN_A}{\sqrt{b^2 - 4ac}} \tag{4.81}$$

Equation (4.81) is a general solution for the heterojunction with a linearly graded layer and a setback layer. For a linearly graded heterojunction without a setback layer ($X_S = 0$), (4.81) reduces to

$$C_J = \frac{qN_A}{\sqrt{\Delta}}$$

where

$$\Delta = \left[\frac{qN_A X_g}{\varepsilon_1} + \frac{qN_A X_g \ln(\varepsilon_2/\varepsilon_1)}{\varepsilon_2 - \varepsilon_1}\right]^2 - 4\left[\frac{qN_D}{2\varepsilon_1}\left(\frac{N_A}{N_D}\right)^2 + \frac{qN_A}{2\varepsilon_2}\right]$$
$$\times \left[\frac{qN_D X_g^2}{2\varepsilon_1} + \frac{qN_D X_g^2}{\varepsilon_2 - \varepsilon_1} - \frac{qN_D X_g^2 \varepsilon_2}{(\varepsilon_2 - \varepsilon_1)^2}\ln\frac{\varepsilon_2}{\varepsilon_1} - (V_{bi} - V)\right]$$

For an abrupt heterojunction with a setback layer ($X_g = 0$), (4.81) reduces to

$$C_J = \sqrt{\frac{1}{2(V_{bi} - V)[(\varepsilon_1 N_D + \varepsilon_2 N_A)/q\varepsilon_1\varepsilon_2 N_A N_D] + X_S^2/\varepsilon_2^2}}$$

For an abrupt heterojunction without a setback layer ($X_g = X_S = 0$), (4.81) reduces to the well-known formula

$$C_J = \sqrt{\frac{q\varepsilon_1\varepsilon_2 N_A N_D}{2(V_{bi} - V)(\varepsilon_1 N_D + \varepsilon_2 N_A)}}$$

The effect of setback layer thickness on the heterojunction capacitance at $V = -1.0$, -0.5, 0, 0.5, and 1.0 V is plotted in Fig. 4.60. In this figure the squares, lines, and triangles represent the analytical predictions and the circles represent the device simulation—MEDICI [69]. Good agreement between the analytical predictions and the device simulation is obtained up to high forward bias. The discrepancy between the model predictions and the device data at high forward biases is because (4.81) does not account for free carriers in a forward-biased junction.

4.3.4 Thermal Effects on f_T and $f_{\max}$

It is known that GaAs HBTs grown on a semi-insulated GaAs substrate exhibit a self-heating effect at high power dissipations. Self-heating degrades not only current gain, but the cutoff frequency of the HBT. Figure 4.61 shows the cutoff frequency versus the collector current density. The increase of cutoff frequency at low collector current density is obviously due to the decrease in emitter charging time ($\propto 1/J_C$). When the collector current density is sufficiently high, the cutoff frequency decreases with J_C due to an increase in junction temperature from self-heating. Self-heating decreases the saturation velocity in the collector–base depletion layer and the electron mobility

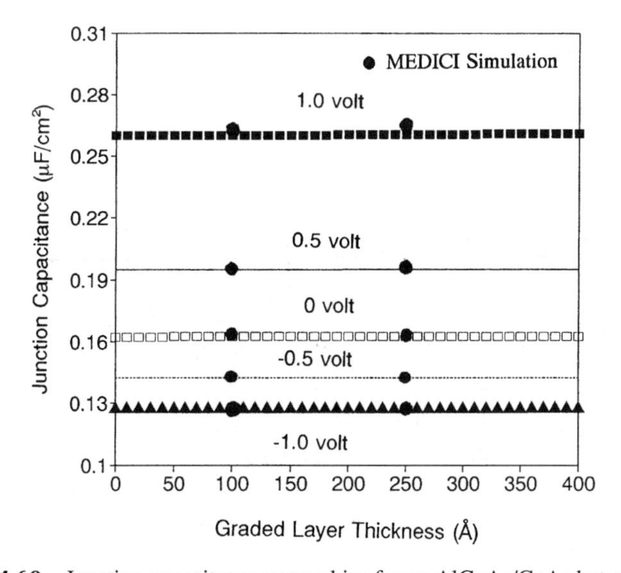

FIGURE 4.60 Junction capacitance versus bias for an AlGaAs/GaAs heterojunction.

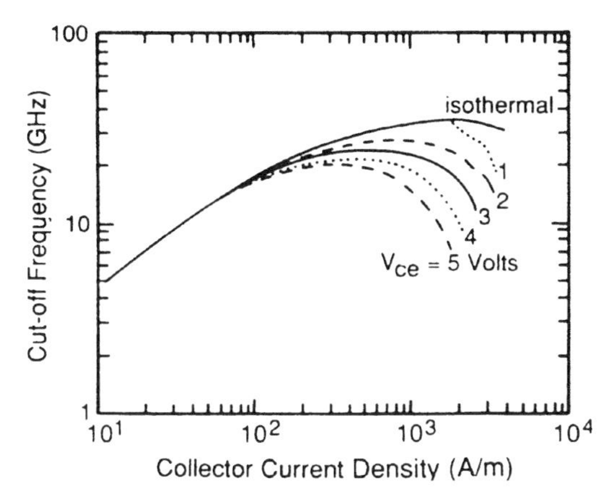

FIGURE 4.61 Cutoff frequency versus collector current for an AlGaAs/GaAs HBT (after Liou et al., Ref. 70 © IEEE).

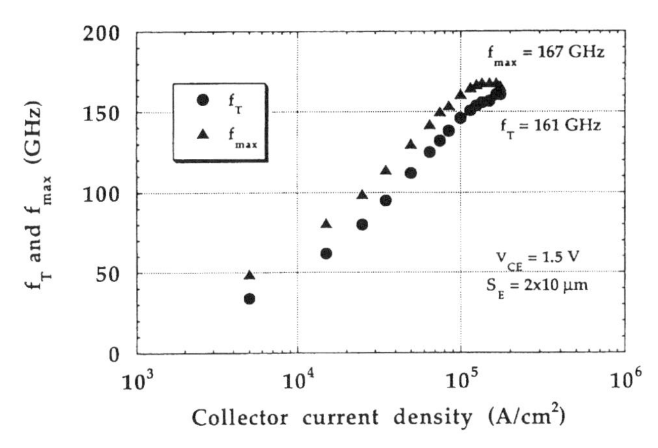

FIGURE 4.62 Cutoff frequency and $f_{\max}$ versus collector current density for an InP/InGaAs HBT (after Shigematsu et al., Ref. 71 © IEEE).

in the base, which increases the collector–base depletion layer transit time and the base transit time, respectively.

Figure 4.62 shows the measured f_T and $f_{\max}$ of MOCVD grown InP/InGaAs HBTs with a base doping concentration of 1×10^{20} cm^{-3} and an emitter area of $2 \times 10\ \mu\text{m}^2$. The measured f_T and $f_{\max}$ reach peak points ($f_T = 161$ GHz, $f_{\max} = 167$ GHz) at $J_C = 1.65 \times 10^5$ A/cm^2. These results indicate the great potential of InP HBTs for ultrahigh-speed applications. No significant self-heating effect is observed due to the high thermal conductivity of the InP substrate.

REFERENCES

1. J. J. Liou, F. A. Lindholm, and J. S. Park, "Forward-voltage capacitance and thickness of p-n junction space-charge regions," *IEEE Trans. Electron Devices*, **ED-34**, 1752 (1987).
2. J. S. Yuan, "Modeling Si/Si$_{1-x}$Ge$_x$ heterojunction bipolar transistors," *Solid-State Electron.*.,**7**, 921 (1992).
3. P. J. Van Wijnen and R. D. Gardner, "A new approach to optimizing the base profile for high-speed bipolar transistors," *IEEE Electron Device Lett.*, **EDL-4**, 149 (1990).
4. K. Suzuki, "Optimum base doping profile for minimum base transit time," *IEEE Trans. Electron Devices*, **ED-39**, 2128 (1991).
5. J. S. Yuan, "Effect of base profile on the base transit time of the bipolar transistor for all levels of injection," *IEEE Trans. Electron Devices*, **ED-41**, 212 (1994).
6. S. Winterton, C. Peters, and N. Tarr, "Composition grading for base transit-time minimization in SiGe base heterojunction bipolar transistors," *Solid-State Electron.*, **36**, 1161 (1993).
7. S. Winterton, S. Searless, C. Peters, N. Tarr, and D. Pulfrey, "Distribution of base dopant for transit time minimization in a bipolar transistor," *IEEE Trans. Electron Devices*, **ED-43**, 170 (1996).
8. A. H. Marshak, "Optimum doping distribution for minimum base transit time," *IEEE Trans. Electron Devices*, **ED-14**, 190 (1967).
9. R. G. Meyer and R. S. Muller, "Charge-control analysis of the collector–base space-charge-region contribution to bipolar-transistor time constant τ_T," *IEEE Trans. Electron Devices*, **ED-34**, 450 (1987).
10. R. Katoh, M. Kurata, and J. Yoshida, "A self-consistent particle simulation for (Al,Ga)As/GaAs HBTs with improved base–collector structures," *IEEE Trans. Electron Devices*, **ED-36**, 846 (1989).
11. Y. Yamauchi and T. Ishibashi, "Electron velocity overshoot in the collector depletion layer of AlGaAs/GaAs HBTs," *IEEE Electron Device Lett.*, **EDL-7**, 655 (1986).
12. J. M. Early, "P-N-I-N and N-P-I-N junction transistor triodes," *Bell Syst. Tech. J.*, **33**, 517 (1954).
13. S. E. Laux and W. Lee, "Collector signal delay in the presence of velocity overshoot," *IEEE Electron Device Lett.*, **EDL-4**, 174 (1990).
14. T. Ishibashi, "Influence of electron velocity overshoot on collector transit times of HBTs," *IEEE Trans. Electron Devices*, **ED-37**, 2103 (1990).
15. J. J. Liou and H. Shakouri, "Collector signal delay time and collector transit time of HBTs including velocity overshoot," *Solid-State Electron.*, **35**, 15 (1992).
16. H. Zhou and D. L. Pulfrey, "Bipolar effects on the signal delay time in HBTs at high currents," *IEEE Trans. Electron Devices*, **ED-40**, 44 (1993).
17. D. L. Harame, J. H. Comfort, J. D. Cressler, E. F. Crabbe, J. Y.-C. Sun, B. S. Meyerson, and T. Tice, "Si/SiGe epitaxial-base transistors: I. Materials, physics, and circuits," *IEEE Trans. Electron Devices*, **ED-42**, 455 (1995).
18. B. Jalali and S. J. Pearton, eds., *InP HBTs: Growth, Processing, and Applications*, Artech House, Norwood, MA (1995).
19. D. A. Sunderland and P. L. Dapkus, "Optimizing N-p-n and P-n-p heterojunction bipolar transistors for speed," *IEEE Trans. Electron Devices*, **ED-34**, 367 (1987).

20. S. Yamahara, Y. Matsuoka, and T. Ishibashi, "High-f_{max} collector-up AlGaAs/GaAs heterojunction bipolar transistors with a heavily carbon-doped base fabricated using oxygen-ion implantation," *IEEE Electron Device Lett.*, **EDL-14**, 173 (1993).

21. J. N. Burghartz, K. A. Jenkins, D. A. Grützmacher, T. O. Sedgwick, and C. L. Stanis, "High-performance emitter-up/down SiGe HBTs," *IEEE Electron Device Lett.*, **EDL-15**, 360 (1994).

22. M. Kawanka, N. Iguchi, and J. Sone, "112-GHz collector-up Ge/GaAs heterojunction bipolar transistors with low turn-on voltage," *IEEE Trans. Electron Devices*, **ED-43**, 670 (1996).

23. Y. Miyamoto, J. M. M. Rios, A. G. Dentai, and S. Chandrasekhar, "Reduction of base–collector capacitance by undercutting the collector and subcollector in GaInAs/InP DHBTs," *IEEE Electron Device Lett.*, **EDL-17**, 97 (1996).

24. W. L. Chen, H. F. Chau, M. Tutt, M. C. Ho, T. S. Kim, and T. Henerson, "High-speed InGaP/GaAs HBTs using a simple collector undercut technique to reduce base–collector capacitance," *IEEE Electron Device Lett.*, **EDL-18**, 355 (1997).

25. Y.-F. Yang, C.-C. Hsu, E. S. Yang, and H.-J. Qu, "A high-frequency GaInP/GaAs heterojunction bipolar transistor with reduced base–collector capacitance using a selective buried sub-collector," *IEEE Electron Device Lett.*, **EDL-17**, 531 (1996).

26. S. C. Jue, D. J. Day, A. Margittai, and M. Svilans, "Transport and noise in GaAs/AlGaAs heterojunction bipolar transistors: II. Noise and gain at low frequencies," *IEEE Trans. Electron Devices*, **ED-36**, 1020 (1989).

27. A. van der Ziel, X. Zhang, and A. H. Pawlikiewicz, "Location of $1/f$ noise sources in BJTs and HBJTs: I. Theory," *IEEE Trans. Electron Devices*, **ED-33**, 1371 (1986).

28. S. C. Jue, D. J. Day, A. Martgittai, and M Svilans, "Transport and noise in GaAs/AlGaAs heterojunction bipolar transistors: II. Noise and gain at low frequencies," *IEEE Trans. Electron Devices*, **ED-36**, 1020 (1989).

29. G. Massobrio and P. Antognetti, *Semiconductor Device Modeling with SPICE*, 2nd ed., McGraw-Hill, New York (1993).

30. J. D. Cressler, L. Vempati, J. A. Babcock, R. C. Jaeger, and D. L. Harame, "Low-frequency noise characteristics of UHV/CVD epitaxial Si- and SiGe-based bipolar transistors," *IEEE Electron Device Lett.*, **EDL-17**, 13 (1996).

31. J. Kilmer, A. van der Ziel, and G. Bossan, "Presence of mobility-fluctuation $1/f$ noise identified in silicon p^+np transistors," *Solid-State Electron.*, **26**, 71 (1983).

32. W. J. Ho, M. F. Chang, A. Sailer, P. Zampardi, D. Deakin, B. McDermott, R. Pierson, J. A. Higgins, and J. Waldrop, "GaInP/GaAs HBTs for high-speed integrated circuit applications," *IEEE Electron Device Lett.*, **EDL-14**, 572 (1993).

33. K. Kitahara, M. Hoshino, and M. Ozeki, "Observation of donor-related deep levels in $Ga_xIn_{1-x}P$ $(0.52 < x < 0.7)$," *Jpn. J. Appl. Phys.*, **27**, L110 (1988).

34. D. Costa and J. S. Harris, Jr., "Low-frequency noise properties of N-p-n AlGaAs/GaAs heterojunction bipolar transistors," *IEEE Trans. Electron Devices*, **ED-39**, 2383 (1992).

35. V. K. Raman, C. R. Viswanathan, and M. E. Kim, "Dc conduction and low-frequency noise characteristics of GaAlAs/GaAs single heterojunction bipolar transistors at room temperature and low temperatures," *IEEE Trans. Microwave Theory Tech.*, **MTT-39**, 1054 (1991).

36. G. Gonzalez, *Microwave Transistor Amplifiers: Analysis and Design*, Prentice Hall, Upper Saddle River, NJ (1984).

37. K. Lu, P. A. Perry, and T. J. Brazil, "A new large-signal AlGaAs/GaAs HBT model including self-heating effects, with corresponding parameter-extraction procedure," *IEEE Trans. Microwave Theory Tech.*, **MTT-43**, 1433 (1995).

38. J. Laskar, R. N. Nottenburg, J. A. Baquedano, A. F. J. Levi, and J. Kolodzey, "Forward transit time delay in $In_{0.53}Ga_{0.47}As$ heterojunction bipolar transistors with nonequilibrium electron transport," *IEEE Trans. Electron Devices*, **ED-41**, 1942 (1993).

39. J. S. Yuan, "Switching-off transient analysis for heterojunction bipolar transistors in saturation," *Solid-State Electron.*, **36**, 1261 (1993).

40. M. Ugajin and Y. Amemiya, "The base–collector heterojunction effect in SiGe-base bipolar transistors," *Solid-State Electron.*, **34**, 593 (1991).

41. B. Mazhari and H. Morkoç, "Minority charge storage characteristics of HBTs in saturation," *Solid-State Electron.*, **36**, 455 (1993).

42. J. Song and J. S. Yuan, "Comment on 'On the base profile design and optimization of Si- and SiGe-epitaxial bipolar technology for 77 K applications. II. Circuit performance issues,' " *IEEE Trans. Electron Devices*, **ED-44**, 915 (1997).

43. J. D. Cressler, E. F. Crabbé, J. H. Comfort, J. M. C. Stork, and J. Y.-C. Sun, "On the profile design and optimization of epitaxial Si- and SiGe-base bipolar technology for 77 K applications: II. Circuit performance issues," *IEEE Trans. Electron Devices*, **ED-40**, 542 (1993).

44. J. Song and J. S. Yuan, "Optimal Ge profile for base transit time minimization of SiGe HBT," *Solid-State Electron.*, **41**, 1957 (1997).

45. D. M. Richey, J. D. Cressler, and A. J. Joseph, "Scaling issues and Ge profile optimization in advanced UHV/CVD SiGe HBTs," *IEEE Trans. Electron Devices*, **ED-44**, 431 (1997).

46. S. R. Stiffler, J. H. Comfort, C. L. Stanis, D. L. Harame, E. de Fresart, and B. S. Meyerson, "The thermal stability of SiGe films deposited by ultrahigh-vacuum chemical vapor deposition," *J. Appl. Phys.*, **70**, 1416 (1991).

47. A. J. Joseph, J. D. Cressler, D. M. Richey, R. C. Jaeger, and D. L. Harame, "Neutral base recombination and its influence on the temperature dependence of Early voltage and current gain–Early voltage product in UHV/CVD SiGe heterojunction bipolar transistors," *IEEE Trans. Electron Devices*, **ED-44**, 404 (1997).

48. M. Ugajin, G.-B. Hong, and J. G. Fossum, "Inverse base-width modulation and collector space-charge-region widening: degrading effects at high current densities in highly scaled BJTs (and HBTs)," *IEEE Trans. Electron Devices*, **ED-41**, 266 (1994).

49. M. Ugajin and J. G. Fossum, "Significant time constants defined by high-current charge dynamics in advanced silicon-based bipolar transistors," *IEEE Trans. Electron Devices*, **ED-41**, 1796 (1994).

50. J. D. Cressler, J. H. Comfort, E. F. Crabbé, G. L. Patton, J. M. C. Stork, and J. Y.-C. Sun, "On the profile design and optimization of epitaxial Si- and SiGe-base bipolar technology for 77 K applications: I. Transistor dc design consideration," *IEEE Trans. Electron Devices*, **ED-40**, 525 (1993).

51. J. H. Lee, Y. J. Park, and J. D. Lee, "Effects of high injection barrier (HIB) position on bipolar transistor characteristics," *IEEE Trans. Electron Devices*, **ED-41**, 102 (1994).

52. J.-H. Lee, W.-G. Kang, J.-S. Lyu, and J. D. Lee, "Modeling of the critical current density of bipolar transistor with retrograde collector doping profile," *IEEE Electron Device Lett.*, **EDL-17**, 109 (1996).

53. A. J. Joseph, J. D. Cressler, D. M. Richey, and D. L. Harame, "Impact of profile scaling on high-injection barrier effects in advanced UHV/CVD SiGe HBTs," *IEDM Tech Dig.*, 253 (1996).

54. G.-B Gao, J.-I. Chyi, J. Chen, and H. Morkoç, "Emitter region delay time of AlGaAs/GaAs heterojunction bipolar transistors," *Solid-State Electron.*, **33**, 389 (1990).

55. M. S. Lundstrom, "Boundary Conditions for pn heterojunctions," *Solid-State Electron.*, **27**, 491 (1984).

56. H. Kroemer, "Two integral relations pertaining to the electron transport through a bipolar transistor with a non-uniform energy gap in the base region," *Solid-State Electron.*, **28**, 1101 (1985).

57. S. M. Sze, ed., *High-Speed Semiconductor Devices*, Wiley, New York (1990).

58. O.-S. Ang and D. L. Pulfrey, "The cut-off frequency of base-graded and junction-graded $Al_xGa_{1-x}As$ DHBTs," *Solid-State Electron.*, **34**, 1325 (1991).

59. D. A. Ahmari, M. T. Fresina, Q. J. Hartmann, D. W. Barlage, P. J. Mares, M. Feng, and G. E. Stillman, "High-speed InGaP/GaAs HBTs with a strained $In_xGa_{1-x}As$ base," *IEEE Electron Device Lett.*, **EDL-17**, 226 (1996).

60. S. C.- M. Ho, and D. L. Pulfrey, "The effect of base grading on the gain and high frequency performance of AlGaAs/GaAs heterojunction bipolar transistors," *IEEE Trans. Electron Devices*, **ED-36**, 2173 (1989).

61. O.-S. Ang and D. L. Pulfrey, "The cut-off frequency of base-graded and junction-graded $Al_xGa_{1-x}As$ DHBTs," *IEEE Trans. Electron Devices*, **ED-34**, 1325 (1991).

62. J. J. Liou, W. W. Wong, and J. S. Yuan, "A study of base built-in field effects on the steady-state current gain of heterojunction bipolar transistor," *Solid-State Electron.*, **33**, 845 (1990).

63. J. Song and J. S. Yuan, unpublished data.

64. Q. M. Zhang, G. L. Tan, J. M. Xu, and D. J. Day, "Current gain and transit-time effects in HBTs with graded emitter and base regions," *IEEE Trans. Electron Devices*, **ED-11**, 508 (1990).

65. Sadao Adachi, "GaAs, AlAs, and $Al_xGa_{1-x}As$: material parameters for use in research and device applications," *J. Appl. Phys.*, **58(3)**, 1 (1985).

66. M. Shur, *GaAs Devices and Circuits*, Plenum, New York (1986).

67. M. E. Hafizi, C. R. Crowell, L. M. Pawlowicz, and M. E. Kim, "Improved current gain and f_T through doping profile selection in linearly graded heterojunction bipolar tranistors," *IEEE Trans. Electron Devices*, **ED-37**, 1779 (1990).

68. J. S. Yuan and J. Ning, "Analysis of abrupt and linearly-graded heterojunction bipolar transistors with or without a setback layer," *IEE Pro. G*, **142**, 254 (1995).

69. *MEDICI*: *Two-Dimensional Device Simulation*, Technology Modeling Associates, Inc., Palo Alto, CA (1993).

70. L. L. Liou, J. L. Ebel, and C. I. Huang, "Thermal effects on the characteristics of AlGaAs/GaAs heteorjunction bipolar transistors using two-dimensional numerical simulation," *IEEE Trans. Electron Devices*, **ED-40,** 35 (1993).

71. H. Shigematsu, T. Iwai, Y. Matsumiya, H. Ohnishi, O. Ueda, and T. Fujii, "Ultrahigh f_T and f_{max} new self-alignment InP/InGaAs HBTs with a highly Be-doped base layer grown by ALE/MOCVD," *IEEE Electron Device Lett.*, **EDL-16**, 55 (1995).

PROBLEMS

4.1. Compare the junction capacitance and the diffusion capacitance versus junction voltage for SiGe- and InP-based HBTs.

4.2. Discuss qualitatively three different overshoot velocity profiles (uniform, linear, and inverse linear) on the collector–base space-charge-layer delay.

4.3. Should the f_T of an InP-based HBT increase or decrease with temperature? Explain.

4.4. Determine the value of base thickness W_B that maximizes $f_{\max}$ of a bipolar transistor. Assume that minority-carrier transport across the base is by diffusion only.

4.5. Derive the base transit time of the HBT with a box Ge profile and an exponential base doping concentration $N_B = N_{B0}\exp(-\alpha x)$. Assume no heavy doping effect and low injection.

4.6. Discuss the effect of linear, box, and trapezoid Ge profiles in the SiGe base on the collector–base space-charge-layer delay under nonequilibrium condition.

4.7. Derive and plot the transconductance versus base–emitter voltage for the SiGe HBT with and without neutral-base recombination.

4.8. Plot the junction capacitance versus setback layer thickness of an AlGaAs/GaAs p-n heterojunction at 0 V. The quasi-neutral p region has uniform doping of 5×10^{17} cm^{-3}, and the quasi-neutral n region has uniform doping of 1×10^{19} cm^{-3}. An intrinsic (setback) layer is inserted between AlGaAs and GaAs layers.

4.9. Discuss qualitatively what bipolar device parameters such as R_B, R_E, β, and so on, will affect S_{11}, S_{12}, S_{21}, and S_{22}.

4.10. Analyze the collector–base junction capacitance when transconductance clipping occurs.

4.11. **(a)** Plot f_T versus J_C for a uniform $Si_{0.8}Ge_{0.2}$ base, a linearly graded $Si_{1-x}Ge_x$ base ($x = 0$ at the emitter–base junction and $x = 0.2$ at the collector–base junction), and a trapezoidal SiGe base ($X_T = 0.5W_B$).
(b) Explain their relative importance in low power applications.
The dopings and device dimensions are given as follows: $N_E = 1 \times 10^{20}$ cm^{-3}, $N_B = 5 \times 10^{18}$ cm^{-3}, $N_C = 2 \times 10^{17}$ cm^{-3}, $W_E = 0.1\ \mu$m, $W_B = 0.08\ \mu$m, and $W_C = 0.5\ \mu$m.

CHAPTER FIVE

HBT Modeling

Device models are commonly used in circuit simulation for the design of integrated circuits. Understanding of physical equations in the derivation of device models can also help process improvement, diagnosis, and new device design. Generally speaking, there are two types of device models: the physics-based model and the empirical model. The physics-based model is derived from device physics. This model is able to account for bandgap discontinuities and other physical effects in the transport equations of the HBT. It can predict device behavior for given device doping profiles and dimensions. The physics-based model could be used for the evaluation of process variation on device behavior. The physics-based model is comprehensive but computationally intensive. The accuracy of physical parameters is important for the determination of terminal currents. The empirical model is developed based on measured I–V and C–V characteristics. The goal of empirical modeling is to obtain analytic formulas that approximate existing data with good accuracy and minimal complexity. Empirical models provide efficient approximation and interpolation. They do not provide insight, predictive capabilities, or encapsulation of theoretical knowledge. The empirical model, however, is compact and suitable for circuit simulation with a large number of transistors on integrated circuits.

5.1 SILICON–GERMANIUM HBT MODELS

In this section we present collector and base current equations for SiGe heterojunction bipolar transistors.

5.1.1 Analytical Collector Current Equation

The electron and hole current densities across the base of the bipolar transistor involve two carrier concentrations and the gradients of the quasi-Fermi levels [1]:

$$J_n = \mu_n n \frac{dE_{Fn}}{dx} \tag{5.1}$$

$$J_p = -\mu_p p \frac{dE_{Fp}}{dx} \tag{5.2}$$

For the bipolar transistor with high current gain, the hole current compared to the electron current can be neglected (i.e., $dE_{Fp}/dx \ll dE_{Fn}/dx$). Therefore,

$$J_n = \mu_n n \left(\frac{dE_{Fn}}{dx} - \frac{dE_{Fp}}{dx} \right) \tag{5.3}$$

Since the pn product for a nondegenerate base is

$$np = n_i^2 e^{(E_{Fn} - E_{Fp})/kT} \tag{5.4}$$

from which one obtains

$$\frac{dE_{Fn}}{dx} - \frac{dE_{Fp}}{dx} = kT \frac{n_i^2}{np} \frac{d}{dx} \left(\frac{np}{n_i^2} \right) \tag{5.5}$$

Inserting (5.5) into (5.3) yields

$$\frac{d}{dx} \left(\frac{np}{n_i^2} \right) = \frac{J_n}{q} \frac{p}{D_n n_i^2} \tag{5.6}$$

Applying the integration on both sides of (5.6) gives

$$\left. \frac{np}{n_i^2} \right|_{W_B} - \left. \frac{np}{n_i^2} \right|_x = \frac{J_n}{q} \int_x^{W_B} \frac{p}{D_n n_i^2} \, dy \tag{5.7}$$

The boundary conditions at $x = 0$ and $x = W_B$ are

$$\left. \frac{np}{n_i^2} \right|_0 = e^{qV_{BE}/kT} \tag{5.8}$$

$$\left. \frac{np}{n_i^2} \right|_{W_B} = e^{qV_{BC}/kT} \tag{5.9}$$

Inserting (5.8) and (5.9) into (5.7) yields the generalized Moll–Ross current relation suitable for SiGe heterojunction bipolar transistors:

$$J_n = -\frac{q \left(e^{qV_{BE}/kT} - e^{qV_{BC}/kT} \right)}{\int_0^{W_B} (p/D_n n_i^2) \, dx} \tag{5.10}$$

The effect of Ge content in the base of the SiGe heterojunction bipolar transistor is accounted for by the position-dependent intrinsic carrier concentration n_i^2 in the denominator of (5.10). The larger the Ge content in the base, the higher the intrinsic concentration and the larger the electron current density.

5.1.2 Generalized Integral Charge-Control Relation for SiGe HBTs

The integral charge-control relation (ICCR) proposed by Gummel [2] gives the relation between the collector current density and the voltages V_{BE} and V_{BC}. The ICCR is no longer valid for heterojunction bipolar transistors. Referring to Gummel [2], the collector current of the BJT is

$$J_C = \frac{kT\left(e^{qV_{BE}/kT} - e^{qV_{BC}/kT}\right)}{\int (p/\mu_n n_i^2) e^{q(V_{BE}-\phi_p)/kT}} \tag{5.11}$$

where ϕ_p is the quasi-Fermi potential for holes.

In practice, the influence of the exponential term on the value of the integral can be neglected [3]. Thus (5.11) is simplified as

$$J_C = \frac{kT\left(e^{qV_{BE}/kT} - e^{qV_{BC}/kT}\right)}{\int (p/\mu_n n_i^2)} \tag{5.12}$$

Equation (5.12) is identical to (5.10). To make (5.10) applicable to compact models, the denominator is separated in parts related to the emitter, base, and collector region [4]:

$$\int \frac{p}{\mu_n n_i^2}\,dx = \int_E \frac{p}{\mu_n n_i^2}\,dx + \int_B \frac{p}{\mu_n n_i^2}\,dx + \int_C \frac{p}{\mu_n n_i^2}\,dx \tag{5.13}$$

The weighting functions $(\mu_n n_i^2)^{-1}$ of the hole densities are taken out of the integrals by introducing average values defined by

$$\frac{1}{\overline{\mu_n n_i^2}} = \frac{\int (p/\mu_n n_i^2)\,dx}{\int p\,dx} \tag{5.14}$$

Thus the denominator in (5.12) can be written as

$$\int \frac{p}{\mu_n n_i^2}\,dx = \frac{1}{\overline{\mu_{nE} n_{iE}^2}} \int_E p\,dx + \frac{1}{\overline{\mu_{nB} n_{iB}^2}} \int_B p\,dx + \frac{1}{\overline{\mu_{nC} n_{iC}^2}} \int_C p\,dx \tag{5.15}$$

In agreement with the ICCR, the averaged products $\overline{\mu_n n_i^2}$ are normalized to that of the base. The generalized integral charge-control relation (GICCR) is obtained [4]:

$$J_C = \frac{(kT/q)\overline{\mu_{nB} n_{iB}^2}}{q_b^+}\left(e^{qV_{BE}/kT} - q^{qV_{BC}/kT}\right) \tag{5.16}$$

where

$$q_b^+ = q_{b0} + q_e + q_c + h_e q_{fe} + q_{fb} + h_c q_{fe}$$
$$h_e = \frac{\overline{\mu_{nB} n_{iB}^2}}{\overline{\mu_{nE} n_{iE}^2}}, \tag{5.17}$$
$$h_c = \frac{\overline{\mu_{nB} n_{iB}^2}}{\overline{\mu_{nC} n_{iC}^2}}$$

Note that q_b^+ in the GICCR is related to q_b in the ICCR as follows:

$$q_b^+ = q_b + (h_e - 1)q_{fe} + (h_c - 1)q_{fe} \tag{5.18}$$

For SiGe HBTs, $h_e \gg 1$ and/or $h_c \gg 1$. For Si BJTs, $h_e \approx h_c \approx 1$. The GICCR reduces to ICCR. In principle, the GICCR is not restricted to SiGe-base HBTs but might also be applicable to HBTs not based on silicon.

The main advantage of the ICCR is that the operating point–dependent terms can be determined from measurements. For example, q_{b0} is determined from the base sheet resistance, and q_e and q_c are determined from the junction capacitances C_{JE} and C_{JC}, respectively. q_f is obtained from the forward transit time τ_F. The ICCR is, however, not suited for HBTs.

The validity of the GICCR is verified for several SiGe HBTs with different doping profiles and Ge contents [4]. Transistor A is designed with a high Ge mole fraction ($x_{\text{Ge}} \approx 0.28$) and a comparatively moderately doped emitter (typical for wide-gap emitter III/V HBTs), as shown in Fig. 5.1*a*. The Gummel plot of transistor A is displayed in Fig. 5.1*b*. The I–V characteristics predicted by GICCR agree with the device simulation results very well. The ICCR, however, overestimates collector current at high base–emitter biases. Device B is designed with respect to a high transit frequency, as shown in Fig. 5.2*a*. The Ge mole fraction in the base is increased from a rather low value near the emitter–base junction to the maximum value ($x_{\text{Ge}} \approx 0.076$) near the collector–base junction, resulting in an accelerating drift field for the electrons in the base. A comparison of GICCR and ICCR for transistor B is shown in Fig. 5.2*b*. Again, the GICCR is in good agreement with device simulation, while ICCR overestimates the collector current at very high base–emitter biases. The ICCR, however, is quite accurate for low injection and moderate injection.

For the compact transistor models, the GICCR has difficulty separating the minority charges q_{fe}, q_{fb}, and q_{fc} by measurements. Note that q_{fe}, q_{fb}, and g_{fc} are small compared to q_{b0}. The weighted charges $h_e q_{fe}$ and $h_c q_{fc}$ together with q_{fb} exceed q_{b0} at medium current densities and thus become the dominating term in q_b^+. At high current densities, $h_e q_{fe} \gg q_{b0}$. The difference between the ICCR and GICCR is significant.

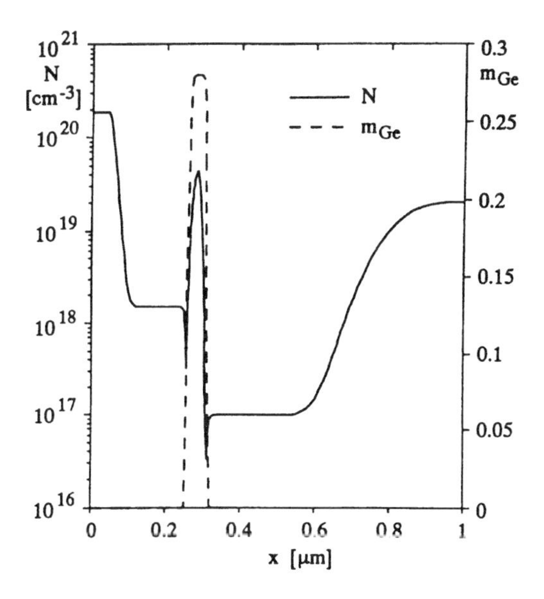

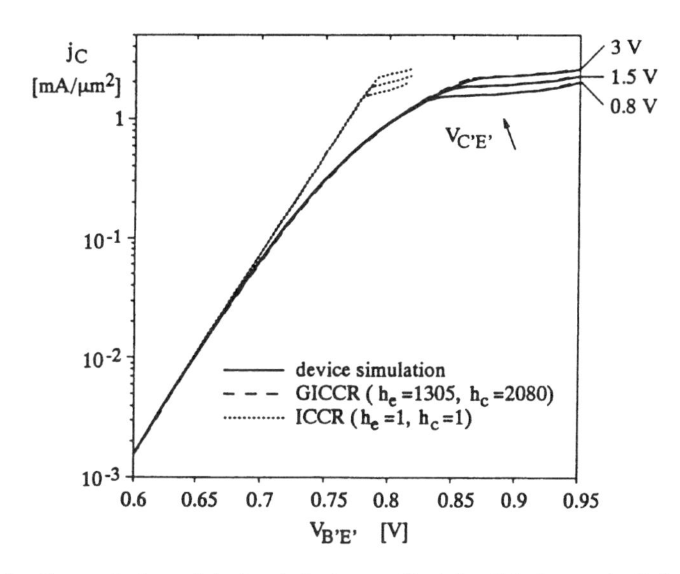

FIGURE 5.1 Gummel plot and device A doping profile (after Schröter et al., Ref. 4 © IEEE).

5.1.3 Base Current of SiGe HBTs

For an n-p-n SiGe bipolar transistor, the base current consists of the hole back-injection into the wide-gap emitter, recombination current in the narrow-gap base, and recombination current in the emitter–base space-charge region. The base current

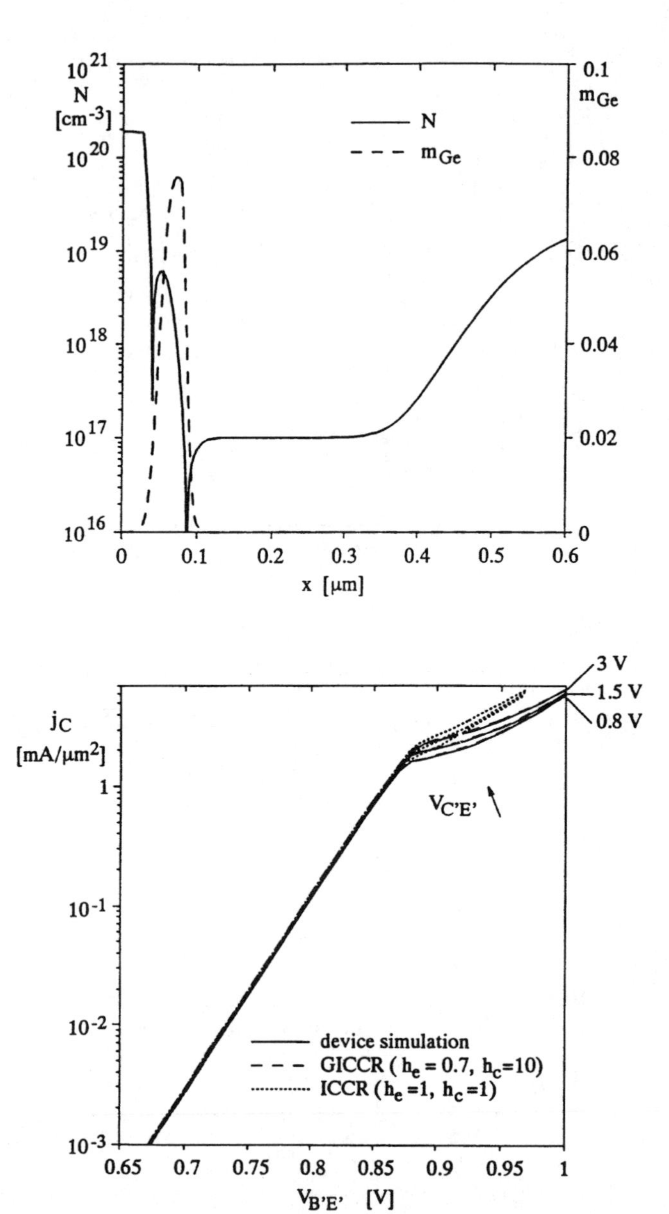

FIGURE 5.2 Gummel plot and device B doping profile (after Schröter et al., Ref. 4 © IEEE).

is approximated as [5]

$$I_B \approx \frac{e^{qV_{BE}/kT}}{\int_0^{W_E} N_E(x)\,dx/qAn_{iE}^2(x)D_E(x)} + \frac{qAn_{iB}^2 W_B}{2N_B\tau_B'} e^{qV_{BE}/kT} e^{\Delta E_V/kT} + \frac{1.7qAn_{iB}^2 X_p}{N_B\tau_B'} e^{qV_{BE}/kT} e^{\Delta E_G/kT} \tag{5.19}$$

where X_p is the space-charge layer width on the SiGe side of the emitter–base heterojunction and τ_B' is the minority-carrier lifetime in the base. Note that the base current above does not include surface recombination and emitter–base tunneling current. The emitter–base tunneling current can be modeled empirically by $I_{St}\exp(qV_{BE}/kT)$, where I_{St} is the tunneling current, when both emitter and base doping densities are on the order of 10^{19} cm^{-3} [6].

5.1.4 High Current Operation

Heterojunction bipolar transistors rely on a difference in bandgap to obtain a selective suppression of carriers from the base into the emitter. Double-heterojunction bipolar transistors such as SiGe HBTs exhibit a unique effect at high current densities. As the collector current is larger than the onset current for base pushout, holes tend to move into the lightly doped collector when the electric field at the collector–base junction changes from negative to positive. The valence-band discontinuity at the base–collector heterojunction, however, prevents holes from moving into the collector. Thus electrons in the collector are not compensated with holes and an electric field is established that creates a potential barrier. This barrier opposes the electron flow into the collector and an increased density at the base–collector heterojunction is required to support the collector current density.

Figure 5.3 shows the collector current versus the base–emitter voltage as a function of temperature. In this plot the symbols represent the experimental data and the solid lines represent the simulation results [7]. It is clear from Fig. 5.3 that the collector current clipping effect occurs at high base–emitter voltages. At a higher temperature the crossover point starts earlier, due to a reduced onset current for the Kirk effect.

The collector–base heterojunction barrier effect is modeled [8]. Poisson's equation in the collector–base depletion region is written as

$$\frac{d\phi^2}{dx^2} = -\frac{q}{\varepsilon_s}\left(N_D - N_A + p - n\right) \tag{5.20}$$

where ϕ is the electron potential, ε_s the collector permittivity, N_D the donor concentration, N_A the acceptor concentration, p the hole concentration, and n the electron concentration. When the bipolar transistor is operating at very high collector current densities, the electron concentration in the collector–base space-charge layer is equal

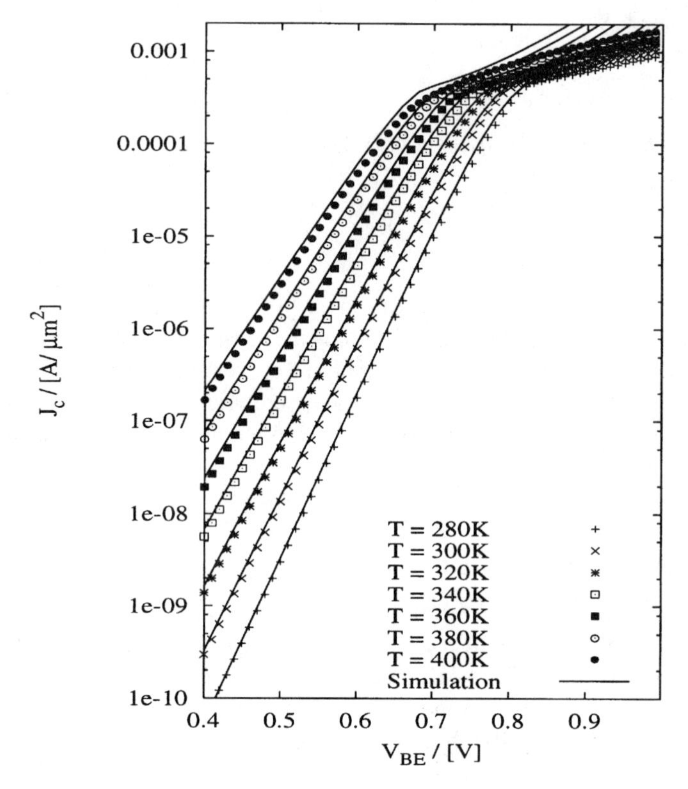

FIGURE 5.3 Collector current versus base–emitter voltage as a function of temperature (after Nuernbergk et al., Ref. 7 © IEEE).

to

$$n = \frac{J_C}{q v_s} \tag{5.21}$$

where v_s is the saturation velocity.

The free carrier hole concentration could be modeled as

$$p = \frac{N_B}{e^{(\Delta E_V - \Delta E_C)/kT}} \tag{5.22}$$

where ΔE_V is the valence-band discontinuity at the collector–base heterojunction and ΔE_C is the conduction-band barrier resulting from the high-current-base pushout effect. In the exponential term in (5.22), $\Delta E_V - \Delta E_C$ can be considered as an effective valence-band discontinuity accounting for high current barrier effect. Combining

(5.20), (5.21), and (5.22) gives

$$\frac{d\phi^2}{dx^2} = -\frac{q}{\varepsilon_s}\left[N_C - \frac{J_C}{qv_s} + \frac{N_B}{e^{(\Delta E_V - \Delta E_C)/kT}}\right] \tag{5.23}$$

where N_C and N_B are the uniform collector and base dopings.

Integration of (5.23) yields

$$V_{bi} + V_{CB} = -\frac{qW^2}{2\varepsilon_s}\left[N_C - \frac{J_C}{qv_s} + \frac{N_B}{e^{(\Delta E_V - \Delta E_C)/kT}}\right] \tag{5.24}$$

where V_{biC} is the built-in voltage in the collector–base heterojunction, V_{CB} the collector–base applied voltage, and W the collector–base depletion layer thickness. Equation (5.24) can be rewritten as

$$\Delta E_C = \Delta E_V + kT \ln\left[\frac{J_C}{qv_sN_B} - \frac{N_C}{N_B} - \frac{2\varepsilon_s(V_{bi} + V_{CB})}{qN_BW^2}\right] \tag{5.25}$$

For a uniformly doped base, the collector current density in written as

$$J_C = \frac{qD_nn_{i0}^2}{W_BN_B}\,\frac{e^{(qV_{BE}+\Delta E_V - \Delta E_C)/kT}}{1 + D_ne^{\Delta E_C/kT}/W_Bv_s} \tag{5.26}$$

Solving (5.25) and (5.26) together gives the conduction-band barrier height as a function of J_C and the collector current density, including the effect of bias-dependent ΔE_C.

Figure 5.4 shows the electron barrier height versus collector current density. The barrier increases with the current density. In Fig. 5.4 the solid line represents the analytical predictions and the solid circles represent the simulation results [9]. Good agreement between the analytical and simulation results is obtained until the collector current density is very high. The barrier effect on the collector current density at 300 and 400 K is shown in Fig. 5.5. At high base–emitter voltages, the collector current density exhibits the current clipping due to the formation of an electron barrier at high collector current densities. The agreement between the model predictions (solid lines) and experimental data (solid circles) in Fig. 5.5 is fairly good.

5.2 III/V COMPOUND HBT MODELS

With the development of III/V compound heterojunction bipolar transistor technology, there has been concomitant progress in analytical modeling of the HBTs. The Ebers–Moll and Gummel–Poon methodologies have both been utilized to model the HBT. Marty et al. [10] derived a charge-control model in which they assumed that current flow across an abrupt heterojunction was by drift and diffusion only. Grinberg et al. [11] derived an Ebers–Moll-like model assuming a thermionic-field-emission current

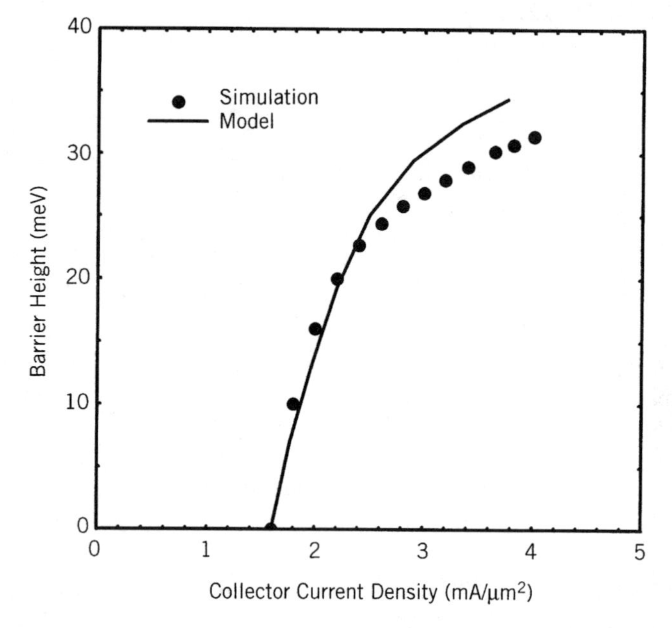

FIGURE 5.4 Electron barrier at the collector–base junction as a function of collector current (after Song and Yuan, Ref. 8 © IEEE).

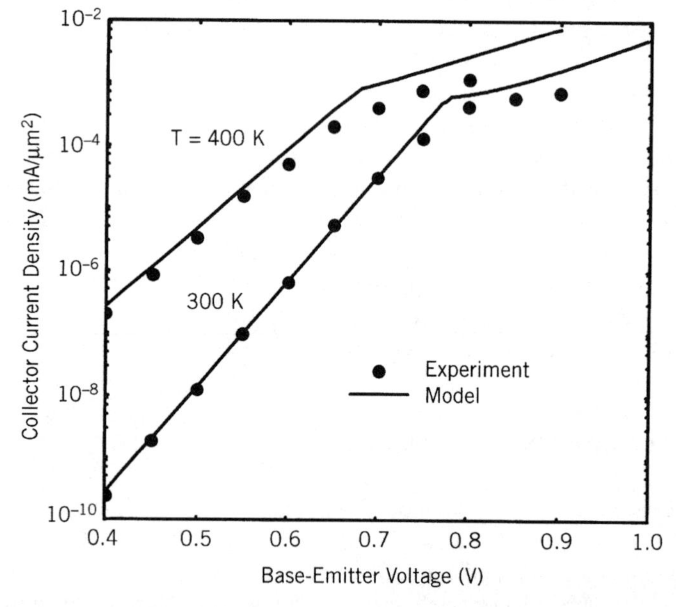

FIGURE 5.5 Collector current versus base–emitter bias (after Song and Yuan, Ref. 8 © IEEE).

flow across the heterojunction interface. Lundstrom [12] proposed that the carrier flow across the heterojunction interface could be modeled by a generalized interface transport velocity. Using such a velocity, he derived an Ebers–Moll-like model [13] for the HBT. Ryum and Adbel-Motaleb [14] unitized Lundstrom's concept of an interface transport velocity to derive a Gummel–Poon-like model for the HBT, taking into account the Early and base pushout effects. Grinberg and Luryi [15] derived exact analytical formulas for the current–voltage characteristics of a double-heterojunction bipolar transistor valid for arbitrary levels of injection and base doping, including the degenerate base. Parikh and Lindholm [16] derived a new charge-control relation for single- and double-heterojunction bipolar transistors valid for arbitrary doping profiles and for all levels of injection in the base. Liou et al. [17] derived an analytical model for the abrupt HBT with a setback layer. Yuan and Ning [18] generalized analytical equations for abrupt and linearly graded heterojunction bipolar transistors with or without a setback layer, including the self-heating effect. Hafizi et al. [19] presented a compact HBT model using Gummel–Poon equations for SPICE modeling. Yuan [20] used the Gummel–Poon model equations to study the thermal and reverse base current effects on heterojunction bipolar transistors. The models discussed above are presented here in a self-consistent manner.

5.2.1 Thermionic-Field-Diffusion Model

Carriers that have the velocity and the energy to inject over the barrier may suffer quantum-mechanical reflections because of the abrupt barrier. Carriers that do not have the requisite energy or velocity to inject over the barrier may still cross into the energy spike by tunneling through the thin barrier. This finite nonzero probability of penetration can make a substantial difference in the magnitude of the current at low temperatures where the thermionic current reduces. This emission behavior is often referred to as *field emission*, and its inclusion in the thermionic emission model leads to the thermionic-field-emission model.

Based on the thermionic emission concept, the electron current density at the heterojunction interface ($x = 0$) can be described as the difference of the two opposing electron fluxes [11]:

$$J_n(0) = q v_n \left[n(0^-) - n(0^+) e^{-\Delta E_C/kT} \right] \tag{5.27}$$

where

$$n(0^-) = n(-X_1) e^{-qV_{B1}/kT} \approx N_E e^{-V_{B1}/kT} \tag{5.28}$$

$$n(0^+) = n(X_2) e^{qV_{B2}/kT} \tag{5.29}$$

To include the current resulting from electron tunneling through the conduction-band spike, the tunneling coefficient γ_n is used. Equation (5.27) is then modified as [11]

$$J_n(0) = qv_n \left[n(0^-) - n(0^+)e^{-\Delta E_C/kT} \right] \gamma_n \tag{5.30}$$

where γ_n is given by

$$\gamma_n = 1 + \frac{q}{kT} e^{qV_{B1}/kT} \int_{V^*}^{V_{B1}} D(x) e^{-qV/kT} \, dV \tag{5.31}$$

For $V_{B1} > \Delta E_C/q$, $V^* = V_{B1} - \Delta E_C/q$; otherwise, $V^* = 0$. $D(X \equiv V/V_{B1})$ is the barrier transparency [21,22]

$$D\left(\frac{V}{V_{B1}}\right) = \exp\left\{ \frac{2V_{B1} m_n^* \varepsilon_E}{\hbar N_E} \left[\sqrt{1-X} + 0.5X \ln X - X \ln(1 + \sqrt{1-X}) \right] \right\} \tag{5.32}$$

Using the diffusion current in the base, the collector current is expressed as

$$J_C \approx -J_n(0) \approx \frac{qD_n n(X_2)}{W_B} \tag{5.33}$$

Using Eqs. (5.28), (5.29), (5.30), and (5.33), we have

$$n(X_2) = \frac{v_n \gamma_n N_E e^{-qV_{B1}/kT}}{(D_n/W_B) + v_n \gamma_n e^{q(V_{B2} - \Delta E_C)/kT}} \tag{5.34}$$

Inserting (5.34) into (5.33) gives

$$J_C = \frac{qv_n \gamma_n N_E e^{-qV_{B1}/kT}}{1 + (W_B/D_n) v_n \gamma_n e^{q(V_{B2} - \Delta E_C)/kT}} \tag{5.35}$$

The collector current equation in (5.35) is the thermionic-field-diffusion model. If the tunneling current is neglected (i.e., $\gamma_n = 1$), the thermionic-field-diffusion model reduces to the thermionic diffusion model. The thermionic-field-diffusion model is more accurate than the thermionic diffusion model when comparing their model predictions to the experimental data [11].

5.2.2 Grinberg–Luryi Physics-Based Collector Current Model

The theory of the minority-carrier transport in heterostructure bipolar transistors is presented with a particular emphasis on the difference between the cases of abrupt and graded emitter–base junctions. Exact analytical formulas are derived [15] for the current–voltage characteristics of a double-heterojunction HBT, valid for arbitrary levels of injection and base doping, including the degenerate case.

As is well known, the minority-carrier transport in the base of a bipolar transistor is adequately described by the drift-diffusion equation, provided that the base width is sufficiently large. The drift-diffusion equation for electrons in a semiconductor with the Fermi level E_{Fn} is given by

$$E_{Fn} = E_C + kT \ln \frac{n}{N_C} \tag{5.36}$$

where the conduction-band edge E_C results from both the electrostatic potential and the affinity [i.e., $E_C = -q(\phi + \chi)$]. Since $J_n/\mu_n n = \nabla E_{Fn}$, equation (5.36) can be rewritten as

$$\frac{J_n}{\mu_n n} = -\frac{d(q\phi + q\chi)}{dx} + kT \frac{d}{dx} \ln \frac{n}{N_C} \tag{5.37}$$

Consider an abrupt heterojunction where $q(\chi_2 - \chi_1) = \Delta$. Integrating (5.37) from the contact $C_1(x < 0)$ to the other contact $C_2(x > 0)$ including the band discontinuity at $x = 0$ gives

$$\int_{C_1}^{C_2} \frac{J_n\,dx}{\mu_n n} = \int_{C_1}^{0^-} \frac{J_n\,dx}{\mu_n n} + \int_{0^+}^{C_2} \frac{J_n\,dx}{\mu_n n} \tag{5.38}$$

Substituting (5.39) into (5.38) gives

$$\int_{C_1}^{C_2} \frac{J_n\,dx}{\mu_n n} = qV_{BE} - \Delta E_{Fn} \tag{5.39}$$

where the electron quasi-Fermi level difference at the abrupt heterojunction is

$$\Delta E_{Fn} \equiv \Delta - kT \ln \left[\frac{n(0^+)}{n(0^-)} \frac{N_C^{(E)}}{N_C^{(B)}} \right] = kT \ln \left[1 - \frac{J_n e^{\Delta/kT}}{qn(0^+)v_R} \right] \tag{5.40}$$

Assuming that at the top of the barrier, the electron concentration $n(x = 0^-)$ is nondegenerate, we have

$$\begin{aligned} n(0^-) &= N_C^{(E)} e^{-[E_C(-\varepsilon) - E_{Fn}]/kT} \\ &= N_C^{(E)} e^{-[E_G^{(B)} + \Delta]/kT} e^{qV_{BE}/kT} e^{[E_{Fp} - E_V(+\varepsilon)]/kT} \\ &= \frac{N_C^{(E)}}{N_C^{(B)} N_V^{(B)}} e^{-\Delta/kT} e^{qV_{BE}/kT} \Im_{1/2}^{-1} \left(\frac{p(0^+)}{N_V^{(B)}} \right) \end{aligned} \tag{5.41}$$

where $\Im_{1/2}^{-1}$ is the inverse transformation to $\Im_{1/2}$ and

$$\Im_{1/2}(\eta) \equiv \frac{2}{\pi^{1/2}} \int_0^\infty \frac{x^{1/2}dx}{1+\eta e^x} \tag{5.42}$$

Using (5.40) and eliminating $n(0^-)$ from (5.41) gives

$$n(0^+)e^{\Delta E_{Fn}/kT} = \frac{n_i^2}{N_V^{(B)}} e^{qV_{BE}/kT} \Im_{1/2}^{-1}\left(\frac{p(0^+)}{N_V^{(B)}}\right) \tag{5.43}$$

For a nondegenerate hole concentration, $\Im(\eta) \approx \eta^{-1}$ and (5.43) reduces to

$$n(0^+)p(0^+)e^{\Delta E_{Fn}/kT} = n_i^2 e^{qV_{BE}/kT} \tag{5.44}$$

Now consider a double-heterostructure transistor in which the emitter–base junction is abrupt while the collector–base junction is graded. The base bandgap and doping are assumed uniform.

The diffusion equation in the base is

$$\frac{\partial n}{\partial z^2} = \frac{n-n_0}{L_D^2} \tag{5.45}$$

where L_D is the electron diffusion length and

$$n_0 \equiv \frac{n_i^2}{N_V^{(B)}} \Im_{1/2}^{-1}\left(\frac{p_0}{N_V^{(B)}}\right) \tag{5.46}$$

The solution of (5.45) has the form

$$n-n_0 = \frac{[n(0^+)-n_0]\sinh[(W_B-z)/L_D]}{\sinh(W_B/L_D)} + \frac{[n(W_B)-n_0]\sinh(z/L_D)}{\sinh(W_B/L_D)} \tag{5.47}$$

Expressed in terms of the boundary conditions, the collector current density is

$$\frac{J_C}{qD_n} = -\frac{[n(0^+)-n_0]/L_D}{\sinh(W_B/L_D)} + \frac{[n(W_B)-n_0]/L_D}{\tanh(W_B/L_D)} \tag{5.48}$$

At the emitter–base and collector–base junctions, we have

$$n(0^+) = \frac{n_i^2}{N_V^{(B)}} e^{(qV_{BE}-\Delta E_{Fn})/kT} \Im_{1/2}^{-1}\left(\frac{p(0^+)}{N_V^{(B)}}\right) \tag{5.49}$$

$$n(W_B) = \frac{n_i^2}{N_V^{(B)}} e^{qV_{BC}/kT} \Im_{1/2}^{-1}\left(\frac{p(W_B)}{N_V^{(B)}}\right) \tag{5.50}$$

Inserting (5.46), (5.49), and (5.50) into (5.48) gives the collector current density of the HBT, valid for arbitrary levels of injection, including the degenerate base case.

5.2.3 New Charge-Control Model

A new charge-control relation is derived for heterojunction bipolar transistors [16]. The relation is valid for arbitrary base doping profiles. It is applicable to both single- and double-heterojunction bipolar transistors for all levels of injection in the base. Figure 5.6 shows a schematic band diagram of a double N-p-N heterojunction bipolar transistor. The band diagram of Fig. 5.6 shows that both the base–emitter and base–collector junctions are under forward bias. The zero of the x-axis is at the base–emitter metallurgical interface.

The key assumptions used in the derivation of the new charge-control model [16] for single- and double-heterojunction bipolar transistors are:

- The gradient of the majority hole quasi-Fermi level is much smaller than that of the minority electron quasi-Fermi level in the base.
- The energy bandgap in the base is uniform (i.e., the intrinsic carrier concentration is not a function of position.)
- The recombination current in the base is neglected.

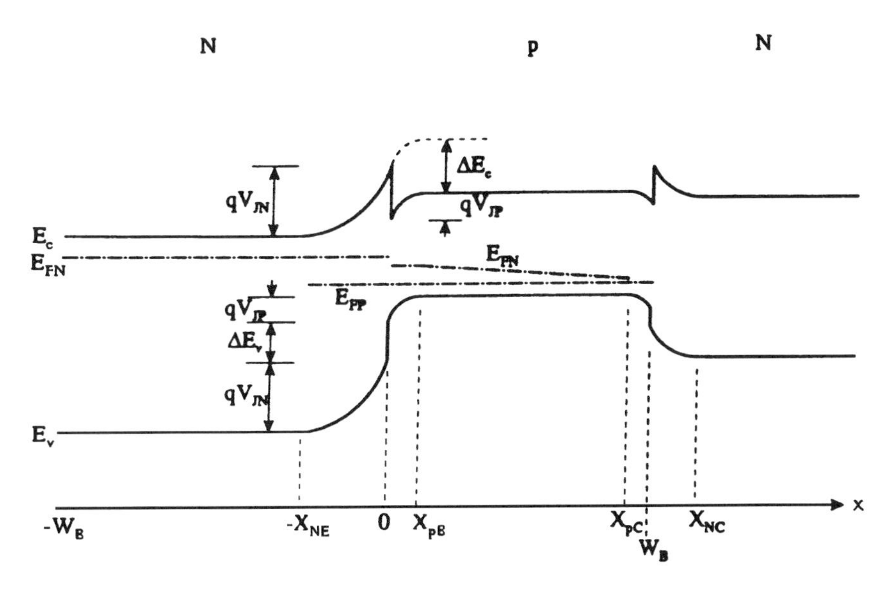

FIGURE 5.6 HBT schematic for the derivation of charge-control model (after Parikh and Lindholm, Ref. 16 © IEEE).

- The quasi-Fermi levels are nearly flat across the space-charge regions except for a possible discontinuity in the electron quasi-Fermi level at the heterojunction interface.
- The device is isothermal.
- The analysis is one-dimensional.

The carrier flow across the heterojunction interface is assumed by thermionic emission. Thus the electron flux F_{EN} across the emitter–base heterojunction interface is the difference of the thermionic electron fluxes in the positive and negative directions:

$$F_{EN} = v_x n(0^-) - v_x n(0^+) e^{-\Delta E_C/kT} \tag{5.51}$$

Equation (5.51) can be rewritten as

$$F_{EN} = v_x n(X_{NE}) e^{-qV_{JN}/kT} - v_x n(X_{PE}) e^{(qV_{JP}-\Delta E_C)/kT} \tag{5.52}$$

where $V_J = V_{JN} + V_{JP}$ is the electrostatic potential difference across the emitter–base space-charge layer. Now defining an effective interface carrier velocity $S_{EN} \equiv v_x \exp([(qV_{JP} - \Delta E_C)/kT]$, one gets

$$F_{EN} = v_x n(X_{NE}) e^{-qV_{JN}/kT} - v_x n(X_{PE}) e^{(qV_{JP}-\Delta E_C)/kT} \tag{5.53}$$

Rearranging (5.53) gives the boundary condition relating the electron carrier densities on both sides of a heterojunction:

$$n(X_{PE}) + \frac{F_{EN}}{S_{EN}} = n(X_{NE}) e^{(-qV_J+\Delta E_C)/kT} \tag{5.54}$$

Similarly, the boundary condition relating the hole carrier densities on both sides of a heterojunction is derived as

$$p(X_{PE}) + \frac{F_{EP}}{S_{EP}} = p(X_{NE}) e^{(-qV_J+\Delta E_V)/kT} \tag{5.55}$$

Multiplying the two equations above and rearranging the resulting equation gives

$$p(X_{pE}) n(X_{pE}) = -\frac{F_{EN}}{S_{EN}} p(X_{pE}) + p(X_{NE}) n(X_n E) e^{(\Delta E_C+\Delta E_V)/kT} \tag{5.56}$$

Since the hole quasi-Fermi level is constant across the emitter–base space-charge layer and hole flow is as in a homojunction, one can write

$$p(X_{NE}) n(X_{NE}) = n_{iE}^2 e^{[E_{FN}(X_{NE})-E_{FP}(X_{NE})]/kT} = n_{iE}^2 e^{qV_{BE}/kT} \tag{5.57}$$

Furthermore, neglecting the changes in the density of states N_C and N_V between the emitter and base, the intrinsic carrier concentration on the two sides of the heterojunction can be related as

$$n_{iE}^2 e^{\Delta E_G/kT} = n_{iB}^2 \tag{5.58}$$

where $\Delta E_G = \Delta E_C + \Delta E_V$. Combining the last three equations results in

$$p(X_{pE})n(X_{pE}) = n_{iB}^2 e^{qV_{BE}/kT} - \frac{F_{EN}}{S_{EN}} p(X_{pE}) \tag{5.59}$$

Equation (5.59) defines the pn product at the space-charge layer boundary of a heterojunction at $x = X_{pE}$. Similarly, the pn product at $x = X_{pC}$ is given by

$$p(X_{pC})n(X_{pC}) = n_{iB}^2 e^{qV_{BC}/kT} - \frac{F_{CN}}{S_{CN}} p(X_{pC}) \tag{5.60}$$

The charge-control relation for any bipolar transistor including HBTs pointed out by Kroemer [1] is

$$J_n = qD_n \frac{(pn)_{X_{pC}} - (pn)_{X_{pE}}}{\int_{X_{pE}}^{X_{pC}} p\, dx} \tag{5.61}$$

Using the boundary conditions in (5.59) and (5.60) and the charge-control relation yields the collector current density

$$J_C = \frac{qD_n n_{iB}^2 (e^{qV_{BE}/kT} - e^{qV_{BC}/kT})}{\int_{X_{pE}}^{X_{pC}} p\, dx + (qD_n/S_{EN})p(X_{pE}) + (qD_n/S_{CN})p(X_{pC})} \tag{5.62}$$

The new charge-control relation in (5.62) is different from the conventional equation in (5.10) by inclusion of emitter–base and collector–base heterojunction interface velocity effects in the denominator of the Gummel number.

5.2.4 Base Recombination Currents

5.2.4.1 Space-Charge-Region Recombination Currents

The space-charge-region recombination current density is [23]

$$J_{SCR} = J_{SCR1} + J_{SCR2} + J_{SCR3} = q\int_{-X_{NE}}^{-\lambda} R\, dx + q\int_{-\lambda}^{0} R\, dx + q\int_{0}^{X_{pE}} R\, dx \tag{5.63}$$

where λ is the edge of graded layer in the emitter–base space-charge region next to the emitter as shown in Fig. 5.7 and R is the Shockley–Read–Hall recombination rate

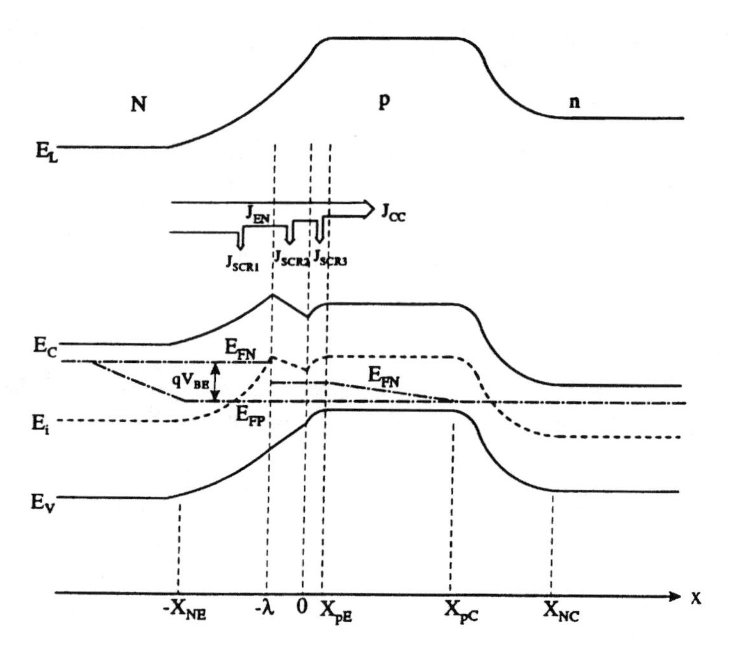

FIGURE 5.7 HBT schematic for the derivation of space-charge recombination (after Parikh and Lindholm, Ref. 23 © IEEE).

and is given by

$$R = \frac{n_i}{\sqrt{\tau_{n0}\tau_{p0}}} \times \frac{\sinh[(E_{Fn} - E_{Fp})/2kT]}{\cosh\left[\frac{2E_i - (E_{Fn} + E_{Fp})}{2kT} + \ln\sqrt{\frac{\tau_{n0}}{\tau_{p0}}}\right] + e^{(E_{Fn}-E_{Fp})/2kT}\cosh\left[\frac{E_T - E_i}{kT} + \ln\sqrt{\frac{\tau_{p0}}{\tau_{n0}}}\right]} \tag{5.64}$$

The intrinsic carrier concentration in the emitter–base space-charge region of an AlGaAs/GaAs HBT is

$$n_i = \begin{cases} n_{iB}e^{-\Delta E_G/2kT} & \text{for } x \geq 0 \\ n_{iB}e^{-1.55y(x)/2kT} & \text{for } \lambda \leq x \leq 0 \\ n_{iB}e^{-1.55y_{\max}/2kT} & \text{for } x \leq \lambda \end{cases} \tag{5.65}$$

The intrinsic energy level E_i for GaAs HBTs can be approximated as [24]

$$E_i(x) \approx -0.6E_G - \frac{\Delta E_G}{2} - qV(x) \tag{5.66}$$

Using (5.64)–(5.66) J_{SCR1} is derived as

$$J_{\mathrm{SCR1}} = \frac{\sinh[(E_{FN} - E_{FP})/2kT]}{\sqrt{\tau_{n0}\tau_{p0}}} n_{iE} \int_{-X_{NE}}^{-\lambda} \frac{dx}{\cosh[u(x) - 0.155 y_{\max} + U_F] + b} \tag{5.67}$$

where

$$u(x) = \frac{q}{kT} V(x)$$

$$U_F = \frac{E_{FN} + E_{FP}}{2kT} - \frac{1}{2} \ln \frac{\tau_{p0}}{\tau_{n0}}$$

$$b = \exp\left(\frac{E_{FP} - E_{FN}}{kT}\right) \cosh\left(\frac{E_T - E_i}{kT} + \frac{1}{2} \ln \frac{\tau_{p0}}{\tau_{n0}}\right)$$

From (5.64)–(5.66), J_{SCR2} is written as

$$\begin{aligned} J_{\mathrm{SCR2}} = {} & \frac{\sinh[(E_{FN} - E_{FP})/2kT]}{\sqrt{\tau_{n0}\tau_{p0}}} n_{iB} \\ & \times \int_{-\lambda}^{0} \frac{\exp[1.55 y(x)/2kT]}{\cosh[0.155 y(x) - u(x) - U_F] + b} dx \end{aligned} \tag{5.68}$$

Similarly, J_{SCR3} is given by

$$\begin{aligned} J_{\mathrm{SCR3}} = {} & \frac{\sinh[(E_{FN} - E_{FP})/2kT]}{\sqrt{\tau_{n0}\tau_{p0}}} n_{iB} \\ & \times \int_{0}^{X_{PE}} \frac{\exp[1.55 y(x)/2kT]}{\cosh[0.155 y(x) - u(x) - U_F] + b} dx \end{aligned} \tag{5.69}$$

Figure 5.8 shows the current gain of the AlGaAs/GaAs HBT as a function of the collector current density for three different grading layer thicknesses. For all three grading widths, the current gain at higher current densities is seen to be approximately the same. This is because at these high current densities, the recombination in the quasi-neutral base region dominates the base current, and thus the SCR recombination characteristics become irrelevant. At lower current densities, the junction with a grading width of 100 Å gives the flat-test characteristics. For the abrupt (1 Å) junction, the current gain is lower due to a lower collector current, which is caused by the conduction-band peak. For the 300-Å grading width device, the current gain is lower than the 100 Å device, due to higher recombination in the emitter SCR.

A more comprehensive space-charge recombination current model, including Schockley–Read–Hall, Auger, and radiative processes, is derived [25]. The recombination currents due to SRH recombination at the emitter and base sides

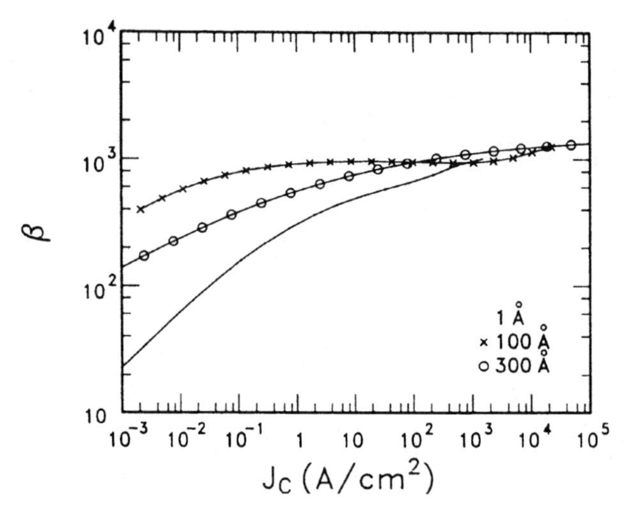

FIGURE 5.8 Current gain for different graded-layer thickness (after Parikh and Lindholm, Ref. 23 © IEEE).

are

$$J_{\mathrm{SRH},E} \approx C_S \frac{N_D n_{i,p}}{2\tau_n} \exp\left(\frac{qV_{BE}}{2kT}\right) \tag{5.70a}$$

$$J_{\mathrm{SRH},B} \approx C_S \frac{N_D n_{i,p}}{\tau_{n0,p} n_{i,n}} \times \exp\left(\frac{\Delta E_i - qN_{\mathrm{rad}}V_{bi}}{kT}\right) \exp\left(\frac{qN_{\mathrm{rad}}V_{BE} - \Delta E_{Fn}}{kT}\right) \tag{5.70b}$$

where $C_S = kT\{2\varepsilon/[qN_B(1 - N_{\mathrm{rad}})V_{bi}]\}^{1/2}$ and $N_{\mathrm{rad}} = N_E/(N_B + N_E)$. The recombination currents due to Auger recombination are

$$J_{\mathrm{Aug},E} \approx C_S n_{i,p}^2 A_{p,p} N_A \exp\left(\frac{qV_{BE}}{kT}\right) \tag{5.71a}$$

$$J_{\mathrm{Aug},B} \approx C_S n_{i,n}^2 A_{n,n} N_D \exp\left(\frac{qV_{BE} - \Delta E_{Fn}}{kT}\right) \tag{5.71b}$$

where A_n and A_p are the electron and hole Auger coefficients. The recombination currents due to radiation are

$$J_{\mathrm{rad},E} \approx \frac{qC_S V_{bi}}{kT} n_{i,n}^2 B_n N_{\mathrm{rad}} \exp\left(\frac{qV_{BE}}{kT}\right) \tag{5.72a}$$

$$J_{\mathrm{rad},B} \approx \frac{qC_S V_{bi}}{kT} n_{i,p}^2 B_p (1 - N_{\mathrm{rad}}) \exp\left(\frac{qV_{BE} - \Delta E_{Fn}}{kT}\right) \tag{5.72b}$$

where B is the radiative recombination coefficient.

Figure 5.9 shows the results for the various SCR current contributions as a function of base–emitter bias. The slopes of the curves are not constant, owing to the voltage dependence of emitter–base space-charge layer thickness. The parameters used in the simulation are $N_E = 5 \times 10^{17}$ cm^{-3}; $N_B = 1 \times 10^{19}$ cm^{-3}, $\tau_{n0} = 5$ ns; $\tau_{p0} = 20$ ns; 30% Al mole fraction; $\Delta E_C = 0.24$ eV; $\Delta E_i = 77.3$ meV; $V_{bi} = 1.671$ V; $n_{\text{in}} = 4.21 \times 10^3$ cm^{-3}; $n_{i,p} = 4.07 \times 10^6$ cm^{-3}; $N_{\text{rad}} = 0.952$; $A_{n,n} = 7.99 \times 10^{-32}$ cm^{6}s^{-1}; $A_{p,p} = 1.12 \times 10^{-30}$ cm^{6}s^{-1}; $B_n = 1.29 \times 10^{-10}$ cm^3; $B_p = 7.82 \times 10^{-11}$ cm^{3}s^{-1}; and $D_n = 30$ cm^{3}s^{-1}.

It is clear from Fig. 5.9 that all the base-side SCR recombination components have about the same ideality factor and that n is considerably less than that of the emitter-side SCR recombination current. Specifically, at $V_{BE} = 1.2$ V, $n_{\text{SCR},E} = 1.90$, and adding all the base-side currents together, $n_{\text{SCR},B} = 1.19$. The width of the SCR on the base side of the heterojunction is much less than that on the emitter side, and this fact alone would make $J_{\text{SCR},B} \ll J_{\text{SCR},E}$ beyond some forward bias. In Fig. 5.9 this occurs around $V_{BE} = 1.45$ V. This transfer from $n \approx 2$ to an $n \approx 1$ slope in the SCR current does not occur in a homojunction device, as there is no spatial change in n_i to inflate the current in the more highly doped side of the junction.

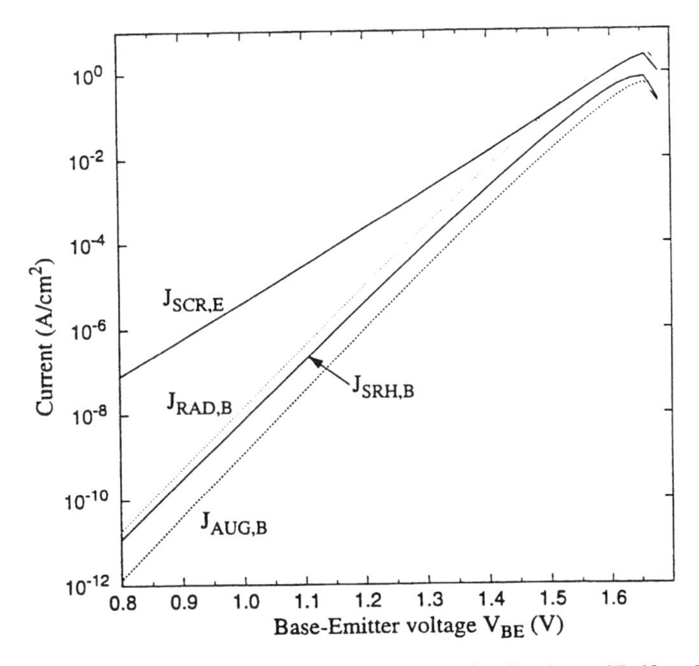

FIGURE 5.9 Base current versus base–emitter voltage (after Searles and Pulfrey, Ref. 25 © IEEE).

5.2.4.2 *Surface Recombination Currents*

Electron–hole recombination at the surface of a GaAs-related compound is more significant than that of Si and InP. This leads to nonnegligible emitter and base surface recombination currents in AlGaAs/GaAs heterojunction bipolar transistors, due to considerable electron–hole recombination taking place at the surface of the emitter and base peripheries. Based on the finite-difference approach and relevant device physics such as Shockley–Read–Hall recombination statistics and Fermi-level pining, a model for emitter and base surface recombination currents is developed [26].

Consider an abrupt N-p$^+$-n AlGaAs/GaAs HBT under normal operation as shown in Fig. 5.10, where the circles represent the surface states at which surface recombination can occur. The effect of the n$^+$ emitter cap region, which is usually placed on top of the emitter for a better ohmic contact is neglected. Since the HBT normally has a very thin base layer, the spreading of the carriers into the horizontal direction in the base is negligible. This suggests that most minority carriers in the base are confined in the region underneath the emitter (intrinsic base region), resulting in a negligibly small recombination current at the surface of the extrinsic base. Thus one can focus on the intrinsic base region and assume that the great majority of surface recombination takes place at the emitter surface and at the intrinsic base surface (or the emitter–base heterointerface), which results in emitter surface recombination current I_{ES} and base surface recombination current I_{BS}, respectively. Note that the extrinsic base surface can be important if the base layer is thick ($>$ 1000 Å) and/or if the base is not graded [27].

The emitter surface recombination current is modeled as

$$J_{ES} = \sum_{i=1}^{N} q S^i p^i \frac{X_{JE} P}{N} \tag{5.73}$$

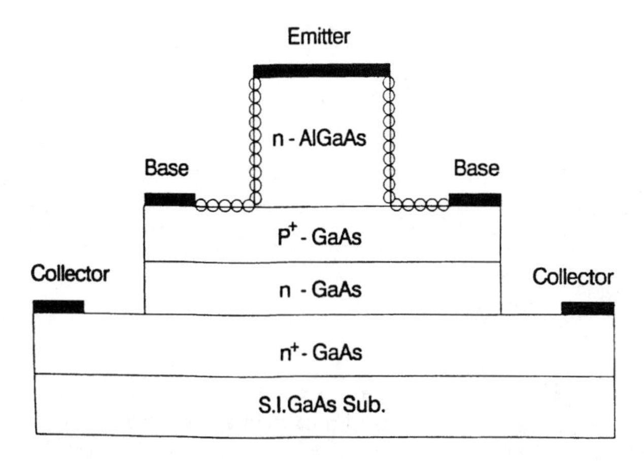

FIGURE 5.10 Schematic of an HBT, including surface recombination.

where S is the surface velocity, X_{JE} the emitter junction depth, P the emitter–base junction periphery, and N the number of regions in the emitter. The effective surface velocity in each region is

$$S^i = \sum_j \frac{N_{Sj}\sigma_{nj}\sigma_{pj}v_n v_p(n_0^i + p_0^i)}{\sigma_{nj}v_n(n^i + n_{Tj}) + \sigma_{pj}v_p(p^i + p_{Tj})} \tag{5.74}$$

where j is the number of types of surface traps, N_S is the surface trapping density (cm^{-2}), σ_n and σ_p are the electron and hole capture cross section, v_n and v_p are the electron and hole thermal velocities, n_0 and p_0 are the equilibrium free-carrier concentrations, and n_{Tj} and p_{Tj} are the electron and hole densities if the equilibrium Fermi level E_F were located at the trap energy E_T:

$$n_{Tj} = N_C e^{-(E_C - E_{Tj})/kT} \tag{5.75a}$$

$$p_{Tj} = N_V e^{-(E_{Tj} - E_V)/kT} \tag{5.75b}$$

$$n_0^i = N_E e^{-(qV_{B1} - \Delta E_C)/kT} e^{[(qV_{B1}/kT)(M+1-i)^2/(M-1-N)^2]} \tag{5.75c}$$

$$p_0^i = N_B e^{-qV_B/kT} e^{[(qV_{B1}/kT)(M+1-i)^2/(M-1-N)^2]} \tag{5.75d}$$

$$n^i = N_E e^{-qV_S/kT} e^{-[(qV_{bi} - V_{BE} - \Delta E_C)/kT](M-1-i)^2/(M+1-N)^2} \tag{5.75e}$$

$$p^i = N_B e^{-(V_B - V_{BE})/kT} e^{qV_S/kT} e^{-[(qV_{bi} - V_{BE})/kT](M-1-i)^2/(M+1-N)^2} \tag{5.75f}$$

$i = M + 1, \ldots, N$ in the space-charge emitter subregions. The values of E_{tj} and N_{Sj} at the emitter surface are needed in calculating S_i and thus I_{ES}. It has been suggested that there are two dominant traps at the oxided n-type GaAs interface [28]: One type of trap has an energy E_{t1} located at 0.73 eV (relative to the conduction-band edge) with a density of $N_{S1} = 10^{11}\ \text{cm}^{-2}$. Assuming that these values also apply to AlGaAs, one can calculate S_i and I_{ES} with $j = 2$ and $i = N$ (N is selected based on accuracy and computation efficiency).

Since the surface states at the heterointerface are arranged along the interface plane, the heterointerface trapping density N_{TI} is in units of number/cm^2. Using SRH recombination statistics, the base interface recombination current I_{BS} is given by

$$I_{BS} = qA \int_{X_{JE}}^{X_E} U_{\text{SRH}} \delta(x - X_{JE})\, dx \approx qAU_{\text{SRH}}(X_{JE}) \tag{5.76}$$

where U_{SRH} is the SRH recombination rate.

For a single trap in the middle of the energy bandgap,

$$I_{BS} = \tfrac{1}{2} qA\sigma v N_{TI} n_{iE} e^{qV_{BE}/2kT} \tag{5.77}$$

where $\sigma = \sigma_n = \sigma_p$ and $v = v_n = v_p$.

Note that the base interface recombination current is proportional to $\exp(qV_{BE}/2kT)$, whereas the emitter surface recombination current follows an $\exp(qV_{BE}/kT)$ dependence. The significance of emitter surface recombination current is proportional to the ratio of emitter–base junction perimeter to emitter area. Both emitter and base surface recombination currents are significant at low base–emitter voltages.

5.2.4.3 *Quasi-Neutral Recombination Currents*

Electrons injected from the emitter contact are recombined with holes in the quasi-neutral base and the emitter. The recombination in the quasi-neutral base is

$$I_{RB} = I_n(0)(1 - \alpha_B) \tag{5.78}$$

where $I_n(0)$ is the electron current at the edge of the emitter–base space-charge region in the base side and α_B is the base transport factor. The base transport factor is expressed as

$$\alpha_B = \frac{1}{\cosh(W_B/\sqrt{D_n \tau_n})} \tag{5.79}$$

τ_n is the electron lifetime in the base. Note that α_B approaches unity if the electron–hole recombination in the quasi-neutral base is negligible. The recombination current in the quasi-neutral emitter can be obtained using the diffusion approximation

$$I_{RE} = \frac{qAN_E D_p}{W_E} e^{-q(V_{B1}+V_{B2}+\Delta E_V)/kT} \tag{5.80}$$

where $V_{B1} + V_{B2} = V_{bi} - V_{BE}$. In general, the quasi-neutral base recombination current is much larger than the quasi-neutral emitter recombination current for the heterojunction bipolar transistor. This is because the base is usually more heavily doped than the emitter. In addition, the valence-band discontinuity ΔE_V suppresses hole injection from base to emitter significantly.

5.2.5 Analytical Collector Current Model, Including the Self-Heating Effect

A more general collector current equation for abrupt and linearly graded HBTs with or without an undoped spacer layer at the emitter–base heterojunction has been derived [18]. A one-dimensional schematic showing various positions at the edges of the space-charge regions is depicted in Fig. 5.11 for the derivation of analytical equations. The collector current, including the space-charge region recombination current, is given by

$$J_C = J_n(-X_g) - (J_{\mathrm{SCR}g} - J_{\mathrm{SCR}s} - J_{\mathrm{SCR3}}) \tag{5.81}$$

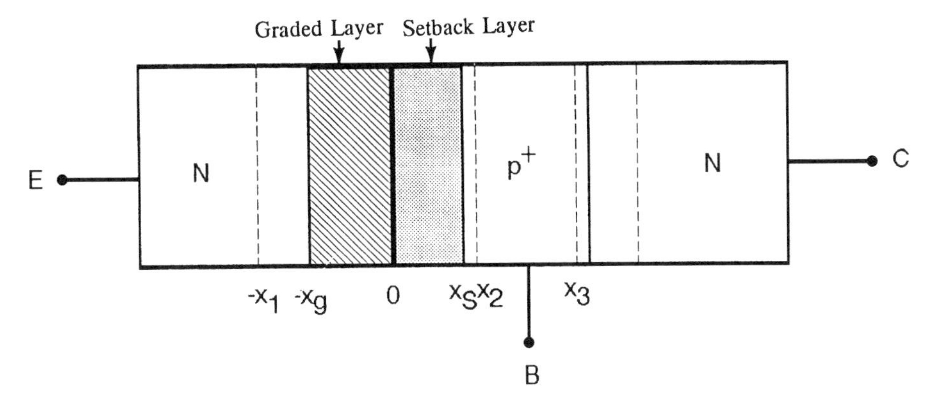

FIGURE 5.11 One-dimensional HBT schematic, including a setback layer.

The electron current density across the heterointerface is

$$J_n(-X_g) = qv_n\gamma_n\left[n(-X_g^-) - n(-X_g^+)e^{-\Delta E_C/kT}\right]$$
$$= qv_n\gamma_n\left[N_E e^{-qV_{B1}/kT} - n(X_2)e^{q(V_{B2}+V_{BS}+V_{BgC})/kT}\right] \tag{5.82}$$

where $V_{B1} = -V_1(-X_g)$, $V_{B2} = -[V_2(X_2) - V_S(X_S)]$, $V_{BS} = -[V_S(X_S) - V_g(0)]$, and $V_{BgC} = -\Delta E_C - V_g(0) + V_1(-X_g)]$ [18].

Since the electron transport in the uniformly-doped base is given by diffusion, the collector current can be written as

$$J_C(X_2) \approx \frac{qD_n n(X_2)}{X_3 - X_2} \tag{5.83}$$

Combining Eqs. (5.60), and (5.78)–(5.80) obtains the electron concentration at $x = X_2$, including the effects of thermionic emission and space-charge recombination:

$$n(X_2) = \frac{v_n\gamma_n N_E e^{-qV_{B1}/kT} - (J_{SCRg} + J_{SCRS} + J_{SCR3})/q}{D_n/(X_3 - X_2) + v_n\gamma_n e^{q(V_{B2}+V_{BS}+V_{BgC})/kT}} \tag{5.84}$$

Inserting (5.81) into (5.80) gives

$$J_C(X_2) = \frac{qv_n\gamma_n N_E e^{-qV_{B1}/kT} - (J_{SCRg} + J_{SCRS} + J_{SCR3})}{1 + v_n\gamma_n e^{q(V_{B2}+V_{BS}+V_{BgC})/kT}(X_3 - X_2)/D_n} \tag{5.85}$$

Equation (5.85) accounts for thermionic emission across the heterojunction interface and space-charge recombination in the emitter–base heterojunction for determining the collector current of the HBT with a uniform bandgap and doping concentration in the base. For an abrupt emitter–base heterojunction with an undoped spacer layer and space-charge-region recombination (i.e., $V_{BgC} = 0$, $V_{BS} = 0$, $J_{SCRg} = 0$, $J_{SCRS} = 0$, $J_{SCRB} = 0$), (5.85) reduces to (5.35).

Heterojunction bipolar transistors using a GaAs semi-insulating substrate exhibit self-heating owing to low thermal conductivity of the GaAs semi-insulating substrate as described in Chapter 2. To model the self-heating effect, the increment of temperature as a function of power dissipation and thermal resistance is determined. Assuming that the heat is dissipated throughout the semi-insulating substrate with a lateral diffusion angle and the thermal conductivity is proportional to $(T/T_0)^{-b}$, the Kirchhoff transformation gives the junction temperature as [29]

$$T = \left[\frac{1}{T_0^{b-1}} - \frac{(b-1)R_{\text{th0}}P_S}{T_0^b} \right]^{-1/b-1} \tag{5.86}$$

where P_S is the power dissipation ($P_S \approx I_C V_{CE}$), R_{th0} is the thermal resistance at room temperature T_0, and $b = 1.22$ [30]. To account for temperature dependence of collector and base currents, temperature-dependent physical parameters must be used. Equations for doping and temperature-dependent diffusion constants and temperature-dependent n_i, E_G, and v_s are included.

Figure 5.12 shows the current gain against graded-layer thickness at $V_{BE} = 1.25$, 1.3, and 1.35 V, respectively. The current gain increases with graded-layer thicknesss at all biases, owing to reduction of the electron potential barrier at the emitter–base heterojunction. The increase in current gain, however, saturates at around 130 Å. Further increase of graded-layer thickness decreases the current gain slightly. This

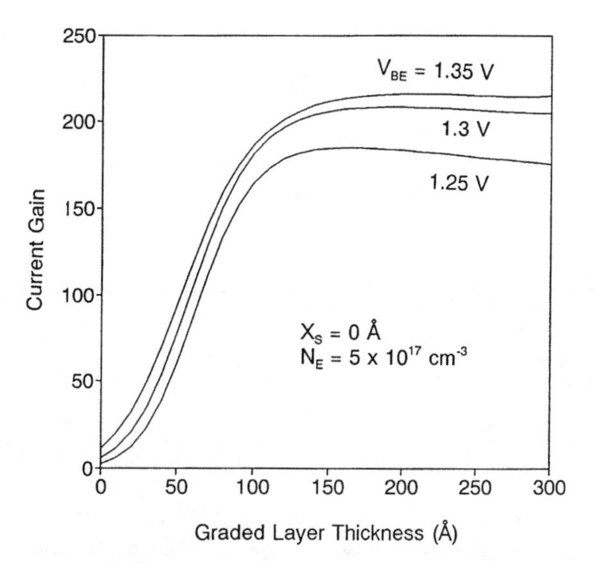

FIGURE 5.12 Current gain versus graded layer thickness at $V_{BE} = 1.25$, 1.3, and 1.35 V (after Yuan and Ning, Ref. 18 © IEE).

is because of the increase of space-charge region recombination in the graded layer at larger X_g. The sensitivity of current gain against setback layer thickness at different graded-layer thicknesses is shown in Fig. 5.13. The current gain is evaluated at $V_{BE} = 1.3$ V. For the abrupt heterojunction the use of a setback layer certainly increases the current gain until it reaches the saturation point. For the linearly graded heterojunction with $X_g = 100$ and 200 Å, however, the current gain decreases with increasing setback layer thickness. This phenomenon is explained as follows. When the heterojunction bipolar transistor has enough junction grading, the emitter–base heterojunction is similar to a homojunction. The injection efficiency cannot be further improved by the use of a setback layer. The use of a setback layer in this case introduces the increment of space-charge recombination current and decreases the current gain of the HBT.

The self-heating effect on the common-emitter current gain of the heterojunction bipolar transistor is shown in Fig. 5.14. The current gain is plotted against the graded-layer thickness for the isothermal model and the nonisothermal model at $V_{BE} = 1.4$ V and $V_{CE} = 3, 6$, and 9 V. The current gain increases with graded-layer thickness and then decreases with X_g when self-heating becomes significant. Self-heating causes the rise of junction temperature, which increases the base and collector currents of the HBT. The increase in the base current, however, is larger than that of the collector current. This results in a decrease of current gain at larger X_g.

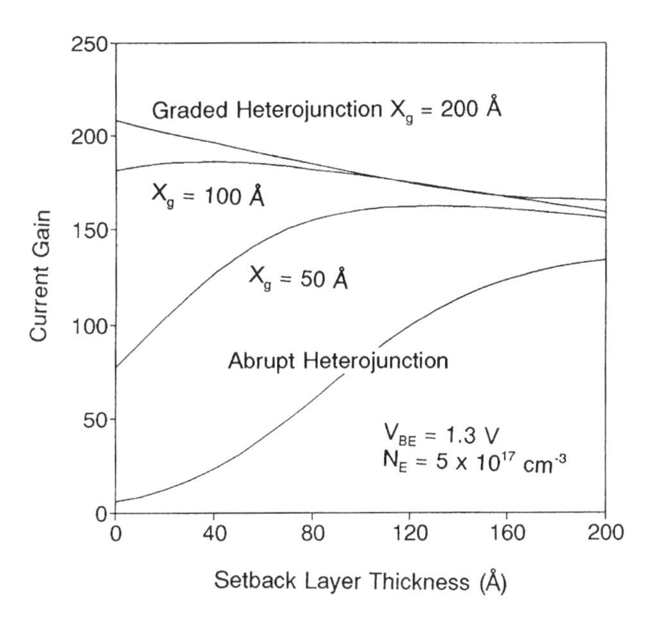

FIGURE 5.13 Current gain versus setback layer thickness as a function of graded layer thickness (after Yuan and Ning, Ref. 18 © IEE).

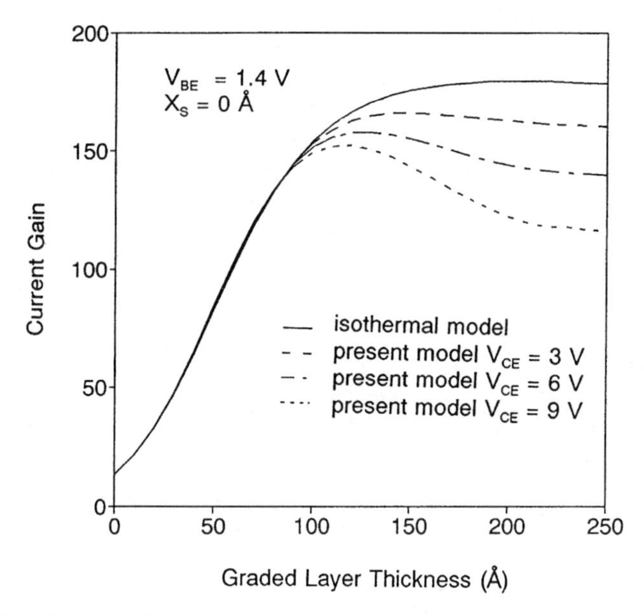

FIGURE 5.14 Current gain versus graded layer thickness for isothermal and nonisothermal models (after Yuan and Ning, Ref. 18 © IEE).

5.2.6 Compact Gummel–Poon Model, Including the Self-Heating Effect

For the heterojunction bipolar transistor, the emitter–base heterojunction can be designed as an abrupt junction or as a graded junction with or without a setback layer. The graded HBT is similar to the homojunction bipolar transistor in that the minority-carrier transport across the emitter–base junction can be described by diffusion. The minority-carrier transport of an abrupt heterojunction bipolar transistor, however, is governed by thermionic emission, tunneling, and diffusion. For a heterojunction bipolar transistor model used for circuit simulation, the model should be universal and can account for various heterojunction systems. Therefore, a compact model suitable for SPICE circuit simulation is needed.

Examine the Gummel–Poon model equations for HBT applications. The collector current of the Gummel–Poon model is given by [31]

$$I_C = \frac{I_S}{q_b}\left(e^{qV_{BE}/n_f kT} - e^{qV_BC/n_r kT}\right) - \frac{I_S}{\beta_R} e^{qV_{BC}/n_r kT} - I_{SC} e^{qV_{BC}/n_c kT} \qquad (5.87)$$

where I_S is the collector saturation current, I_{SC} the preexponential base–collector leakage current, q_b the normalized base charge, β_R the inverse current gain, and n_f, n_r, and n_c are the ideality factors in the collector and base currents. The normalized base charge q_b accounts for base width modulation and the high injection effect by the Early voltage V_A, late voltage V_B, forward knee current I_{KF}, and inverse knee current I_{KR} as follows:

$$q_b = \frac{q_1}{2} + \sqrt{\left(\frac{q_1}{2}\right)^2 + q_2} \tag{5.88}$$

where

$$q_1 = 1 + \frac{V_{BE}}{V_B} + \frac{V_{BC}}{V_A} \approx \frac{1}{1 - V_{BE}/V_B - V_{BC}/V_A}$$

$$q_2 = \frac{I_S}{I_{KF}}\left(e^{qV_{BE}/n_f kT} - 1\right) + \frac{I_S}{I_{KR}}\left(e^{qV_{BC}/n_R kT} - 1\right).$$

AlGaAs/GaAs heterojunction bipolar transistors grown by MBE show a nonnegligible collector–emitter offset voltage. The offset voltage can be computed from the Gummel–Poon model as [19]

$$V_{CE\text{offset}} \approx \frac{kT}{q} \ln\left[\left(\frac{I_B}{I_S}\right)^{n_f}\left(\frac{I_{SC}}{I_B}\right)^{n_c}\right] \tag{5.89}$$

For typical GaAs heterojunction bipolar transistors, the base current is the recombination current in the quasi-neutral base, in the emitter–base space-charge region, at the emitter–base heterojunction interface, and the emitter and extrinsic base surfaces. Taking each contribution into account, the base current of the HBT can be modeled empirically using the Gummel–Poon equation

$$\begin{aligned} I_B = {} & \frac{I_S}{\beta_F}\left(e^{qV_{BE}/n_f kT} - 1\right) + I_{SE}\left(e^{qV_{BE}/n_e kT} - 1\right) \\ & + \frac{I_S}{\beta_R}\left(e^{qV_{BC}/n_r kT} - 1\right) + I_{SC}\left(e^{qV_{BC}/n_c kT} - 1\right) \end{aligned} \tag{5.90}$$

where I_{SE} is the preexponential emitter–base leakage current and n_e is the ideality factor in I_{SE}. The first and third terms in (5.90) represent the bulk recombination in the base and the second and fourth terms correspond to recombination in the emitter–base space-charge region and extrinsic base surface region in the forward and inverse active modes, respectively.

Figure 5.15 shows the collector current versus collector–emitter voltage at three different base–emitter voltages. In this plot the solid lines represent the experimental data and the dashed lines represent the model predictions. The AlGaAs/GaAs HBT has about a 300-meV collector–emitter offset voltage due to an imbalance between the built-in voltages at the emitter–base and collector–base junctions. Since the turn-on voltage at the collector–base homojunction is smaller than that at the emitter–base heterojunction, the collector current is negative for $V_{CE} < V_{CE\text{offset}}$. The model parameters used in this figure are $I_S = 6.2 \times 10^{-22}$ A, $n_f = 1.133$, $n_r = 1.1$, $I_{SE} = 2.5 \times 10^{-18}$ A, $I_{SC} = 5.9 \times 10^{-14}$ A, $n_e = 1.8$, $n_c = 1.97$, $V_A = 116$ V, and

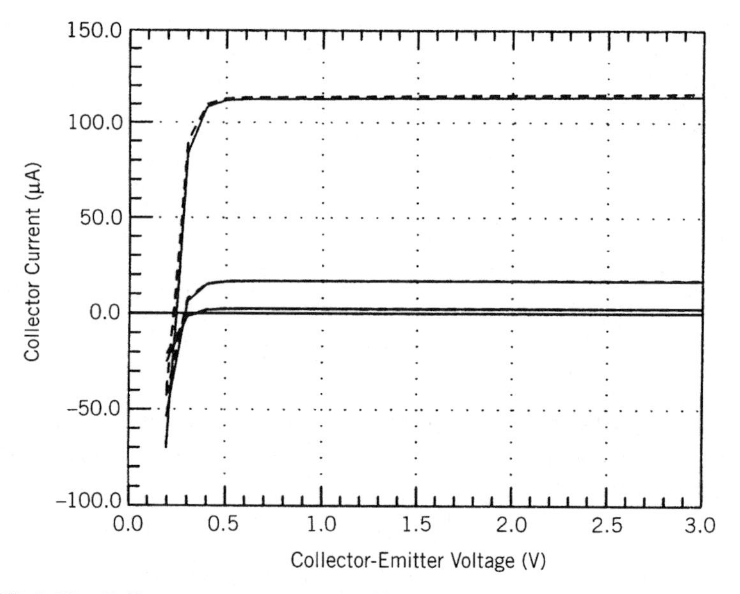

FIGURE 5.15 Collector current versus collector–emitter voltage (after Yuan, Ref. 20 © IEEE).

$\beta_F = 52$. The agreement between the predicted collector currents and measurement is fairly good. The present model also is able to predict the offset voltage with good accuracy. The comparison of the collector and base currents versus the base–emitter voltage for the model predictions (dotted lines) and measurement (solid lines) is shown in Fig. 5.16. The HBT exhibits a higher turn-on voltage ($\approx$ 1.3 V) than that of the silicon bipolar transistor, due to the bandgap discontinuity at the emitter–base heterojunction. At low base–emitter bias, the experimental data show significant leakage current, attributed to device packaging. The agreement between the model predictions and measurement is excellent for the range $V_{BE\text{on}}$ to the onset bias for base pushout ($V_{BE} \approx 1.45$ V). In general, the Gummel–Poon model is accurate enough to model the Gummel plot (I_C and I_B versus V_{BE}), beta plot (current gain versus I_C), and transfer curves (I_C versus V_{CE}) as long as the self-heating effect is negligible.

When the HBT has self-heating, the collector current decreases with increasing collector–emitter voltage at a given I_B. The negative slope is not accounted for in the Gummel–Poon model. By incorporating the thermal resistance from power dissipation, the thermal effects on the HBT can be accounted for.

The increase of junction temperature due to self-heating is modeled as

$$\Delta T = \theta(I_C V_{CE} + I_B V_{BE}) \tag{5.91}$$

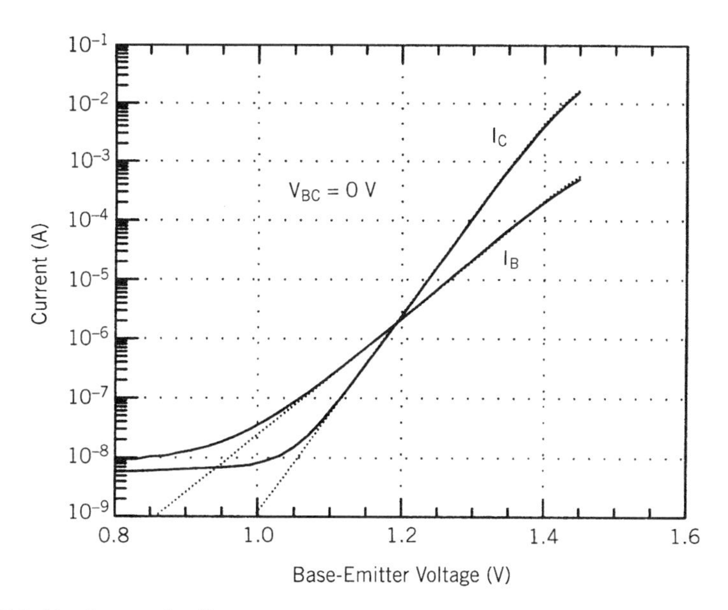

FIGURE 5.16 Base and collector currents versus base–emitter voltage (after Yuan, Ref. 20 © IEEE).

where θ is the thermal resistance. The thermal resistance can be obtained using the following expression [32]:

$$\theta = \frac{(V_{BE1} - V_{BE2})/(P_{S1} - P_{S2})}{(V_{BE1} - V_{BE3})/(T_1 - T_2)} \tag{5.92}$$

where V_{BE1} and P_{S1} are the base–emitter voltage and transistor power dissipation at fixed collector-emitter voltage V_{CE1} and temperature T_1. V_{BE2} and P_{S2} are the base–emitter voltage and transistor power dissipation at V_{CE2} and T_1. V_{BE3} is the base–emitter voltage at V_{CE1} and T_2.

To account for temperature-dependent collector and base currents, the following empirical equations are used [33]:

$$I_S = \left(\frac{T}{T_0}\right)^{X_{TI}} e^{(-E_G/kT)+(E_G/kT_0)} \tag{5.93}$$

$$I_{SE} = I_{SE0}\left(\frac{I_S}{I_{S0}}\right)^{1/n_e} \frac{\beta_F}{\beta_{F0}} \tag{5.94}$$

$$I_{SC} = I_{SC0}\left(\frac{I_S}{I_{S0}}\right)^{1/n_c} \frac{\beta_F}{\beta_{F0}} \tag{5.95}$$

$$\beta_F = \beta_{F0}\left(\frac{T}{T_0}\right)^{a} e^{(E_\infty/kT)-(E_\infty/kT_0)} \tag{5.96}$$

where I_{S0}, I_{SE0}, I_{SC0}, and β_{F0} are measured at a nominal temperature, E_G is the bandgap, and X_{TI}, a, and E_∞ are fitting parameters.

To model the thermal resistance of the substrate, the following nonlinear relation is used:

$$\theta = \theta(T_0)\left(\frac{T}{T_0}\right)^{1.15} \tag{5.97}$$

Figure 5.17 shows the nonlinear thermal resistance versus temperature. In Fig. 5.17 the solid line represents the present model prediction and the solid squares represent the extracted parameter at temperatures 23.3, 49, and 74°C. From 23 to 73°C, the thermal resistance increases by 16%, due to a decrease of GaAs substrate thermal conductivity at high temperature. Figure 5.18 shows the base current versus collector–base voltage. In this figure the solid dots represent the experimental data, the dotted line represents the isothermal model, and the solid line represents the nonisothermal model. Examining the experimental data in Fig. 5.18, the base current increases with increasing collector–base voltage and then decreases with collector–base voltage at high V_{CB}. The initial increase in the base current is attributed to the junction temperature rise due to self-heating, and the precipitous drop of the base current at higher V_{CB} is because of the exponential dependence of the impact ionization coefficient on the electric field. The details of impact ionization are given in Chapter 6.

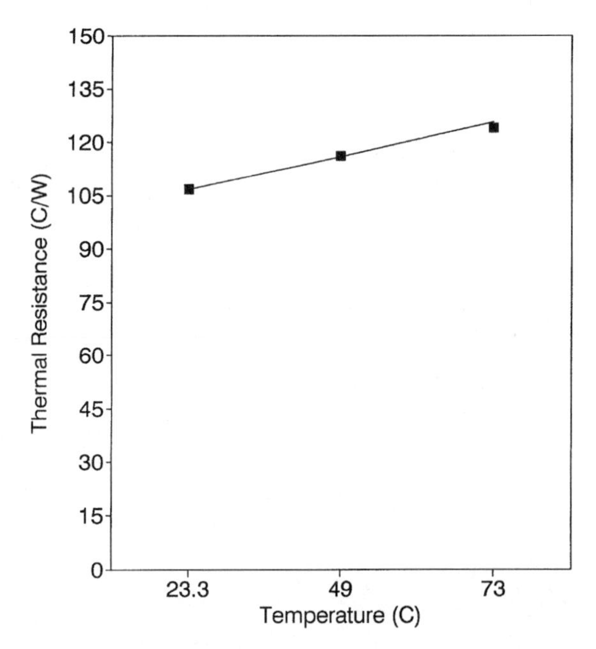

FIGURE 5.17 Thermal resistance as a function of temperature (after Yuan, Ref. 20 © IEEE).

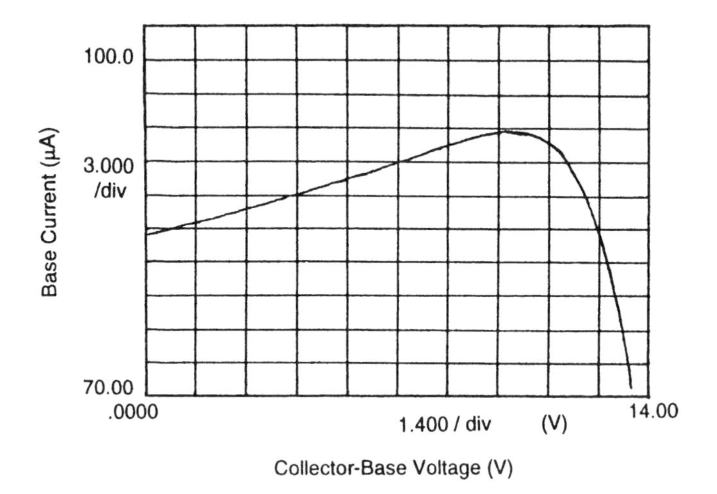

FIGURE 5.18 Base current versus collector–base voltage (after Yuan, Ref. 20 © IEEE).

5.3 LARGE- AND SMALL-SIGNAL MODELS FOR RF APPLICATIONS

The physics-based models presented in Sections 5.2.1 to 5.2.5 are comprehensive but not good for circuit simulation in RF applications. The Gummel–Poon large-signal model is generally used in the bipolar circuit simulation, as shown in Fig. 5.19. In this figure R_B is the base resistance, R_E the emitter resistance, R_C the collector resistance, C_{JE} the emitter–base junction capacitance, C_{JC} the base–collector junction capacitance, C_{DE} the emitter–base diffusion capacitance, and C_{DC} the base–collector diffusion capacitance. For RF applications, packaging capacitances and inductances could be added to the Gummel–Poon model.

The emitter–base and collector–base heterojunction junction capacitances can be modeled empirically as

$$C_{JE,C} = \begin{cases} \dfrac{C_{JE,C}(0)}{\left(1 - V_{BE,C}/V_{biE,C}\right)^{m_{E,C}}} & \text{for } V_{BE,C} \leq FC \times V_{biE,C} \\ \dfrac{C_J(0)}{F_2}\left(F_3 + \dfrac{mV}{V_{bi}}\right)^{-m} & \text{for } V_{BE,C} \geq FC \times V_{biE,C} \end{cases} \tag{5.98}$$

where

$$F_2 = (1 - FC)^{1+m_{E,C}}$$

$$F_3 = 1 - FC(1 + m_{E,C})$$

and $C_J(0)$ is the junction capacitance at zero junction bias, V_{bi} the junction built-in potential, and m the junction grading coefficient. Subscripts E and C

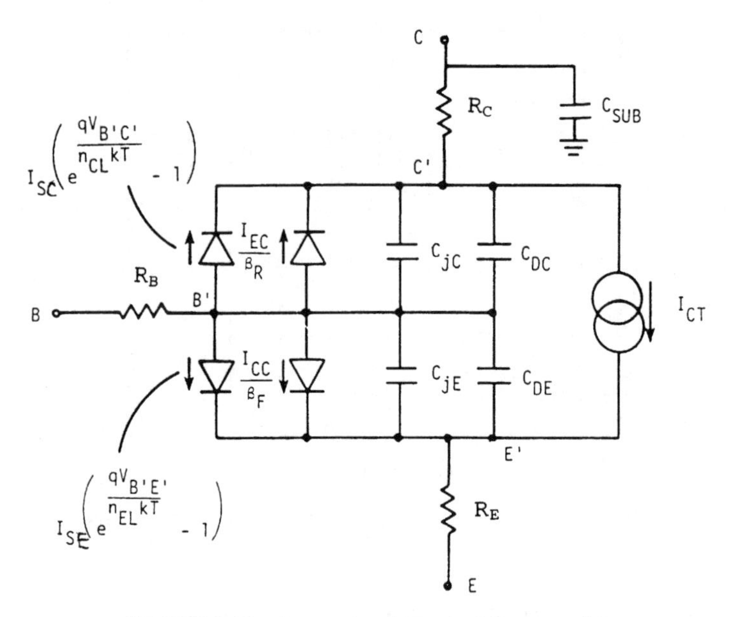

FIGURE 5.19 Large-signal Gummel–Poon model.

represent the emitter–base and collector–base junctions, respectively; $m = \frac{1}{2}$ for the heterojunction with uniform dopings on both sides of the junction.

The diffusion capacitance results from the change of minority-carrier charge with respect to applied bias. For heterojunction bipolar transistors, the diffusion capacitance is usually smaller than the junction capacitance because the base is heavily doped. The diffusion capacitance becomes important when the junction voltage is relatively high. The emitter–base and collector–base diffusion capacitances are derived as

$$C_{DE,C} \equiv \frac{dQ_{BE}}{dV_{BE,C}} = \frac{\tau_{F,R} I_C}{kT/q} \tag{5.99}$$

where τ_F is the forward transit time and τ_R is the reverse transit time.

The small-signal equivalent circuit of heterojunction bipolar transistors for RF applications is shown in Fig. 5.20. The input resistance r_π is given by

$$r_\pi \equiv \frac{\partial V_{BE}}{\partial I_B} = \frac{I_S}{\beta_F} \frac{q}{n_f kT} e^{qV_{BE}/n_f kT} + \frac{q I_{SE}}{n_E kT} e^{qV_{BE}/n_f kT} \tag{5.100}$$

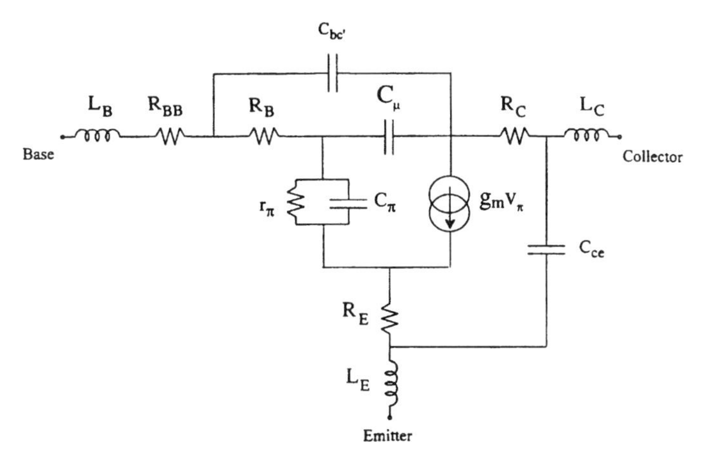

FIGURE 5.20 Small-signal equivalent-circuit model.

The transconductance g_m is given by

$$g_m = g_{m0} \frac{\sin(\omega\tau_C)}{1 + j\omega\tau_B} e^{-j\omega\tau_C} \tag{5.101}$$

where g_{m0} is derived as

$$\begin{aligned} g_{m0} &\equiv \frac{\partial I_C}{\partial V_{BE}} \\ &= \frac{(qI_S/n_f kT)e^{qV_{BE}/n_f kT} - (I_S/q_b)\left(e^{qV_{BE}/n_f kT} - e^{qV_{BC}/n_r kT}\right) dq_b/dV_{BE}}{q_b} - g_o \end{aligned} \tag{5.102}$$

In (5.97) the derivative of q_b respect to V_{BE} is

$$\frac{dq_b}{dV_{BE}} = \frac{(1 + q_1)/V_B + (qI_s e^{qV_{BE}/n_f kT}/n_f kT I_{KF})}{\sqrt{q_1^2 + 4q_2}} \tag{5.103}$$

and the output conductance $g_o (= 1/r_o)$ is given by

$$\begin{aligned} g_o &= \frac{\partial I_C}{dV_{CE}} \approx \frac{dI_C}{dV_{CB}} \\ &= \frac{(qI_S/n_r kT)e^{qV_{BC}/n_r kT} - (I_S/q_b)\left(e^{qV_{BE}/n_f kT} - e^{qV_{BC}/n_r kT}\right) dq_b/dV_{BC}}{q_b} \end{aligned} \tag{5.104}$$

where

$$\frac{dq_b}{dV_{BC}} = \frac{(1+q_1)/V_A + qI_s e^{qV_{BC}/n_r kT}/n_r kT I_{KR}}{\sqrt{q_1^2 + 4q_2}} \tag{5.105}$$

The input capacitance C_π is equal to $C_{JE} + C_{DE}$ and the input–output coupling capacitance C_μ is $C_{JC} + C_{DC}$. C_{p1} and C_{p2} are parasitic capacitances. L_E is the emitter lead inductance, L_B the base lead inductance, and L_C the collector lead inductance.

Modeling thermal effects is particularly important for III/V compound HBTs. In addition to power devices, thermal effects have a significant impact on the performance of HBT comparators. While InP has a 50% better thermal conductivity than GaAs, the high power density as well as narrow-bandgap collector necessitate thermal modeling. A coupled electrical and thermal model of a device can be made by adding to an electrical model a current source I_{TH}, of magnitude equal to the power dissipated in a device, connected to the series combination of a local thermal resistance R_{TH} and a local thermal capacitance C_{TH} [34–36]. In the equivalent network of the electrical/thermal model for an HBT, the junction temperature obtained from the thermal circuit will update the model parameters of the electrical circuit.

A recently published large-signal HBT model [36], including the self-heating effect, is shown in Fig. 5.21. The parasitic elements, such as series resistances R_E, R_B, and R_C and lead inductances L_E, L_B, and L_C, are considered bias independent. Five diodes are used in the model to account for the currents flowing through the two p-n junctions and those caused by parasitic effects. D_E, D_{CI}, and D_{CX} represent the emitter junction, and intrinsic and extrinsic collector junctions, respectively. D_L and D_R represent the base–emitter leakage and the recombination at the device surface and in the emitter-space-charge region. The current-controlled current source I_{CE} is calculated from

$$I_{CE} = \beta_F I_{EE} - \beta_R I_{CC} \tag{5.106}$$

which is similar to that in the Gummel–Poon model. In the circuit model, $V_{J,T}$ can be effectively modeled by using a temperature-controlled voltage source as

$$V_{J,T} = \left.\frac{\partial V_{BE}}{\partial T}\right|_{I=\text{const}} (T - T_0) \tag{5.107}$$

The voltage source V_B is given by

$$V_B(V_{BE}) = V_{BE} n_0 \left[\frac{1}{n_0} - \frac{1}{n(V_{BE})}\right] \tag{5.108}$$

where n_0 is the ideality factor without the conduction-band barrier effect in the collector current and $n(V_{BE})$ is the bias-dependent ideality factor due to the conduction-band barrier effect. $n(V_{BE})$ is obtained from the measured I_C–V_{BE} curve.

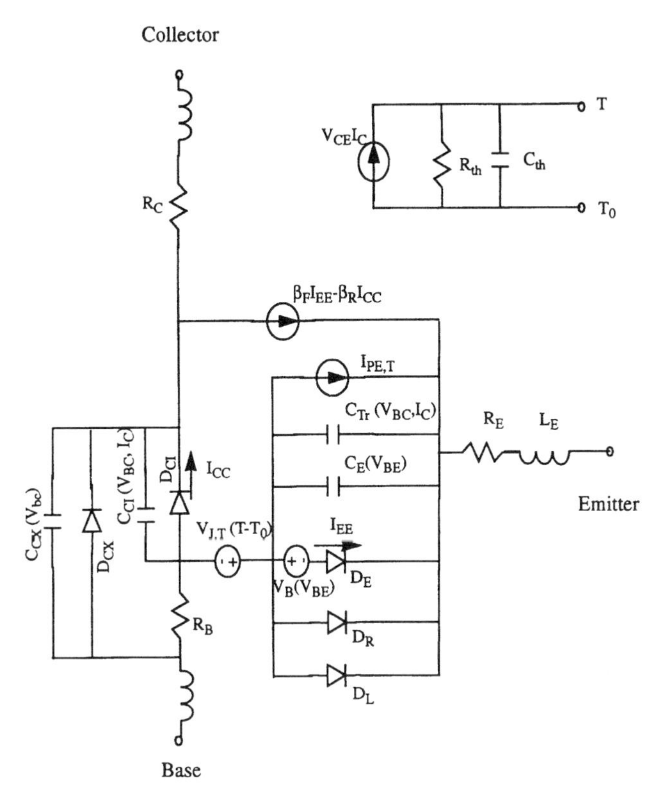

FIGURE 5.21 New large-signal model for RF applications (after Zhang et al., Ref. 36 © IEEE).

The hole current injected into the emitter is modeled as

$$I_{PE,T} = I_{PE}(T) - I_{PE}(T_0) = I_{PE}(T_0) \sum_{n=1} \frac{1}{n!} x^n \tag{5.109}$$

with

$$x = \frac{qV_{J,T}}{nkT_0} \frac{\Delta E_V}{kT_0} \frac{T - T_0}{T}$$

This expression uses a polynomial function of temperature rise to simulate the increase in the backward hole injection current.

Capacitors used to model the charge storage and transit time effects include C_E, C_{CI}, and C_{CX} at the base–emitter, the intrinsic and extrinsic base–collector junctions, and C_{Tr} parallel to C_E. C_{CX} and C_E are treated as classical depletion capacitances.

C_{CI} is modeled as (5.92) to account for the current-dependent capacitance effect:

$$C = C(V)\left(1 - \eta_1 I\right)^{\eta_2} \tag{5.110}$$

where $C(V)$ is the depletion capacitance at the zero current, and η_1 and η_2 are parameters determined by the electron velocity, the junction area, and the junction doping profile.

The transit-time capacitance C_{Tr} is similar to the diffusion capacitance in the Gummel–Poon model. C_{Tr} is modeled as

$$C_{\mathrm{Tr}} = \left[b_0 + b_1\left(1 + \frac{V_{CB}}{V_J}\right)^{b_2}\right] I_C^{b_3} \tag{5.111}$$

where V_J is the built-in potential of the collector junction and b_0 to b_3 are the parameters determined by the device physical structure. In (5.111)

$$\left[b_0 + b_1\left(1 + \frac{V_{CB}}{V_J}\right)^{b_2}\right]$$

is the voltage-dependent term and $I_C^{b_3}$ is the current-dependent one. If τ_{Tr} is independent of current, $b_3 = 1$. In this case, b_0 is the base transit time and

$$b_1\left(1 + \frac{V_{CB}}{V_J}\right)^{b_2}$$

is the collector transit time.

The predicted collector currents as a function of the collector–emitter voltage are compared with the experimental data in Fig. 5.22. The negative differential resistance at high collector currents is correctly accounted for by the model. For RF applications S-parameters from the model predictions are presented. Figure 5.23 shows S_{11}, S_{21}, S_{22}, and S_{12} versus frequency for the HBT biased at $V_{BE} = 1.35$ and $V_{CE} = 3.6$ V. The solid lines represent the model predictions and the symbols represent the experimental data. Good agreement between the model predictions, and measurements in both magnitude and phase have been obtained. Figure 5.24 shows these S-parameters versus frequency at $V_{BE} = 1.65$ and $V_{CE} = 1.65$ V. The modeling results agree very well with the experimental data.

The simulated output power versus input power at 2 GHz is displayed with the measured data in Fig. 5.25. The HBT is biased at $I_B = 0.1$ mA and $V_{CE} = 3.0$ V. The model is very accurate in predicting the first, second, and third powers as a function of input power. The measured and simulated power gain and power-added efficiency versus input power are shown in Fig. 5.26. Again, an excellent agreement between the model predictions and the experimental data is obtained.

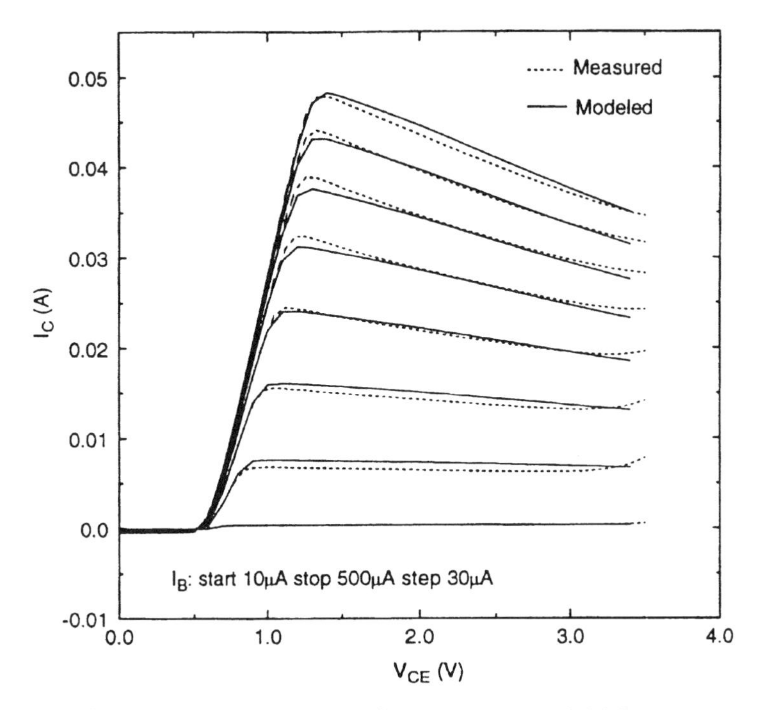

FIGURE 5.22 I_C versus V_{CE} (after Zhang et al., Ref. 36 © IEEE).

5.4 PARAMETER EXTRACTION

Accurate circuit simulation requires accurate model parameters to be used in the simulation. This section presents the Gummel–Poon model parameter extraction for heterojunction bipolar circuit simulation. A setup of the measurement system for parameter extraction is shown in Fig. 5.27.

The saturation current I_S, forward current gain β_F, base–emitter leakage saturation current I_{SE}, the forward knee current I_{KF}, and the ideality factor n_e can be obtained in the Gummel plot as shown in Fig. 5.28. I_S locates the collector current I_C at $V_{BE} = 0$ V. The knee current is the transition point in the I_C curve between low injection and high injection. I_{SE} locates the base current I_B at $V_{BE} = 0$ V. The intercept of I_B at the y-axis I_S/β_F is used to find the forward current gain β_F. The base–emitter leakage current ideality factor is obtained from the base current slope $q/n_e kT$ at low base–emitter voltages. Similarly, the inverse Gummel plot gives the reverse current gain β_R, emitter–base leakage saturation current I_{SC}, the inverse knee current I_{KR}, and the base–collector leakage current ideality factor n_c.

The Early voltage V_A models the effect on the base width modulation in the collector–base junction. It is always a positive number. The simplest way to obtain V_A is from the slope of the I_C versus V_{CE} plot in the linear region. The base–emitter voltage should be constant and the HBT should be biased in the forward-active mode. The

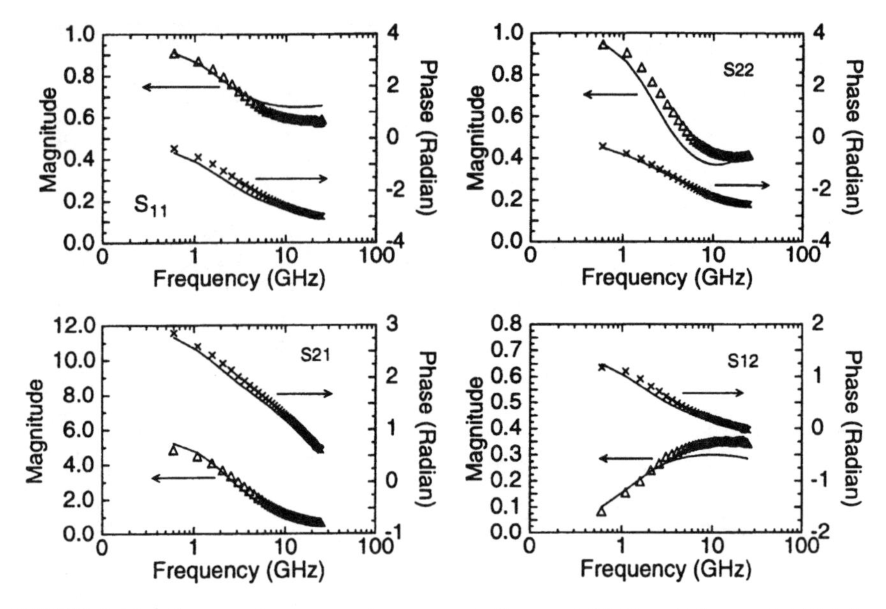

FIGURE 5.23 *S*-parameters versus frequency at $V_{BE} = 1.35$ V and $V_{CE} = 3.6$ V (after Zhang et al., Ref. 36 © IEEE).

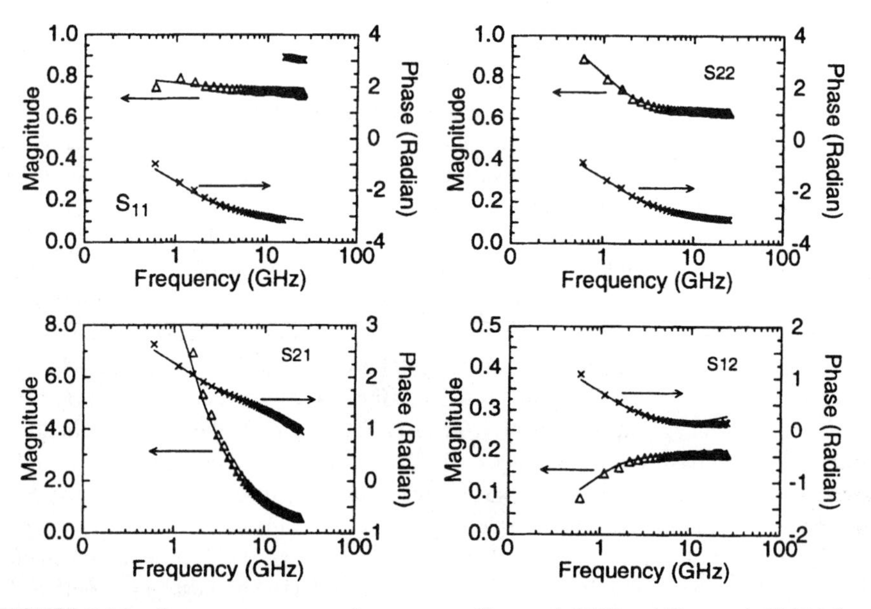

FIGURE 5.24 *S*-parameters versus frequency at $V_{BE} = 1.65$ V and $V_{CE} = 1.65$ V (after Zhang et al., Ref. 36 © IEEE).

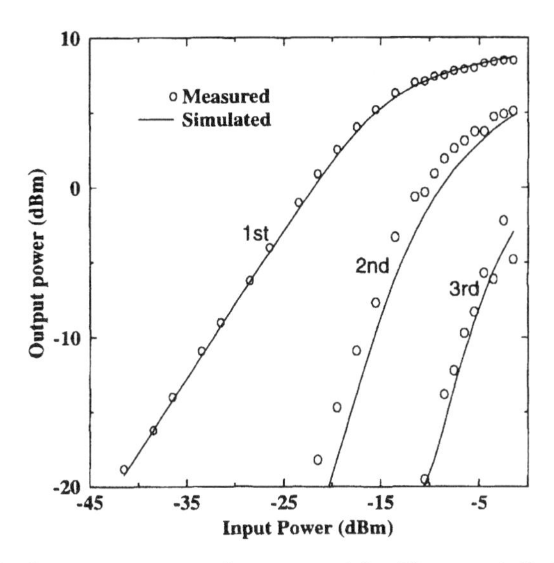

FIGURE 5.25 Output power versus input power (after Zhang et al., Ref. 36 © IEEE).

extrapolation can be performed graphically directly from the curve tracer or parameter analyzer display. However, the experimental error associated with this approach can be very high.

A more accurate technique is the use of the ln I_C versus V_{BE} curve. This technique for finding V_A is the determination of the ln I_C versus V_{BE} characteristics at two different values of V_{BC}. The Early voltage is then determined from the ratio of the

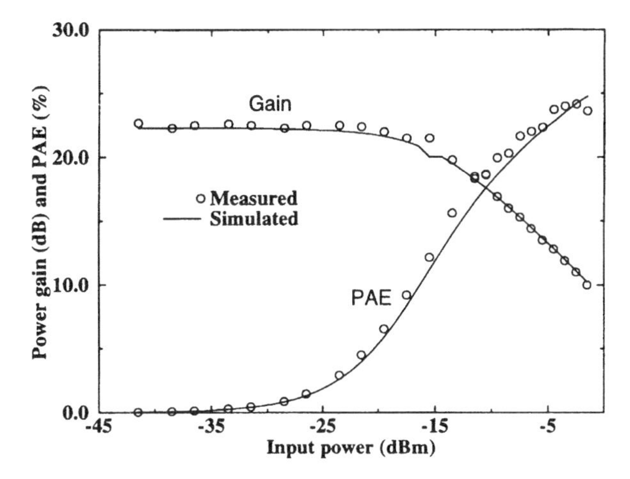

FIGURE 5.26 Power gain and power-added efficiency versus input power (after Zhang et al., Ref. 36 © IEEE).

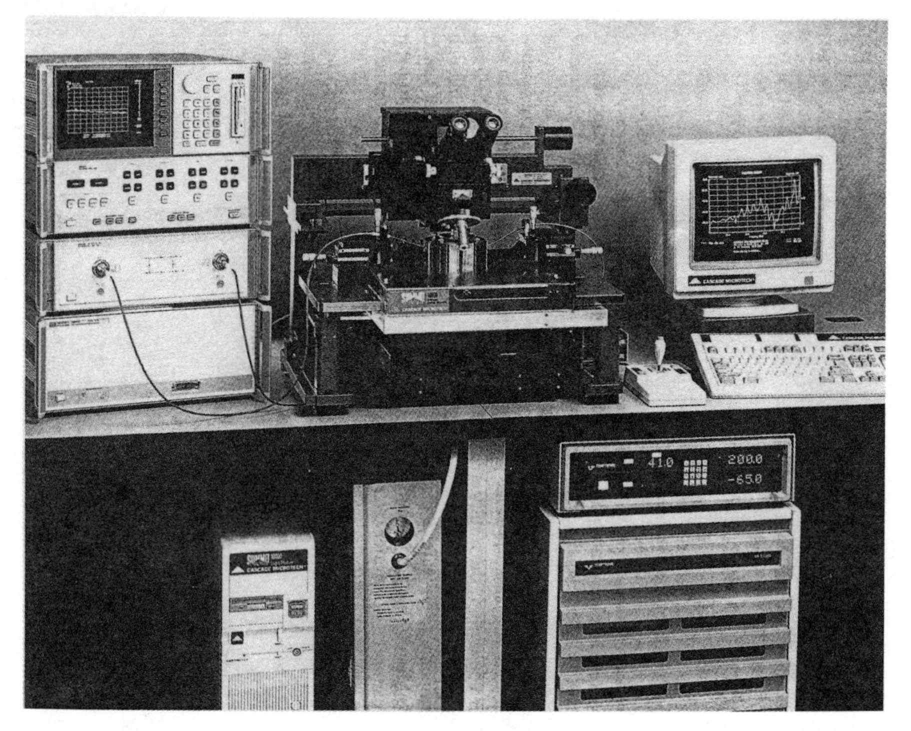

FIGURE 5.27 On-wafer parameter extraction setup (courtesy of Cascade Microtech, Inc.).

(extrapolated) I_S values or the ratio of the values of two I_C values at the same V_{BE}.

$$\frac{I_C(V_{BC1})}{I_C(V_{BC2})} = \frac{1 + V_{BC2}/V_A}{1 + V_{BC1}/V_A} \tag{5.112}$$

The inverse Early voltage V_B models the effect of base-width modulation at the emitter–base junction. V_B is always a positive number. There is a complicating factor in measuring V_B in a process analogous to that used for measuring V_A. The measurement V_A assumed that the variation in the width of the emitter–base depletion region had a negligible effect on the transistor characteristics in the normal active region (i.e., $V_{BE}/V_B \ll 1$). The equivalent assumption that would be necessary if the methods for measuring V_A were to be used directly for measuring V_B is that V_{BC}/V_A is much less than unity. However, this assumption is not always true.

From ln I_C versus V_{BE} in the active region and from ln I_E versus V_{BE} in the inverse region, one obtains

$$\frac{I_C(0)}{I_C(V_{BC1})} = \frac{1 + V_{BE1}/V_B + V_{BC1}/V_A}{1 + V_{BE1}/V_B} \tag{5.113}$$

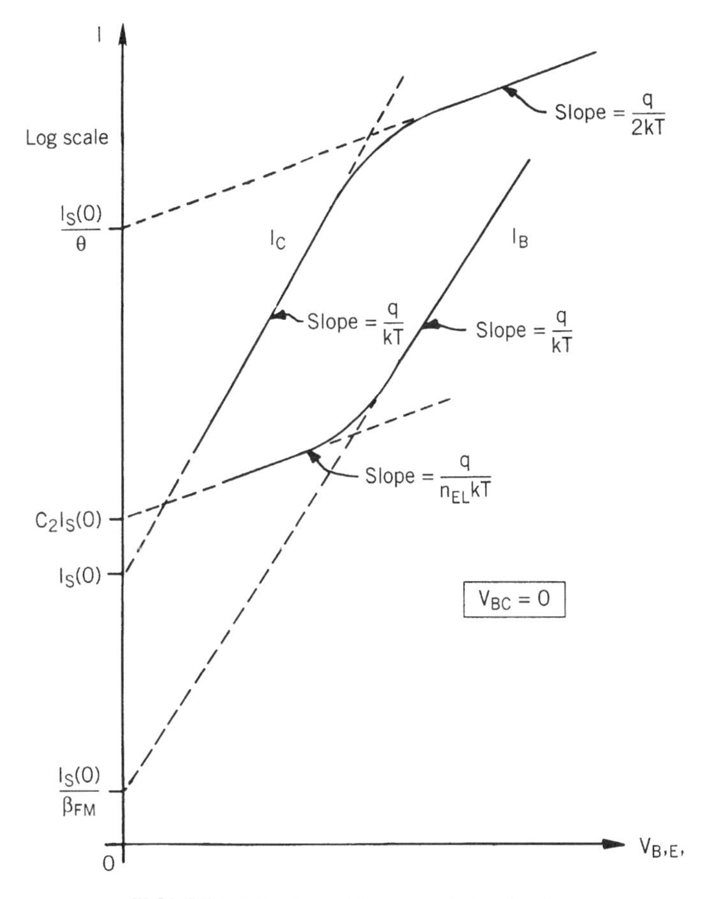

FIGURE 5.28 I_C and I_B versus intrinsic V_{BE}.

$$\frac{I_E(0)}{I_E(V_{BE2})} = \frac{1 + V_{BE2}/V_B + V_{BC2}/V_A}{1 + V_{BE2}/V_B} \tag{5.114}$$

Equations (5.113) and (5.114) can be solved for V_A and V_B.

The base resistance R_B models the resistance between the base region and the base contact. The value obtained from R_B depends strongly on the measurement technique used as well as the transistor's operating condition. The base resistance can be extracted from the ln I_B versus V_{BE} characteristics in Fig. 5.29. Other techniques include noise measurement [37,38], pulse measurement [39], input impedance circle method [40], the phase-cancellation technique [41], the two-port network method [31], and the *h–y* ratio technique [42]. The input impedance circle method involves corrections to account for emitter resistance and parasitic capacitance. It loses accuracy at low collector currents when the diameter of the circle is large. The phase cancellation technique is quick and is relatively unaffected by emitter resistance.

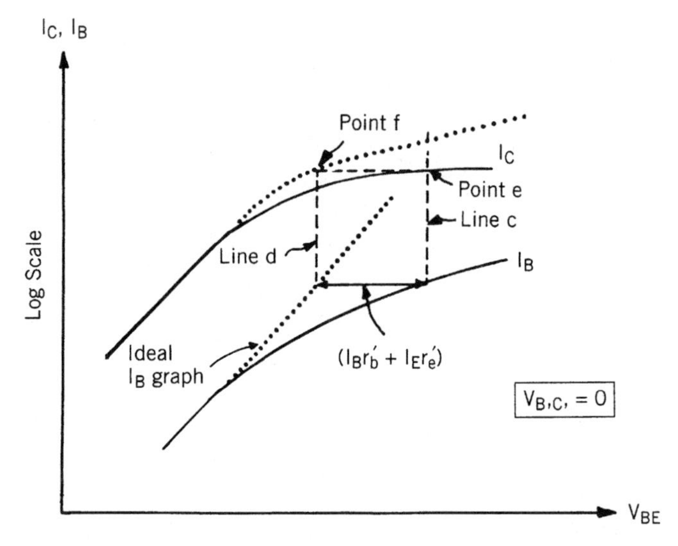

FIGURE 5.29 Gummel plot for base resistance extraction.

However, it can only be used for devices with current gain greater than approximately 10. The two-port network measurement technique is simple, being performed at one frequency. For greatest accuracy, the frequency of the measurement should be between f_β and f_T so that the base resistance is not found from the subtraction of two larger numbers. The *h–y* ratio method is a technique that is quite insensitive to emitter resistance. It requires determination of the full frequency response of the two measurements to verify the assumption of a dominate pole for each case. The noise measurements require the use of very-high-gain amplifiers whose gain is stable with time, as well as extensive shielding to prevent excessive RF interference and 60 Hz pickup. The use of noise measurements presents a number of problems to anyone unfamiliar with noise work and should not be attempted by a beginner. Once a noise measurement system is set up, however, the method can be quite convenient.

For heterojunction bipolar transistors with high base doping, the base resistance is not sensitive to current modulation. When the base doping is relatively low ($< 5 \times 10^{17}$ cm^{-3}), the base resistance becomes current-dependent. The current dependent base resistance can be modeled as [43]

$$R_B = R_{BM} + 3\,(R_B - R_{BM})\,\frac{\tan z - z}{z \tan^2 z} \tag{5.115}$$

where R_{BM} is the minimum base resistance that occurs at high currents and z is a variable given by

$$z = \frac{-1 + \sqrt{1 + 144 I_B/\pi^2 I_{rB}}}{(24/\pi^2)\sqrt{I_B/I_{rB}}} \tag{5.116}$$

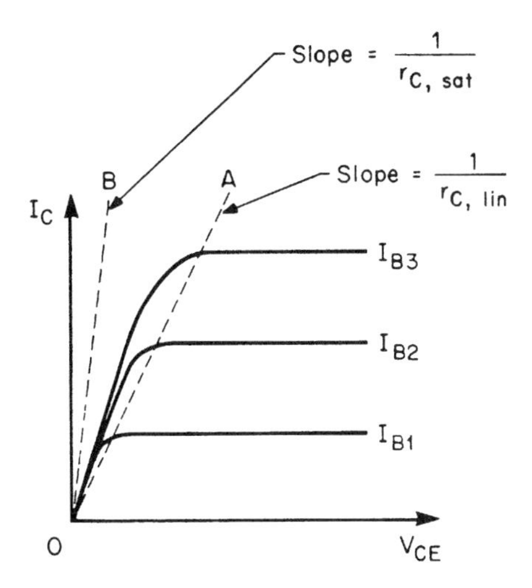

FIGURE 5.30 Collector current versus collector–emitter voltage.

I_{rB} is the base current where the base resistance falls halfway to its minimum value.

The collector resistance R_C models the resistance between the collector region and the collector contact. R_C is calculated from the slope of the I_C versus V_{CE} plot when β_F is forced to 1 in the measurement. Figure 5.30 shows the two limiting values of R_C in the I_C versus V_{CE} characteristics at contact I_B. The collector resistance could vary with the collector current when the bipolar transistor is operated in the high-current quasi-saturation region [44].

The emitter resistance R_E is a constant resistance that models the resistance between the emitter region and the emitter contact. The value of R_E can be obtained by observing the base current I_B as a function of collector–emitter voltage V_{CE} when the collector is open ($I_C = 0$). The open-collector method is shown in Fig. 5.31.

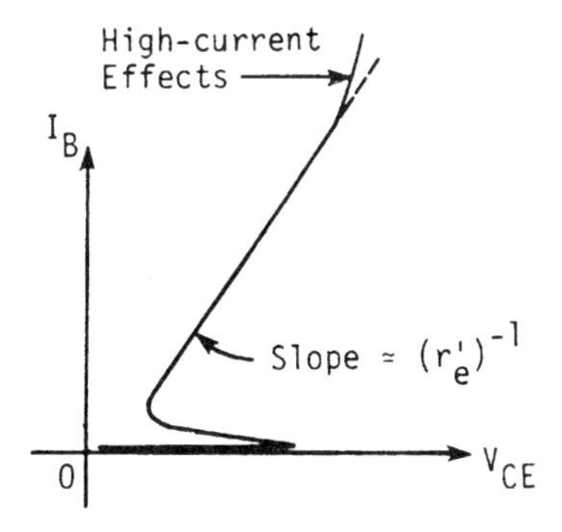

FIGURE 5.31 Open-collector method.

The forward transit time is used for modeling the excess charge stored in the emitter and base when the emitter–base junction is forward biased and $V_{BC} = 0$ V. It is needed to calculate the diffusion capacitance. Generally, τ_F is determined from the f_T versus ln I_C characteristics.

In the region where f_T is constant, τ_F is given by

$$\tau_F = \frac{1}{2\pi f_{T,\max}} - C_{JC} R_C \tag{5.117}$$

where $f_{T,\max}$ is the peak value of f_T and R_C is the collector resistance in the linear region $R_{C,\text{lin}}$. When there is no constant f_T region, τ_F is obtained by plotting $1/f_T$ as a function of $1/I_C$, as shown in Fig. 5.32. The resultant curve can then be extrapolated to obtain τ_F. The intercept (noted by $1/f_A$) of the extrapolated straight line at $1/I_C = 0$ is related to τ_F by

$$\tau_F = \frac{1}{2\pi} \frac{1}{f_A} - C_{JC} R_C \tag{5.118}$$

The reverse forward transit time τ_R is used for modeling the excess charge stored in the base and collector when the base–collector junction is forward biased and $V_{BE} = 0$ V. τ_R can be determined from the f_T versus ln I_E curve using the technique presented in τ_F extrapolation.

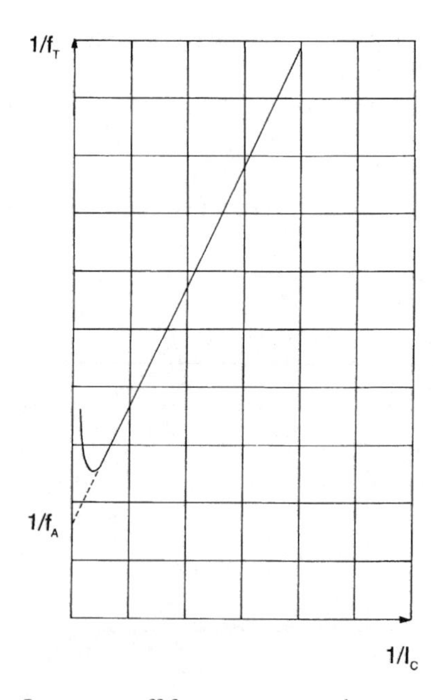

FIGURE 5.32 Inverse cutoff frequency versus inverse collector current.

The junction capacitance at zero bias $C_J(0)$, the junction potential ϕ, and the junction grading coefficient m are the three parameters describing the junction capacitance due to the fixed charge in the junction depletion region. In the junction capacitance model, when the appropriate junction voltage is less than or equal to half of the junction potential, the junction capacitance is modeled by

$$C_{JE,C}(V) = \frac{C_{JE,C}(0)}{(1 - V_{BE,C}/\phi_{E,C})^{m_{E,C}}} \tag{5.119}$$

Subscripts E and C are used to represent the emitter–base and the collector–base junctions, respectively. For example, the emitter–base junction capacitance C_{JE} is a function of the internal base–emitter voltage V_{BE}, and the parameters are $C_{JE}(0)$, ϕ_E, and m_E. Similarly, for the collector–base junction, C_{JC} is expressed in terms of $C_{JC}(0)$, ϕ_C, and m_C.

Both junction capacitances can be obtained as a function of voltage by means of a bridge. The two junction contacts are connected to the bridge and the third contact is left open. The measurement frequency is normally low enough that it can be assumed that the ohmic resistances have a negligible effect.

A complicating factor is the extra capacitance C_k, caused mainly by pin capacitance, stray capacitance, and pad capacitance. C_k is normally assumed to be constant. The capacitance measured by the bridge is

$$C_{\text{meas}} = \frac{C_J(0)}{(1 - V/\phi)^m} + C_k \tag{5.120}$$

C_k can be determined by measurement with a dummy can, by a computer parameter optimization procedure, or by graphical techniques. The graphical technique consists of making an initial guess for ϕ and C_k and then plotting the resultant value of $(C_{\text{meas}} - C_k)$ as a function of $(\phi - V)$ on an ln–ln plot. If a straight line (with a slope between 0.5 and 0.33) results, the chosen values are assumed correct. If the plotted line is not straight, a second guess is made for C_k and/or ϕ and the plot repeated. This process continues until the appropriate straight line is obtained. Since the slope of the straight line is equal to $-m$, the values of ϕ, m, C_k, and $C_J(0)$ can be determined from this plot.

On-wafer dc and RF (S-parameters and noise parameters) are typically performed for HBT modeling and characterization. The measurement system provide a complete wafer mapping of device dc and RF parameters. The on-wafer measurements can be used to screen semiconductor devices effectively and provide correlation of the material and processing parameters to the device performance. Device I–V parameters are obtained using the parameter analyzer. Device C–V parameters are obtained using the impedance analyzer. Small-signal S-parameters are obtained using the network analyzer. Figure 5.33 shows the on-wafer noise and S-parameter measurement setup. The system is comprised of a network analyzer, a radiometer for noise power detection, an electron tuner, wafer probes and an automatic probe station, a control computer, microwave switches, bias supply, and CAD software. Specifically, the

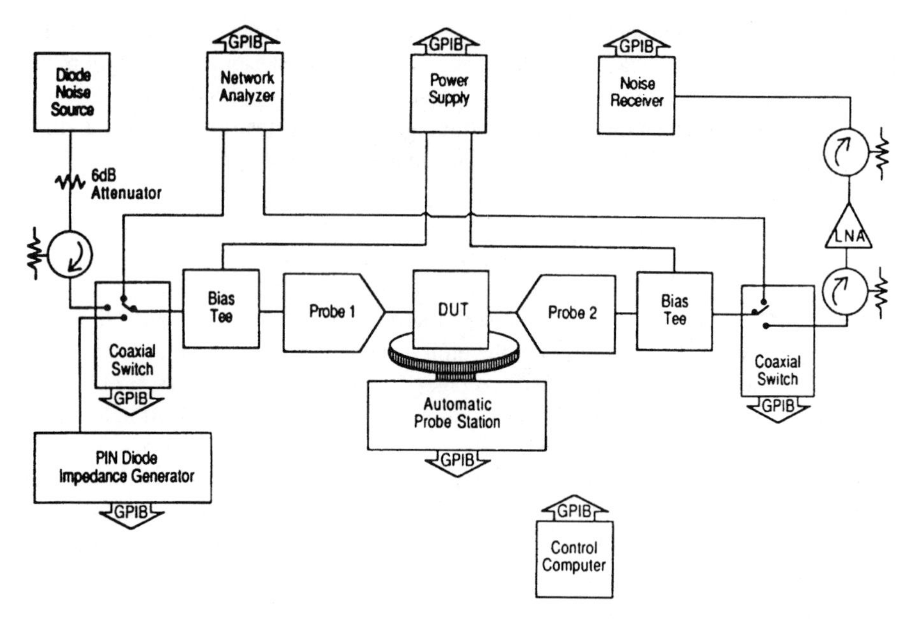

FIGURE 5.33 *S*-parameter and noise parameter measurement setup (after Ali and Gupta, Ref. 45 © Artech House).

system can be used to measure the *S*-parameters, noise parameters, constant noise circles, and gain parameters and gain circles. For RF applications the small-signal model is used to fit the measured *S*-parameters at high frequencies at various bias points. The parasitic element values can thus be extracted based on the small-signal equivalent-circuit fitting results.

REFERENCES

1. H. Kroemer, "Two integral relations pertaining to the electron transport through a bipolar transistor with a nonuniform energy gap in the base region," *Solid-State Electron.*, **28**, 1101 (1985).
2. H. K. Gummel, "A self-consistent iterative scheme for one-dimensional steady-sate transistor calculations," *IEEE Trans. Electron Devices*, **ED-11**, 455 (1964).
3. H.-M. Rein and M. Schröter, "A compact physical large-signal model for high-speed bipolar transistors at high current densities: I. One dimensional model," *IEEE Trans. Electron Devices*, **ED-34**, 1741 (1987).
4. M. Schröter, M. Friedrich, and H.-M. Rein, "A generalized integral charge-control relation and its application to compact models for silicon-based HBTs," *IEEE Trans. Electron Devices*, **ED-40**, 2036 (1993).

5. C. A. King, J. L. Hoyt, and J. F. Gibbons, "Bandgap and transport properties of $Si_{1-x}Ge_x$ by analysis of nearly ideal $Si/Si_{1-x}Ge_x/Si$ HBTs," *IEEE Trans. Electron Devices*, **ED-36**, 2093 (1989).

6. A. G. Chynoweth, W. L. Feldmann, and R. A. Logan, "Excess tunnel current in silicon Esaki junction," *Phys. Rev.*, **121**, 684 (1961).

7. D. Nuernbergk, H. Förster, F. Schwierz, J. S. Yuan, and G. Paasch, "On the temperature behavior of Si/SiGe/Si-HBT: comparison between measurements and numerical simulation," *2nd IEEE International Caracas Conference on Devices, Circuits, and Systems*, Margarita Island, Venezuela, March 1–3, 1998.

8. J. Song and J. Yuan, "Collector–base heterojunction barrier effect on the SiGe heterojunction bipolar transistor," *IEEE Southeastcon'98*, Apr. (1998).

9. A. J. Joseph, J. D. Cressler, D. M. Richey, and D. L. Harame, "Impact of profile scaling on high-injection barrier effects in advanced UHV/CVD SiGe HBTs," *IDEM Tech. Dig.*, 253 (1996).

10. A. Marty, G. Rey, and J. P. Bailbe, "Electrical behavior of an NpN GaAlAs/GaAs heterojunction bipolar transistor," *Solid-State Electron.*, **22**, 549 (1979).

11. A. A. Grinberg, M. S. Shur, R. J Fischer, and H. Morkoc, "An investigation of the effect of graded layers and tunneling on the performance of AlGaAs/GaAs heterojunction bipolar transistors," *IEEE Trans. Electron Devices*, **ED-31**, 1758 (1984).

12. M. S. Lundstrom, "Boundary conditions for pn heterojunctions," *Solid-State Electron.*, **27**, 491 (1984).

13. M. S. Lundstrom, "An Ebers–Moll model for the heterostructure bipolar transistor," *Solid-State Electron.*, **29**, 1173 (1986).

14. B. R. Ryum and I. M. Abdel-Motaleb, "A Gummel–Poon model for abrupt and graded heterojunction bipolar transistors (HBTs)," *Solid-State Electron.*, **33**, 896 (1990).

15. A. A. Grinberg and S. Luryi, "On the thermionic-diffusion theory of minority transport in heterostructure bipolar transistors," *IEEE Trans. Electron Devices*, **ED-40**, 859 (1993).

16. C. D. Parikh and F. A. Lindholm, "A new charge-control model for single- and double-heterojunction bipolar transistors," *IEEE Trans. Electron Devices*, **ED-9**, 1303 (1992).

17. J. J. Liou, C. S. Ho, L. L. Liou, and C. I. Huang, "An analytical model for current transport in AlGaAs/GaAs abrupt HBTs with a setback layer," *Solid-State Electron.*, **36**, 819 (1993).

18. J. S. Yuan and J. Ning, "Analysis of abrupt and linearly graded heterojunction bipolar transistors with or without a setback layer," *IEE Proc G*, **142**, 254 (1995).

19. M. E. Hafizi, C. R. Crowell, and M. E. Grupen, "The dc characteristics of GaAs/AlGaAs heterojunction bipolar transistors with application to device modeling," *IEEE Trans. Electron Devices*, **ED-37**, 2121 (1990).

20. J. S. Yuan, "Thermal and reverse base current effects on heterojunction bipolar transistors and circuits," *IEEE Trans. Electron Devices*, **ED-42**, 789 (1995).

21. R. Stratton, "Theory of field emission from semiconductors," *Phys. Rev.*, **125**, 67 (1969).

22. M. Shur, *GaAs Devices and Circuits*, Plenum, New York (1987).

23. C. D. Parikh and F. A. Lindholm, "Space-charge region recombination in heterojunction bipolar transistors," *IEEE Trans. Electron Devices*, **ED-39**, 2197 (1992).

24. M. S. Lundstrom and R. J. Schuelke, "Modeling semiconductor heterojunctions in equilibrium," *Solid-State Electron.*, **25**, 683 (1982).

25. S. Searles and D. L. Pulfrey, "An analysis of space-charge-region recombination in HBTs," *IEEE Trans. Electron Devices*, **ED-41**, 476 (1994).

26. J. J. Liou and J. S. Yuan, "Surface recombination current of AlGaAs/GaAs heterojunction bipolar transistors," *Solid-State Electron.*, **35**, 805 (1992).

27. S. Tiwari and D. J. Frank, "Analysis and operation of GaAlAs/GaAs HBTs," *IEEE Trans. Electron Devices*, **ED-36**, 2105 (1989).

28. H. H. Wieder, "Surface Fermi level of III-V compound semiconductor–dielectric interfaces," *Surf. Sci.*, **132**, 390 (1983).

29. W. B. Joyce, "Thermal resistance of heat sink with temperature-dependent conductivity," *Solid-State Electron.*, **10**, 161 (1967).

30. D. P. Maycock, "Thermal conductivity of silicon, germanium, III-V compound and II-V alloys," *Solid-State Electron.*, **10**, 161 (1967).

31. I. E. Getreu, *Modeling the Bipolar Transistor*, Tektronix, Inc., Beaverton, OR (1976).

32. D. E. Dawson, A. K. Gupta, and M. L. Salib, "CW measurement of HBT thermal resistance," *IEEE Trans. Electron Devices*, **ED-39**, 2235 (1992).

33. H. Wang, C. Algani, A. Konczykowska, and W. Zuberek, "Temperature dependence of dc currents in HBT," *Proc. 1992 IEEE MTT Symp.*, 731 (1992).

34. P. C. Grossman and J. Choma, Jr., "Large signal modeling of HBTs including self-heating and transit time effects," *IEEE Trans. Microwave Theory Tech.*, **MTT-40**, 449 (1992).

35. K. Lu, P. A. Perry, and T. J. Brazil, "A new large-signal AlGaAs/GaAs HBT model including self-heating effects, with corresponding parameter-extraction procedure," *IEEE Trans. Microwave Theory Tech.*, **MTT-43**, 1433 (1995).

36. Q. M. Zhang, H. Hu, J. Sitch, R. K. Surridge, and J. M. Xu, "A new large signal HBT model," *IEEE Trans. Microwave Theory Tech.*, **MTT-44**, 2001 (1996).

37. R. C. Jaeger and A. J. Brodersen, "Low-frequency noise sources in bipolar junction transistors," *IEEE Trans. Electron Devices*, **ED-17**, 128 (1970).

38. S. T. Hsu, "Noise in high-gain transistors and its application to the measurement of certain transistor parameters," *IEEE Trans. Electron Devices*, **ED-18**, 425 (1971).

39. P. Spiegel, "Transistor base resistance and its effect on speed switching," *Solid-State Design*, **5**, 15 (1965).

40. J. Lindamayer, "Power gain of transistor at high frequencies," *Solid-State Electron.*, **5**, 171 (1962).

41. W. M. C. Sansen and R. G. Meyer, "Characterization and measurement of the base and emitter resistances of bipolar transistors," *IEEE J. Solid-State Circuits*, **SC-7**, 492 (1972).

42. C. M. Meijer and H. J. A. DeRonde, "Measurement of the base resistance of bipolar transistors," *Electron. Lett.*, **11**, 249 (1975).

43. G. Massobrio and P. Antognetti, *Semiconductor Device Modeling with SPICE*, 2nd ed., McGraw-Hill, New York (1993).

44. J. S. Park, A. Neugroschel, V. de la Torre, and P. J. Zdebel, "Measurement of collector and emitter resistances in bipolar transistors," *IEEE Trans. Electron Devices*, **ED-38**, 365 (1991).

45. F. Ali and A. Gupta, *HEMTs and HBTs: Devices, Fabrication, and Circuits*, Artech House, Norwood, MA (1991).

PROBLEMS

5.1. Derive the collector current of the linearly graded SiGe base.

5.2. Derive the normalized sensitivity of the Early voltage with respect to collector current $(\partial I_C/\partial V_A)(V_A/I_C)$ using the Gummel–Poon equations (5.87) and (5.88). Repeat the normalized sensitivity for $(\partial I_C/\partial I_{KF})(I_{KF}/I_C)$.

5.3. Derive the diffusion capacitance of the HBT in saturation.

5.4. Derive the collector–base heterojunction capacitance, including the free carrier capacitance at high collector current.

5.5. Use the hybrid-π small-signal equivalent circuit to derive the f_T expression.

5.6. Derive the transconductance equation, including the series resistances R_E and R_B.

5.7. Plot J_C versus V_{BE} using the drift-diffusion, thermionic-diffusion, and thermionic-drift-diffusion models.

5.8. Use the analytical collector models in Section 5.2.5 and the f_T equation to plot f_T versus I_C at $V_{CE} = 3$, 5, and 9 V.

5.9. Can the analytical models in Sections 5.2.2 and 5.2.3 be used to describe the ideality factor $n (n > 1)$ of HBTs?

5.10. Plot C_{JE} versus V_{BE} using the depletion approximation and (5.98). Compare the depletion capacitance model, with the junction capacitance model, including the free carrier effect.

5.11. In the parameter extraction, the Early voltage extrapolated from one or two I_C–V_{CE} curves usually contains measurement errors. Explain.

5.12. The forward transit time is a strong function of collector current at high current densities. Explain.

5.13. Explain why the collector resistance decreases with collector current at high collector current densities. Derive a current-dependent R_C model using the excess charge Q_{epi}.

5.14. The abrupt heterojunction bipolar transistor has a uniform base doping of 5×10^{18} cm^{-3} and a base thickness of 0.1μm. Its lightly doped collector has a doping of 2×10^{17} cm^{-3} and collector thickness of 0.3μm. At $V_{BE} = 1.5$ V and $V_{CE} = 2.0$ V, will the HBT enter the quasi-saturation mode of operation?

CHAPTER SIX

Heterojunction Device Simulation

Device simulation provides the ability to perform computer simulation experiments without using semiconductor wafers. The simulation is transparent to semiconductor technologies and independent of expensive fabrication equipment. The use of device simulation also provides physical insight into device operation for model development and reduces the design cycle for new transistor design. Figure 6.1 shows the flowchart of device and circuit design using device simulation. For example, device physics provides analytical equations and physical parameters for device simulation. The device simulator solves the fundamental semiconductor equations numerically. The dc, ac, and transient simulation results are used for device design and device model development. The device model is implemented in the circuit simulator. The circuit simulation results are then used for the design of integrated circuits. A more detailed flowchart is illustrated in Fig. 6.2, which demonstrates the relationship of device, process, and circuit simulation in technology CAD. The semiconductor process time, temperature, dopants, and so on, are specified in a two-dimensional process simulator such as SUPREM IV [1]. The process simulation produces the doping profile in two dimensions. This doping profile is loaded into a two-dimensional device simulator such as MEDICI [2]. The device simulation generates I–V and C–V data of semiconductor devices. Using a parameter extraction program, SPICE [3] circuit parameters are obtained. Based on these circuit parameters, SPICE simulation predicts the circuit performance for IC design.

6.1 BOLTZMANN TRANSPORT EQUATION

The Boltzmann transport equation (BTE) describes the time-dependent position and momentum of carrier transport in crystals. The BTE is a pseudoclassical equation that illustrates the statistical behavior of electrons or holes in solids. The total time rate of change of the distribution function $f(k, r, t)$, which describes the occupancy

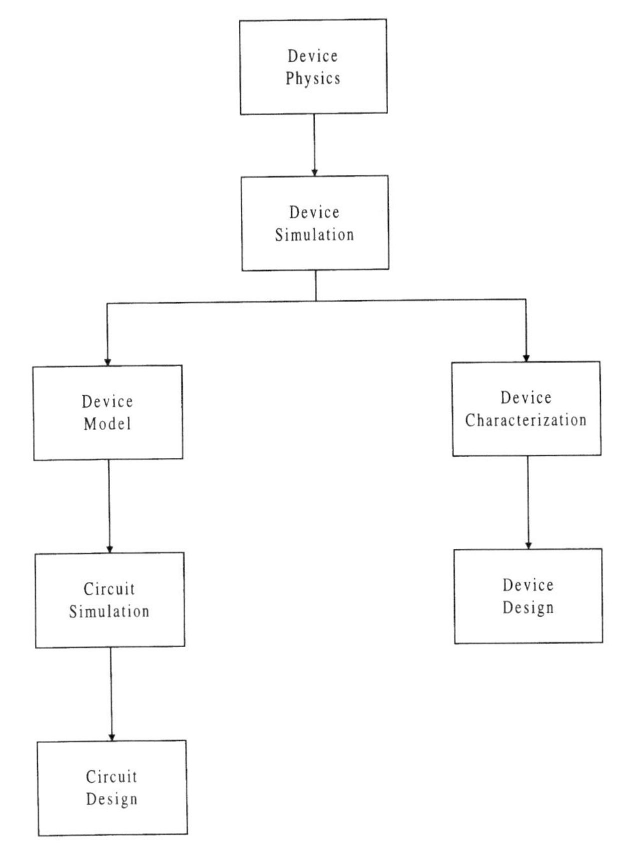

FIGURE 6.1 Role of device simulation in device and circuit design.

of allowed energy states involved in transport processes, is

$$\frac{df}{dt} = \left.\frac{\partial f}{\partial t}\right]_{\text{field}} + \left.\frac{\partial f}{\partial t}\right]_{\text{diffusion}} + \left.\frac{\partial f}{\partial t}\right]_{\text{scattering}} \tag{6.1}$$

For an external force F, one can write

$$\begin{aligned} \left.\frac{\partial f}{\partial t}\right]_{\text{field}} &= -\frac{\partial k}{\partial t} \nabla_k f \\ &= -\frac{1}{\hbar} F \cdot \nabla_k f \end{aligned} \tag{6.2}$$

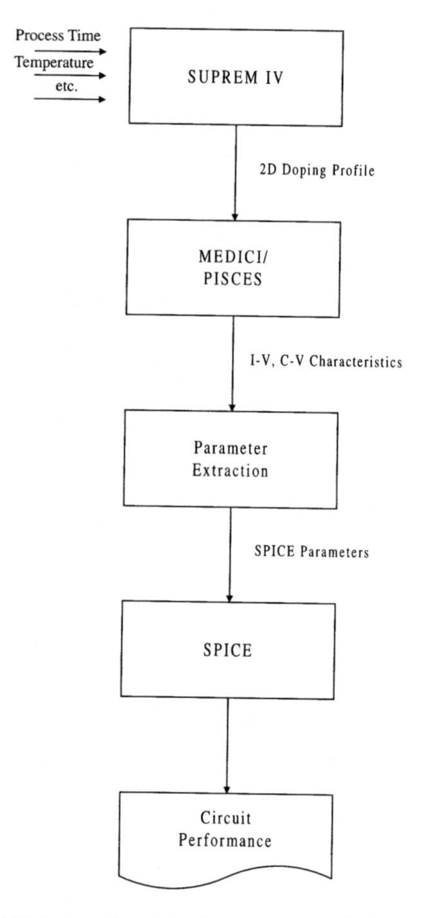

FIGURE 6.2 Relationship of device, process, and circuit simulations.

For the diffusion process, we have

$$\left.\frac{\partial f}{\partial t}\right]_{\text{diffusion}} = -\frac{dr}{dt}\nabla_r f = -v \cdot \nabla_r f \tag{6.3}$$

where v is the velocity. If the collision processes are elastic and the scattering is random, the scattering can be described by using a relaxation time τ:

$$\left.\frac{\partial f}{\partial t}\right]_{\text{scattering}} = \frac{f(k, r) - f_0(E)}{\tau} \tag{6.4}$$

Bearing in mind that the momentum for a periodic structure is quantized according to $p = \hbar k$, we have the following equations for each band β:

$$\frac{dr}{dt} = v_\beta = \frac{\partial H_\beta}{\partial p} = \frac{1}{\hbar}\nabla_k \xi_\beta \tag{6.5}$$

$$\hbar\frac{dk}{dt} = F_\beta = \frac{\partial H_\beta}{\partial r} = \nabla_k E_\beta \tag{6.6}$$

Equations (6.5) and (6.6) specify the motion of an electron between collisions. The result (6.6) also reads

$$F_e = -\nabla E_C - \nabla W_e \qquad \text{(conduction-band)} \tag{6.7}$$

$$F_h = -\nabla E_V + \nabla W_h \qquad \text{(valence band)} \tag{6.8}$$

where W_e is the electron kinetic energy and W_h is the hole kinetic energy. These equations give the total force experienced by an electron in a band. The total force experience by a hole is $F_h = -F_e$.

The derivative of the electron distribution function is

$$\frac{df}{dt} = \frac{\partial f}{\partial t} + \frac{\partial f}{\partial r}\frac{dr}{dt} + \frac{\partial f}{\partial k}\frac{dk}{dt} \tag{6.9}$$

Combining the equations above, we have for electrons in the conduction-band,

$$\frac{\partial f}{\partial t} + v_e \cdot \nabla f - \frac{1}{\hbar}(\nabla E_C + \nabla W_e)\cdot \nabla_k f_e = -\frac{f_e - f_e^0}{\tau_e} \tag{6.10}$$

where τ_e refers to scattering in the conduction-band. For hole distribution f_h of the valence band, one finds that

$$\frac{\partial f_h}{\partial t} + v_h \cdot \nabla f_h - \frac{1}{\hbar}(\nabla E_V + \nabla W_h)\cdot \nabla_k f_h = -\frac{f_h - f_h^0}{\tau_h} \tag{6.11}$$

Note that like current density and carrier density, macroscopic variables for electrons and holes, can be calculated once f_e and f_h in (6.10) and (6.11) are solved.

The solution of the stationary transport equation is discussed as follows. Writing $f_e = f_e^0 + f_e^1$, one finds the perturbation solution for $f_e^1 \ll f_e^0$ by retaining only f_e^0 in the streaming terms and f_e^1 in the collision term of (6.10); thus

$$f_e^1 = -\tau_e\left[v \cdot \nabla f_e^0 - \frac{1}{\hbar}(\nabla E_C + \nabla W_e)\cdot \nabla_k f_e^0\right] \tag{6.12}$$

The local equilibrium distribution is given by the Fermi function:

$$f_e^0 = \left\{\exp\left[\frac{E(k,r) - E_{Fn}(r)}{kT(r)}\right] + 1\right\}^{-1} = f_e^0(E, E_{Fn}, T) \tag{6.13}$$

Equation (6.13) gives the fraction of quantum states having total energy that are occupied by electrons (or equivalently, the probability that a state of energy E is occupied by an electron). For all energies higher than about $2.3kT$ above E_{Fn}, (6.13) reduces to

$$f_e^0 \approx \exp\left(\frac{E_{Fn} - E}{kT}\right) \tag{6.14}$$

Thus if the quasi-Fermi level is more than $2.3kT$ below the edge of the conduction-band, the electron gas obeys Boltzmann statistics and the material is nondegenerate.

For holes, $f_h = f_h^0 + f_h^1$. Again in streaming terms, we retain f_h^0 and in the collision term, f_h^1. For f_h^1 one finds a result similar to (6.12). The local equilibrium distribution is given by

$$f_h^0 = \left\{\exp\left[\frac{E_{Fp}(r) - E(k, r)}{kT(r)}\right] + 1\right\}^{-1} = f_h^0(E, E_{Fp}, T) \tag{6.15}$$

Note that $f_h^0 = 1 - f_e^0$. For all energies lower than about $2.3kT$ below E_{Fp}, (6.15) reduces to

$$f_h^0 \approx \exp\left(\frac{E - E_{Fp}}{kT}\right). \tag{6.16}$$

6.2 MONTE CARLO SIMULATION

The Monte Carlo simulation [4] is a probabilistic approach based on sampling of the parameter distributions of the Boltzmann transport problem using a random distribution. Parameters of the problem are given by probability distributions. Whether a carrier undergoes lattice scattering, defect scattering, or other scattering depends on possible occurrences of these events as represented by their probability. Therefore, if a random distribution is used to sample these events in a transport problem and sufficient statistics are collected, one should obtain realistic distributions of parameters of interest. Examples of these parameters may be carrier distribution as a function of energy, momentum, and position in a sample. In general, this function need not be a Maxwell–Boltzmann distribution in energy, and the Monte Carlo approach can determine what it is. As a statistical method, Monte Carlo is natural for obtaining solutions to transient and large-signal operation conditions. As with any technique, inaccurate solutions can be obtained if the problem is described incompletely.

The Monte Carlo technique involves evaluation of the trajectories of a sufficient number of carriers inside the semiconductor to evaluate a statistically meaningful description of the transport. These trajectories occur under the influence of applied and built-in forces (electric fields) in the semiconductor and due to scattering events such as the numerous mechanisms of defect and lattice scattering. These things can all lead to a change in energy and momentum of the carrier. The technique evaluates both the scattering event and the time between scattering events, a period during which the carrier moves ballistically. These events are accounted for stochastically according to the relative probability of their occurrence.

For a steady-state homogenous problem, the behavior of the motion of the particle in these large scattering events would be representative of the behavior of the particle gas. Consider an example of a carrier's motion and scattering as it drifts in an electric field in momentum and space in Fig. 6.3. Free flight is shown as a solid line and scattering as a dashed line. Corresponding to each scattering event, a change in momentum is shown via the dashed line in k-space, and the particle moves again under the influence of the electric field in the z-direction. If this simulation was continued for many more scattering events and hence in time, the final z-position of the particle as a function of time would vary nearly linearly with time for constant effective mass (i.e., the velocity would be a constant).

The equation of motion of this particle follows from our description of particle motion. The momentum of the particle is and it energy is $\hbar^2k^2/2m^*$ for the simple free particle. Under the influence of electric field E, the equation of motion of the particle is

$$\frac{\partial(\hbar k)}{\partial t} = -qE \tag{6.17}$$

If the initial momentum corresponds to a wave vector k_i, the momentum at an instant in time t during free flight corresponds to a wave vector k, where

$$k(t) = k_i - \frac{qE}{\hbar}t \tag{6.18}$$

Since the motion is by drift, momentum changes only in the direction of the electric field (or z-direction).

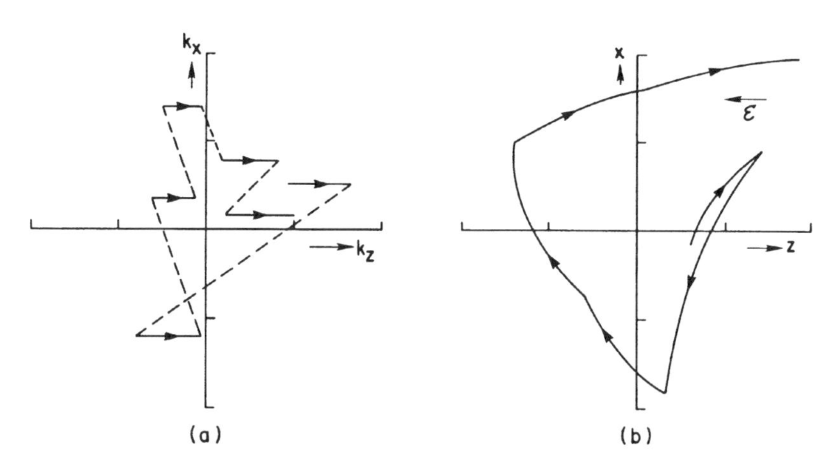

FIGURE 6.3 Carrier scattering in (a) momentum and (b) space.

For a transient problem, where one is interested in the behavior of a particle of gas on a time scale during which too few scattering events occur, as well as for a spatially nonhomogeneous problem such as transport in the base–collector space-charge region of a bipolar transistor, many more particles have to be considered to obtain meaningful results. The test is the standard derivation of the parameter of interest in order to determine its uncertainty. It is usually accomplished by dividing the accumulated information into sufficiently large equal time intervals and determining the parameter of interest (e.g., the velocity). The average of this information is the mean of interest, and its standard derivation is a measure of the uncertainty.

A fully self-consistent Monte Carlo simulation has been used to investigate electron transport through the base–collector region of various Ge_xSi_{1-x} base heterojunction bipolar transistors [5]. For electron transport in Si and SiGe, six anisotropic, nonparabolic valleys in the model of the conduction-band are included. The four valleys that exhibit low effective masses in the growth direction [100] are designated as or transverse valleys. The other two valleys are designated as $\Delta_\perp$ or longitudinal valleys. Both heavy and light holes are included in the simulation, but their respective bands are approximated as isotropic in k-space even under the influence of strain. The carrier scattering mechanisms included in the transport model are acoustic and optical phonons, hole phasmas, positive ionized impurities, and alloy disorder. The scattering by acoustic phonons is assumed elastic and described in the commonly used equipartition approximation. The nonpolar optical phonon scattering is inelastic and its effect on carrier wave vector and energy is included in the simulation. The intervalley phonon scattering of electrons in Si and related materials are treated separately in the simulation. Scattering of electrons by hole plasmas is important in the heavily doped p-type base. The ionized impurity scattering is treated in the Coinwell–Weisskopf approximation [6]. Alloy scattering has been included for electrons by assuming a random alloy described by the virtual crystal approximation and an array of short-range potentials to represent the effects of chemical disorder.

Figure 6.4*a* shows the average drift velocity of electrons as they cross the base–collector region of the $Ge_{0.3}Si_{0.7}$ device for a current density of 1×10^7 A/cm^2. Transient velocity overshoot effect is seen as electrons pass from the base into the collector at $x = 0.025\ \mu$m, where they experience the large electric field (see Fig. 6.4*b*) and are accelerated to a maximum average drift velocity of 1.78×10^7 cm/s. As the electrons transverse the depletion region, their average drift velocity falls as the electrons rapidly come into equilibrium with the field. The rapid increase in average velocity close to the collector contact is due to electrons being absorbed into the subcollector at $x = 0.325\ \mu$m, which is assumed perfectly absorbing. The same SiGe device has been investigated for a current density of 5×10^8 A/cm^2 in order to examine the high current transport effect. Figure 6.5*a* shows the average velocity of the electrons. The redistribution of electric field shown in Fig. 6.5*a* has a marked effect on the average velocity of the electrons. The electrons now reach a velocity of 0.5×10^7 cm/s at $x = 0.025\ \mu$m. Due to the extended field in the collector, though, electrons continue to accelerate and attain a velocity of 1×10^7 cm/s at $x = 0.3\ \mu$m. The increase of average velocity in the collector–base depletion region

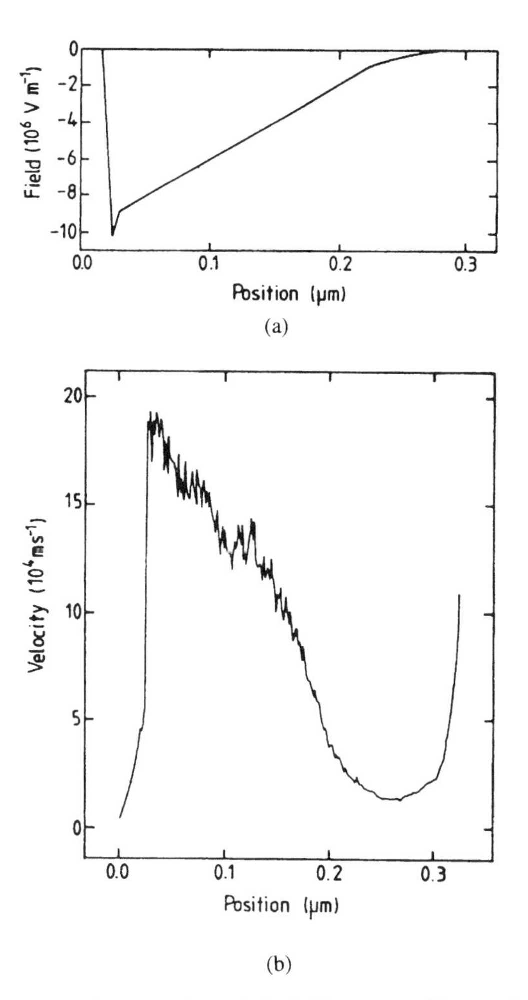

FIGURE 6.4 (*a*) Average drift velocity and (*b*) field versus position at $J_C = 1 \times 10^7$ A/cm^2 (after Hughes et al., Ref. 5 © IEEE).

decreases the collector–base transit time from 9.3 ps at $J_C = 1 \times 10^7$ A/cm^2 to 5.7 ps at $J_C = 5 \times 10^8$ A/cm^2.

A model-based comparison of AlInAs/GaInAs and InP/GaInAs HBTs using one-dimensional self-consistent Monte Carlo simulation was reported [7]. The band structure, electron affinity, effective mass, dielectric constant, and other various physical parameters appearing in the scattering rates were determined by linear interpolation of the data for binary alloy systems such as GaAs, InAs, AlAs, and InP [8–10]. Figure 6.6 shows the J_C–V_{BE} plot for the InP/InGaAs and InAlAs/InGaAs HBTs. Significant difference in turn-on voltage between these two transistors was observed. The electron energy distributions for the two HBTs are shown in Fig. 6.7,

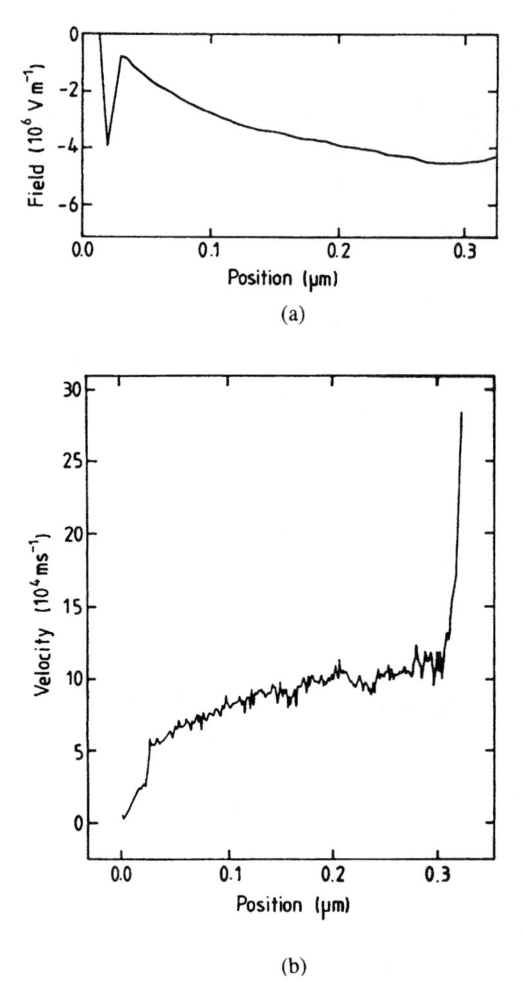

FIGURE 6.5 (*a*) Average drift velocity and (*b*) field versus position at $J_C = 5 \times 10^8$ A/cm^2 (after Hughes et al., Ref. 5 © IEEE).

where the base–emitter voltage was chosen to yield an identical collector current density ($\approx 5 \times 10^4$ A/cm^2). In the emitter region, since the Γ–L band separation energy ($\Delta E_{\Gamma-\mathrm{L}}$) for the InP/InGaAs HBT was by far larger than that for the InAs/InGaAs HBT, there were no electrons in the upper valley of the InP/InGaAs HBT. In the base region, the inclination of the conduction-band for the InAlAs/InGaAs HBT was steeper than for the InP/InGaAs HBT, because of the larger bandgap difference of the AlInAs/GaInAs heterojunction compared with the InP/InGaAs heterojunction. Therefore, the electrons in the InAlAs/InGaAs gained a higher energy within the base region than in the InP/InGaAs HBT. In the collector region, there can be seen a large amount of L-valley electrons for both transistors, thus reducing the average electron velocity,

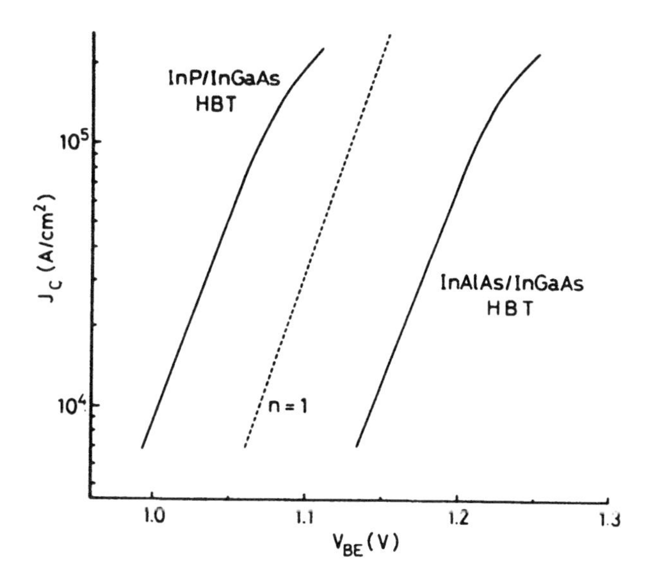

FIGURE 6.6 J_C–V_{BE} for InAlAs/InGaAs and InP/InGaAs HBTs (after Katoh and Kurata, Ref. 7 © IEEE).

even though GaInAs and InP have larger $\Delta E_{\Gamma-\mathrm{L}}$ values than GaAs. In the collector depletion region, a larger number of Γ-valley electrons can be seen in the InAlAs HBT because of a larger $\Delta E_{\Gamma-\mathrm{L}}$. The Γ-valley electrons were transferred from the L-valley via intervalley scattering, thus having a large energy with a random-direction wave number vector. In the collector neutral region, there existed numerous L-valley electrons for the InAlAs HBT, while only a small number of L-valley electrons were found for the InP HBT. This difference in the L-valley population was attributed to the difference in the magnitude of the transfer rate by intervalley scattering from the L-valley to the Γ-valley. Since the intervalley scattering rate is proportional to the power of the electron effective mass in the transferred Γ-valley, the transfer rate in the InP collector is larger than in the GaInAs.

Figure 6.8 shows the average electron velocity of InP and InAlAs HBTs. In the base region, the electron velocity for the InAlAs HBT was about 1.5 times larger that that of the InP HBT, resulting in a smaller base transit time. In the collector region, the peak overshoot velocity for the InP HBT was a little larger because of the smaller velocity in the base region. The overshoot distance at a lower base–emitter bias was about 750 Å for both HBTs. At higher base–emitter voltages, the electron velocity of both HBTs began to increase in a wider range of the collector, thus decreasing the collector transit time. This phenomenon was attributed to the relaxation of the electric field at the onset of the collector high injection. Although the peak overshoot velocity

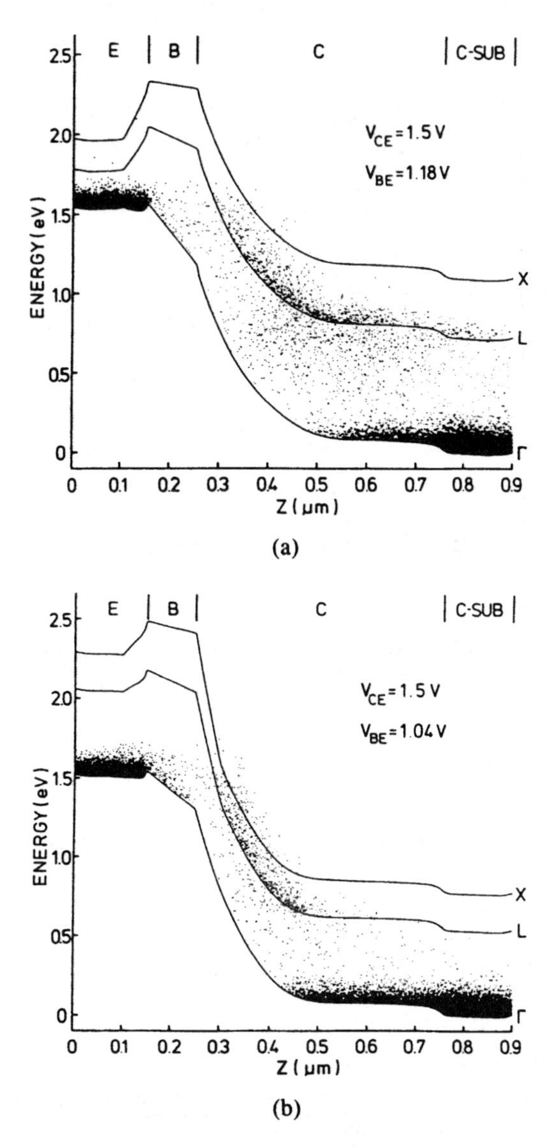

FIGURE 6.7 Electric energy distribution of (*a*) the InAlAs/InGaAs HBT and (*b*) the InP/InGaAs HBT (after Katoh and Kurata, Ref. 7 © IEEE).

decreased markedly as V_{BE} increased, it was insensitive to the magnitude of collector transit time since the overshoot velocity was inherently large.

Nonequilibrium effects become important in heterojunction bipolar transistors because the hot carriers can occur in the collector–base space-charge region, where the electric field is maximum and changes suddenly over scales that are very short. It can also occur in the base if carriers are injected hot using an abrupt heterostructure as

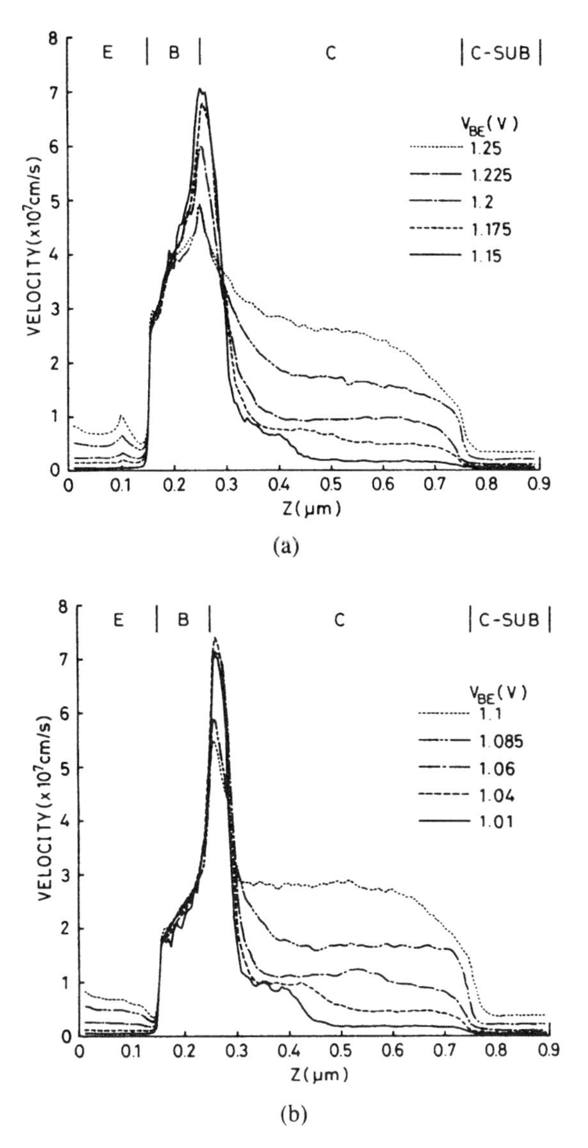

FIGURE 6.8 Average drift velocity versus position for (*a*) the InAlAs/InGaAs HBT and (*b*) the InP/InGaAs HBT (after Katoh and Kurata, Ref. 7 © IEEE).

an emitter or accelerated by quasielectric fields produced by base bandgap grading. Katoh et al. [11] used Monte Carlo self-consistent particle simulation examining effects of different heterojunction structures. Electrons as a function of energy in a $Al_xGa_{1-x}As$/GaAs HBT for a uniformly doped base, an abrupt injection barrier, a grade mole fraction base, and a graded base with higher electron quasielectric field

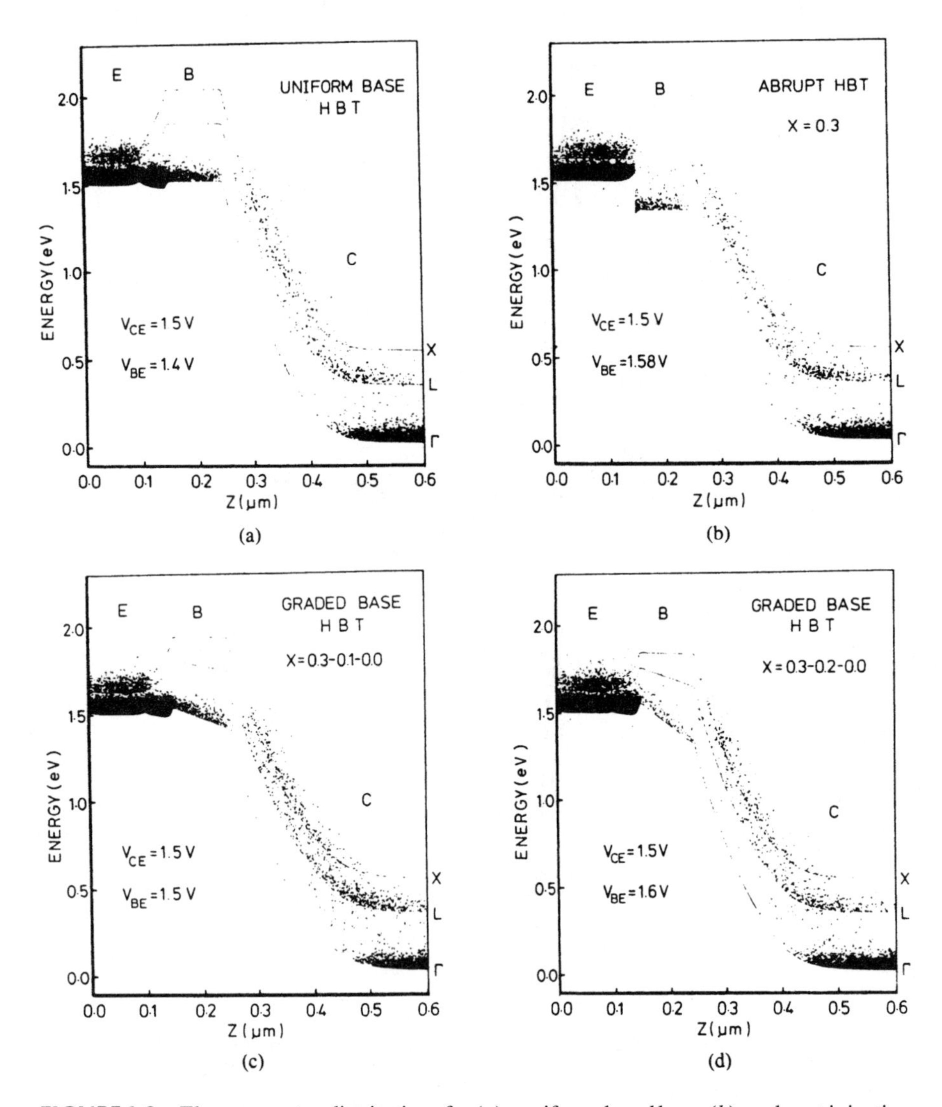

FIGURE 6.9 Electron energy distritutions for (*a*) a uniform doped base, (*b*) an abrupt injection barrier, (*c*) a grade mole fraction base, and (*d*) a graded base AlGaAs/GaAs HBTs (after Katoh et al., Ref. 11 © IEEE).

are shown in Fig. 6.9. A number of interesting features can be seen in these plots. In the $Al_{0.3}Ga_{0.7}As$ emitter, the electrons populate all the Γ, L, and X valleys. While the Γ valley is the lowest, it has a low density of states. Higher L and X valleys have a heavier effective mass and a size of eight equivalent minimum, resulting in a significantly higher density of states. This results in large carrier populations in the

L and X valleys also. In the AlGaAs emitter, the carriers are almost evenly divided between the valleys. In the base, the Γ, L, and X valleys are farther apart, and the barriers to the X and L valleys are large. Figure 6.9*a* shows that electron are higher up in the Γ valley in the emitter at the injecting emitter–base junction. These carriers may enter the base hot if an abrupt heterojunction transition is used, as in part (*b*). Compared to the other cases, there are many more hot electrons at the emitter–base junction with the use of an abrupt barrier. Carriers can pick up energy from the electron quasielectric field in the base, as in parts (*c*) and (*d*), where as they approach the collector end of the base, there is a fair fraction of hot electrons.

The collector depletion region has large electron fields in it resulting from the p-n junction, and at the base end it rises rapidly because of the heavy base doping. This results in rapid carrier heating in the depletion region in all the examples over a short distance, on the order of 500 Å. Carriers heat up in the valley, rapidly increasing in energy and velocity, and when they have sufficient energy they scatter into the X and L valleys, finally resulting in transport at the saturation velocity.

Electron drift velocity as a function of position for the structures of cases (a)–(d) is shown in Fig. 6.10. The velocity in the base varies from the drift-diffusion-like velocity of 2 to 4 $\times$ 10^6 cm/s in the uniformly doped base to higher velocities in the abrupt and graded base cases. The velocity in the collector rises very rapidly due to velocity overshoot and subsequently relaxes to approximately 10^7 cm/s saturation velocity of GaAs. The peak velocities are near the maximum group velocity.

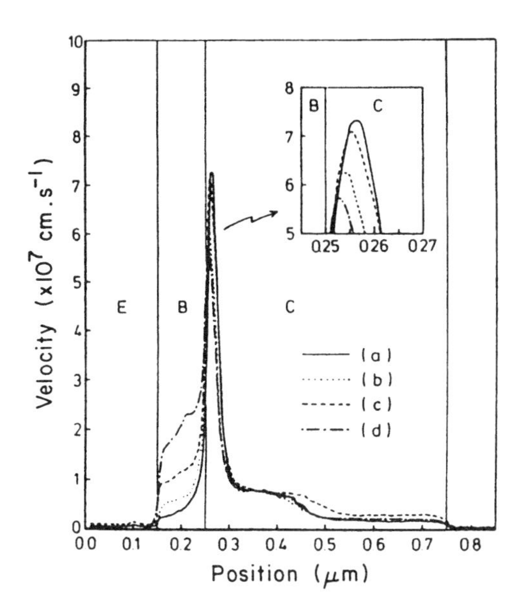

FIGURE 6.10 Electron drift velocity as a function of position (after Katoh et al., Ref. 11 © IEEE).

Electron energy as a function of position for a collector doping of 1×10^{17} cm^{-3} (a), 2×10^{17} cm^{-3} (b), a thickness collector of 5×10^{16} cm^{-3} with a p-extension of the base (c), and a 1×10^{17} cm^{-3} collector with a p-extensions of the base (d) is displayed in Fig. 6.11. For cases (a) and (b) in Fig. 6.11, the depletion region shrinks at the higher doping in the collector, with a corresponding increase in the electric

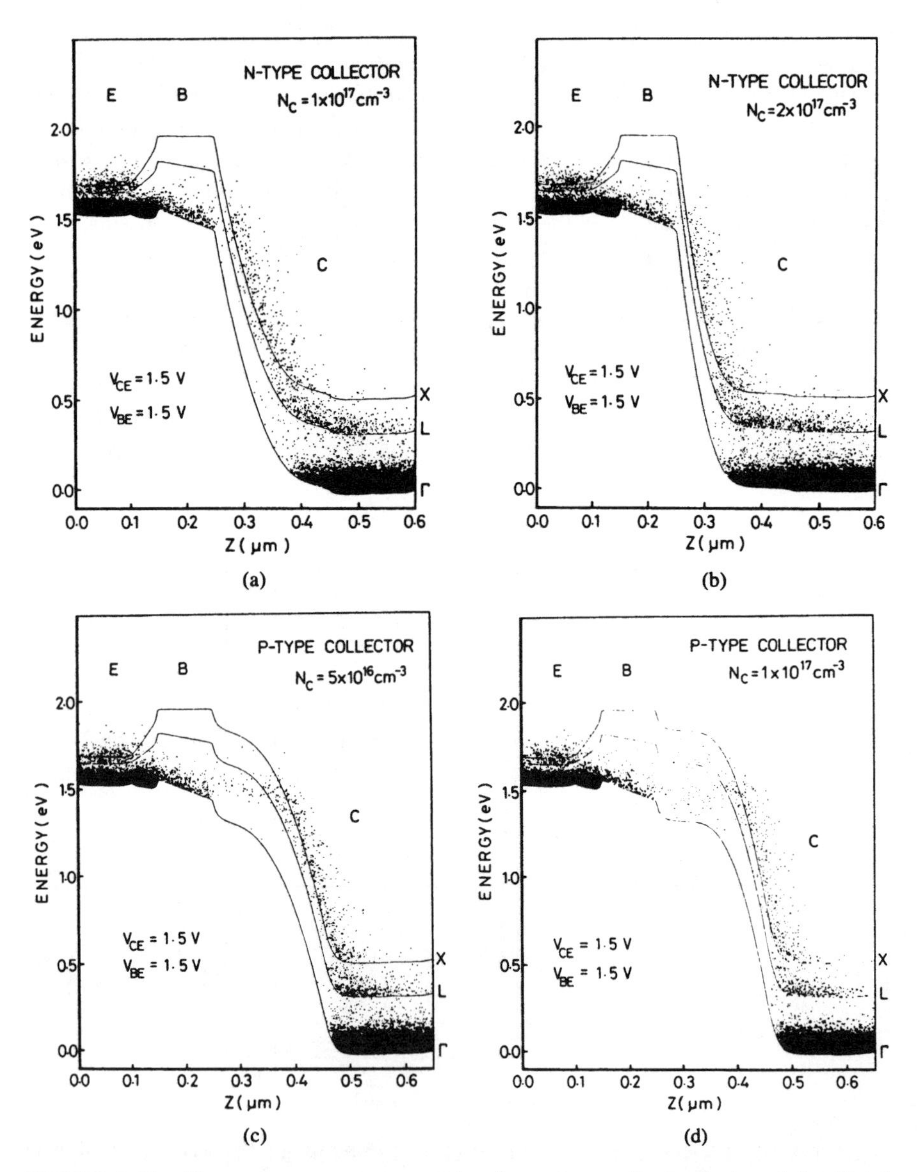

FIGURE 6.11 Electron energy versus position for various collector designs (after Katoh et al., Ref. 11 © IEEE).

field. This leads to a transfer of electrons to the L and X valleys occurring over a shorter distance. Cases (c) and (d) are of more interest because of the decrease in the electric field in the collector depletion region as well as the rapid change in the field by employing a p-type extension of the base region. The decrease in electric field, as a result, allows the carriers to overshoot over a longer distance because they do not pick up enough energy to transfer to the L and X valleys during the approximately 1000 Å of travel. In these structures, during the transit through the collector depletion region, the average velocity can be much larger, and the corresponding transit delay significantly shorter.

Overshoot effects occur because the relaxation rate of momentum is larger than that of energy, and a major cause of these large rates is the large scattering rate resulting from the secondary valley transfer. $Ga_{0.47}In_{0.53}As$, which has secondary valleys placed much higher than the primary valleys, may show stronger nonequilibrium effects than GaAs. The maximum possible velocities are related to this Γ–L separation and the effective mass of electrons. InP, which has a larger bandgap and hence a larger breakdown voltage. InAs is another contrasting choice because it has a very large secondary valley separation, an unusually low scattering rate and effective mass, but a small bandgap and hence a small breakdown voltage.

Consider the switching in heterojunction bipolar transistors. Figure 6.12 shows transient evolution of the potential (a), the kinetic energy (b), the carrier density (c), and the velocity (d) for a switching in V_{BE} from 1.2 V to 1.4 V, with V_{BC} maintained at 0 V of a GaAs HBT [12]. The collector doping is 1×10^{17} cm^{-3} with an expitaxial thickness of 0.175 μm, a compromise between excess storage delay due to the Kirk effect at the 1.4-V input bias and collector signal delay. Indeed, at the high input bias, evidence of the Kirk effect is observable in the conduction-band edge and electron density profiles of Fig. 6.12. The electron transit time, as a function of time, is observable in the collector space-charge region. However, the time taken for this transit (the collector transit time) is larger than the delay time of the collector current in the quasi-neutral collector (the collector signal delay), as explained in Section 4.2.6. Note that in Fig. 6.12 a broader velocity overshoot region develops, due to the presence of a larger electron density in the collector space charge at the higher bias. To generate this, consider the InP steady-state characteristics shown in Fig. 6.13. The peak velocity at the high bias is smaller than at the lower bias. This large bias case is actually quite similar to that of an extended lower doped base. With the switching to a higher bias at the base–emitter junction, the field in the base region decreases because of the higher electron density, and the high-field regions shift toward the collector–collector interface. The charge storage is larger. This occurs despite the collector transit time being smaller in this device. The collector signal delay is still larger. Therefore, both the analysis of the steady-state behavior alone and a strict emphasis on broadness of the overshoot can be deceiving. Similar comments also hold regarding small-signal behavior.

Steady-state behavior, however, does provide an adequate low-order description of the device behavior and is a good tool to compare the behavior of other materials and their behavior versus GaAs. Figure 6.14 shows the steady-state behavior of the conduction-band edge energy, the kinetic energy, and the velocity in GaAs, InP,

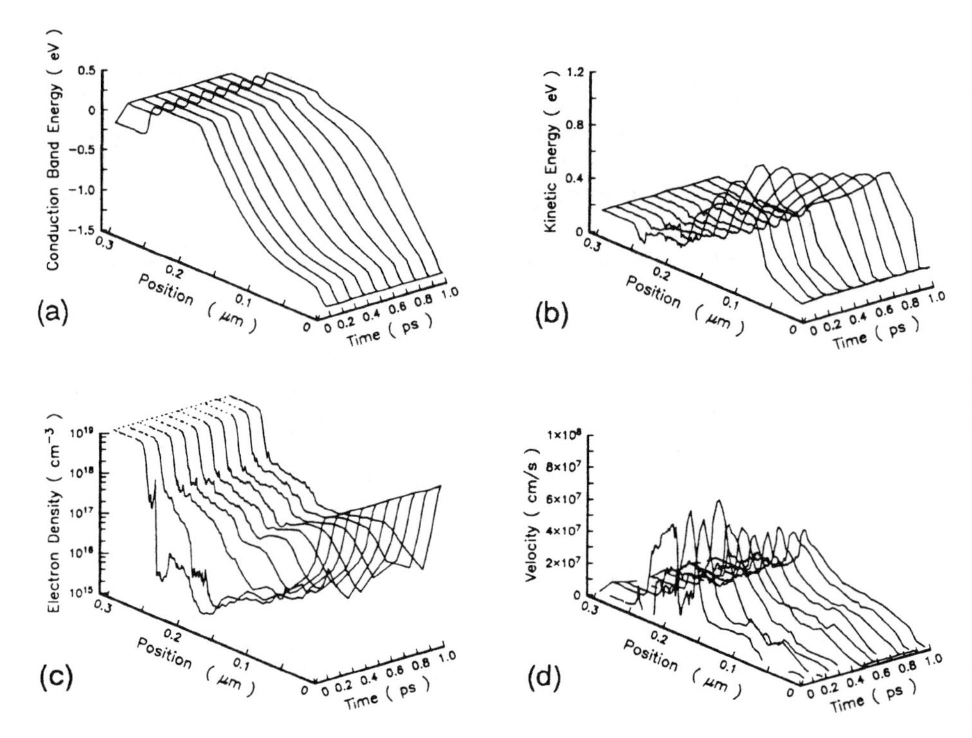

FIGURE 6.12 (*a*) Conduction-band energy, (*b*) kinetic energy, (*c*) electron density, and (*d*) velocity versus position as a function of time (after Tiwari et al., Ref. 12 © IEEE).

$Ga_{0.47}In_{0.53}As$, and InAs under conditions of low Kirk effect. InP is of particular interest because of its larger secondary valley separation. This results in the broader but still peak overshoot behavior. $Ga_{0.47}In_{0.53}As$ HBTs have a secondary valley separation comparable to the bandgap, a smaller fraction of carrier transfer at this bias condition, and a broader overshoot. InAs is an extreme example in this comparison. It has a low scattering rate at low energies, a significantly larger secondary valley separation than the bandgap ($>$ 1 eV separation compared to $\approx$ 0.4 eV bandgap), and hence only a few carriers transfer to the secondary valleys. Consequently, InAs bipolar transistors show the broadest and highest overshoot features of all these devices.

6.3 DRIFT AND DIFFUSION EQUATIONS

Standard drift and diffusion simulation is useful for semiconductor device analysis because of its computation efficiency. The drift–diffusion model assumes isothermal (or constant lattice temperature) and solves the five basic semiconductor equations:

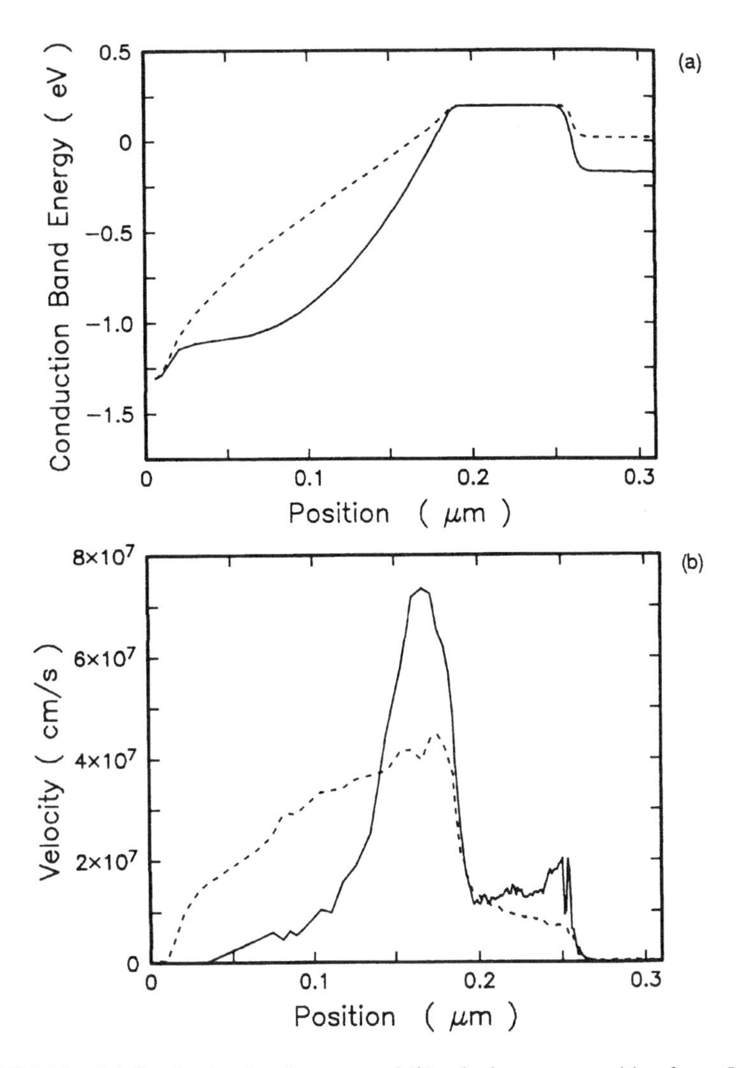

FIGURE 6.13 (*a*) Conduction-band energy and (*b*) velocity versus position for an InP HBT in low and high injection levels (after Tiwari et al., Ref. 12 © IEEE).

namely, Poisson's equation,

$$\varepsilon \nabla^2 \psi = -q(p - n + N_D^+ - N_A^-) - \rho_F \tag{6.19}$$

the continuity equations for electrons and holes,

$$\frac{\partial n}{\partial t} = \frac{1}{q} \nabla J_n - U_n \tag{6.20}$$

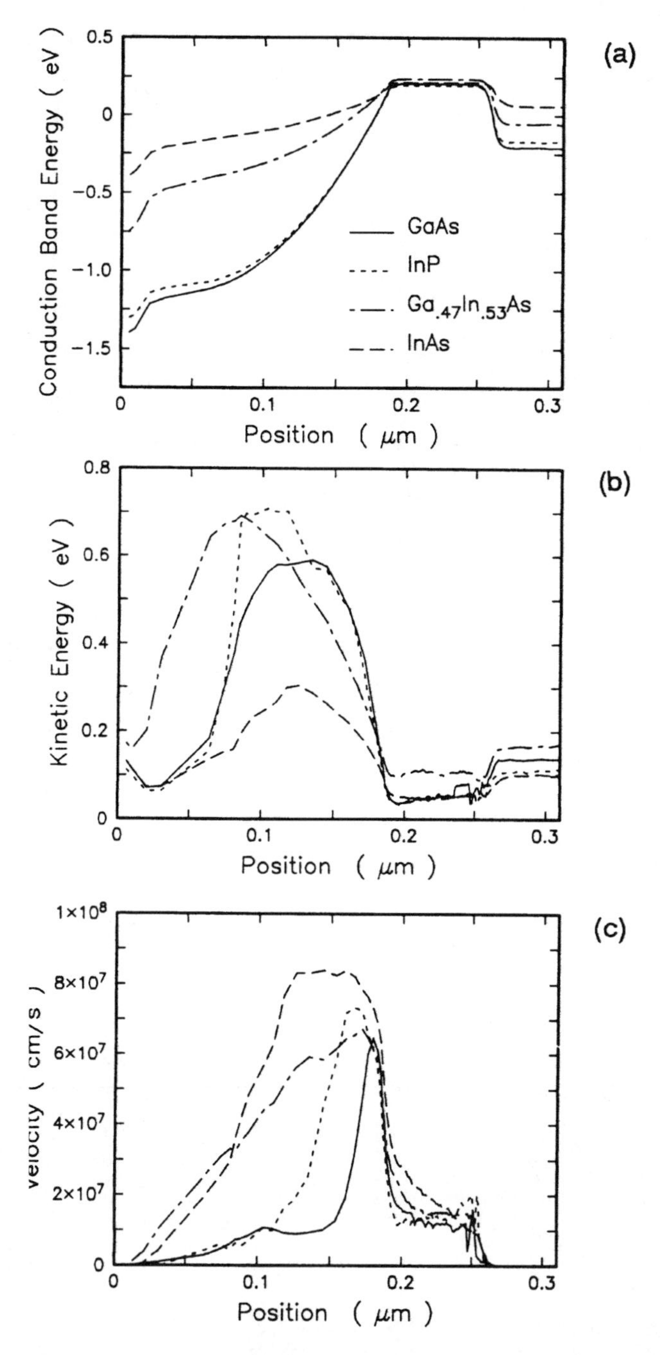

FIGURE 6.14 (*a*) Conduction-band energy, (*b*) kinetic energy, and (*c*) velocity of GaAs, $Ga_{0.47}In_{0.53}As$, and InAs.

$$\frac{\partial p}{\partial t} = -\frac{1}{q}\nabla J_p - U_p \tag{6.21}$$

and the electron and hole current density equations,

$$J_n = q\mu_n nE + qD_n\nabla n \tag{6.22}$$

$$J_p = q\mu_p nE - qD_p\nabla p \tag{6.23}$$

In (6.19)–(6.23) ε is the permittivity, ψ the electrostatic potential, N_D^+ the ionized donor concentration, N_A^- the ionized acceptor concentration, ρ_F the fixed charge density that may be present due to fixed charge in insulating materials or charged interface states, t the time, J_n the electron current density, J_p the hole current density, U_n the electron recombination rate, and U_p the hole recombination rate.

When the semiconductor is lightly doped, the use of Boltzmann statistics is adequate in determining the Fermi level and carrier concentration:

$$n = N_C e^{(E_F - E_C)/kT} \tag{6.24}$$

$$p = N_V e^{(E_V - E_F)/kT} \tag{6.25}$$

where N_C and N_V are the density of states for conduction-band and valence band, E_C and E_V are the energy levels at the bottom of the conduction-band and at the top of the valence band, respectively, and E_F is the Fermi level.

It is, however, necessary to use Fermi–Dirac statistics to calculate the Fermi level when the semiconductor is heavily doped ($> 10^{18}$ cm^{-3}):

$$n = N_c F_{1/2}\frac{E_F - E_C}{kT} \tag{6.26}$$

$$p = N_V F_{1/2}\frac{E_V - E_F}{kT} \tag{6.27}$$

where $F_{1/2}$ is the Fermi–Dirac integral of order $\frac{1}{2}$.

$$F_{1/2}(\eta_{n,p}) = \frac{2}{\sqrt{\pi}}\int_0^\infty \frac{x^{1/2}\,dx}{1 + e^{x-\eta_{n,p}}}$$

and $\eta_n = (E_{Fn} - E_C)/kT$ and $\eta_p = (E_V - E_{Fp})/kT$. Although for most practical cases, full impurity ionization is assumed (i.e., $N_D^+ = N_D$ and $N_A^- = N_A$), the device simulators are able to model the incomplete ionization due to carrier freeze-out effect

$$N_D^+ = \frac{N_D}{1 + g_c e^{(E_{Fn} - E_D)/kT}} \tag{6.28}$$

$$N_A^- = \frac{N_A}{1 + g_d e^{(E_A - E_{Fp})/kT}} \tag{6.29}$$

where g_c is the degeneracy factor for electrons, g_d the degeneracy factor for holes, E_{Fn} the electron quasi-Fermi level, E_{Fp} the hole quasi-Fermi level, E_D the donor energy level, and E_A the acceptor energy level.

In the continuity equations for electrons and holes, Shockley–Read–Hall and Auger recombination models are used:

$$U_{\mathrm{SRH}} = \frac{pn - n_{ie}^2}{\tau_n[n + n_{ie}e^{-(E_T - E_i)/kT}] + \tau_p[p + n_{ie}e^{(E_T - E_i)/kT}]} \tag{6.30}$$

$$U_{\mathrm{Auger}} = c_n(pn^2 - nn_{ie}^2) + c_p(np^2 - pn_{ie}^2) \tag{6.31}$$

where E_i is the intrinsic energy level, E_T the trap density energy level, τ_n the electron lifetime, and τ_p the hole lifetime.

The intrinsic concentration, energy bandgap, effective density of states for the conduction-band, and density of states for the valence band are temperature dependent

$$n_i^2 = N_C(T)N_V(T)e^{-E_G(T)/kT} \tag{6.32}$$

$$N_C(T) = N_C(300)\left(\frac{T}{300}\right)^{3/2} \tag{6.33}$$

$$N_V(T) = N_V(300)\left(\frac{T}{300}\right)^{3/2} \tag{6.34}$$

$$E_G(T) = E_G(300) + \alpha\left(\frac{300^2}{300 + \beta} - \frac{T^2}{T + \beta}\right) \tag{6.35}$$

where $\alpha = 4.73 \times 10^{-4}$ for Si and 5.405×10^{-4} for GaAs and $\beta = 636$ for Si and 204 for GaAs [13]. A list of α and β for various compound semiconductors is given in Table 6.1.

TABLE 6.1 Values of α and β in (6.35) for Various III/V Compound Semiconductors

	$E_G(0)$ (eV)	$\beta(10^{-4}\mathrm{eV/K})$	α (K)
AlAs	2.239	6	408
AlSb	1.687	4.97	213
InP	1.421	3.63	162
InAs	0.420	2.50	75
GaAs	1.519	5.405	204
GaP	2.338	5.771	372
GaSb	0.810	3.78	94
$Ga_{0.47}In_{0.53}As/InP$	0.822	4.5	327

Source: Data from Casey and Panish, Ref. 14. © Academic Press

The electron and hole mobilities as a function of concentration and temperature are expressed as [2]

$$\mu_n = 55.24 + \frac{1429.2T_n^{-2.3} - 55.24}{1 + T_n^{-3.8}(N/1.072 \times 10^{17})^{0.73}} \tag{6.36}$$

$$\mu_p = 49.70 + \frac{479.37T_n^{-2.2} - 49.70}{1 + T_n^{-3.7}(N/1.606 \times 10^{17})^{0.70}} \tag{6.37}$$

for Si and

$$\mu_n = \frac{8500T_n^{-1.0}}{1 + (N/1.69 \times 10^{17})^{0.436}} \tag{6.38}$$

$$\mu_p = \frac{400T_n^{-2.1}}{1 + (N/2.75 \times 10^{17})^{0.395}} \tag{6.39}$$

for GaAs.

The diffusion coefficients are related by the Einstein relation

$$D_n = \frac{kT}{q}\mu_n \tag{6.40}$$

$$D_p = \frac{kT}{q}\mu_p \tag{6.41}$$

for nondegenerate semiconductors and

$$D_n = \frac{kT}{q}\mu_n \frac{F_{1/2}(\eta_n)}{F_{-1/2}(\eta_n)} \tag{6.42}$$

$$D_p = \frac{kT}{q}\mu_p \frac{F_{1/2}(\eta_p)}{F_{-1/2}(\eta_p)} \tag{6.43}$$

for degenerate semiconductors.

When solving for structures that contain heterojunctions, the intrinsic Fermi potential is in general not a solution to Poisson's equation. The vacuum level, however, is a solution to Poisson's equation. The intrinsic Fermi potential and vacuum level are related by

$$\varphi = \psi - \theta \tag{6.44}$$

where ψ is the intrinsic Fermi potential and θ is the band structure parameters given by

$$\theta = \chi + \frac{E_g}{2q} + \frac{2kT}{q} \ln \frac{N_C}{N_V} \tag{6.45}$$

Note that if the band structure parameter is spatially constant, ψ will be a solution to Poisson's equation. However, this is seldom the case in structures containing heterojunctions due to differences in bandgap, electron affinity, and densities of states in adjacent materials. Thus Poisson's equation must be written in the form

$$\nabla \cdot \varepsilon \nabla(\psi - \phi) = -q(p - n + N_D^+ - N_A^-) - \rho_F \tag{6.46}$$

The form of the continuity equations remains unchanged for the heterojunction except that the electric field terms $\mathbf{E}_n$ and $\mathbf{E}_p$ in the transport equations must account for gradients in conduction- and valence-band edges:

$$\mathbf{E}_n = -\nabla\left(\psi + \frac{kT}{q}\ln n_{ie}\right) = -\nabla\left(\frac{E_C}{q} + \frac{kT}{q}\ln N_C\right) \tag{6.47}$$

$$\mathbf{E}_p = -\nabla\left(\psi - \frac{kT}{q}\ln n_{ie}\right) = -\nabla\left(-\frac{E_V}{q} - \frac{kT}{q}\ln N_V\right) \tag{6.48}$$

The material parameters available for describing the properties of the heterojunction are just the usual parameters available for describing the properties of the materials that meet at the heterojunction. Some of these include the energy bandgap parameters, electron affinity, and densities of states, as well as the various parameters for describing recombination, mobility, and so on.

6.4 HYDRODYNAMIC EQUATIONS

The drift–diffusion approximation can lead to unacceptable inaccuracies in the predicted electrical characteristics of modern devices. For example, when the base thickness of the heterojunction bipolar transistor is made very thin, electrons injected into the base encounter only a few collisions before reaching the collector. High-energy electrons transverse across the thin base ballistically. Conventional drift and diffusion mechanisms used to describe carrier transport become invalid. To describe the movement of electrons accurately in the thin-base bipolar transistor, the use of Monte Carlo simulation is essential. Monte Carlo simulations, however, are extremely time consuming and not useful for studying device terminal currents. Hydrodynamic or energy balance models have become popular as a way of accounting for nonlocal transport effects to emulate the Monte Carlo simulation behavior. The solutions of the standard hydrodynamic equation are approximations to the lowest three moments of the Boltzmann transport equations.

Furthermore, in device simulation it is typically assumed that the carrier transport is always isothermal, and the carriers are always in a quasi-static state such that a direct relation between the carrier energy and the local electrical field exists. The first assumption eliminates the carrier flux component driven by the temperature gradient, while the second allows the use of local field-dependent relationships for carrier transport parameters such as the carrier mobility. Since only potential and carrier concentrations are included in the solutions, it is assumed implicitily that dissipated power density inside devices is so low that no significant lattice temperature increase occurs. These assumptions, however, begin to fail for heterojunction semiconductor devices on the GaAs semi-insulating substrate, where self-heating is significant. Moreover, hot carrier nonlocal transport such as velocity overshoot begins to emerge due to the large field gradient and the finite-energy relaxation rate. Consequently, a full thermodynamic (nonisothermal) solution is needed to account for nonuniform lattice temperature for more accurate device performance prediction.

The complete thermodynamic system consists of a set of additional partial differential equations in conjunction with standard drift and diffusion equations. These additional equations are the particle energy balance equations for electrons and holes:

$$\nabla S_n = J_n \nabla \left(\frac{E_C}{q} \right) - W_n \tag{6.49}$$

$$\nabla S_p = J_p \nabla \left(\frac{E_V}{q} \right) - W_p \tag{6.50}$$

and the lattice energy-balance equation

$$\nabla (\kappa \nabla T_L) = -H \tag{6.51}$$

where S_n and S_p are the energy flux densities associated with electrons and holes, W_n and W_p are the energy density loss rates for electrons and holes, and H is the heat source for the lattice.

The hydrodynamic set of equations uses generalized expressions for the electron and hole current densities [2]:

$$\begin{aligned} \mathbf{J}_n &= kT_n \mu_n \left[n\mathbf{E} + \nabla \left(\frac{kT_n n}{q} \right) \right] \\ &= q\mu_n n\mathbf{E} + \mu_n \nabla (kT_n n) + kT_n n \frac{\partial \mu_n(T_n)}{\partial T_n} \nabla \left(\frac{T_n}{q} \right) \end{aligned} \tag{6.52}$$

$$\begin{aligned} \mathbf{J}_p &= kT_p \mu_p \left[n\mathbf{E} + \nabla \left(\frac{kT_p p}{q} \right) \right] \\ &= q\mu_p p\mathbf{E} + \mu_p \nabla (kT_p p) + kT_p p \frac{\partial \mu_p(T_p)}{\partial T_p} \nabla \left(\frac{T_p}{q} \right) \end{aligned} \tag{6.53}$$

Figure 6.15 shows the Gummel plot of the heterojunction simulation using MEDICI. It is obvious from Fig. 6.15 that self-heating increases the collector and base currents at high base–emitter biases. The nonuniform temperature distribution of the HBT biased at $V_{BE} = 1.5$ V is shown in Fig. 6.16. The effect of self-heating affects the Early voltage of the bipolar transistor significantly. Figure 6.17 shows the simulated Early voltage of a bipolar device using drift diffusion and the energy balance equation versus the collector current. The standard drift-diffusion simulation predicts an increase of

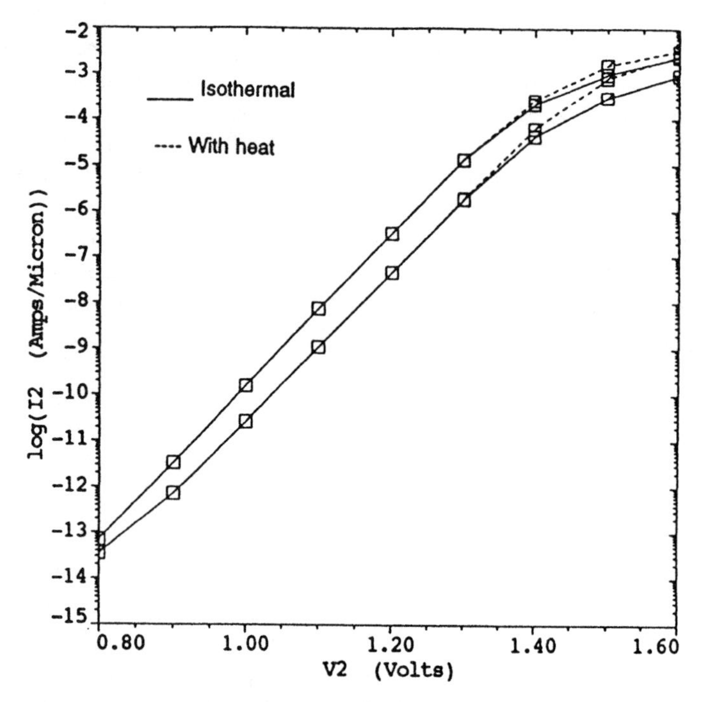

FIGURE 6.15 Simulated collector and base currents versus base–emitter bias.

the Early voltage at high collector currents. The energy balance simulation, however, gives a decrease of the Early voltage at high collector currents. The latter simulation accounts for device self-heating, which reduces the Early voltage of the bipolar device.

The coupling effect of the AlGaAs/GaAs HBT between nonlocal charge transport and nonisothermal behavior was investigated [16]. The $Al_{0.3}Ga_{0.7}As$ emitter has a width of 0.15 μm, with the last 30 nm graded. The GaAs base, n^- collector, and n^+ collector widths are 0.1, 0.5, and 0.15 μm, respectively. The doping concentrations in the emitter, base n^- collector, and n^+ collector are 5×10^{17}, 10^{19}, and 10^{17}m and 10^{18} cm^{-3}, respectively. The lengths of the emitter and collector are 1 and 4 μm. A thermal resistor with a value of $R_{\text{th}} = 6.7 \times 10^{-4}$ K·cm/W connects the bottom of the device to a heat sink held at 300 K. Figure 6.18 shows the cutoff frequency of the HBT simulated by drift–diffusion (DD), nonisothermal drift–diffusion (NDD), energy balance (EB), and nonisothermal energy balance (NEB). Significant differences between the results calculated using the DD and EB models were observed. This is because the EB model accounts for velocity overshoot, which reduces the base–collector transit time, thereby increasing the cutoff frequency. Inclusion of the nonisothermal model reduces the cutoff frequency at high collector current compared to that of the EB model. The nonisothermal energy-balance model predicts significant negative differential resistance (NDR) as shown in Fig. 6.19. The drift–diffusion model predicts no NDR. The nonisothermal drift–diffusion model predicts much less

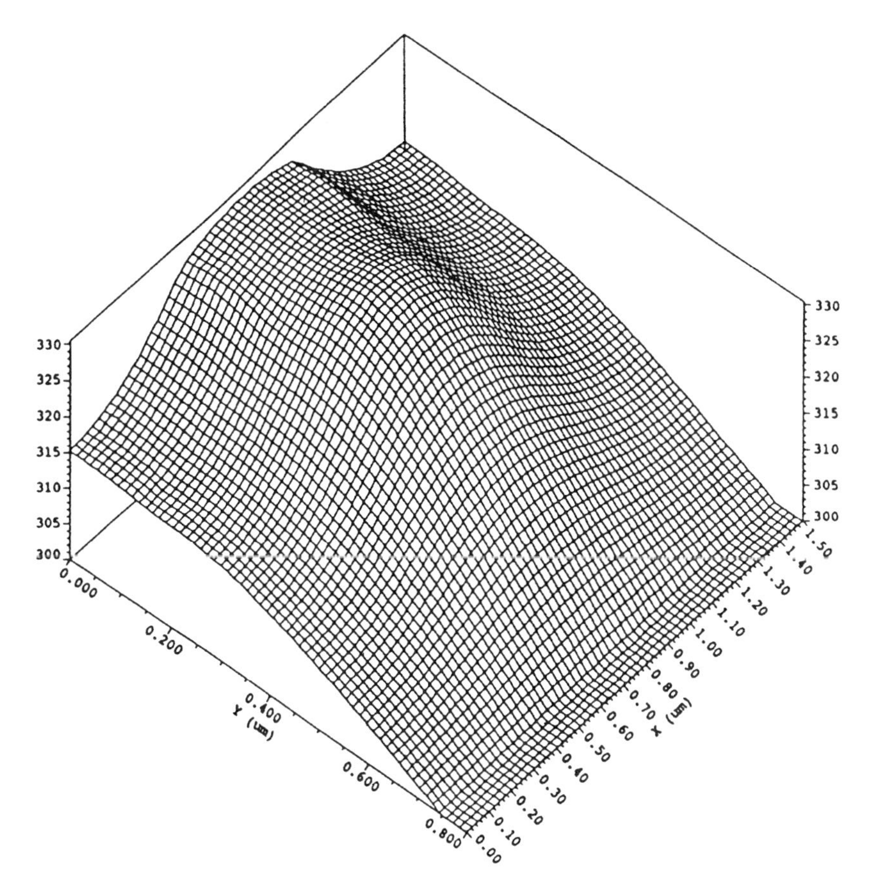

FIGURE 6.16 Temperature distribution of an HBT in MEDICI simulation.

pronounced NDR. The energy-balance model predicts NDR over only a limiting range of bias voltage.

The Monte Carlo, energy balance, and drift–diffusion simulations for a SiGe HBT were compared [17]. Figure 6.20 shows the electric field, energy, velocity, and electron concentration versus position of the SiGe HBT. The electric field is virtually the same as that predicted by these three simulations. However, the energy distribution predicted by the energy balance simulation is closer to the Monte Carlo simulation, while the drift–diffusion simulation predicts a uniform energy distribution, as shown in Fig. 6.20*b*. The velocity overshoot profiles between the MC and EB simulations are pretty much the same, while the drift–diffusion simulation projects a uniform saturation velocity. In Fig. 6.20*c* the electron velocity for 150 nm $< x <$ 300 nm predicted by the MC and EB simulations is lower that of the drift–diffusion result. This is due to an enhancement of electron concentration in that region, as shown in Fig. 6.20*d*. For a given set of boundary conditions (V_{BE} = 0.7 V and V_{CE} = 1.0 V), the

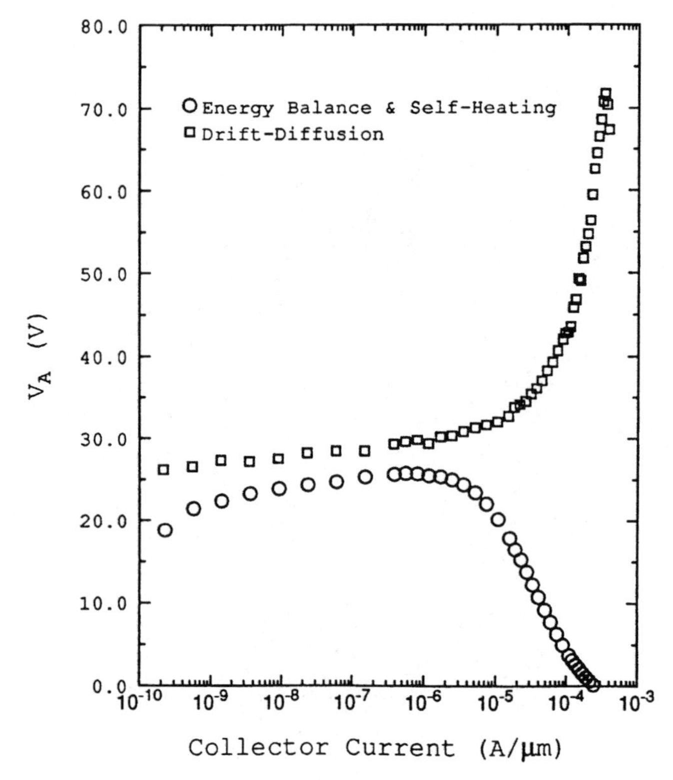

FIGURE 6.17 Early voltage versus collector current (after Liang and Law, Ref. 15 © IEEE).

electron velocity is reduced when the electron concentration is increased to maintain a constant collector current density.

6.5 TRANSISTOR DESIGN USING HETEROJUNCTION DEVICE SIMULATION

Device simulators are powerful tools that model the two- or three-dimensional distributions of potential and carrier concentrations as well as current vectors in a device and predict the device electrical characteristics for any bias condition. Simulators are useful not only for majority-carrier devices, which involve a single carrier type such as the MESFET and MODFET, but also for minority-carrier devices, which involve both carriers (electrons and holes) such as HBTs. The device can be simulated under steady-state, small-signal, and transient operating conditions.

In general, device simulation provides the ability to perform computer simulation experiments without using semiconductor wafers. The simulation is transparent to semiconductor technologies and independent of expensive fabrication equipment. In addition, the turnaround design time of advanced or prototype devices is reduced. Effi-

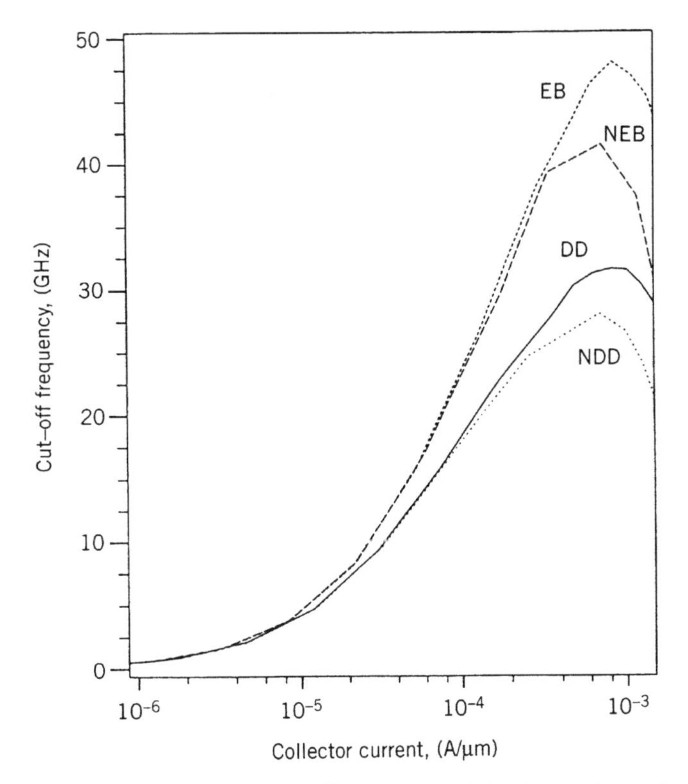

FIGURE 6.18 Cutoff frequency versus collector current (after Apanovich et al., Ref. 16 © IEEE).

cient use of device simulators, however, requires knowledge of semiconductor device physics for the insightful interpretation of device simulation results. The accuracy of device simulation relies heavily on the physical parameters used in the simulator. The simulation results are sensitive to the design of mesh points, especially for non-planar device structures. In addition, the accuracy of the doping profile is important, especially for small-geometry devices. Nevertheless, device simulation provides a relatively accurate model of device response with respect to boundary conditions. It also provides insight into device operation and doping profile optimization.

There are many device simulators available, such as SEDAN [18], PISCES [19], BIPOLE [20], BAMBI [21], MINIMOS [22], MEDICI [2], DAVINCI [23], and ATLAS [24]. Here we use the device simulator MEDICI for illustration purposes. MEDICI accurately describes the behavior of deep submicron devices by providing the ability to solve the electron and hole energy balance equations consistently with the other device equations. Effects such as carrier heating and velocity overshoot are accounted for. Many physical models are incorporated into the program. Among these are models for recombination, mobility, and lifetime. Both Boltzmann

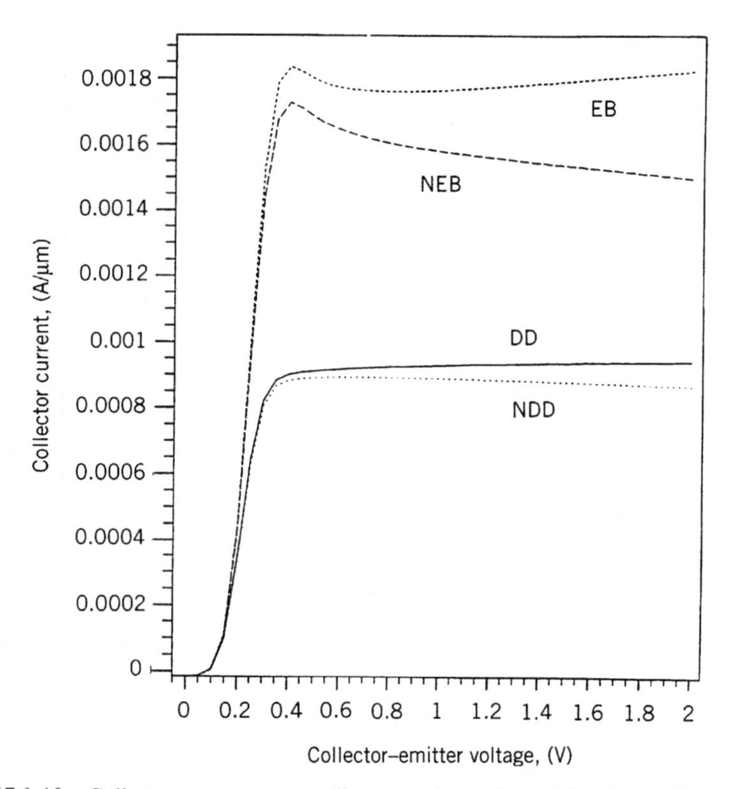

FIGURE 6.19 Collector current versus collector–emitter voltage (after Apanovich et al., Ref. 16 © IEEE).

and Fermi–Dirac statistics, including the incomplete ionization of impurities and effect of carrier freeze-out, are included. The simulator is able to attach lumped resistors, capacitors, and inductors to contacts as well as distributed contact resistances. Both voltage and current boundary conditions during a simulation can be specified. In addition, a light source can be specified for examining the photoelectrical response of photodiodes and phototransistors. Advanced application modules (AAMs) are available. The lattice temperature AAM is able to solve the lattice heating equation self-consistently with the other device equations. The heterojunction device AAM simulates the heterojunction device behavior.

With its nonuniform triangular simulation grid, MEDICI can model arbitrary device geometries with both planar and nonplanar surface topologies. From the grid structure of the device and impurity doping profile, MEDICI used the finite element method for solving the five basic semiconductor equations. The finite element method lends itself to structures that are far from rectangular. Since they employ elements that can be quite different in size for different regions, the method is also ideal for having small elements in the vicinity of junctions, where the potential is rapidly changing. The boundary conditions are incorporated as integrals in a function that is minimized,

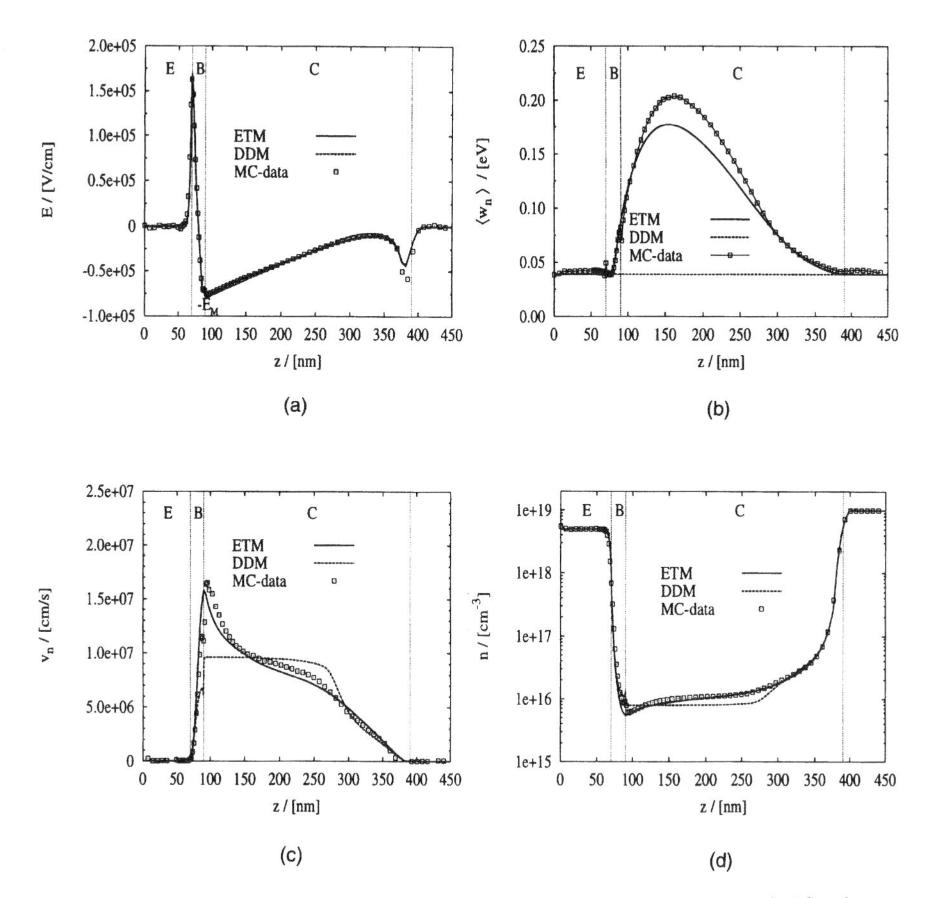

FIGURE 6.20 (*a*) Electric field, (*b*) electron energy, (*c*) velocity, and (*d*) electron concentration versus position for a SiGe HBT (after Nuernbergk et al., Ref. 17 © IEE).

and the scheme is independent of the specific boundary conditions; this provides a good degree of flexibility, particularly since elements of different sizes may be added without increasing the complexity.

MEDICI uses the box method to discretize the differential operators on a general triangular grid. Each equation is integrated over a small polygon enclosing each node. The integration equates the flux into the polygon with the sources and sinks inside it so that conservation current and electric flux are built into the solution. The integral involved is performed on a triangle-by-triangle basis, which leads to a simple and elegant way of handling general surface and boundary conditions.

The discretization of the semiconductor device equations gives rise to a set of coupled nonlinear algebraic equations which must be solved by a nonlinear iteration method starting from an initial guess. Two approaches are widely used: Gummel's method and Newton's method. In Gummel's method, the equations are solved sequentially. Poisson's equation is solved assuming fixed quasi-Fermi potentials; since Poisson's equation is nonlinear, it is solved by an inner Newton loop. Then the new potential is substituted into the continuity equations, which are linear and can be solved directly. The new carrier concentrations are substituted back into the charge term of Poisson's equation and another cycle begins. At each state only one equation is being solved, so the matrix has N rows and N columns, regardless of the number of carriers being solved. This is a decoupled method; one set of variables is held fixed while another set is solved. The success of the method therefore depends on the degree of coupling between the equations. In Newton's method, all of the variables in the problem are allowed to change during each iteration, and all of the coupling between variables is taken into account. Due to this, the Newton method is very stable and the solution time is nearly independent of bias condition even at high levels of injection. Each approach involves solving several large linear systems of equations. The number of equations in each system is on the order of one to three times the number of grid points, depending on the number of carriers being solved for. The nonlinear iteration usually converges either at a linear rate or at a quadratic rate. In the former, the error decreases by about the same factor at each iteration. In a quadratic method, the error should go down at each iteration, giving rise to rapid convergence. For accurate solutions, it is advantageous to use a quadratic method. Newton's method is quadratic, whereas Gummel's method is linear in most cases.

A solution is considered converged, and iterations will terminate, when either the X norm falls below a certain tolerance or the right-hand-side (RHS) norm falls below a certain tolerance. The X norm measures the error as the size of the updates to the device variables at each iteration. For the X norm, the default tolerance is 10^{-5} kT/q for the potentials and 10^{-5} relative change in the concentrations. The RHS norm measures the error as a function of the difference between the left- and right-hand sides of the equations. For the RHS norm, the tolerance is 10^{-26} C/μm for the continuity equations. By default, MEDICI uses a combination of the X norm and RHS norm to determine convergence, and therefore alleviates the difficulty of choosing one method or the other. Basically, the program assumes that a solution is converged when either the X norm or RHS norm tolerances are satisfied at every node in the device. This often greatly reduces the number of iterations required to obtain a solution, compared to when either the X norm or RHS norm is used alone, without sacrificing the accuracy of the solution.

The correct allocation of the grid is a crucial issue in device simulation. The number of nodes in the grid has a direct influence on the simulation time. Because the different parts of a device have very different electrical behavior, it is usually necessary to allocate a fine grid in some regions and a coarse grid in others to maintain reasonable simulation time. Another aspect of grid allocation is the accurate representation of small-device geometries. To model the carrier flow correctly, the grid must be a reasonable fit to the device shape. This consideration becomes more and more important as smaller, more nonplanar devices are simulated. MEDICI supports a general irregular grid structure. This permits the analysis of arbitrarily shaped devices and allows the refinement of particular regions with minimum impact to others.

Because of the extremely rigid mathematical nature of Poisson's equation and the electron and hole current continuity equations, strong stability requirements are placed on any proposed transient integration scheme. Additionally, it is most convenient to use one-step integration methods so that only the solution at the most recent time step is required. MEDICI has made use of the first-order (implicit) backward difference in time formula; that is, electron and hole concentrations are discretized as

$$\frac{n_j - n_{j-1}}{\Delta t_j} = F_n(\psi_n, n_k, p_k) = F_n(j) \tag{6.54}$$

$$\frac{p_j - p_{j-1}}{\Delta t_j} = F_p(\psi_p, n_k, p_k) = F_p(j) \tag{6.55}$$

where $\Delta t_j = t_j - t_{j-1}$ and ψ_j, n_j, and p_j denotes the potential, electron concentration, and hole concentration at time t_j, respectively. The backward Euler method is known to be both A- and L-stable, but suffers from a large local truncation error (LTE), which is proportional to the size of the time steps taken.

The backward Euler method is a one-step method. An alternative to the backward Euler method is the second-order backward difference formula

$$\frac{1}{t_j - t_{j-2}}\left[\frac{(2-\gamma)n_j}{1-\gamma} - \frac{n_{j-1}}{1-\gamma} + \frac{(1-\gamma)n_{j-2}}{\gamma}\right] = F_n(j) \tag{6.56}$$

$$\frac{1}{t_j - t_{j-2}}\left[\frac{(2-\gamma)p_j}{1-\gamma} - \frac{p_{j-1}}{1-\gamma} + \frac{(1-\gamma)p_{j-2}}{\gamma}\right] = F_p(j) \tag{6.57}$$

where $\gamma = (t_{j-1} - t_{j-2})/(t_j - t_{j-2})$. At a given point of time, MEDICI checks the local truncation error for both the first- and second-order backward difference methods and selects the method that yields the largest time step. By default MEDICI will select time steps so that the local truncation error matches the user-specified criteria. Specifying a larger tolerance will result in a quicker but less accurate simulation. MEDICI also places additional restrictions on the time steps:

(1) The time step size is allowed to increase at most by a factor of 2.
(2) If the new time step is less than half of the preceding step, the preceding time step is recalculated.
(3) If a time point fails to converge, the time step is reduced by a factor of 2 and the point is recalculated.

In addition to steady-state and transient simulation, MEDICI also allows small-signal analysis to be performed. An input of given amplitude and frequency can be applied to a device structure from which sinusoidal terminal currents and voltages are calculated. In MEDICI, the use of small-signal device analysis published by Laux [25] is adopted. An ac sinusoidal voltage bias is applied to an electrode i such that

$$V_i = V_{i0} + \tilde{v}_i e^{j\omega t} \tag{6.58}$$

where V_{i0} is the dc bias, $\tilde{v}_i$ the small-signal amplitude, ω the angular frequency, and V_i the actual bias (sum) to be simulated. Rearranging the basic partial differential equations in (6.19)–(6.21), we obtain

$$F_\psi(\psi, n, p) = \varepsilon\nabla^2\psi + q(p - n + N_D^+ - N_A^-) + \rho_F = 0 \tag{6.59}$$

$$F_n(\psi, n, p) = \frac{\nabla \mathbf{J}_n}{q} - U_n = \frac{\partial n}{\partial t} \tag{6.60}$$

$$F_p(\psi, n, p) = -\frac{\nabla \mathbf{J}_p}{q} - U_p = \frac{\partial p}{\partial t} \tag{6.61}$$

The ac solution to the equations above can be written as

$$\psi_i = \psi_{i0} + \tilde{\psi}_i e^{j\omega t} \tag{6.62}$$

$$n_i = n_{i0} + \tilde{n}_i e^{j\omega t} \tag{6.63}$$

$$p_i = p_{i0} + \tilde{p}_i e^{j\omega t} \tag{6.64}$$

where ψ_{i0}, n_{i0}, and p_{i0} are the dc potential and carrier concentrations at node i, while $\tilde{\psi}$, $\tilde{n}_i$, and $\tilde{p}_i$ are the respective ac values. By substituting (6.62)–(6.64) into (6.59)–(6.61) and expanding as a Taylor series of first order only, one obtains nonlinear equations of the form for each of the three partial differential equations

$$F(\psi, n, p) = F(\psi_0, n_0, p_0) + \tilde{\psi} e^{j\omega t}\frac{\partial F}{\partial \omega} + \tilde{n} e^{j\omega t}\frac{\partial F}{\partial n} + \tilde{p} e^{j\omega t}\frac{\partial F}{\partial p} \tag{6.65}$$

The heterojunction device simulation can be used to investigate different device design [26–28]. Liou [28] used the MEDICI simulation to study three different HBTs: the emitter island (E-island), emitter–base island (E/B-island), and emitter–base/collector (E/B/C-island) structure as shown in Fig. 6.21*a*–*c*, respectively. Device dc and RF characteristics were examined. Through the analysis, all three HBT structures have N-p$^+$-n $Al_{0.3}Ga_{0.7}As$/GaAs three-finger HBTs with the following practical makeup: a 5×10^{18} cm^{-3} and 1000-Å emitter, 300-Å graded layer, 10^{19} cm^{-3} and 1000-Å base, 5×10^{16} cm^{-3} and 7000-Å collector, 4×10 μm^2 emitter finger area, and 10-μm finger spacing.

Figure 6.22 shows the current gain versus the collector current of the three HBTs simulated at $V_{CE} = 2$ and 5 V, respectively. It is clear that the E-island HBT has the highest current gain for a wide range of collector currents. This is due to the fact that the lattice temperature in such a device is the lowest among the three HBTs, as evidenced by the lattice temperature contours in Fig. 6.23. The current gain at high collector current ($> 10^{-4}$ A/μm) decreases with increasing I_C and V_{CE} due to self-heating. Note that the intrinsic HBT is located at $0 < y < 0.93$ μm and 234 μm $< x <$ 266 μm, and the lattice temperature decreases rapidly toward the sidewalls and semi-insulating substrate. Also, the middle finger is hotter than

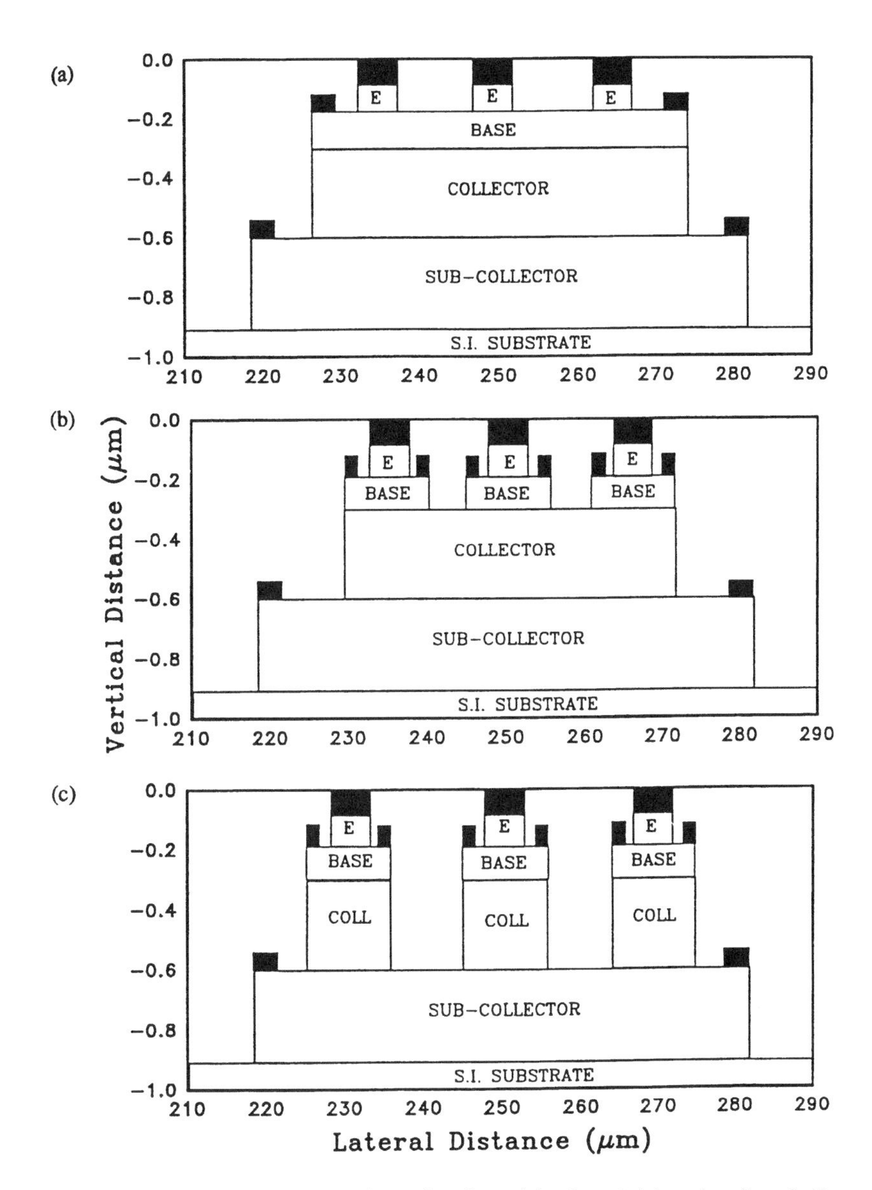

FIGURE 6.21 (*a*) Emitter-island, (*b*) emitter/base-island, and (*c*) emitter/base/collector-island HBTs (after Liou, Ref. 28 © Artech House).

the outer fingers, due to the thermal coupling among the fingers. At low collector current ($< 10^{-6}$ A/μm) the largest current gain in the E-island HBT results from the continuous base structure, which makes the electron–hole recombination in the base less prominent and hence reduces the base current. Intuitively, one

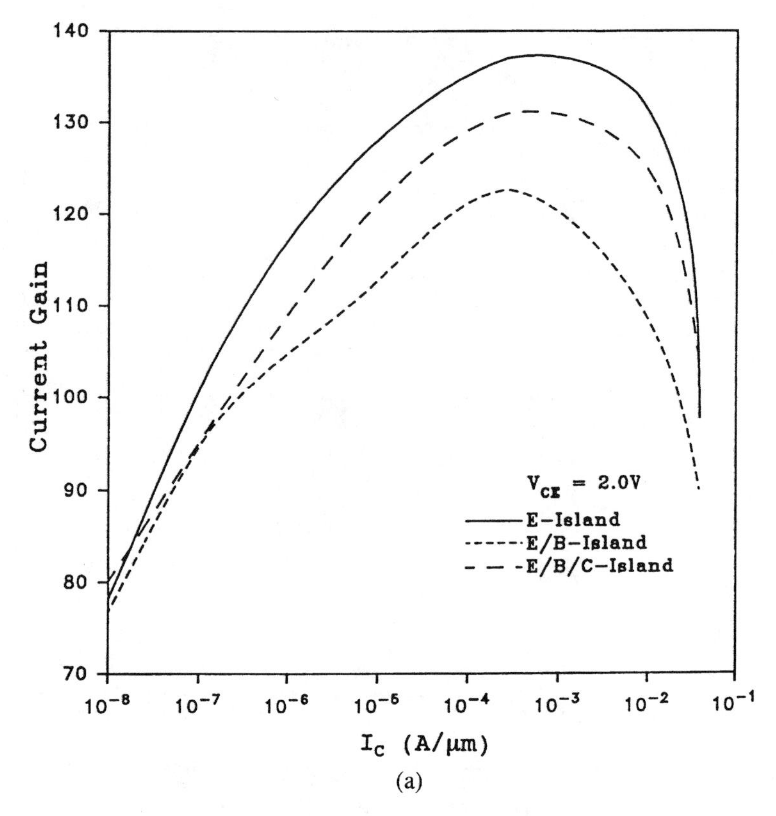

FIGURE 6.22 (*a*) Current gain versus collector current at $V_{CE} = 2$ V (after Liou, Ref. 28 © Artech House).

expects that the E/B-island HBT has a lower lattice temperature than the E/B/C-island HBT, due to a more uniform collector formation. This is not the case, however, because the contiguous collector in the E/B-island HBT results in a more uniform current contour in the collector and thus a higher collector current. The higher collector current consequently gives rise to a more significant thermal effect and a higher lattice temperature in the E/B-island HBT than in its E/B/C-island counterpart.

Figure 6.24 shows the cutoff frequency of the three HBTs simulated at $V_{CE} = 2$ and 5 V, respectively. Like the current gain, the high-current f_T degradation at larger V_{CE} is due to thermal effect. At low collector current, however, f_T is slightly higher at larger V_{CE}, due to a larger electric field in the base–collector junction and a smaller quasi-neutral base thickness as V_{CE} increased. Furthermore, the simulation results show that the E/B/C-island HBT has the lowest peak f_T, attributed to the fact that the E/B/C-island HBT has the least uniform electric field in the base–collector junction among the three devices. It has been shown that the free-carrier transit time across the base–collector space-charge region is often the limiting factor for the HBT cutoff

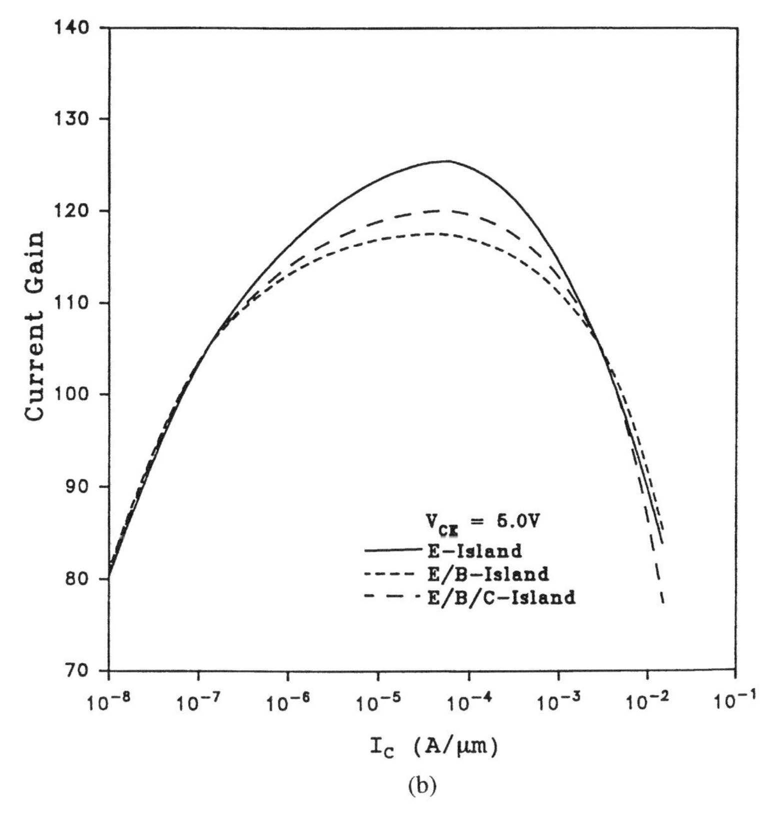

FIGURE 6.22 (*b*) Current gain versus collector current at $V_{CE} = 5$ V (after Liou, Ref. 28 © Artech House).

frequency. Thus the lower field in the regions between fingers found in the E/B- and E/B/C-island HBTs increases the overall transit time and decreases the cutoff frequency of the HBTs.

Advanced epitaxial growth of strained SiGe into a Si substrate enhances the free form for designing high-speed bipolar transistors. The use of device simulation helps the design of an optimized Ge profile in the base to achieve the best electrical characteristics, such as current gain and cutoff frequency. Hueting et al. [29] used the one-dimensional device simulator HeTRAP to study the optimization of SiGe-base bipolar transistors. The effect of the position of the leading edge of the Ge profile relative to the emitter–base metallurgical junction (distance d) on the HBT electrical performance was investigated. The maximum cutoff frequency and the current gain as a function of the distance d is shown in Fig. 6.25. Three regions of d are defined in Fig. 6.25. In region 1 the Ge edge is in the emitter–base space-charge region. Here the Ge causes extra bandgap narrowing and increased storage time, which reduces $f_{T,\max}$. Shifting the profile edge to region 2 increases $f_{T,\max}$ but decreases the current

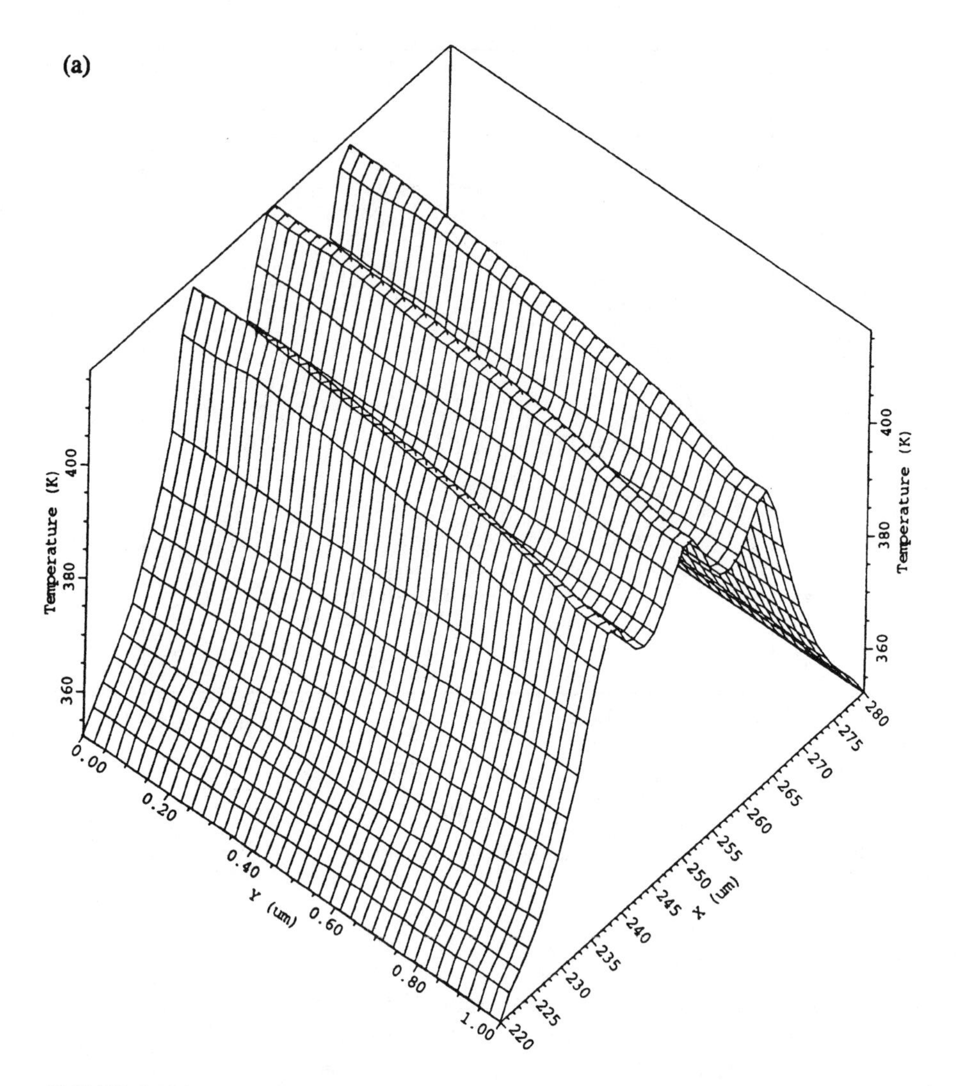

FIGURE 6.23 (a) Lattice temperature contours in the E-island HBTs at $V_{BE} = 1.5$ V and $V_{CE} = 2.0$ V (after Liou, Ref. 28 © Artech House).

gain. In this region the base Gummel number is very sensitive to V_{BE}. Increasing the distance d causes an enhancement of the base Gummel number. This reduces the collector current density. Consequently, the current gain is reduced. In region 3 the profile edge enters deeper into the neutral base region, causing a considerable reduction in the collector current density and current gain. The peak cutoff frequency $f_{T,\max}$, however, decreases only slightly with increasing distance.

Note that the slopes of both curves in Fig. 6.25 depend on the device structure, temperature, collector–base voltage, and Ge percentage. Only regions 1 and 3 have

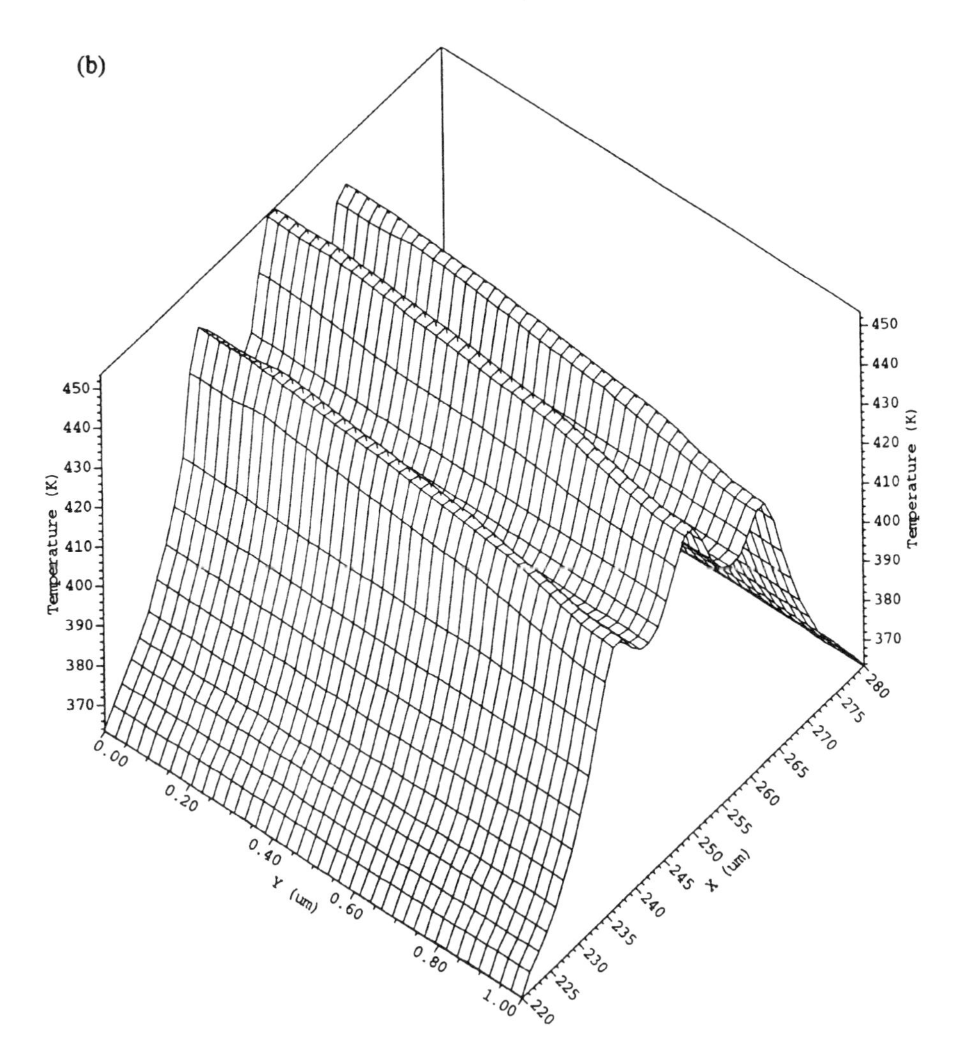

FIGURE 6.23 (*b*) Lattice temperature contours in the E/B-island HBTs at $V_{BE} = 1.5$ V and $V_{CE} = 2.0$ V (after Liou, Ref. 28 © Artech House).

ideal Gummel plots. In region 2 the Ge profile is rather close to the emitter–base space charge, and this causes a parasitic energy barrier that strongly varies with the emitter–base bias. This modulates the base Gummel number and introduces strong nonideal I_C–V_{BE} characteristics. If the Ge profile reaches beyond region 3, a much larger SiGe layer thickness would be needed. This increases the risk of misfit dislocations in the material. Therefore, the optimal position of the Ge profile is in region 3.

The various transit times between the Si BJT and SiGe HBT are compared [30]. The comparison is based on the charge partitioning method (quasistatic approximation) presented by van den Biesen [31]. From the integration of charge of two very similar

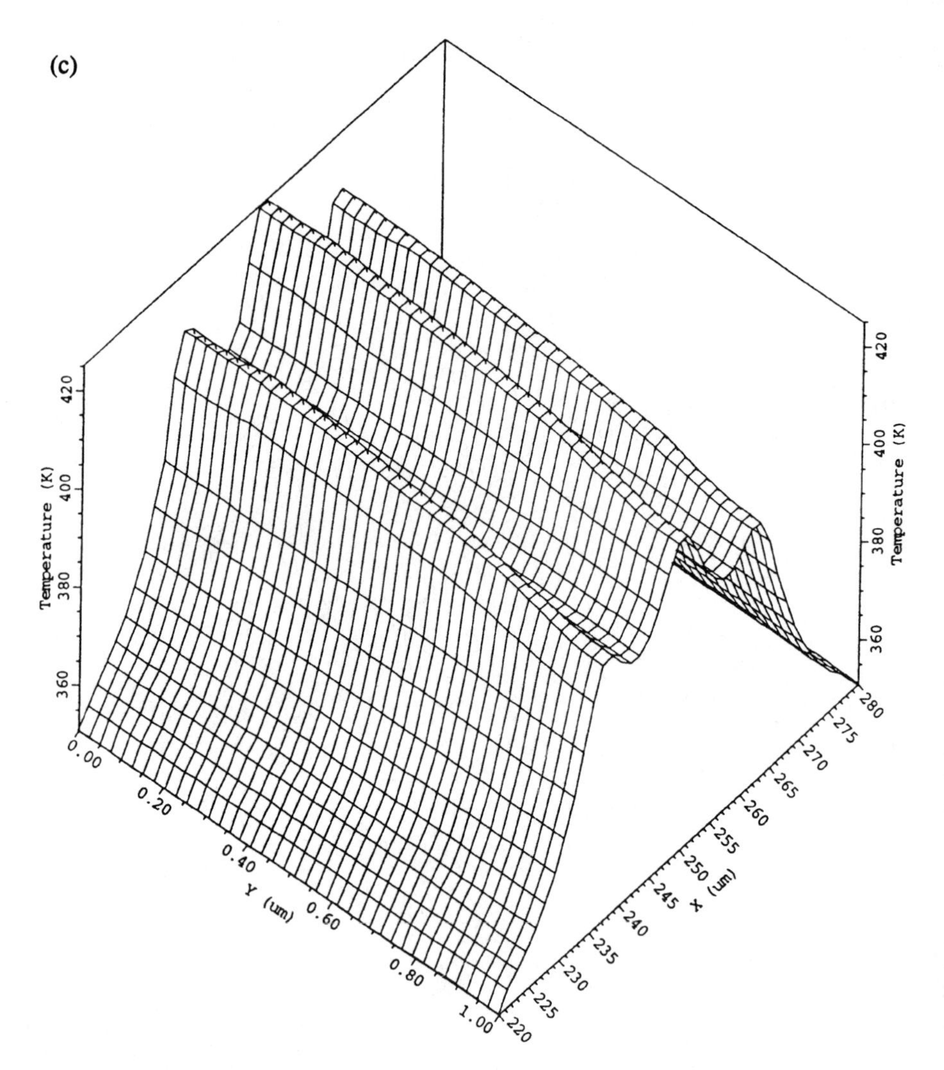

FIGURE 6.23 (*c*) Lattice temperature contours in the E/B/C-island HBTs at $V_{BE} = 1.5$ V and $V_{CE} = 2.0$ V (after Liou, Ref. 28 © Artech House).

operating points, one gets the emitter-to-collector transit time

$$\tau_{EC} = q \left| \int_0^L \frac{dp}{dJ_C} dx \right|_{V_{CE}=0} = q \left| \int_0^L \frac{dn}{dJ_C} dx \right|_{V_{CE}=0} \tag{6.66}$$

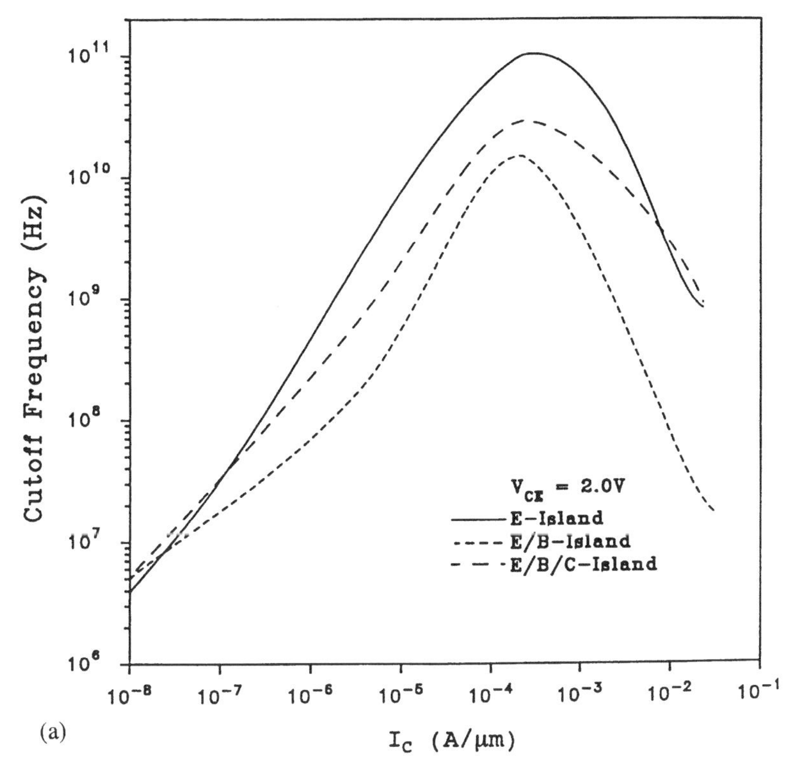

FIGURE 6.24 (*a*) Cutoff frequency versus collector current at $V_{CE} = 2$ V (after Liou, Ref. 28 © Artech House).

while the transit time within the different regions of a device is given by

$$\tau_E = q \left| \int_0^{x_{eb}} \frac{dp}{dJ_C} dx \right|_{V_{CE}=0} \tag{6.67}$$

$$\tau_{EB} = q \left| \int_0^{x_{eb}} \left(\frac{dn}{dJ_C} - \frac{dp}{dJ_C} \right) dx \right|_{V_{CE}=0} \tag{6.68}$$

$$\tau_B = q \left| \int_{x_{eb}}^{x_{bc}} \frac{dn}{dJ_C} dx \right|_{V_{CE}=0} \tag{6.69}$$

$$\tau_{BC} = q \left| \int_{x_{bc}}^{L} \left(\frac{dn}{dJ_C} - \frac{dp}{dJ_C} \right) dx \right|_{V_{CE}=0} \tag{6.70}$$

$$\tau_C = q \left| \int_{x_{cb}}^{L} \frac{dp}{dJ_C} dx \right|_{V_{CE}=0} \tag{6.71}$$

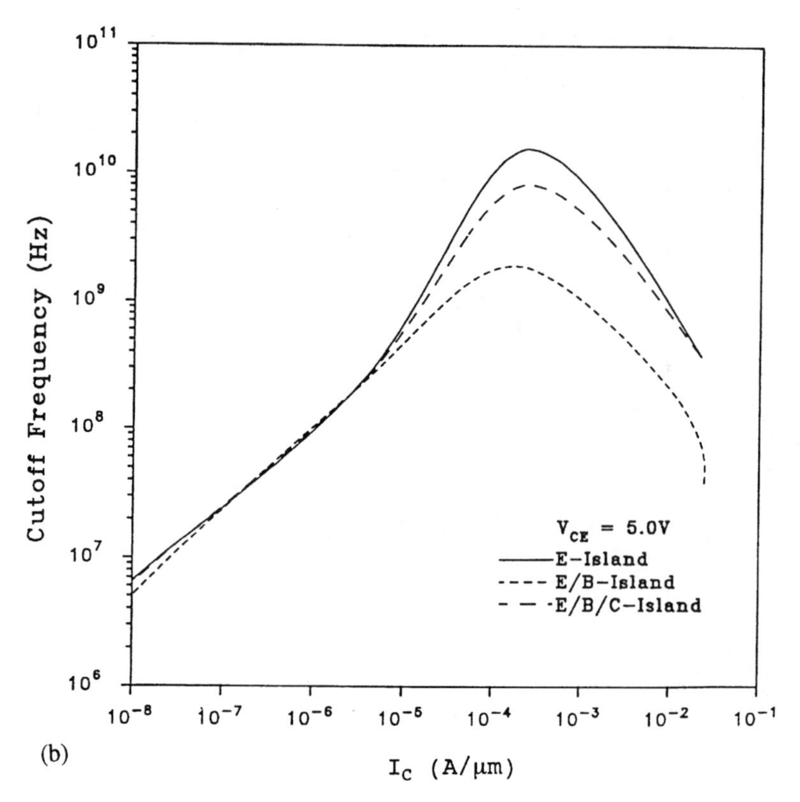

FIGURE 6.24 (*b*) Cutoff frequency versus collector current at $V_{CE} = 5$ V (after Liou, Ref. 28 © Artech House).

Note that τ_E, τ_B, and τ_C are influenced by the neutral, mobile charge stored in the emitter, base, and collector. τ_{EB} and τ_{BC} are the charging times for the space-charge regions. The boundary conditions x_{be} and x_{bc} are determined from the intersection of the curves $|dn/dJ_C|_{V_{CE}=0}$ and $|dp/dJ_C|_{V_{CE}=0}$. Therefore, x_{be} and x_{bc} are not the usual boundaries for the space-charge regions.

Figure 6.26 shows a comparison of various transit times for a Si BJT and a SiGe HBT. As shown in Fig. 6.26, the collector–emitter transit time has been reduced significantly for the SiGe HBT, due to reduction of base transit time and emitter transit time resulting from the introduction of Ge in the base.

6.6 MULTIEMITTER SIMULATION

The heterojunction bipolar transistors use multiemitter fingers to achieve excellent RF performance while maintaining dc current driving capability. The emitter finger width and spacing effects could only be studied by using three-dimensional analysis.

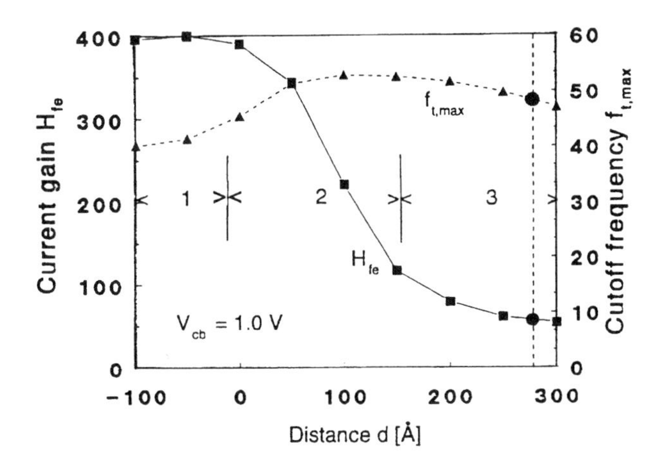

FIGURE 6.25 Peak cutoff frequency and current gain versus distance (after Hueting et al., Ref. 29 © IEEE).

Figure 6.27 shows the schematic diagram of conventional HBT chip with multiemitter fingers fabricated on the top surface of a GaAs substrate. Each emitter finger is further divided into a number of unit areas. Each unit area is assumed to be a heat source with constant temperature, with the exception of the substrate bottom surface, where the temperature is kept at heat sink temperature T_0 and the heat sources on the top.

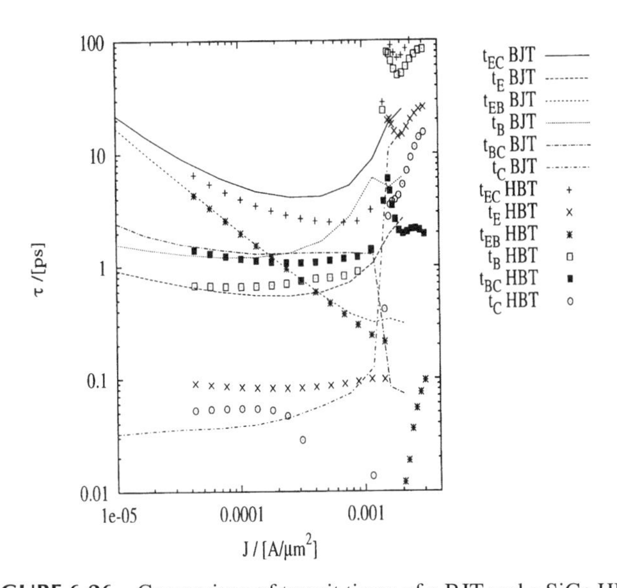

FIGURE 6.26 Comparison of transit times of a BJT and a SiGe HBT.

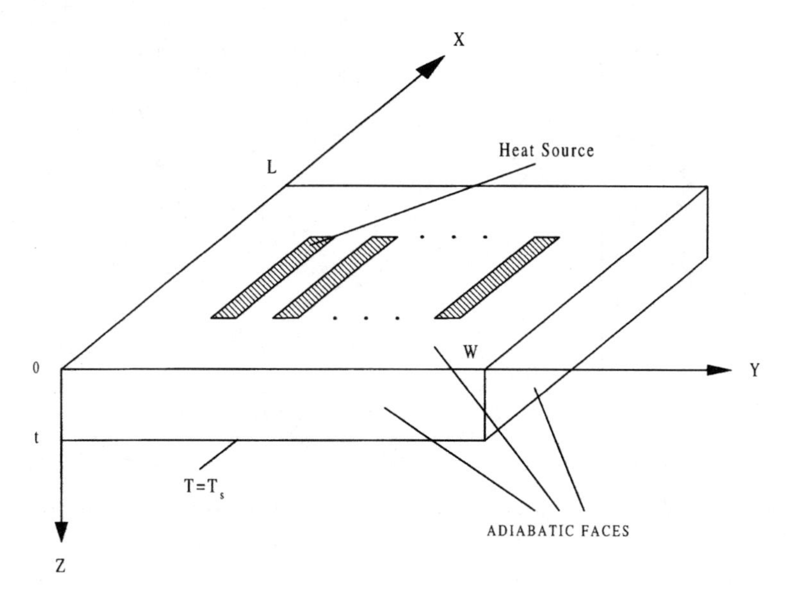

FIGURE 6.27 Multiemitter fingers on the top surface of a GaAs substrate.

The remaining surfaces are assumed adiabatic. Using these boundary conditions

$$\left.\frac{\partial T}{\partial x}\right|_{x=0,L} = 0$$

$$\left.\frac{\partial T}{\partial y}\right|_{y=0,W} = 0$$

$$T|_{x=t} = T_s$$

the temperature distribution at the top surface (i.e., $z = 0$) can be obtained by solving the heat transfer equation

$$\nabla\kappa(T)T = 0 \tag{6.72}$$

If thermal conductivity is temperature independent [$\kappa(T) = \kappa_0$], the solution is given by [32,33]

$$\theta(x, y, 0) = T_0 + \sum_{i=1}^{N}\left[\frac{q_i\Delta_x\Delta_y d}{L_x L_y \kappa_0} + \sum_{\mu,v=0}^{\infty}\left(C_{\mu v} q_i \cos\frac{\mu\pi x_i}{L_x}\cos\frac{v\pi y_i}{L_y}\frac{\tanh\gamma_{\mu v} d}{\kappa_0\gamma_{\mu v}}\right) \times \cos\frac{\mu x}{L_x}\sin\frac{v y}{L_y}\right] \tag{6.73}$$

where

$$\gamma_{\mu v} = \pi \sqrt{\left(\frac{\mu}{L_x}\right)^2 + \left(\frac{v}{L_y}\right)^2}$$

$$C_{\mu v} = \begin{cases} \dfrac{16}{\mu v \pi^2} \sin(0.5\mu\pi x)\sin(0.5v\pi y) & \mu \neq 0,\ v \neq 0 \\ \dfrac{4y}{\mu\pi L_y} \sin\dfrac{0.5v\pi x}{L_x} & \mu \neq 0,\ v = 0 \\ \dfrac{4x}{\mu\pi L_x} \sin\dfrac{v\pi y}{L_y}, & \mu = 0,\ v \neq 0 \end{cases}$$

$q_i \Delta x \Delta y$ is the heat generated at ith unit area with the center located at (x_i, y_i), $\Delta x \Delta y$ the unit area, N the number of the unit area on the surface, κ_0 the thermal conductivity at the heat sink temperature, and L_x, L_y, and d are the x, y, and z-dimensions, respectively. Kirchhoff's transformation can be used to account for the temperature-dependent thermal conductivity. Assuming that the thermal conductivity is proportional to $(T/T_0)^{-1.22}$, the corrected temperature is

$$T = \left[\frac{1}{T_0^{0.22}} - \frac{0.22(\theta - T_0)}{T_0^{1.22}}\right]^{-4.5454} \tag{6.74}$$

where θ is as given in (6.73).

The effect of chip thickness on HBT thermal behavior is shown in Fig. 6.28. Both the active region and substrate are assumed to have a square shape. L_a/L_c is the ratio of side lengths between the active region and the substrate. If this ratio is equal to 0.8, which is typical of conventional Si power transistors, the peak junction temperature decreases almost linearly with the decrease in chip thickness. However, if the ratio is less than 0.5, thinning of the substrate has little impact on peak junction temperature based on the result of the three-dimensional thermal analysis. Where the dimension of the active region is much less than the chip size, the heat flow is distributed in a cone shape, and the junction temperature is determined primarily by the lateral size of the device rather than the chip thickness.

Analytical and experimental results used to investigate thermal behavior depending on the emitter-finger area and substrate thickness in multifinger HBTs were also published [34]. As the substrate thickness decreases from 100 μm to 30 μm under the fixed power dissipation condition, the peak temperature of the large-emitter-area HBT (Fig. 6.29) decreases from 373 K to 346 K, while that of the small-emitter-area HBT (Fig. 6.30) decreases to 367 K. Thermal resistance R_{th} ($= T_j - T_0$/dissipate power) of the large-emitter-area HBT decreases from 40°C/W to 25°C/W, while the R_{th} values of the small-emitter-area HBT decreases from 200°C/W to 186°C/W. A reduction in R_{th} is obtained in the large-emitter-area HBT, while only a small reduction of 7% is obtained in the small-emitter-area HBT.

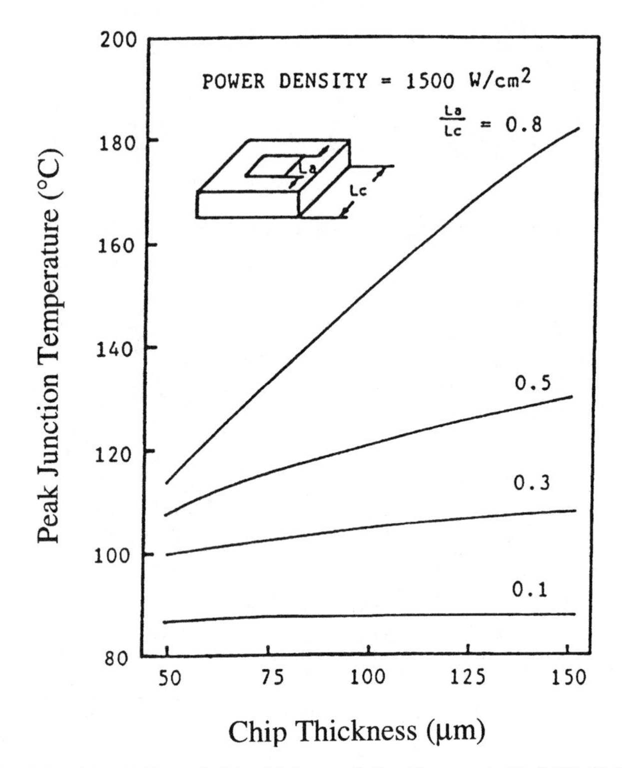

FIGURE 6.28 Effect of chip thickness (after Gao et al., Ref. 32 © IEEE).

Clearly, the junction temperature in the larger-emitter-area HBT decreases significantly when the substrate is thinned. This can be explained in term of the three-dimensional effect of the heat flow. As the emitter-finger area becomes smaller, the three-dimensional effect becomes dominant, and the heat concentrates in the vicinity of the emitter fingers. In this case, the junction temperature decreases very gradually with the thinning of the substrate. On the other hand, as the emitter-finger area becomes larger, the three-dimensional effect weakens, and the heat spreads over the entire chip. In this case the junction temperature decreases significantly with thinning of the substrate.

Three-dimensional transmission-line matrix simulation was used to evaluate the thermal behavior of power HBTs with multiple-emitter fingers [35]. The power device consists of five cells. Each cell has two emitter fingers grouped closely together by the emitter–base self-alignment process. The temperature distribution is simulated at a dissipated power density of 2 mW/μm^2 under steady state. Figure 6.31 shows the isometric view of temperature distribution of five emitter fingers, and Fig. 6.32 is the contour plot with interval of 15°C. The isometric projection and the corresponding temperature contours clearly display that two peak values of junction temperature occur at the centers of the two central emitter fingers. The junction temperature drops sharply around the central cell, with an average temperature gradient as high

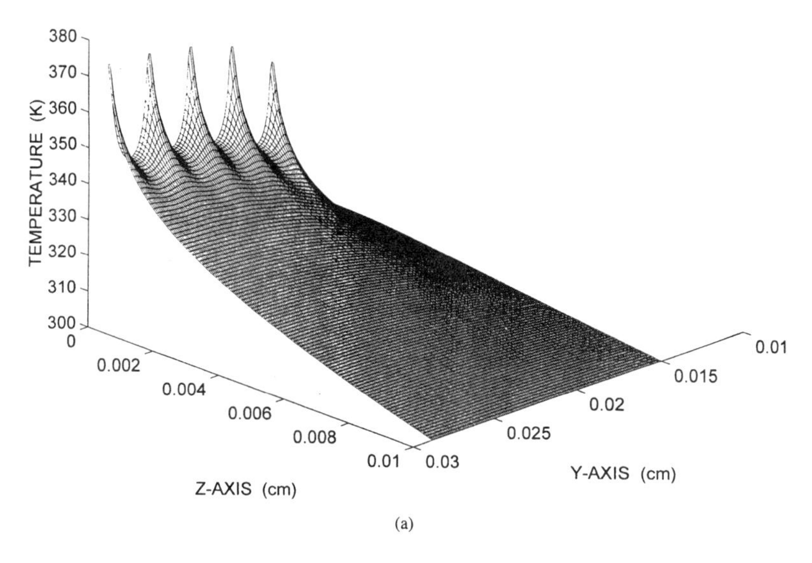

(a)

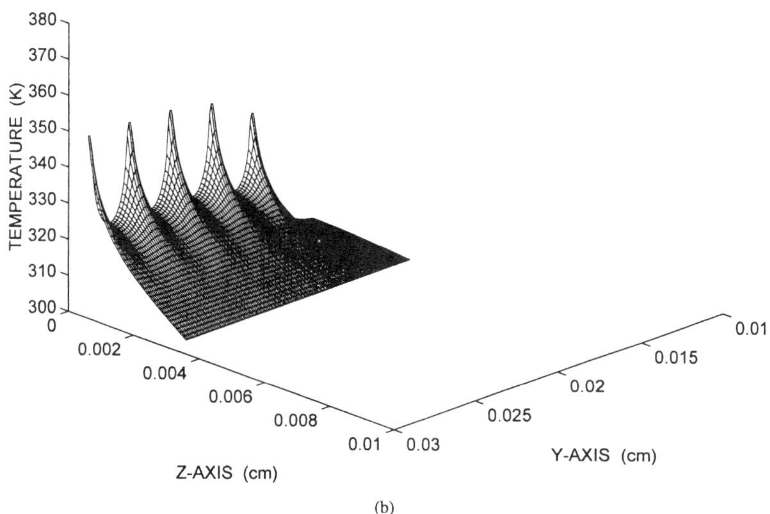

(b)

FIGURE 6.29 Three-dimensional temperature distribution of a $40 \times 4\ \mu m^2$ HBT at a substrate thickness of (*a*) 100 μm and (*b*) 30 μm; the dissipated power is fixed at 1.7 W (after Kim et al., Ref. 34 © IEEE).

as 9°C/μm in these areas. From a device-reliability point of view, the peak junction temperature is the most important parameter for the device thermal design and should be used to characterize the device thermal behavior.

The nonuniform junction temperature distribution of HBTs for various substrate materials is depicted in Fig. 6.33. The peak junction temperatures are 180, 140, and 101°C for the HBTs on GaAs, InP, and Si substrates, respectively. The HBT on a Si substrate obviously operates better than others for the same power dissipation.

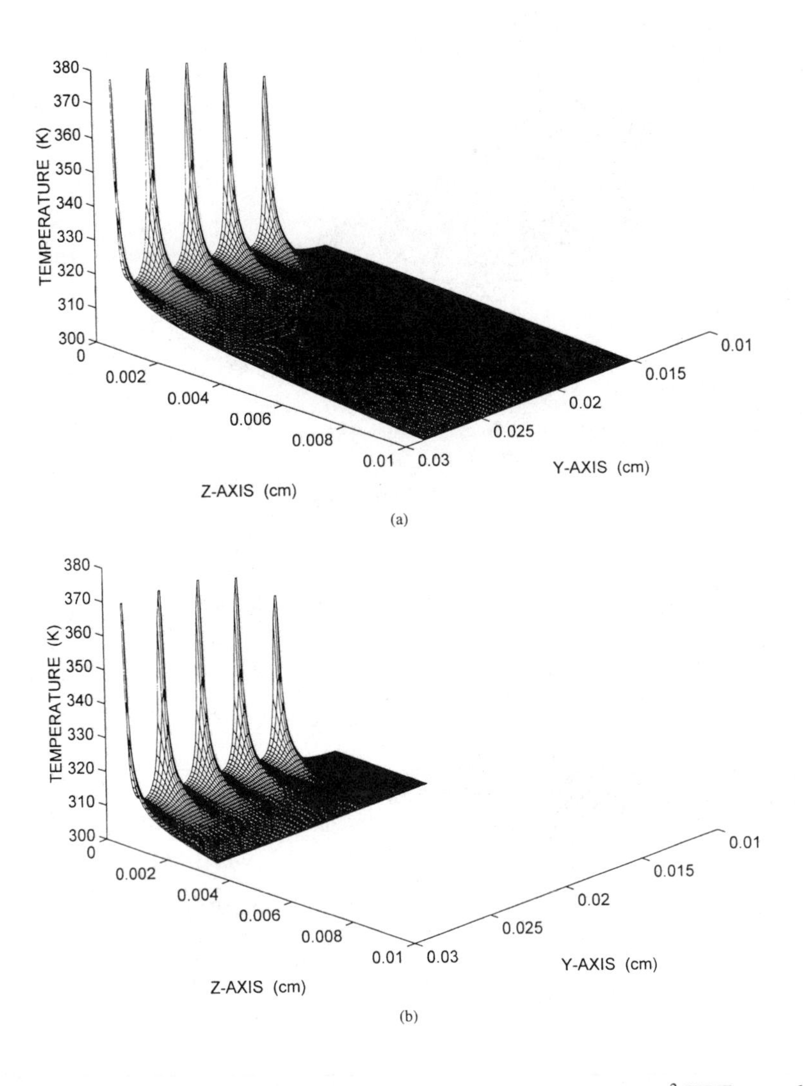

FIGURE 6.30 Three-dimensional temperature distribution of a $2.7 \times 4\ \mu m^2$ HBT at a substrate thickness of (*a*) 100 μm and (*b*) 30 μm; the dissipated power is fixed at 0.34 W (after Kim et al., Ref. 34 © IEEE).

In the design of power heterojunction bipolar transistors, the emitter-finger length is determined primarily by the emitter metallization resistance and the junction temperature nonuniformity along the emitter length. In general, the length and the length/width ratio of an emitter finger should not exceed 30 μm and have a 20:1 ratio in order to minimize the junction temperature and the voltage drop along the length of the finger. From thermal considerations, a power HBT with a longer emitter length results in a higher junction temperature and in turn causes a reduction in the output power density.

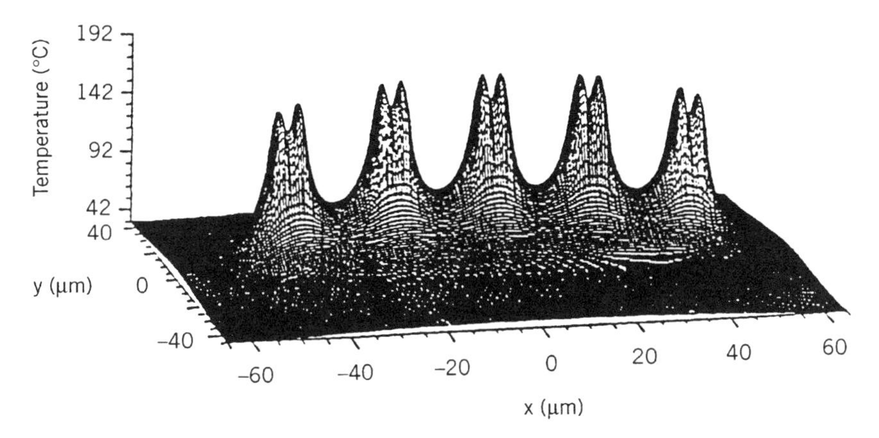

FIGURE 6.31 Isometric view of temperature distribution of five emitter fingers (after Gui et al., Ref. 35 © IEEE).

The total emitter area is determined by the desired output power or the maximum peak collector current. For silicon bipolar transistors, the peak current is strongly dependent on the emitter periphery, due to current crowding at high currents. For HBTs with high base doping concentration, the emitter crowding effect is reduced significantly. The calculated emitter utilization factor as a function of operating frequency and emitter size is depicted in Fig. 6.34. Clearly, an emitter width of 2 μm is sufficient for an operating frequency of 10 GHz. In this case, the emitter utilization

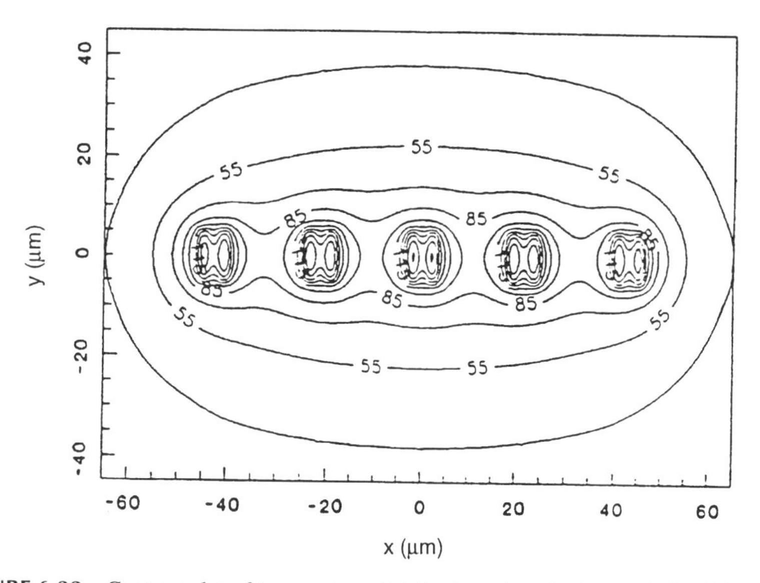

FIGURE 6.32 Contour plot of temperature distribution (after Gui et al., Ref. 35 © IEEE).

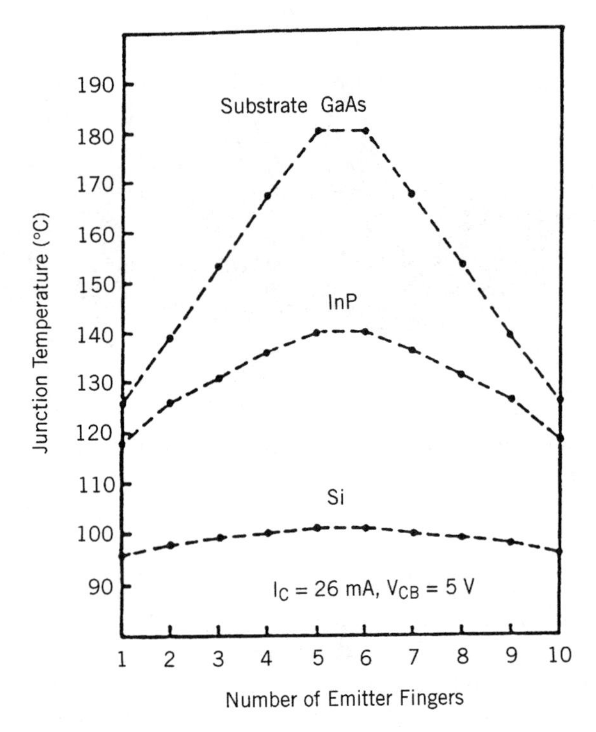

FIGURE 6.33 Junction temperature distribution for different substrate materials (after Gao et al., Ref. 36 © IEEE).

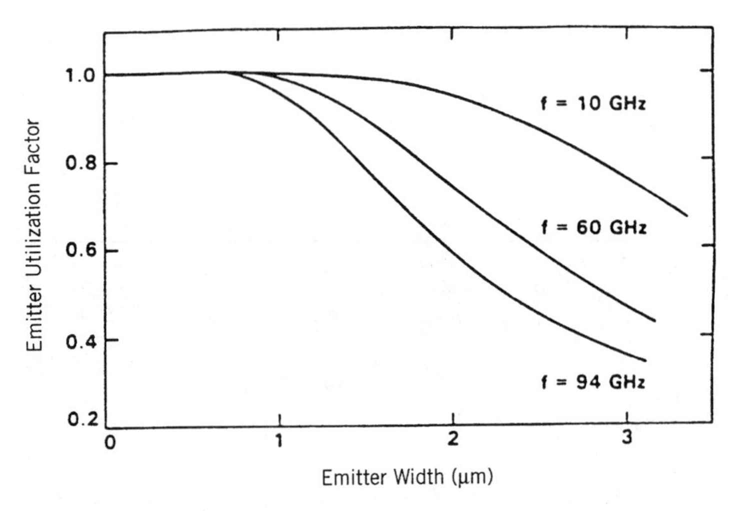

FIGURE 6.34 Emitter utilization factor as a function of frequency (after Gao et al., Ref. 36 © IEEE).

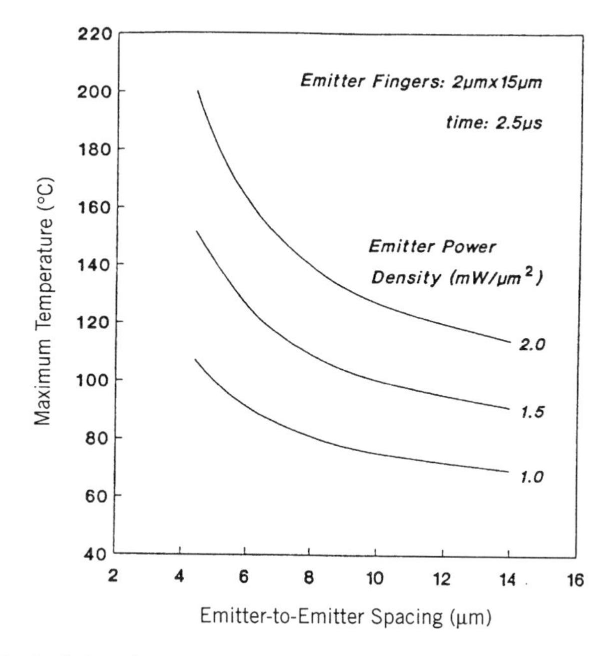

FIGURE 6.35 Peak junction temperature versus emitter spacing (after Gao et al., Ref. 36 © IEEE).

factor should be greater than 95%. This allows the current capability and the output power for HBTs to be dependent on the emitter area rather than the emitter periphery.

Another very important parameter for multiple-emitter-finger HBT design is the emitter spacing, which is defined as the center distance between two adjacent emitter fingers. A larger spacing results in a higher base resistance and capacitance, both of which degrade the $f_{\max}$ or the power gain. On the other hand, a larger spacing reduces the peak value of junction temperature at a given power dissipation, as seen in Fig. 6.35. When the spacing increases from 4.4 μm to 6, 10, and 14 μm, the peak value of junction temperature is reduced by 17.5, 35.0, and 42.5%, respectively, for a power density of 2 mW/μm^2.

REFERENCES

1. *SUPREM IV*: *Two-Dimensional Semiconductor Process Simulation*, Technology Modeling Associates, Inc., Palo Alto, CA (1993).
2. *MEDICI*: *Two-Dimensional Semiconductor Device Simulation*, User's Manual, Technology Modeling Associates, Inc., Palo Alto, CA (1993).
3. L. W. Nagel, *SPICE2*: *A Computer Program to Simulate Semiconductor Circuits*, Electronics Research Laboratory Report ERL-M520, University of California, Berkeley, CA (1975).

4. M. V. Fischett and S. E. Laux, "Monte Carlo analysis of electron transport in small semiconductor devices including band structures and space charge effects," *Phys. Rev. B*, **38**, 9721 (1988).

5. D. T. Hughes, R. A. Abram, and R. W. Kelsall, "An investigation of graded and uniform base Ge_xSi_{1-x} HBTs using a Monte Carlo simulation," *IEEE Trans. Electron Devices*, **ED-42**, 201 (1995).

6. B. K. Ridley, *Quantum Processes in Semiconductors*, University Press, Oxford, Oxford (1993).

7. R. Katoh and M. Kurata, "A model-based comparison of AlInAs/GaInAs and InP/GaInAs HBTs: a Monte Carlo study," *IEEE Trans. Electron Devices*, **ED-37**, 1245 (1990).

8. S. Adachi, "Material parameters of $In_{1-x}Ga_xAs_yP_{1-y}$ and related binaries," *J. Appl. Phys.*, **53**, 846 (1989).

9. S. Adachi, "GaAs, AlAs, and $Al_xGa_{1-x}As$: material parameters for use in research and device applications," *J. Appl. Phys.*, **58**, R1 (1985).

10. K. Brennan and K. Hess, "High field transport in GaAs, InP, and InAs," *Solid-State Electron.*, **27**, 237 (1984).

11. R. Katoh, M. Kurata, and J. Yoshida, "Self-consistent particle simulation for (Al,Ga)-As/GaAs HBTs with improved base–collector structures," *IEEE Trans. Electron Devices*, **ED-36**, 846 (1989).

12. S. Tiwari, M. Fischetti, and S. E. Laux, "Transient and steady-state overshoot in GaAs, InP, $Ga_{0.47}In_{0.53}As$, and InAs bipolar transistors," *IEDM Tech. Dig.*, 435 (1990).

13. S. Sze, *Physics of Semiconductor Devices*, 2nd ed., Wiley, New York (1981).

14. H. C. Casey, Jr., and M. B. Panish, *Heterostructure Lasers*: *Part B. Materials and Operating Characteristics*, Academic Press, San Diego, CA (1978).

15. M. Liang and M. E. Law, "Influence of lattice self-heating and hot-carrier transport on device performance," *IEEE Trans. Electron Devices*, **ED-41**, 2391 (1994).

16. Y. Apanovich, P. Blakey, R. Cottle, E. Lyumkis, B. Polsky, A. Shur, and A. Tcherniaev, "Numerical simulation of submicrometer devices including coupled nonlocal transport and nonisothermal effects," *IEEE Trans. Electron Devices*, **ED-42**, 890 (1995).

17. D. M. Nuernbergk, H. Forster, F. Schwierz, J. S. Yuan, and G. Paasch, "Comparison of Monte Carlo, energy transport, and drift–diffusion simulations for a Si/SiGe/Si HBT," *High Performance Electron Devices for Microwave and Optoelectronics Applications*, London, Nov. 24–25 (1997).

18. *TMA SEDAN-2*: *One-Dimensional Device Analysis Program*, *User's Manual*, Technology Modeling Associates, Inc., Palo Alto, CA (1984).

19. *PISCES*: *Two-Dimensional Device Analysis Program*, *User's Manual*, Technology Modeling Associates, Inc., Palo Alto, CA (1989).

20. *BIPOLE3*: *Bipolar Semiconductor Device Simulation*, *User's Manual*, Technology Modeling Associates in conjunction with Electrical and Computer Engineering Department, University of Waterloo (1993).

21. A. F. Franz, G. A. Franz, W. Kausel, G. Nanx, P. Dickinger, and C. Fischer, *BAMBI 2.1 User's Guide*, Institute for Microelectronics, Technical University, Vienna (1989).

22. C. Fischer, P. Habas, O. Heinreichsberger, H. Kosina, Ph. Lindorfer, P. Pichler, H. Potzl, C. Sala, A. Schutz, S. Selberherr, M. Stiftinger, and M. Thurner, *MINIMOS 6.0*, *User's Guide*, Institute for Microelectronics, Technical University, Vienna (1994).

23. *DAVINCI, Three-Dimensional Semiconductor Device Simulation*, Technology Modeling Associates, Inc., Palo Alto, CA (1993).

24. *ATLAS User Manual*, Device Simulation Software, Silvaco International, Inc., Santa Clara (1996).

25. S. E. Laux, "Techniques for small-signal analysis of semiconductor devices," *IEEE Trans. Electron Devices*, **ED-32**, 2028 (1985).

26. A. J. Kager, J. J. Liou, L. L. Liou, and C. I. Huang, "A semi-numerical model for multiemitter finger AlGaAs/GaAs HBTs," *Solid-State Electron.*, **37**, 1825 (1994).

27. W. Liu, S. Nelson, D. Hill, and A. Khatibzadeh, "Current gain collapse in microwave multifinger heterojunction bipolar transistors operated at very high power density," *IEEE Trans. Electron Devices*, **ED-40**, 1917 (1993).

28. J. J. Liou, *Principles and Analysis of AlGaAs/GaAs Heterojunction Bipolar Transistors*, Artech House, Norwood, MA (1996).

29. J. E. Hueting, J. A. Slotboom, A. Pruijmboom, W. B. de Boer, C. E. Timmering, and N. E. B. Cowern, "On the optimization of SiGe-base bipolar transistors," *IEEE Trans. Electron Devices*, **ED-43**, 1518 (1996).

30. D. M. Nuernbergk and J. S. Yuan, unpublished data.

31. J. J. H. van den Biesen, "A simple regional analysis of transit times in bipolar transistors," *Solid-State Electron.*, **29**, 529 (1986).

32. G.-B Gao, M.-Z. Wang, X. Gui, and H. Morkoç, "Thermal design studies of high-power heterojunction bipolar transistors," *IEEE Trans. Electron Devices*, **ED-36**, 854 (1989).

33. L. L. Liou and B. Bayraktaroglu, "Thermal stability analysis of AlGaAs/GaAs heterojunction bipolar transistors with multiple emitter fingers," *IEEE Trans. Electron Devices*, **ED-41**, 629 (1994).

34. C.-W. Kim, N. Goto, and K. Honjo, "Thermal behavior depending on emitter finger and substrate configurations in power heterojunction bipolar transistors," *IEEE Trans. Electron Devices*, **ED-45**, 1190 (1998).

35. X. Gui, G.-B. Gao, and H. Morkoç, "Simulation study of peak junction temperature and power limitation of AlGaAs/GaAs HBTs under pulse and CW operation," *IEEE Electron Device Lett.*, **EDL-13**, 411 (1992).

36. G.-B Gao, H. Morkoç, and M.-C Chang, "Heterojunction bipolar transistor design for power applications," *IEEE Trans. Electron Devices*, **ED-39**, 1987 (1992).

PROBLEMS

6.1. Consider heavily doped n^+ silicon. A model for minority hole transport in this material is based on the assumption that the effective doping density is constant; that is,

$$N_{D(\text{eff})} \approx K_D \neq f(N_D)$$

Using Nilsson's empirical approximation,

$$\Im_{1/2}^{-1}(u) \approx \ln u + \frac{u}{(64 + 3.6u)^{0.25}}$$

derive an approximate expression for the phenomenological bandgap narrowing $\Delta E_g(N_D)$ commensurate with a nearly constant $N_{D(\mathrm{eff})}$.

6.2. Show that in moderately doped but nondegenerate n-type semiconductor, if $E_C - E_D$ is fixed,

$$n_0 \approx \sqrt{\frac{N_D N_C}{g_D}} \exp\left(\frac{E_D - E_C}{2kT}\right)$$

Why might this prediction of particle ionization of the dopant impurities be invalid?

6.3. Perform a device simulation for an $Al_{0.3}Ga_{0.7}As$/GaAs abrupt HBT with $N_E = 5 \times 10^{17}\ \mathrm{cm}^{-3}$, $N_B = 1 \times 10^{19}\ \mathrm{cm}^{-3}$, and $N_C = 1 \times 10^{17}\ \mathrm{cm}^{-3}$. Set up a grid and determine the current gain. Divide the grid spacing by 2. Simulate the current gain again. Is the current gain a function of the grid spacing?

6.4. How would you design the mesh points for a SiGe HBT with a Gaussian profile in the emitter, a Gaussian profile in the base, and a uniform profile in the collector?

6.5. Use a device simulator to simulate E_{Fn} versus V_{BE} of an AlGaAs/GaAs HBT.

6.6. Use Table 6.1 to plot E_G as a function of temperature for InP, GaAs, InAs, and Si materials.

6.7. Use a device simulator to simulate the current gain versus I_C using the drift–diffusion and hydrodynamic models.

6.8. Simulate the relationship between $f_{\max}$, d, and current gain for a SiGe HBT.

CHAPTER SEVEN

Breakdown and Thermal Instability

7.1 AVALANCHE BREAKDOWN

For heterojunction bipolar transistors the collector doping concentration is on the order of $10^{17}cm^{-3}$. As a result of higher collector doping, the electric field in the base–collector space-charge region increases accordingly. The resulting peak electric field can be as high as 10^5 V/cm and impact ionization-induced current can become prominent even under normal operating conditions. Understanding the avalanche breakdown mechanism will help device designers and circuit designers make reliable, high-performance transistors and integrated circuits.

The breakdown phenomenon occurs in the collector–base depletion region, where the electric field is highest. A high electric field accelerates the carriers and imparts to them a large kinetic energy. Indeed, this kinetic energy can become sufficiently large to cause an electron, or even more than one electron under the right conditions, to be removed from the valence band and transferred to the conduction-band. Take the case of the GaAs energy-band diagram shown in Fig. 7.1. The electron- and hole-initiated transition, involving a single electron–hole pair generation, consists of a transfer of energy from the hot carrier to an electron in the valence band, which jumps into the conduction-band leaving a hole behind [1]. In Fig. 7.1 the ionization process initiated by a hot electron is shown by solid lines. The ionization process initiated by a hot hole from the split-off band is shown by dashed lines. For small-bandgap semiconductors, even multiple pairs may be generated.

The ionization process is characterized in a semiconductor by the parameter ionization rate. The ionization rate is the relative increase in carrier density per unit length of carrier travel. For GaAs, the impact ionization occurs largely from carriers that are not in the lowest bands. Electrons cause ionization from higher bands, and holes generally cause it from the split-off band. At a low field, the ionization rate is also orientation dependent because of a lack of randomization. For most compound semiconductors with greater than 1 eV of bandgap, the hole ionization rate is larger

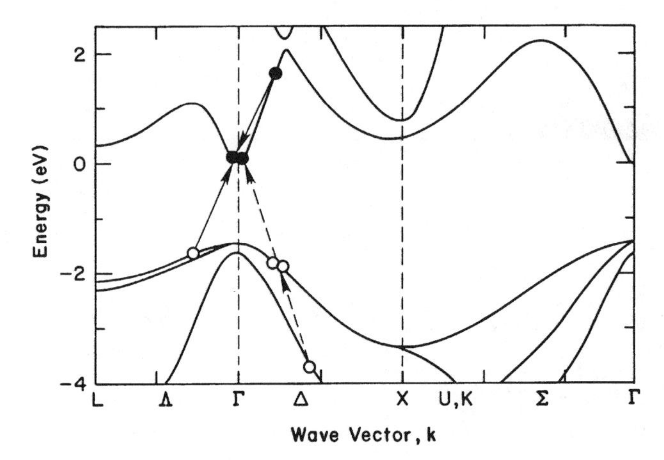

FIGURE 7.1 GaAs energy-band diagram for illustration of impact ionization (after Tiwari, Ref. 1 © Academic Press).

than the electron ionization rate. For smaller-bandgap semiconductors such as InAs and $Ga_{0.47}In_{0.53}As$, the hole ionization rate is always low because the threshold energy for ionization is lower. Figure 7.2 shows the impact-ionization coefficient for electrons and holes for GaAs, InP, and $Ga_{0.47}In_{0.53}As$ as a function of the inverse of the electric field. The electron and hole impact ionization coefficients for Si, Ge, $Al_{0.48}In_{0.52}As$, and InAs are displayed in Fig. 7.3.

It is interesting to note that impact ionization does not necessarily need an electric field to show a large generation rate for electrons and holes. In compound semiconductor heterostructures, there exist many small-bandgap compounds that can be lattice matched to other semiconductors with a bandgap discontinuity for a conduction or valence band that is larger than the bandgap. Even in the absence of an electric field, high-energy carriers injected into the smaller-bandgap semiconductor can cause impact ionization. This process is often referred to as *Auger generation*.

When impact ionization occurs in the presence of an electric field, such as in a p-n junction, both electrons and holes are accelerated in opposite directions, and both cause further impact ionization. The process of electron–hole pair creation, acceleration of carriers, and creation of more electron–hole pairs sets up an avalanche breakdown. Assume that a current I_{p0} is incident at the left-hand side of the depletion region. When the electric field in the depletion region is high enough, free electron–hole pairs are created by impact ionization. The hole current I_p will increase with distance through the depletion region and reaches $M_p I_{p0}$ at the end of the depletion region ($x = W$). Similarly, the electron current I_n will increase from $x = W$ to $x = 0$. The total current $I (= I_p + I_n)$ should be constant at steady state. The incremental hole current at x equals the number of electron–hole pairs generated per second in the distance dx:

$$d\left(\frac{I_p}{q}\right) = \frac{I_p}{q}\alpha_p\,dx + \frac{I_n}{q}\alpha_n\,dx \tag{7.1}$$

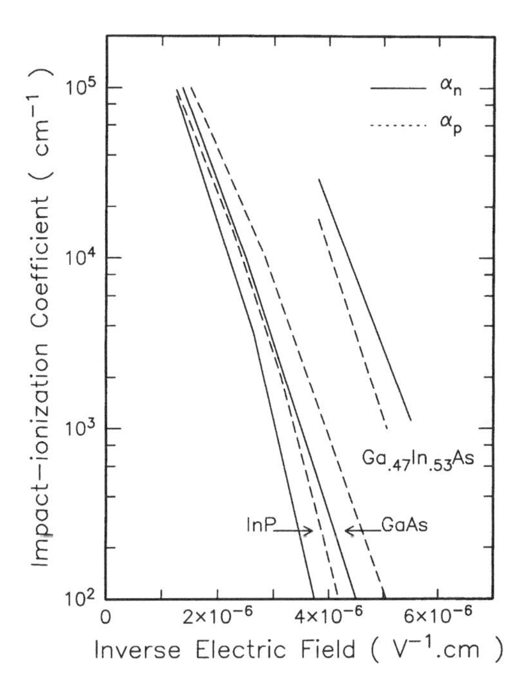

FIGURE 7.2 Impact ionization rate for electrons and holes for GaAs, InP, and $Ga_{0.47}In_{0.53}As$ (after Tiwari, Ref. 1 © Academic Press).

or

$$\frac{dI_p}{dx} - (\alpha_p - \alpha_n)I_p = \alpha_n I \tag{7.2}$$

The solution of (7.2) with the boundary condition of $I_p(W) = M_p I_{p0}$ is given by

$$I_p(x) = \frac{I\left\{1/M_p + \int_0^x \alpha_n \exp\left[-\int_0^x (\alpha_p - \alpha_n)\,dy\right]dx\right\}}{\exp\left[-\int_0^x (\alpha_p - \alpha_n)\,dy\right]} \tag{7.3}$$

where M_p is defined as the multiplication factor:

$$M_p \equiv \frac{I_p(W)}{I_p(0)} \tag{7.4}$$

Equation (7.3) can be rewritten as

$$1 - \frac{1}{M_p} = \int_0^W \alpha_p \exp\left[-\int_0^x (\alpha_p - \alpha_n)\,dy\right]dx \tag{7.5}$$

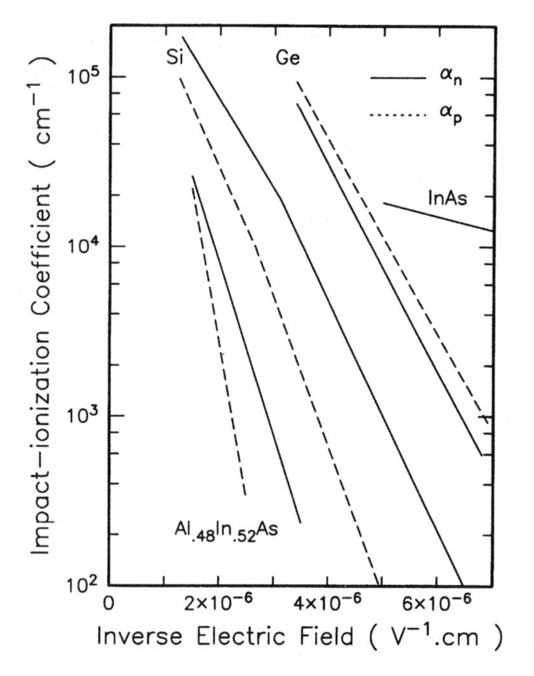

FIGURE 7.3 Impact ionization rate for electrons and holes for Si, Ge, $Al_{0.48}In_{0.52}As$, and InAs (after Tiwari, Ref. 1 © Academic Press).

The condition for breakdown caused by electron-initiated avalanche is the condition where M_p goes to infinity. This occurs when the integral in this equality goes to unity. Avalanche breakdown therefore occurs when

$$\int_0^W \alpha_p \exp\left[-\int_0^x (\alpha_p - \alpha_n)\, dy\right] dx = 1 \tag{7.6}$$

The condition for breakdown caused by electron-initiated avalanche can be found following a similar analysis. The expression for the electron multiplication factor is given as

$$1 - \frac{1}{M_n} = \int_0^W \alpha_n \exp\left[-\int_W^x (\alpha_n - \alpha_p)\, dy\right] dx \tag{7.7}$$

and hence breakdown is said to have occurred when

$$\int_0^W \alpha_n \exp\left[-\int_x^W (\alpha_n - \alpha_p)\, dy\right] dx = 1 \tag{7.8}$$

Note that equations (7.6) and (7.8) are equivalent. The breakdown conditions depend on what is happening in the depletion region and not on the carriers that initiate the avalanche process.

Figure 7.4 shows the avalanche breakdown voltage versus impurity concentration for one-sided abrupt junctions in Si, Ge, GaAs, and GaP. The dashed line indicates the maximum doping beyond which the tunneling mechanism will dominate the voltage breakdown characteristics. The breakdown voltage decreases with increasing impurity concentration. At given doping concentrations, GaP has the highest breakdown voltage, followed by GaAs, Si, and Ge. The avalanche breakdown voltage versus the impurity gradient for linearly graded junctions in Si, Ge, GaAs, and GaP is shown in Fig. 7.5. The dashed line indicates the maximum gradient beyond which the tunneling mechanism will dominate. Again, GaP has the highest breakdown voltage, followed by GaAs, Si, and Ge for given impurity gradients.

7.1.1 Reverse Base Current Phenomenon

In the collector–base depletion region for substantial reverse bias, the electric field intensity is fairly high. Free carriers can be accelerated by the field to a point where they acquire sufficient energy to create additional carriers when collisions occur. If there is a sufficient distance for the carriers to be accelerated to a high-enough velocity, generation of additional carriers during collisions will continue to sustain the avalanche process. Electrons and holes generated in the base–collector space-charge region drift to the quasi-neutral regions under the influence of the electric field. For an n^+-p-n-n^+ bipolar transistor, impact-ionization-induced electrons are

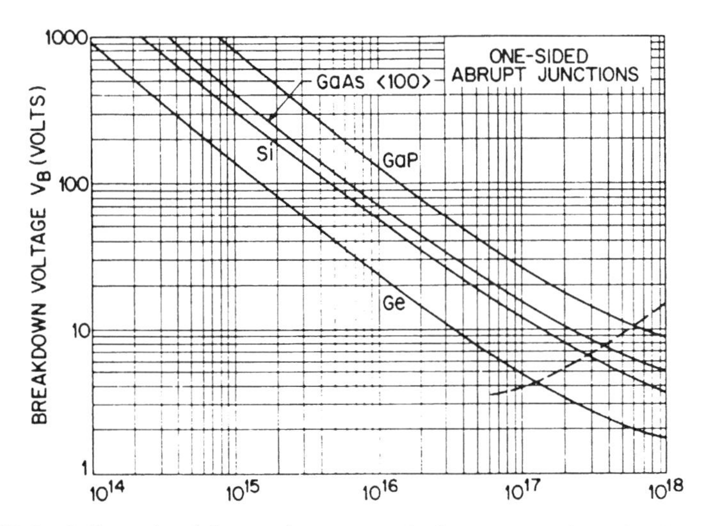

FIGURE 7.4 Collector breakdown voltage versus doping concentration (after Sze, Ref. 2 © John Wiley & Sons).

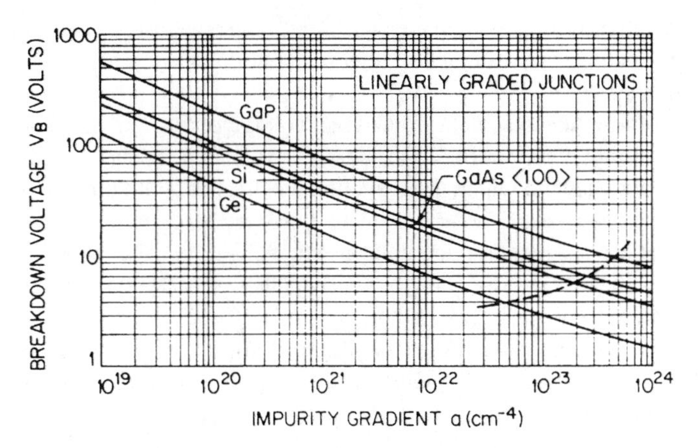

FIGURE 7.5 Collector breakdown voltage versus doping gradient (after Sze, Ref. 2 © John Wiley & Sons).

swept to the collector and increase the collector current:

$$I_C = (\xi + 1)I_C^0 \tag{7.9}$$

where ξ is the avalanche multiplication rate ($\xi = M - 1$) and I_C^0 is the collector current in the absence of avalanche multiplication.

Impact-ionization-induced holes also drift to the quasi-neutral base, as shown in Fig. 7.6. Since the hole current from the base to the emitter is fixed by the base–emitter voltage, excess holes are forced to flow to the base terminal, as shown in Fig. 7.6. The external base current decreases as [3]

$$I_B = I_B^0 - \xi I_C \tag{7.10}$$

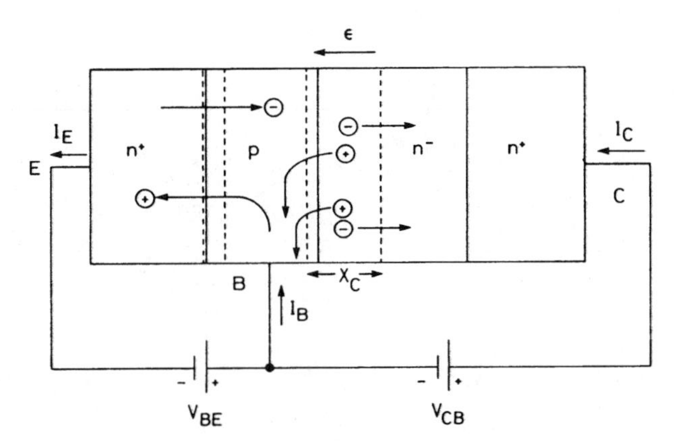

FIGURE 7.6 Hole current flow in avalanche breakdown.

where I_B^0 is the base current in the absence of avalanche multiplication.

If the avalanche current ξI_C is larger than I_B^0, the terminal current will reverse its sign. However, the base current can be made positive at higher base–emitter bias. This is because the hole diffusion current across the base–emitter junction increases at higher base–emitter bias so that I_B^0 is larger than ξI_C. The base–collector avalanche multiplication has strong dependence on the collector current density.

Avalanche multiplication has strong dependence on the collector current density. The avalanche multiplication rate decreases at higher base–emitter bias because it reduces the electric field in the collector–base depletion region. This is explained as follows. When the collector–base voltage V_{BE} is increased, the electric field in the collector–base space-charge region increases and the space-charge layer expands. The electric field distribution in the bipolar transistor as a function of the collector current for a given V_{CB} is depicted in Fig. 7.7. The peak electric field is located at the collector–base metallurgical junction. From the local field model, the electron and hole impact ionization coefficients are highest at the collector–base metallurgical junction, as evidenced by the impact-ionization-rate plot in Fig. 7.8. When the base–emitter voltage increases, the peak electric field at the collector base junction decreases gradually, due to electron injection from the emitter (curve 2 in Fig. 7.7). The peak electric field decreases until a uniform distribution is attained (curve 3). At this point, impact ionization at the collector–base junction reaches a minimum. As injection increases further, the peak electric field moves toward the collector region (curve 4). The peak field starts to rise again at higher base–emitter voltages and collector current densities.

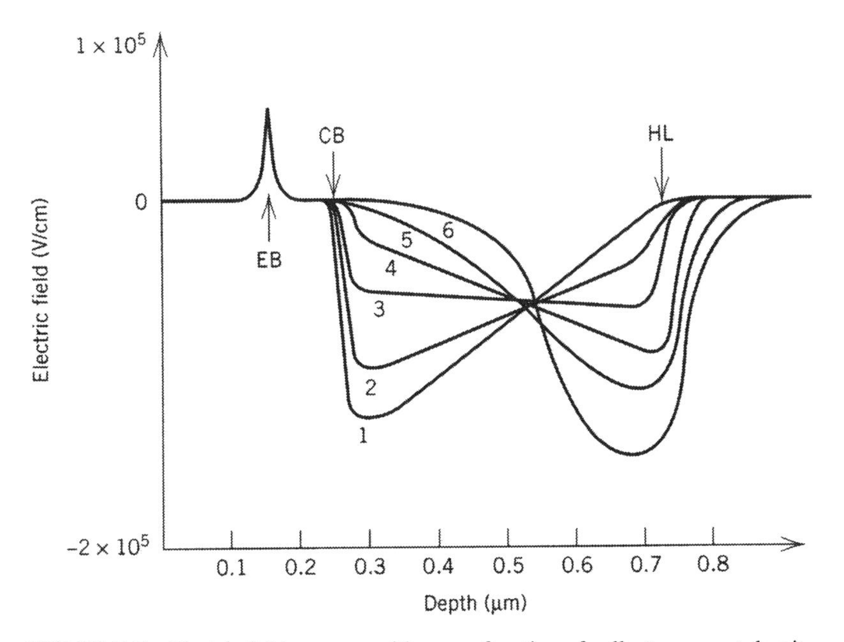

FIGURE 7.7 Electric field versus position as a function of collector current density.

When the collector current density approaches the onset current density for the Kirk effect J_0, base pushout occurs. The effective base region is then widened and the peak electric field is pushed into the collector high–low junction (curve 5). When $J_C > J_0$, the impact ionization rate increases again, due to increases in the peak electric field at the collector high–low junction.

The qualitative explanation of injection-modulated impact ionization is evidenced by the experimental data [4,5], as shown in Fig. 7.9, where the avalanche multiplication

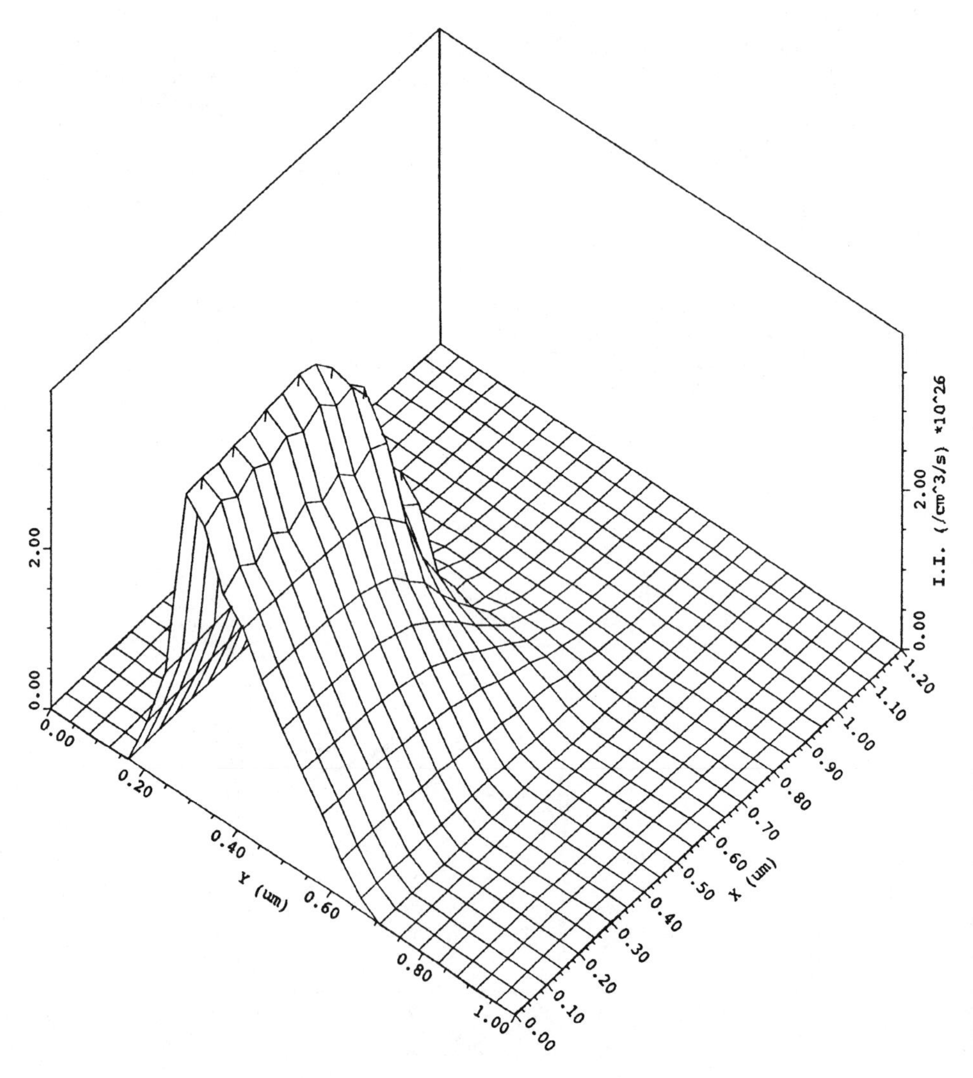

FIGURE 7.8 Impact ionization rate versus position from device simulation.

rate ξ is extracted by using

$$\xi = \frac{I_B(V_{BE}, V_{BC} = 0) - I_B^0(V_{BE}, V_{BC})}{I_C(V_{BE}, V_{BC})} \tag{7.11}$$

The reverse base current was observed in Si bipolar transistors first [6,7]. A negative base current at a moderate base–emitter bias has also been seen in AlGaAs/GaAs heterojunction bipolar transistors [8]. The physical description of impact ionization in silicon bipolar transistors [9] is suitable for explaining base current reversal in AlGaAs/GaAs heterojunction bipolar transistors. In GaAs heterojunction bipolar transistors, however, the thermal conductivity of the GaAs semi-insulating substrate is low. The AlGaAs/GaAs heterojunction bipolar transistor experiences severe self-heating. Since self-heating increases the junction temperature, avalanche multiplication is reduced.

Figure 7.10 shows the measured collector and base currents versus emitter current for an AlGaAs/GaAs HBT biased at various collector–base voltages. At $V_{CB} = 0$, impact ionization does not occur and I_B is always positive. On increasing V_{CB} and $|I_E|$, the number of electron–hole pairs generated by impact ionization increases. The electrons generated are swept in the collector, contributing a positive term to I_C, while the holes generated are collected at the base electrode, contributing a negative term to I_B. In particular, at $V_{CB} = 6$ V this contribution is so high that I_B reverses it sign

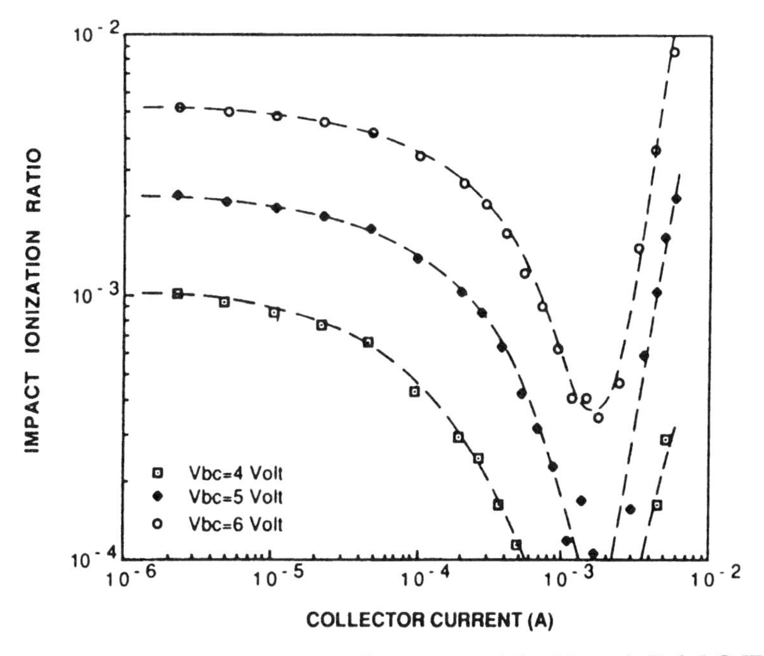

FIGURE 7.9 Impact ionization versus collector current (after Liu et al., Ref. 5 © IEEE).

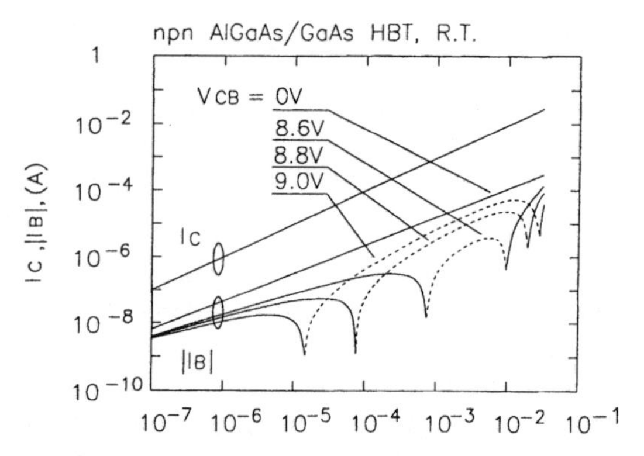

FIGURE 7.10 Gummel plot of an AlGaAs/GaAs HBT (after Zanoni et al., Ref. 8 © IEEE).

and becomes negative. In general, I_B is negative if $\beta(M-1) > 1$. The value of $|I_E|$ at which the reversal of I_B occurs becomes lower and lower with increasing V_{CB}. On increasing $|I_E|$ further, different effects contribute to a decrease in the generation rate and I_B recovers to a positive value, as occurs for $|I_E| > 10$ mA at $V_{CB} = 8.6$ V (see Fig. 7.10). These effects are voltage drops due to R_C and R_B parasitic resistances, device self-heating, and electric field lowering in the base–collector junction due to the presence of mobile carriers. Figure 7.11 shows the effect of reducing device self-heating by pulsing I_E with a duty cycle as low as 0.2% and correcting voltage drops on parasitic resistances (the extracted values being $R_B = 180\ \Omega$, $R_C = 10\ \Omega$) [10]. At a given $|I_E|$ in Fig. 7.11, the difference between dc and pulsed response is caused by the self-heating effect. The effect of parasitic resistances on the avalanche multiplication factor is seen by the pulsed curves with and without R_B and R_C correction.

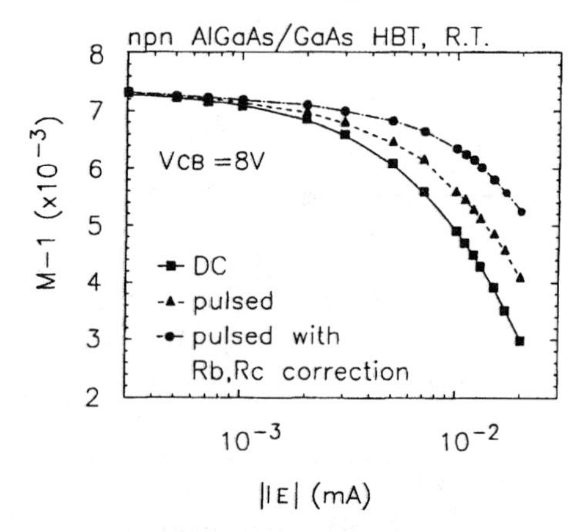

FIGURE 7.11 Self-heating effect on ionization multiplication factor (after Zanoni et al., Ref. 8 © IEEE).

7.1.2 Nonlocal Avalanche Effect

The collector doping profile directly influences impact-ionization effects and breakdown voltage, which are important factors for power applications of HBTs. High-speed performance and breakdown voltage cannot be optimized independently. Due to the strong electric fields present in the base–collector space-charge region of HBTs and to the presence of hot electron phenomena, conventional drift-diffusion analysis, which assumes locality of ionization events, is inadequate to describe and predict the avalanche breakdown. The local model implies that the impact-ionization coefficients at the base–collector metallurgical junction are the highest. Since electrons must travel a nonnegligible distance within the collector before attaining sufficient energy to initiate impact ionization, ionization events do not occur in correspondence to maximum electric field. Such a phenomenon is referred to as a *nonlocal* or *dead-space effect* [11] since the electron energy usually lags the electric field. For an n-p-n bipolar transistor, the peak electron impact-ionization coefficient shifts from the base–collector metallurgical junction into the collector, and the peak hole impact-ionization coefficient shifts from the base–collector metallurgical junction into the base, as evidenced by Monte Carlo simulations [10]. The local electric field model thus overestimates the avalanche multiplication rate compared to that obtained from the energy model, which uses mean carrier energy. Furthermore, as the collector–base voltage approaches the breakdown voltage, the contribution of secondary holes must be accounted for. Experimentally, the secondary holes cause a steep increase in the avalanche rate at a peak electric field of 9×10^5 V/cm [11,12].

The nonlocal model for impact-ionization current may be implemented as follows [13]. For a given $E(x)$ in the collector–base depletion region, the electron temperature $T_e(x)$ can be approximated by [14]

$$\frac{d\left[T_e(x) - T_0\right]}{dx} + \frac{T_e(x) - T_0}{\lambda_e} = \frac{2qE(x)}{5k} \tag{7.12}$$

where λ_e is the energy relaxation length of the electron. The solution of the equation is given by

$$T_e(x) - T_o = -\frac{2q}{5k}\int_0^x \exp\left(\frac{u - x}{\lambda_e}\right) E(u)\, du \tag{7.13}$$

For an abrupt n-p junction with uniform doping concentration in the quasi-neutral regions,

$$E(x) = \begin{cases} \dfrac{qN_D}{\varepsilon}x & \text{for } 0 \le x \le x_a \\[2ex] \dfrac{qN_A}{\varepsilon}x_b - \dfrac{qN_A}{\varepsilon}x & \text{for } x_a \le x \le x_b \end{cases}$$

where $x = 0$ is at the edge of the depletion layer next to the quasi-neutral n region, $x = x_a$ is at the metallurgical junction, and $x = x_b$ is at the edge of the depletion layer next to the quasi-neutral p region. Inserting the electric field into (7.13) gives

$$T_e(x) - T_0 = \begin{cases} \dfrac{2q\lambda_e}{5k}\dfrac{qN_D}{\varepsilon}\left\{x - \lambda_e\left[1 - \exp\left(-\dfrac{x}{\lambda_e}\right)\right]\right\} & \text{for } 0 \le x \le x_a \\ \dfrac{2q\lambda_e}{5k}\left\{\dfrac{qN_A x_b}{\varepsilon}\left[1 - \exp\left(\dfrac{x_a - x}{\lambda_e}\right)\right]\right. & \\ \qquad \left. - \dfrac{qN_A}{\varepsilon}\left[x - x_a \exp\left(\dfrac{x_a - x}{\lambda_e}\right)\right]\right\} & \\ -\dfrac{2q\lambda_e}{5k}\left\{\dfrac{qN_A\lambda_e}{\varepsilon}\left[1 - \exp\left(\dfrac{x_a - x}{\lambda_e}\right)\right]\right. & \\ \qquad \left. - \exp\left(-\dfrac{x}{\lambda_e}\right) Y_1\right\} & \text{for } x_a \le x \le x_b \end{cases} \tag{7.14}$$

where

$$Y_1 = \frac{qN_D}{\varepsilon}\left\{x_a - \lambda_e\left[1 - \exp\left(-\frac{x_a}{\lambda_e}\right)\right]\right\}\exp\left(\frac{x_a}{\lambda_e}\right)$$

For $x > x_b$,

$$T_e(x) - T_0 = \frac{2q\lambda_e}{5k}(Y_1 + Y_2)\exp\left(-\frac{x}{\lambda_e}\right) \tag{7.15}$$

where

$$Y_2 = \frac{qN_A x_b}{\varepsilon}\left[1 - \exp\left(\frac{x_a - x_b}{\lambda_e}\right)\right] - \frac{qN_A}{\varepsilon}\left\{x_b - x_a\exp\left(\frac{x_a - x_b}{\lambda_e}\right) - \lambda_e\left[1 - \exp\left(\frac{x_a - x_b}{\lambda_e}\right)\right]\right\}\exp\left(\frac{x_b}{\lambda_e}\right)$$

A graphical representation of the normalized electric field and the electron temperature distributions is displayed in Fig. 7.12. The solid lines in this figure represent the nonlocal model and the dashed lines represent the local model. It is clear from this plot that the nonlocal model shows the energy lag effect. Figure 7.13 shows experimental data for $M - 1$ obtained at $|I_E| = 50$ nA compared with values calculated by the local model and the energy model with and without the contribution of secondary holes to ionization. In the local model, $\alpha_{n,p}(E(x))$ is modeled as $\alpha_{n,p}\exp[-\beta_{n,p}/E(x)]$. In Fig. 7.13, $\alpha_n = 5 \times 10^6$ cm^{-1} and $\beta_n = 1.6 \times 10^6$ V/cm. When holes are

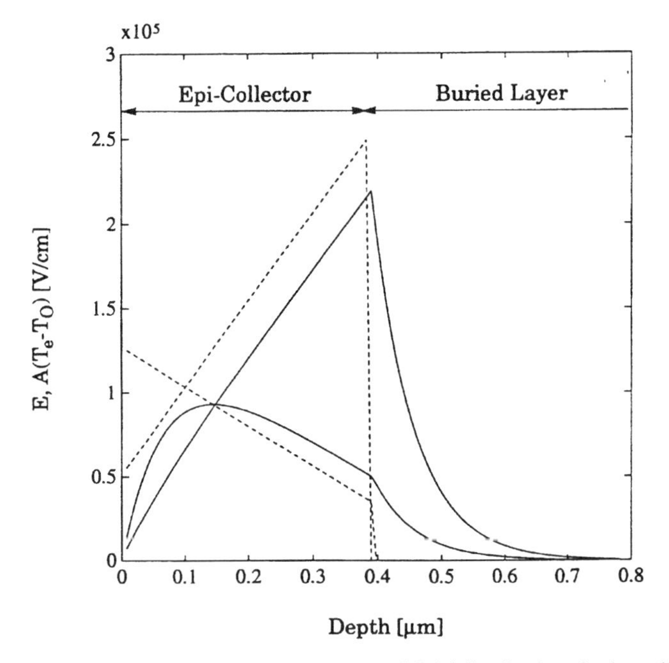

FIGURE 7.12 Normalized electron temperature and field distributions in the collector (after Hong and Fossum, Ref. 13 © IEEE).

considered (dashed lines in Fig. 7.13), $\alpha_p = 1.01 \times 10^6$ cm^{-1} and $\beta_p = 2.11 \times 10^6$ V/cm. For the energy model, $\lambda_e = 70$ nm and $\lambda_h = 100$ nm. At low V_{CB} the simulated curves with and without the contribution of holes are superimposed. The inclusion of secondary holes gives more accurate $M - 1$ predictions at high V_{CB}. To improve the predictability of the avalanche multiplication model at very high electric field intensity, incorporation of a delay in the conventional current equation has been proposed [11]. This simple analytical approach provides a steep increase in the avalanche multiplication rate for $\xi > 10$.

The multiplication factor is defined by the ratio of electron current density $J_n(x_b)$ where electrons enter the space-charge region to $J_n(0)$, where electrons leave the space-charge region:

$$M_N \equiv \frac{J_n(x_b)}{J_n(0)} \approx \frac{1}{1 - \int_0^{x_b} \alpha_n \, dx} \tag{7.16}$$

To model the avalanche current correctly, the impact-ionization rate as a function of electron temperature must be used. For example,

$$\alpha_n \left(T_e\left(x\right)\right) = A_0 \exp\left[-\frac{\beta_0}{T_e\left(x\right)}\right] \tag{7.17}$$

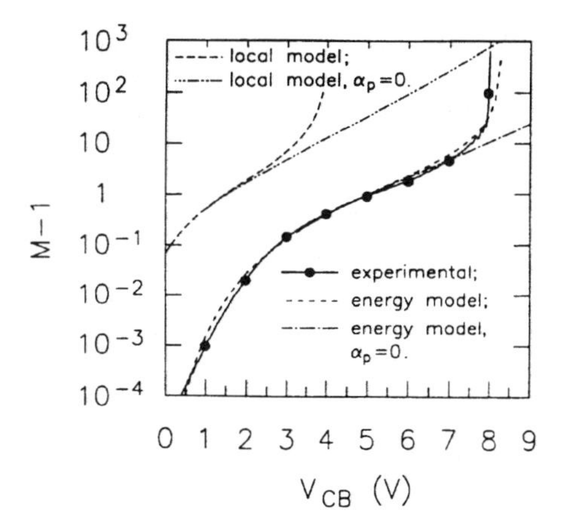

FIGURE 7.13 Comparison of local and nonlocal model in impact ionization characteristics (after Zanoni et al., Ref. 12 © IEEE).

where A_0 and β_0 are impact ionization coefficients and $T_e(x)$ is given in (7.14) and (7.15).

7.1.3 Influence of the Base Thickness on the Collector Breakdown

When the base thickness is less than the electron mean free path, electrons enter the collector region with excess energy and initiate impact ionization after acquiring additional kinetic energy from the electric field. Figure 7.14 illustrates the energy-band diagram of an AlInAs/InGaAs HBT showing the collector impact ionization caused by high-energy injection from the emitter under nonequilibrium transport. In the impact-ionization process, an energetic electron undergoes a transition to a lower state, during which another electron is excited across the bandgap.

A key to understanding the effect of high-energy injection from the emitter and ballistic base transport on impact ionization in the collector is the threshold energy for electron-induced impact ionization. The threshold energy can be calculated from knowledge of energy-band structures. For example, consider an initial electron in a conduction-band with a wave vector k_i making a transition to a lower state with a wave vector k_1. This will then excite an electron from a state at k_2 in the valence band to k_3 in the conduction-band. Conservation of total energy and momentum together with the condition for minimum energy requires that at threshold

$$v_{g1} = v_{g2} = v_{g3} \tag{7.18}$$

where $v_{gi} = 2\pi/h\nabla_k E(k_i)$ is the group velocity of a charge carrier with a wave vector k_i. Using the condition in (7.18) with the method of Anderson and Crowell [16], the threshold for impact ionization along the direction of electron transport in

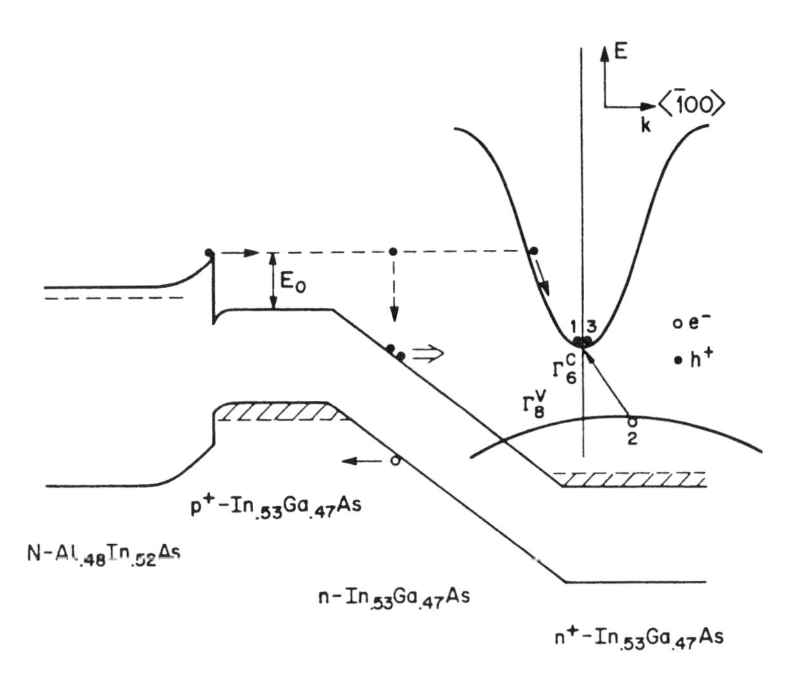

FIGURE 7.14 Energy-band diagram of an AlInAs/InGaAs HBT (after Jalali et al., Ref. 15 © IEEE).

the collector occurs for an electron with $E_i = 0.83$ eV above the conduction-band minimum.

Comparison of the collector ionization threshold $E_i = 0.83$ eV and the emitter injection energy $E_0 = 0.48$ eV indicates that hot electron injection enhances impact ionization in the collector if electrons traverse the base before dissipating their energy via inelastic collisions. Figure 7.15 shows the measured multiplication factor using (7.11) for AlInAs/InGaAs transistors with a base thickness of 200, 400, 700, 1200, and 4000 Å. The common-base breakdown voltage is BV_{CB0} 6.5 V for all devices. High collector–base junction breakdown is desirable for high device linearity and high-power application. The breakdown voltage is governed by the collector doping and width because the base is highly doped. Thicker and lower-doped collectors increase the breakdown voltage at the expense of increased collector depletion-layer transit time. The collector–emitter breakdown voltage is governed by collector–base breakdown coupled with a feedback multiplication factor of the current gain. BV_{CB0} is related to BV_{CE0} as follows:

$$BV_{CE0} = \frac{BV_{CB0}}{\beta^{1/n}} \tag{7.19}$$

where n is an empirical constant between 2 and 6. In Fig. 7.15 the base–emitter voltage

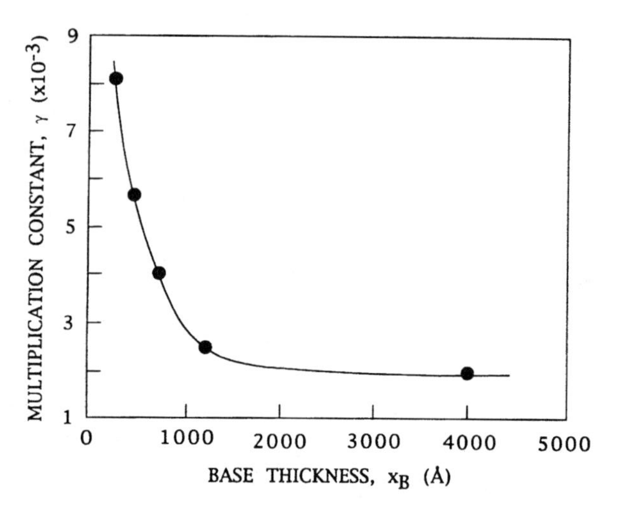

FIGURE 7.15 Measured multiplication factor versus base thickness (after Jalali et al., Ref. 15 © IEEE).

is kept the same for all measurements to ensure the same injection energy. In the thin base region, charge transport can be described by ballistic or quasiballistic motion. The multiplication factor has been increased significantly.

Since the base and collector transport are coupled, there is a trade-off between the need for short base and collector transit times and adequate collector breakdown voltage. This is more important for AlInAs emitter HBTs than those with InP emitters because of the 0.48-eV injection energy compared to 0.24 eV for the latter. The optimum base design must result in a small intrinsic delay with reasonable transistor output impedance.

7.1.4 Avalanche Effect on the Collector-Base Junction Capacitance

The free carrier capacitance concept has been applied to the collector–base junction capacitance where free carriers in the collector–base depletion region are injected from the emitter. For silicon bipolar transistors, free carrier capacitance becomes important when base pushout occurs. For GaAs heterojunction bipolar transistors, base pushout is suppressed due to high collector doping and valence-band discontinuity at the collector–base heterojunction. However, free carriers generated by avalanche multiplication can become important. The collector–base junction capacitance of the HBT, including the impact-ionization effect, is written as [17]

$$C'_{JC} = C_{JC} + C_F \left(\equiv \left| \frac{\partial \Delta Q_p}{\partial V_{BC}} \right| \right) \tag{7.20}$$

where ΔQ_p is the hole mobile charge in the collector–base depletion region ($\Delta Q_p = \Delta Q_n$).

For an N-p-n heterojunction bipolar transistor, impact-ionization-induced electrons in the collector depletion region are swept to the collector and impact-ionization-induced holes drift to the base terminal. The ionization-induced hole charge is associated with the incremental base current density $\Delta J_B[= (M - 1)J_C]$ and can be expressed by using the charge-control model:

$$Q_p = \tau(M - 1)J_C \tag{7.21}$$

where τ is the lifetime and

$$M = \frac{1}{1 - \left(V_{CB}/BV_{CB0}\right)^n} \tag{7.22}$$

Using (7.21) and (7.22), the free carrier capacitance per unit area due to impact ionization is expressed as

$$C_F = (M - 1)\left|\frac{\partial(\tau J_C)}{\partial V_{BC}}\right| + \tau J_C \left|\frac{\partial(M - 1)}{\partial V_{BC}}\right| \tag{7.23}$$

Since the Early voltage of the heterojunction bipolar transistor is very large, the base-width modulation effect on the collector current $\partial J_C/\partial V_{BC}$ is very small. For an injection-level independent lifetime, (7.23) is approximated as

$$C_F \approx \frac{n\tau J_C V_{CB}^{n-1}}{BV_{CB0}^n\left[1 - \left(V_{CB}/BV_{CB0}\right)^n\right]^2} \tag{7.24}$$

The analytical prediction of collector–base junction capacitance including avalanche effect is compared against MEDICI simulation. The $Al_{1-x}Ga_xAs$/GaAs heterojunction bipolar transistor has an abrupt emitter–base heterojunction with an aluminum mole fraction $x = 0.3$, emitter doping 2×10^{17} cm^{-3}, base doping 10^{19} cm^{-3}, collector doping 1×10^{17} cm^{-3}, emitter thickness 0.1 μm, base thickness 0.1 μm, and collector thickness 0.6 μm. The physical models used in MEDICI include Shockley–Read–Hall recombination, Auger recombination, bandgap narrowing, impact ionization, and concentration- and field-dependent mobilities. Figure 7.16 shows the electron and hole concentrations in the collector obtained from MEDICI for the HBT at $V_{BE} = 1.4$ V and $V_{CB} = 6$ and 12 V. The electron and hole concentration and depletion region thickness at $V_{CB} = 12$ V are much larger than those at $V_{CB} =$ 6 V. The difference is obviously due to avalanche multiplication at the large reverse-biased junction. Impact-ionization-induced holes drift to the base terminal, which reduce the base current at high V_{CB}, as shown in Fig. 7.17. In this plot the line represents the model prediction using (7.10) and the squares represent the MEDICI simulation results. The agreement between the model predictions and simulation data is excellent. The agreement allows us to obtain empirical parameters BV_{CB0} and n accurately. From the base current, $BV_{CB0}(= 13$ V) and $n(= 1.4)$ are extracted and

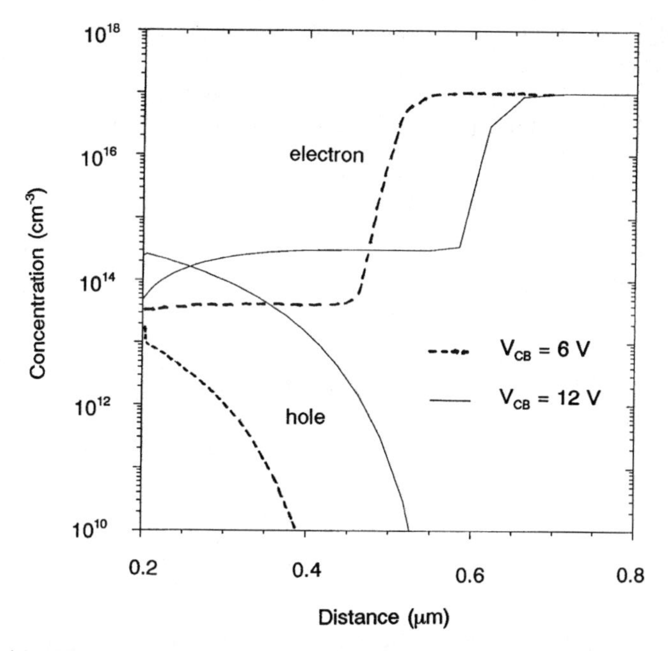

FIGURE 7.16 Electron and hole concentrations in the collector (after Yuan and Ning, Ref. 17 © Elsevier Science).

then used to predict the collector–base junction capacitance at high collector–base bias. The results are illustrated in Fig. 7.18. The collector–base junction capacitance decreases with collector–base voltage and then increases with V_{CB} when the device is in close proximity to avalanche breakdown. An abrupt increase of the collector junction capacitance at $V_{CB} > 12$ V is seen in Fig. 7.18, which is consistent with base current reversal when V_{CB} is approaching BV_{CB0}.

7.1.5 Avalanche Effect on the Output Conductance

Avalanche multiplication also changes the output conductance of the HBT. The output conductance of the bipolar transistor is defined as

$$g_o \equiv \frac{\partial J_C}{\partial V_{CE}}$$

$$= \frac{\partial J_C}{\partial M}\frac{\partial M}{\partial V_{CE}} \approx \frac{\partial J_C}{\partial M}\frac{\partial M}{\partial V_{CB}} \tag{7.25}$$

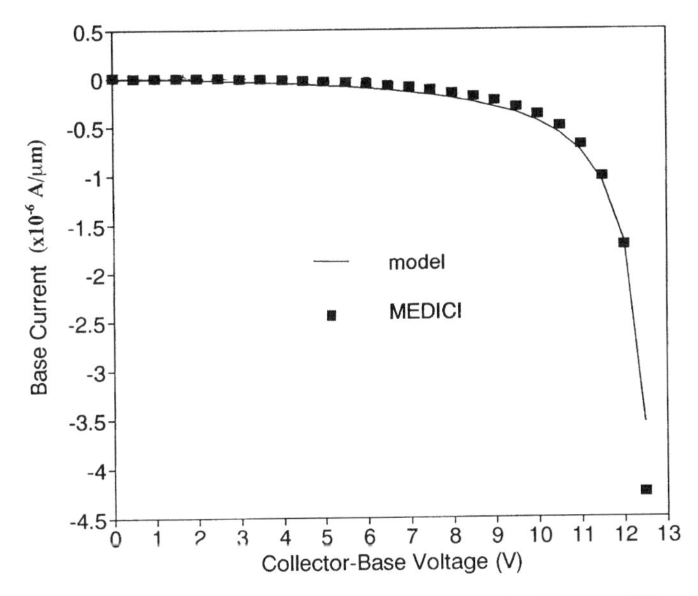

FIGURE 7.17 Base current versus collector–base voltage (after Yuan and Ning, Ref. 17 © Elsevier Science).

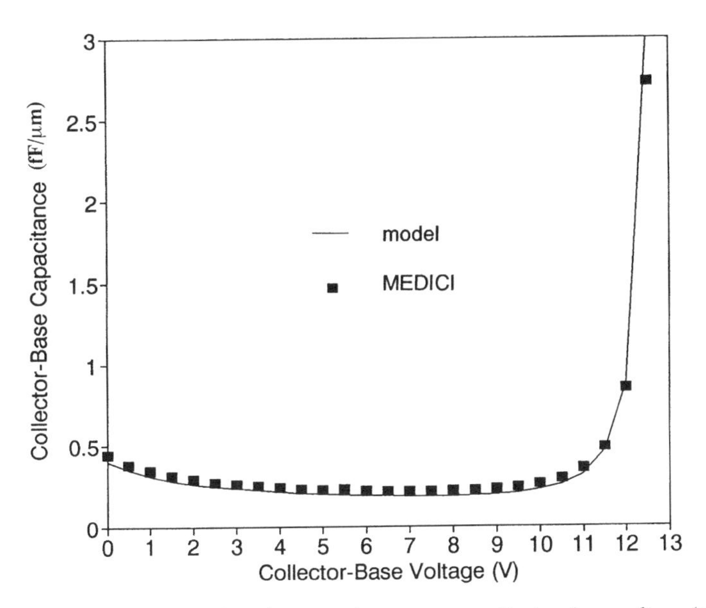

FIGURE 7.18 Collector–base junction capacitance versus collector–base voltage (after Yuan and Ning, Ref. 17 © Elsevier Science).

Using (7.9) and (7.22), the output conductance per unit area is derived as

$$g_o = J_C \frac{nV_{CB}^{n-1}}{BV_{CB0}^{n-1}\left[1 - \left(V_{CB}/BV_{CB0}\right)^n\right]^2} \tag{7.26}$$

7.1.6 Breakdown and Speed Considerations in InGaAs HBTs

In the III/V compound heterojunction bipolar transistors, AlGaAs/GaAs and AlInAs/GaInAs HBTs exhibit different temperature behavior in collector breakdown voltage [18]. Figure 7.19 shows the common-emitter I–V characteristics at 25 and 125°C for the GaAs and $In_{0.53}Ga_{0.47}As$ HBTs. The dc current gains are $\beta(\text{GaAs}) = 60$ and $\beta(In_{0.53}Ga_{0.47}As) = 130$ for $I_C = 1$ mA. It is evident that there are significant differences in the I–V characteristics of the transistors. For the GaAs HBT, the collector current saturates completely with negligible output conductance. The collector breakdown voltage BV_{CEX} is greater than 10 V at 25°C and actually increases at 125°C. In contrast, the $In_{0.53}Ga_{0.47}As$ HBT has poor saturation characteristics with high output conductance. The collector breakdown voltage, BV_{CEX}, is very low, less than 2.5 V at 25°C, which is related to the high current gain in the transistor. The collector breakdown voltage is also seen to decrease at 125°C. The poor saturation characteristics and low collector breakdown voltages in the $In_{0.53}Ga_{0.47}As$ collector are caused by thermally generated leakage current at the collector–base junction.

It has been noted by many researchers that InGaAs-collector HBTs grown on InP substrates exhibit low common-emitter collector breakdown BV_{CE0} and high output conductance [18,19]. This is believed to be related to the low bandgap in the InGaAs collector. The use of δ-doping techniques for the improvement of breakdown characteristics [20] may not be applicable to the InP/InGaAs due to the high electric fields associated with δn^+ or δp^+ layers. Although the collector–base breakdown voltage increases with collector thickness, the collector transit time becomes larger in this case and degrades the overall speed performance. To increase the collector breakdown voltage, the use of InP rather than InGaAs collectors is attractive. InP has a larger bandgap energy (1.35 eV) and greater saturation velocity (1.3×10^7 cm/s) than InGaAs. The conduction spike at the collector–base heterojunction, however, tends to degrade the electron transport.

The study of breakdown–speed characteristics has been carried out using single and double heterostructures with InGaAs and InP collectors [21]. The design of different-collector HBTs is shown in Fig. 7.20. The thickness, doping, and composition of the emitter, base, and subcollector are kept the same for all designs in order to have a fair comparison. All collectors have a thickness of 3000 Å and are either doped at 5×10^{16} cm^{-3} or left undoped. For single HBTs, the InGaAs collectors are chosen based on (A) n^-, (B) p^-, (C) i, and (D) p^--n^- design schemes. The first three structures correspond to the conventional, inverted field, and modified (without δp^+ toward the subcollector side) undoped collector designs. Design (E) uses a wider bandgap InP collector to improve the breakdown characteristics. This collector results in a double heterojunction with an abrupt transition from the InGaAs base to the InP collector.

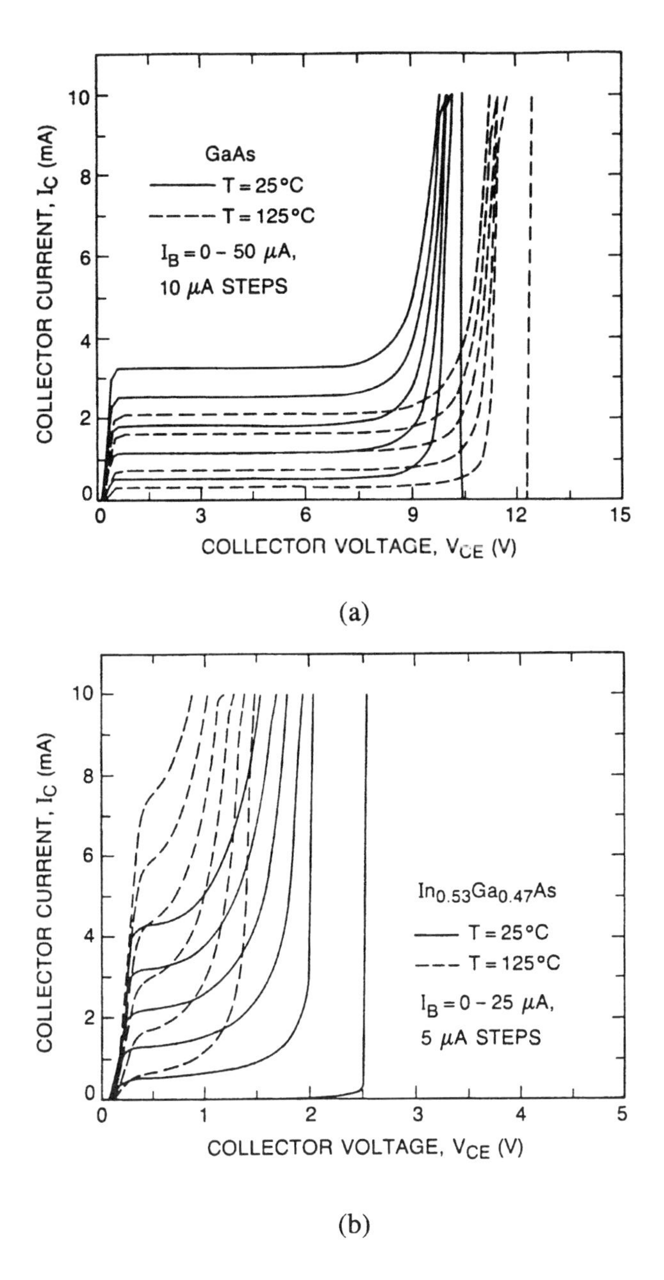

FIGURE 7.19 Collector current versus collector–emitter voltage for (*a*) GaAs and (*b*) InP HBTs (after Malik et al., Ref. 18 © IEEE).

2000 Å	n^+	$In_{0.53}Ga_{0.47}As$	2×10^{19} cm^{-3}
1500 Å	n^+	InP	1.6×10^{19} cm^{-3}
700 Å	n	InP	5×10^{17} cm^{-3}
100 Å	i	$In_{0.53}Ga_{0.47}As$	undoped
500 Å	p^+	$In_{0.53}Ga_{0.47}As$	1×10^{19} cm^{-3}
3000 Å		Collector	5×10^{16} cm^{-3} or undoped
5000 Å	n^+	$In_{0.53}Ga_{0.47}As$	2×10^{19} cm^{-3}
	S. I.	InP	

Collector Options

(A) n^- InGaAs (conventional)
(B) p^- InGaAs (inverted field)
(C) i InGaAs (undoped)
(D) p^-(top)-n^-(bottom) InGaAs 1500 Å each
(E) n^- InP (abrupt)
(F) n^- InGaAsP/InP 1500 Å each (parabolically graded)

FIGURE 7.20 Design of different-collector HBTs (after Chau et al., Ref. 21 © IEEE).

The base–collector heterojunction spike can be smoothed out by using composition grading in design (F), where a parabolically graded profile is used.

The collector breakdown voltage BV_{CB0} of InP/InGaAs single HBTs as a function of collector current density is depicted in Fig. 7.21. The breakdown voltage was obtained from one-dimensional device simulation. The simulation results underestimate the breakdown characteristics compared to the experimental data. However, the expected trends of the simulation results are correct and therefore serve as guidelines

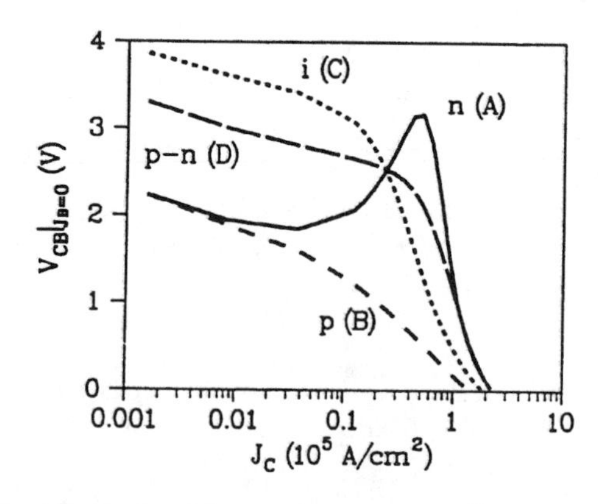

FIGURE 7.21 Base–collector breakdown voltage versus collector current (after Chau et al., Ref. 21 © IEEE).

for comparing the relative performance of various designs rather than estimating the absolute values of the breakdown voltages. In Fig. 7.21 the breakdown voltage is highest for doping profile (C) and smallest for (A) and (B) at low current densities. At high collector current densities, devices (B), (C), and (D) show monotonic decrease in BV_{CB0} with J_C, whereas device (A) shows a peak in the breakdown voltage–J_C characteristics. This makes BV_{CB0} of device (A) larger than the other three devices at high J_C. The difference arises because of the change in the peak electric field with current. In the n^- collector (A), E_{peak} occurs at the base–collector interface, whereas in (B) and (C), it occurs at the collector–subcollector interface. Since high injection at the collector changes the position and magnitude of E_{peak} in design (A), the breakdown voltage increases with J_C initially and then decreases rapidly at high collector current densities. In the case of p^- and i collectors, E_{peak} always occurs at the collector–subcollector interface, resulting in a rapid decrease of BV_{CB0} with J_C under all current-level injections.

Figure 7.22 shows the sum of the base transit time and the collector space-charge layer transit time of the four HBTs as a function of collector current density. The p^- collector (device B) has the shortest total transit time followed by i (C), p^--n^- (D), and n^- (A) collector devices. Device A is the slowest due to the high electric field at the base end of the collector SCR region which results in electron excitation to the upper energy valleys. The other three HBTs [devices (B), (C), and (D)] all show broad velocity overshoot characteristics, which give rise to shorter transit time. The figure of merit

$$V_{CB|J_B=0}/\ (\tau_B + \tau_{SCR})\big|_{V_{CE}=V_{CE,br}/2}$$

is plotted as a function of collector current density in Fig. 7.23. The figure of merit for all four devices decreases with the collector current density. Among the four HBTs, the i-collector design (C) has the highest breakdown voltage and the p^--collector

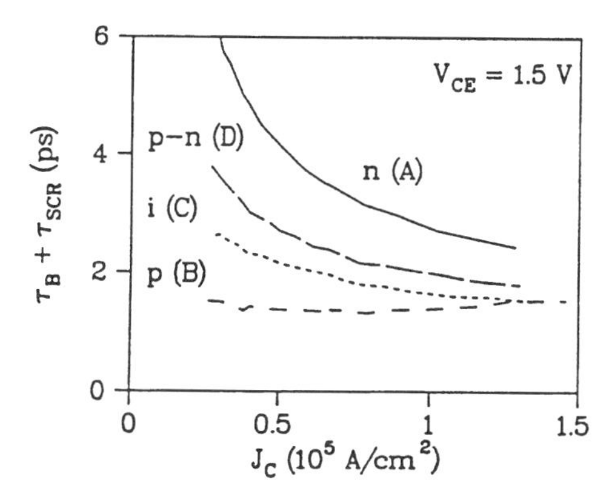

FIGURE 7.22 $\tau_B + \tau_{SCR}$ versus collector current density (after Chau et al., Ref. 21 © IEEE).

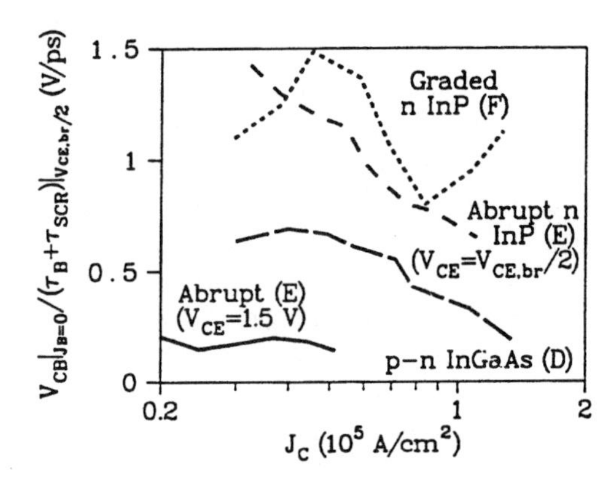

FIGURE 7.23 Figure of merit versus collector current density (after Chau et al., Ref. 21 © IEEE).

design (B) has the shortest total transit time. In terms of speed–breakdown trade-offs, the results of Fig. 7.23 show that p^--n^- collectors have the best performance, while design p^- has the worst performance. Thus optimum collector designs such as (B), which normally give the highest speed, do not necessarily combine good power characteristics.

7.2 THERMAL INSTABILITY

A three-dimensional transmission-line matrix simulation was used to evaluate the thermal behavior of power HBTs with multiple emitter fingers [22]. The power device consists of five cells. Each cell has two emitter fingers grouped closely together by the emitter–base self-alignment process. The temperature distribution at a dissipated power density of 2 mW/μm^2 under steady state is simulated. Figure 7.24*a* shows the isometric view of the temperature distribution of five emitter fingers and Fig. 7.24*b* is the contour plot with intervals of 15°C. The isometric projection and the corresponding temperature contours clearly display that two peak values of the junction temperature occur at the centers of the two central emitter fingers. The junction temperature drops sharply around the central cell, with an average temperature gradient of as high as 9°C/μm in these areas. From a device-reliability point of view, the peak junction temperature is the most important parameter for the device thermal design and should be used to characterize the device thermal behavior.

Despite the fact that the multifinger HBT has become increasingly important and popular in high-power microwave applications, efforts to model such a device analytically have been limited due to the complicated nature of thermal coupling between the neighboring emitter fingers. Consider a three-finger-pattern HBT. The temperature of the two outer fingers, which have identical thermal properties due to the symmetrical

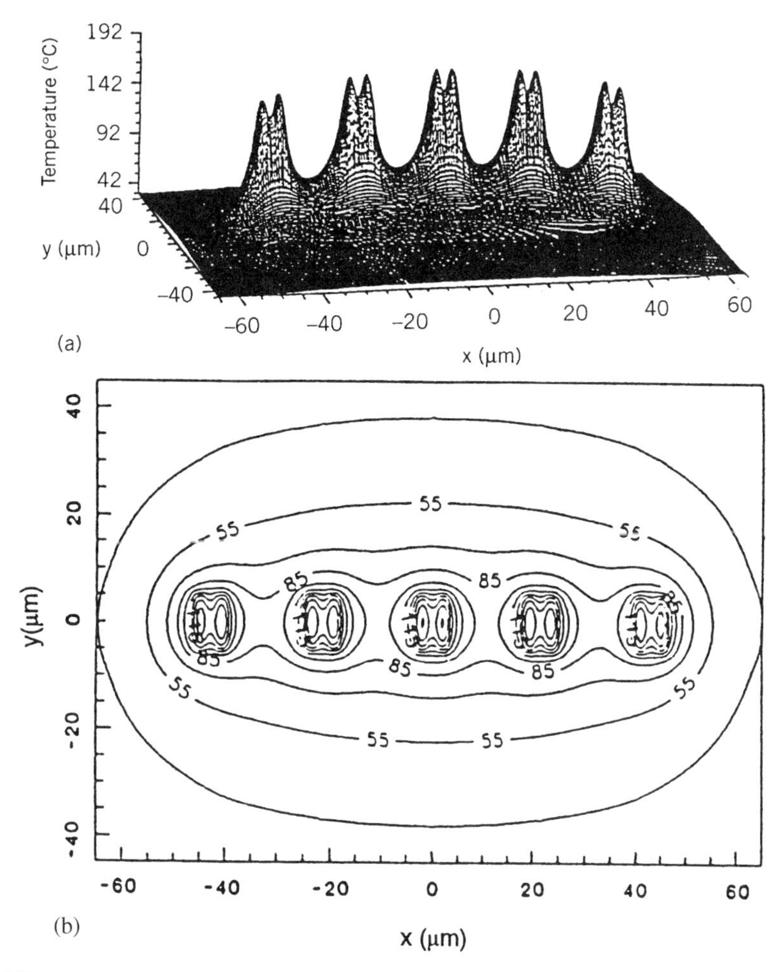

FIGURE 7.24 (*a*) Isothermal and (*b*) contour plot of a five-finger HBT (after Gui et al., Ref. 22 © IEEE).

geometry, and the temperature at the center finger are given by

$$T_S = T_0 + P_{s,S} R_{\text{th},S} + P_{s,C} R_{cp,C} \tag{7.27}$$

$$T_C = T_0 + P_{s,C} R_{\text{th},C} + 2P_{s,S} R_{C,S} \tag{7.28}$$

where the subscripts S and C denote outer and center fingers, respectively. R_{th} is the thermal resistance due to self-heating in the unit HBT, and R_{cp} is the thermal resistance due to heating from the neighbor emitter elements.

R_{th} can be determined using (3.45). The value of R_{cp} depends on the geometry of the emitter fingers and the process including the emitter mesa etching and metallization. For typical mesa-etch HBTs, R_{cp} has been determined empirically as

$$R_{cp} = \delta R_{\text{th}} \left(\frac{10}{S_f} \right)^{1.5} \tag{7.29}$$

where δ is a fitting parameter with a typical value of 0.25, and S_f is the emitter-finger spacing in μm. Decreasing S_f will increase R_{cp}.

7.2.1 Emitter Collapse Phenomenon

Power heterojunction bipolar transistors are designed to deliver large amounts of power at high frequencies. Because these HBTs are operated at high power densities, the ultimate limit on their performance is imposed by thermal considerations. Recently, a thermal phenomenon was observed for the case when a multifinger GaAs HBT was operating at high power densities [23]. This phenomenon, referred to as *emitter collapse*, occurs when one finger of a multifinger HBT suddenly conducts most of the device current. This leads to an abrupt decrease in current gain. The collapse does not cause the HBTs to fail immediately, and the device can be biased in and out of the collapse by adjusting the level of power dissipation. This collapse, however, degrades the device performance significantly and should be avoided.

The collapse phenomenon is illustrated by the common-emitter I–V characteristics measured at constant base currents. As the collector–emitter voltage increases, the power dissipation in the HBT increases. This increase in power dissipation elevates the junction temperature above the ambient temperature and causes an increase in base and collector currents. At high V_{CE}, impact ionization increases I_C. At a certain collector–emitter bias, the collapse of collector current occurs, as shown in Fig. 7.25. At an even higher collector–emitter voltage, the collector current goes to infinity due to the domination of avalanche breakdown again.

Consider the case of a multifinger HBT. If one finger becomes slightly warmer than the others, this particular finger conducts more current at a given base–emitter bias. This increased collector current, in turn, increases the power dissipation in the junction and raises the junction temperature and collector current even further. The collapse occurs when the junction temperature in one finger becomes much hotter than the rest of the fingers.

Note that the collapse does not cause the HBTs to fail immediately. In fact, the collapse is reversible and the same device can be measured for several experiments. This is markedly different from thermal runaway in silicon BJTs, which causes the devices to fail irreversibly. Both thermal runaway in BJTs and the collapse of HBTs result from an increase in junction temperature. Since the current gain in Si bipolar transistors increases rather than decreases with temperature, the increased current in that finger of the BJT will draw more and more current. This is in contrast to the collapse when I_C decreases after the critical condition occurs.

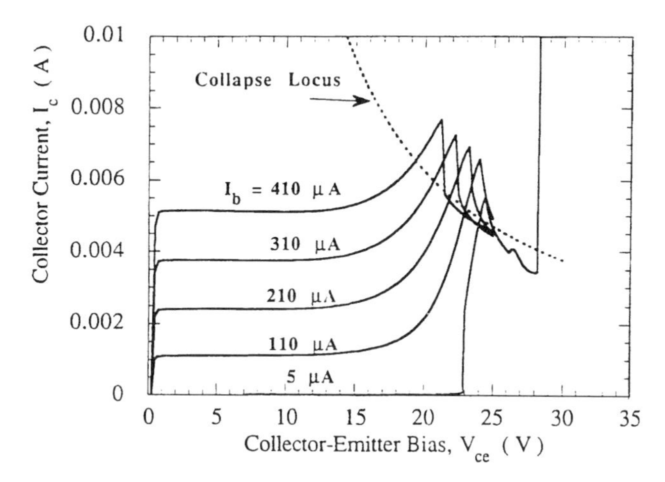

FIGURE 7.25 Collector current versus collector–emitter voltage (collapse phenomenon).

The collector current collapse (or crush) effect has also been observed by using numerical electrothermal simulation [24]. The contour plots of the temperature distribution on the top surface of a three-finger HBT under the bias of $V_{CE} = 2, 3, 5,$ and 6 V are shown in Fig. 7.26. The first two operation points are before the current crush. At this point the temperature variation is small across the three emitter fingers. The last two operating points occur after the current crush. The temperature distribution here shows a drastic difference between the center and the other two emitter fingers. As V_{CE} increases, the center finger will eventually conduct the largest base current, and the other two fingers become nearly inactive. The collector current will continue to decrease due to the smaller current gain at higher junction temperatures.

7.2.2 Relation Between Emitter Collapse and Avalanche Breakdown

The interdependence between collapse phenomenon and avalanche breakdown is discussed in detail in this section. The measured and calculated I–V characteristics for $I_B = 5\ \mu$A are shown in Fig. 7.27. The calculated collector current of a six-finger HBT flowing in the hot finger is I_{C1} and in one of the cold fingers is I_{C2}. The calculated total current is $I_{C1} + 5I_{C2}$. Significant avalanche multiplication starts at $V_{CE} \approx 22.5$ V for both curves. The collector current reaches a maximum value of 5.3 mA before the collapse occurs and the collector current decreases. In Fig. 7.27, at point A, which corresponds to $V_{CE} \approx 22.5$ V and $I_C \approx 0$ A, the collector current level is well below the collapse locus. Therefore, the total device current at point A is uniformly distributed among the six fingers. As V_{CE} increases slightly past point A, impact ionization in the base–collector junction becomes noticeable. The product of the avalanche multiplication factor M_n and the current transfer ratio α_0 approaches unity, and causes the collector current to increase rapidly toward the collapse locus. This well-known trend of increasing I_C, however, stops abruptly upon initiation of

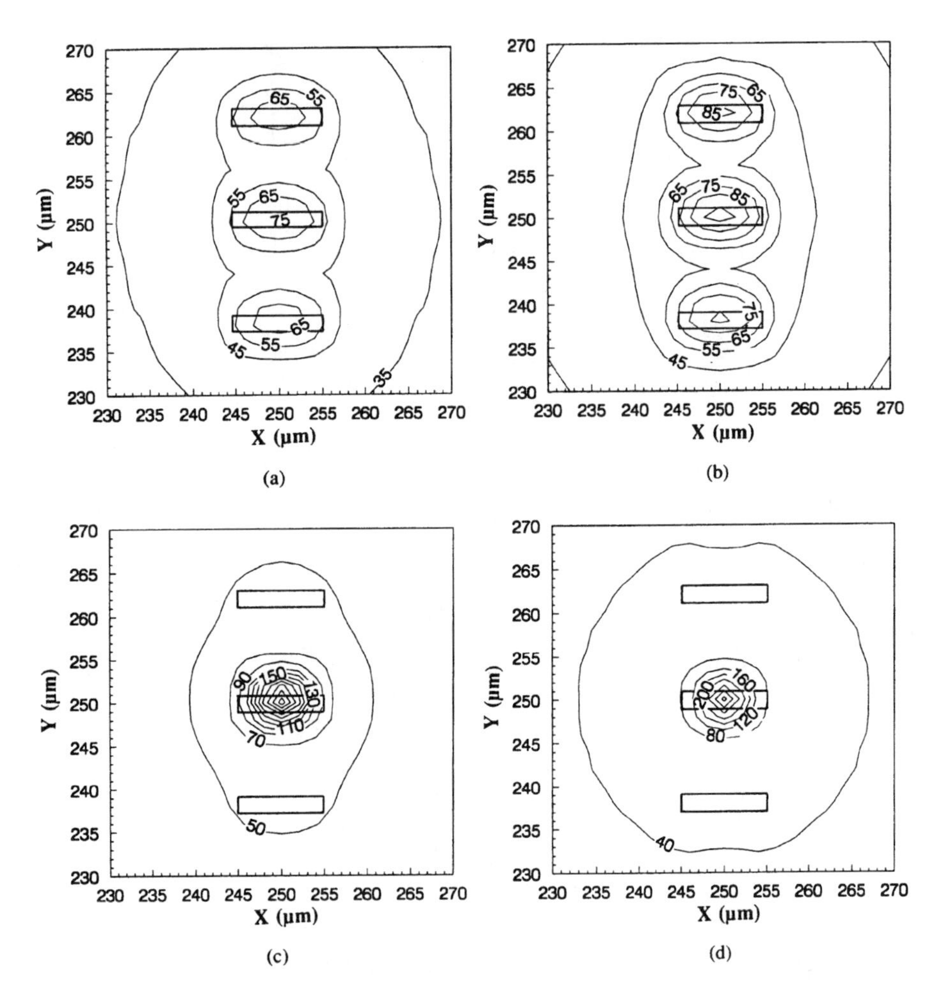

FIGURE 7.26 Contour plot of temperature distribution for (*a*) $V_{CE} = 2$ V, (*b*) $V_{CE} = 3$ V, (*c*) $V_{CE} = 5$ V, and (*d*) $V_{CE} = 6$ V (after Liou and Bayraktaroglu, Ref. 24 © IEEE).

the collapse at point B, corresponding to $V_{CE} \approx 24.1$ V. Beyond this point, a further increase in collector–emitter voltage results in a greater proportion of the collector current being conducted in the hot finger, until all of the available device current is conducted through the hot finger.

Between point B, which marks the onset of the collapse, and point D, where the device burns out, device operation can be separated into two phases according to the features in the junction temperature of the hot finger (T_1). Between points B and C, T_1 quickly increases toward some drastically higher temperature than that prior to collapse. The rapid increase of T_1 in the hot finger corresponds to the rapid decrease of T_2 in the cold finger. During this phase, the incremental increase of I_{C1} with respect

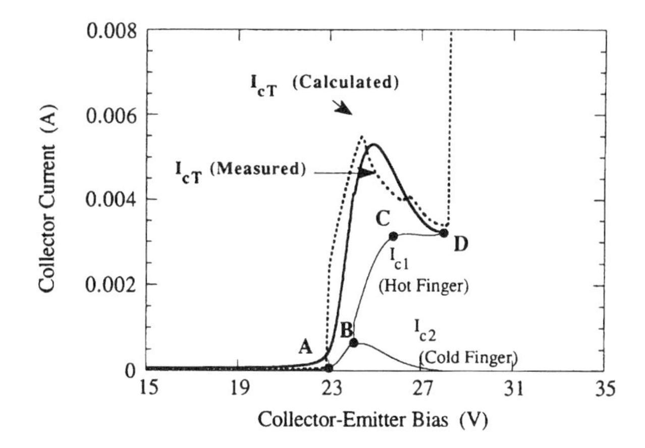

FIGURE 7.27 Measured and calculated I–V characteristics (after Liu, Ref. 25 © IEEE).

to V_{CE} is high since a greater amount of current flows into the hot finger. The surge in power density further increases the hot finger junction temperature. In contrast, the cold finger loses more and more of its current with increasing V_{CE} and the cold finger temperature plummets. Between points C and D, most of the available current has been directed to the hot finger. The six-finger HBT is effectively a one-finger HBT, with the hot finger conducting the entire device current.

The classification of the two distinct phases in the T_1 characteristics is critical to the understanding of the collapse–avalanche breakdown interaction. Figure 7.28 shows both electron ionization multiplication factor M_n and zero-ionization current gain β_0 as a function of V_{CE}. The two phase boundaries are marked in Fig. 7.28 as dashed lines. All six fingers share current equally before phase I. Hence M_n and β_0 of the six fingers are indistinguishable. In phase I, when the current redistributes among the fingers, T_1 increases dramatically while T_2 decreases toward room temperature. Because in general both M_n and β_0 decrease with temperature, M_{n1} and β_{01} of the hot finger decrease from their values prior to phase I. In contrast, M_{n2} and β_{02} of the cold fingers increase with V_{CE} because the cold finger temperature decreases. In phase II, T_1 remains relatively constant and T_2 continues to decrease slowly toward room temperature. Because T_2 no longer increases dramatically, M_{n1} reverses its decreasing trend and increases gradually with V_{CE} because of the increased ionization rate with higher electric field. Similarly, β_{01} no longer decreases dramatically with V_{CE} but stays relatively constant. Meanwhile, M_{n2} and β_2 continue to increase, as in phase I.

7.2.3 InP HBT Thermal Instability

For AlGaAs/GaAs heterojunction bipolar transistors, the collapse of current gain occurs when the thermal instability condition is met. Because the current gain decreases

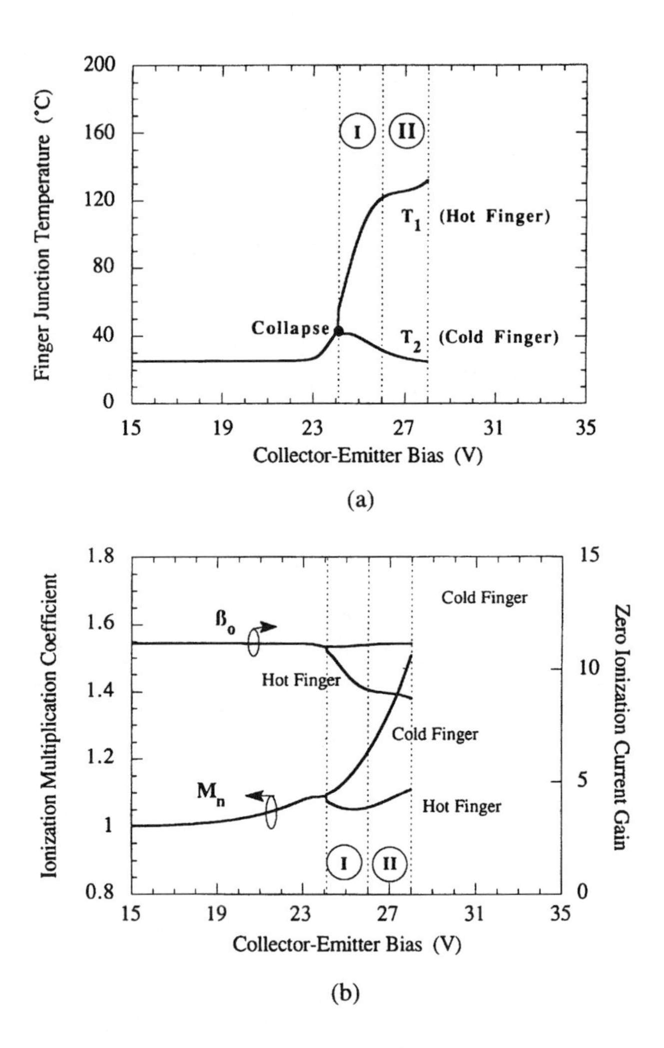

FIGURE 7.28 Calculated (*a*) junction temperature and (*b*) electron ionization multiplication factor and current gain as a function of V_{CE} (after Liu, Ref. 25 © IEEE).

with temperature in the AlGaAs/GaAs HBT, the increasing junction temperature during thermal instability causes the current gain to decrease dramatically leading to the collapse. For InP heterojunction bipolar transistors, the current gain is essentially independent of the junction temperature. Therefore, further examination is required to see how the InP HBT characteristics are modified upon entering the thermal instability condition.

Figure 7.29 shows the measured I–V characteristics of a two-finger InP HBT. Each finger has an area of $2 \times 10\ \mu m^2$. The total collector current of these two fingers

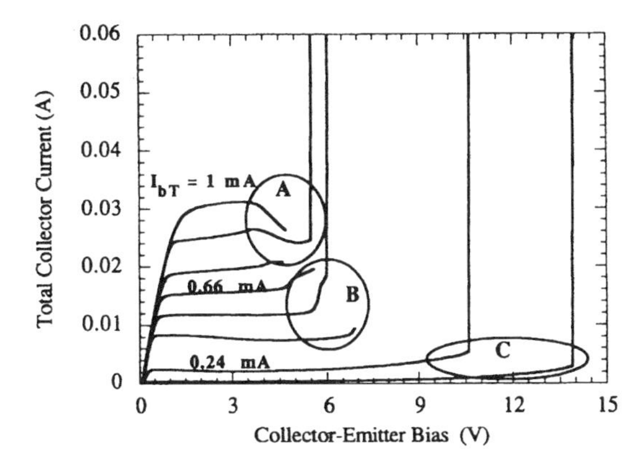

FIGURE 7.29 Measured I–V characteristics of a two-finger InP HBT (after Liu et al., Ref. 26 © IEEE).

is plotted against V_{CE} as a function of the total base current from 0.07, 0.24, 0.44, 0.56, 0.66, 0.76, 0.86 to 1 mA. The emitter resistance is measured to be 8 Ω per finger. The familiar decrease of I_C with V_{CE} appears only at high current levels in the region marked A in Fig. 7.29. At medium current levels (region B), the collector current increases abruptly at certain values of V_{CE}, opposite to what is observed in region A. At low current levels, the collector current increases toward infinity similarly to the breakdown mechanism.

Figure 7.30 displays the individual collector currents corresponding to the I_C–V_{CE} plot in Fig. 7.29. The dashed and dotted lines in Fig. 7.30 represent the hot finger and cold finger, respectively. In contrast to Fig. 7.25, the individual currents clearly demonstrate that thermal instability occurs at some bias conditions in all three regions.

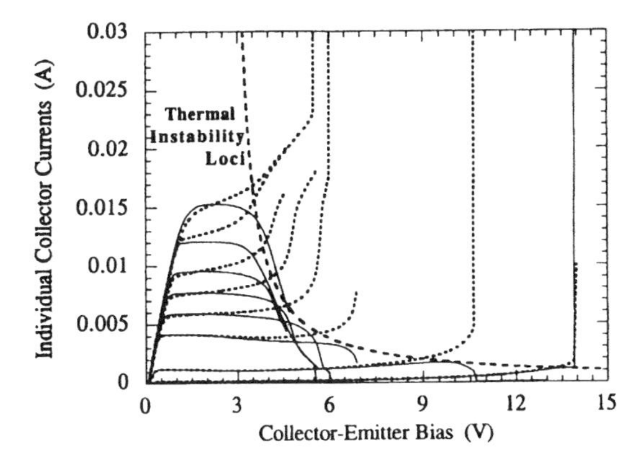

FIGURE 7.30 Individual collector current corresponding to the I–V of Fig. 7.29 (after Liu et al., Ref. 26 © IEEE).

Before the thermal instability loci, both fingers conduct relatively identical currents. Beyond the instability loci, a current and junction temperature increase in the hot finger occurs. When the hot finger takes away the cold finger current, the hot-finger current gain decreases, resulting in the collapse of the overall collector current. The collapse observed in a multifinger InP HBT is in principle similar to that in the AlGaAs/GaAs HBT. The major difference is that the gain collapses in the InP HBT are due to its current gain dependence on collector current, while the gain collapses in the AlGaAs HBT are due to its current gain dependence on temperature. Therefore, the collapse in the AlGaAs/GaAs HBT occurs over the entire range of current levels and the collapse in the InP HBT occurs only at high current levels.

7.2.4 Modeling the Emitter Collapse Loci

The critical current level at the collapse loci separating the normal region and the collapse region is modeled as [27]

$$I_{\text{critical}} = N \frac{n_f k T_A}{q} \frac{1}{\phi_{fb} R_{\text{th}} V_{CE} - R_e} \tag{7.30}$$

where N is the number of emitter fingers and ϕ_{fb} is the base–emitter junction voltage feedback coefficient.

Figure 7.31 plots the collapse loci of the I–V characteristics at the four substrate temperatures. In this figure, solid lines represent the measured collapse loci and the dashed lines are the calculated collapse loci. The temperature dependence of all the parameters was evaluated. The ideality factor n_f of the collector currents are 1.165, 1.110, 1.070, and 1.041 as T_A increases from 25, 100, 175, and 250°C, respectively. The emitter resistances measured with an open-collector method [28] are found to be

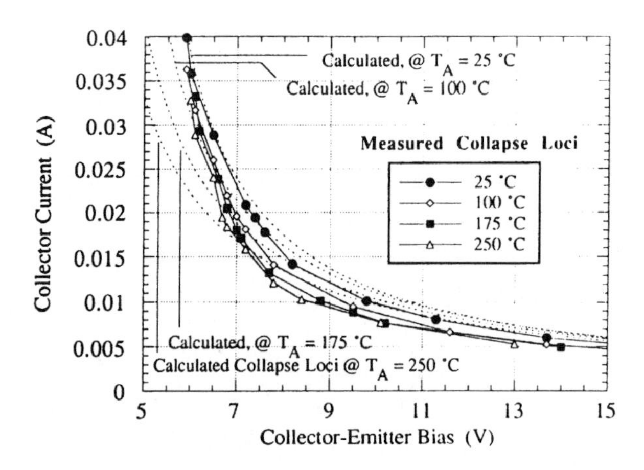

FIGURE 7.31 Collapse loci at the four different substrate temperatures (after Liu and Khatibzadeh, Ref. 27 © IEEE).

4.0 Ω per finger throughout the various ambient temperatures. The thermal-voltage feedback coefficients are determined by measuring the base–emitter turn-on voltage at $I_C = 60$ mA at various substrate temperatures between 25 and 300°C. The base–collector bias was maintained at 0 V during such measurements, so that the actual junction temperature was well approximated by the substrate temperature. Similar to the emitter resistance, ϕ_{fb} does not vary with ambient temperature, remaining roughly at 1.0 mV/°C. Among all the parameters, the thermal resistance is the one that manifests the strongest dependence on T_A. The thermal resistance per finger, estimated from measurements, is found to be 930, 1050, 1200, and 1370°C/W as T_A increases from 25, 100, 175, and 250°C, respectively. The increase in thermal resistance with respect to T_A is caused by the reduction of the GaAs thermal conductivity at higher temperatures. As seen in Fig. 7.31, the model predictions are in good agreement with the experimental data. Equation (7.30) is generally correct and applicable at various elevated temperatures.

The thermal instability criterion is the S-factor proposed for silicon bipolar transistors [29]. The S-factor is given by

$$S = R_{\text{th}} V_{CE} \left. \frac{\partial I_C}{\partial T} \right|_{V_{BE}=\text{const}} \tag{7.31}$$

The transistor is said to be thermally unstable when $S > 1$. From the plot of collector current versus junction temperature rise, the S-factor can be plotted. Figure 7.32 shows the S-factor as a function of junction temperature rise at different base–emitter voltages. The curve connecting the points where $S = 1$ is the S-factor loci. Note that in practice S-factor values obtained are very close to, but not exactly equal to, unity. Thus the S-factor loci are the points having the highest S-factor values. In fact, the collapse loci and the S-factor loci are governed by the same equation, namely, (7.31).

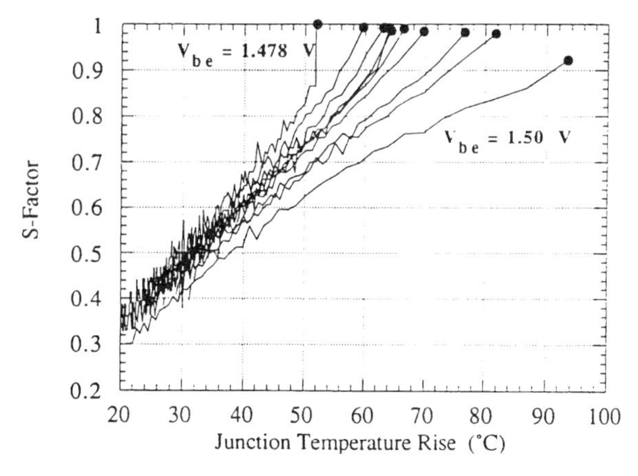

FIGURE 7.32 S-factor versus junction temperature (after Liu and Khatibzadeh, Ref. 27 © IEEE).

For the S-factor loci, one can write

$$V_{BE} = \frac{n_f kT}{q} \ln \frac{I_C}{I_S} + R_e I_C + \frac{E_{G0}}{q} - \frac{\beta^*}{q} T \tag{7.32}$$

Taking the derivative of (7.32) with respect to T, we have

$$0 = \frac{\partial I_C}{\partial T} \left(\frac{n_f kT}{q} \frac{1}{I_C} + R_e \right) + \frac{n_f k}{q} \ln \frac{I_C}{I_S} - \frac{\beta^*}{q} \tag{7.33}$$

where β^* is the coefficient that measures the amount of bandgap shrinkage as T increases. Solving for $\partial I_C/\partial T$ and setting $S = 1$, one then deduces the same equation for the S-factor loci as (7.31).

If the temperature variation of R_{th} is considered, the S-factor defined in (7.31) should be revised as

$$\begin{aligned} S &= \frac{d}{dT} (R_{\text{th}} V_{CE} I_C)|_{V_{BE}=\text{const}} \\ &= R_{\text{th}} V_{CE} \left(\frac{\partial I_C}{\partial T} + \frac{I_C}{R_{\text{th}}} \frac{\partial R_{\text{th}}}{\partial T} \right) \Bigg|_{V_{BE}=\text{const}} \end{aligned} \tag{7.34}$$

From (7.34) the governing equation of the collapse loci, taking into account the R_{th} variation with temperature, can be derived:

$$I_{\text{critical}} = N \frac{n_f k T_A}{q} \frac{1}{\phi_{fb} R_{\text{th}} V_{CE} \left\{1 - [(T - T_A)/R_{\text{th}}](\partial R_{\text{th}}/\partial T)\right\}^{-1} - R_e} \tag{7.35}$$

7.3 DESIGN IN THERMAL STABILITY

7.3.1 Emitter Ballasting Resistors

To reduce the likelihood of the emitter collapse phenomenon, and to ensure a more uniform current distribution, emitter ballasting (EB) resistors can be inserted in series with each emitter finger. However, the determination of the value of ballasting resistance is not simple. There are a number of conflicting requirements. Too large a resistance will cause degradation of power gain while too small a resistance will not effectively protect the HBT from thermal instability. The optimal value of the total ballasting resistance is determined by [30]

$$R_E = \frac{-\phi_T V_{CE} R_T - (R_{EC} + nkT/qI_E)[1 - (T_j - T_c)\varsigma]}{1 + (T_j - T_c)(\zeta - \varsigma)} \tag{7.36}$$

where R_T is the total thermal resistance from the active device area to the point where the case temperature T_c is measured, $T_j = I_C V_{CE} R_T + T_c$, R_{EC} is the emitter contact resistance, $\phi_T = \partial V_{BE}/\partial T$, $\zeta = 1/R_e(dR_e/dT)$, and $\varsigma = 1/\kappa(T)(d\kappa/dT)$.

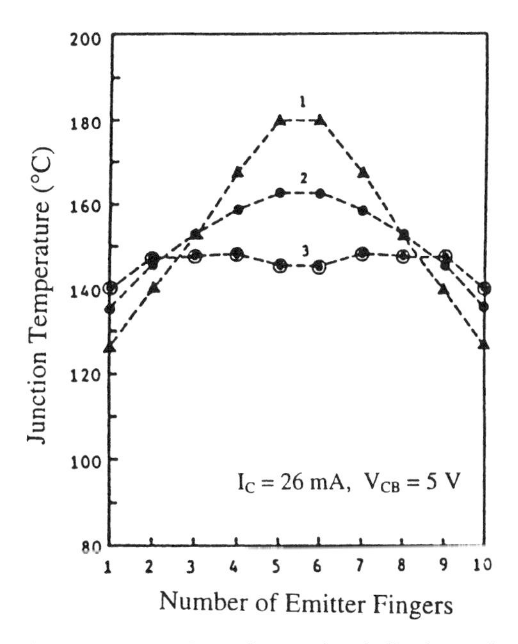

FIGURE 7.33 Junction temperature for various emitter ballasting resistors (after Gao et al., Ref. 30 © IEEE).

The effect of the emitter ballasting resistors on the junction temperature of the AlGaAs/GaAs HBT is shown in Fig. 7.33. In the figure, curve 1 represents the HBT without ballasting resistors, curve 2 represents the HBT with equally valued emitter ballasting resistors, and curve 3 represents the HBT with unequally valued emitter ballasting resistors. As seen in Fig. 7.33, the nonuniform junction temperature distribution with multiple emitter fingers over the entire active region can be improved by using ballasting resistors, especially unequally valued emitter resistors. It is anticipated that highly thermostable HBTs with homogeneous junction temperature can be fabricated by this technique.

The effects of emitter-finger layout, emitter resistance, and substrate thermal conductivity on collector current crush are also demonstrated by the numerical simulation results [25]. The dependence of current instability on the spacing between emitter fingers is shown in Fig. 7.34*a*. The maximum temperature on the chip as a function of V_{CE} is also given. The case of zero spacing corresponds to an HBT with only one emitter finger with an area of $10 \times 6\ \mu\text{m}^2$. As the spacing between the emitter fingers increases, V_{CE} shifts slightly to a smaller value, while I_C shifts slightly to a larger value, at the current crush point. This result shows that the threshold power current instability is only minimally improved by increasing the emitter-finger spacing beyond 10 μm and not increasing the number of emitter fingers. The dependence of current crush due to thermal instability on the emitter specific resistance is shown in Fig. 7.34*b*. As the emitter specific resistivity ρ_e increases, threshold power increases. This result

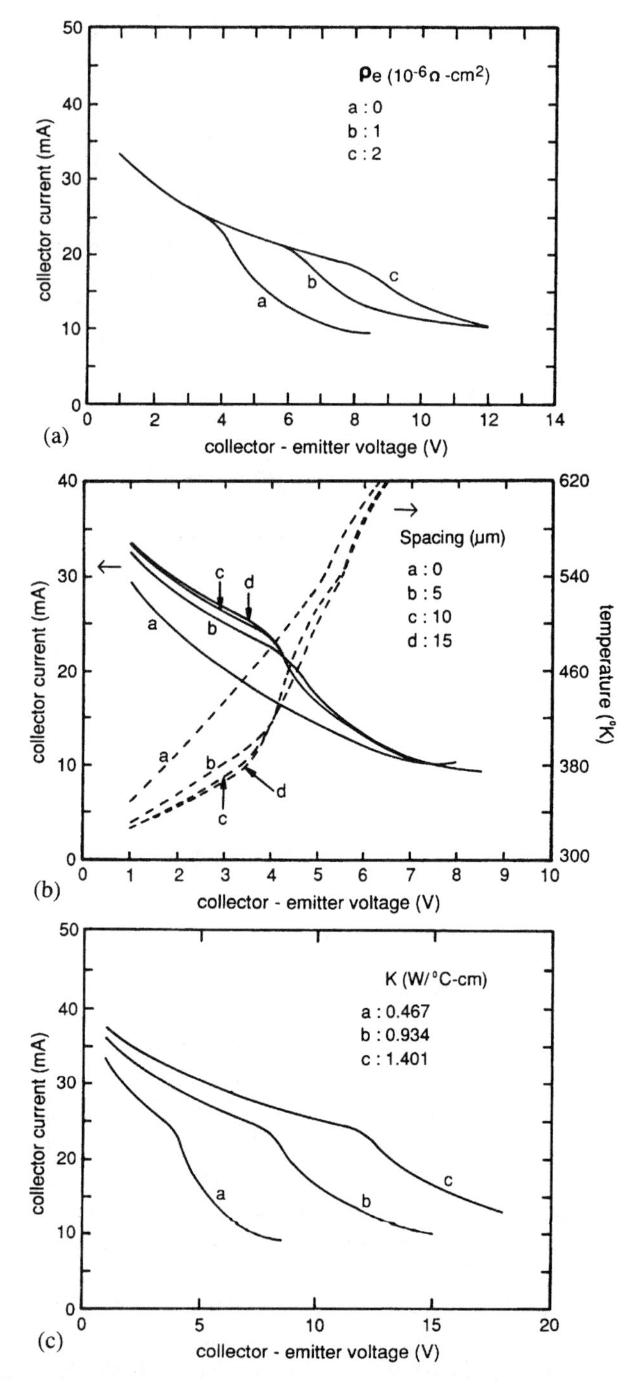

FIGURE 7.34 I_C–V_{CE} for various (*a*) emitter resistivities, (*b*) spacing, and (*c*) substrate conductivities (after Liou and Bayralktaroglu, Ref. 24 © IEEE).

verifies the reduction in thermal instability by implementing a ballast resistor at each emitter node. The dependence of current crush on the substrate thermal conductivity is shown in Fig. 7.34*c*. A reduction in the thermal resistance due to an increase in the thermal conductivity results in a smaller temperature rise and a larger critical voltage for current crush. Improvement in thermal instability by increasing substrate conductivity is the most efficient way of increasing power capability without encountering high-frequency limitations and consuming extra semiconductor area. This suggests that the power HBT on Si substrate provides a smaller negative differential resistance and a larger current crush voltage and is more thermally stable than the HBT on GaAs substrate.

7.3.2 Emitter Thermal Shunt

Recently developed thermal shunt technology [31,32] has the benefit of improving the thermal stability of the HBT while the device cutoff frequency and efficiency are not reduced. Thermal-shunted HBTs utilize a thermal stabilization feature to minimize problems due to nonuniform junction temperature distribution, without the addition of ballast resistors. In thermal-shunted devices, the emitter elements are connected by a thick metal bridge (thermal shunt). Temperature differences between emitter elements are reduced since the emitters are now thermally coupled to each other. To reduce temperature nonuniformities efficiently, the thermal shunt thickness needs to be larger than 10 μm, which is significantly larger than the conventional emitter metallization thickness. In addition, the shunt channels transfer heat to spreading areas on the substrate that further lowers the thermal resistance compared to conventional device geometries. The heat-spread areas are referred to as *thermal lenses*. Figure 7.35 shows a typical layout for a thermally shunted device. The thermal lenses are the areas of the thermal shunts that anchor onto the emitter pads. Bar shaped emitter elements (device B1) are shown in Fig. 7.35*a* and dot-shaped emitter elements (device D1) are displayed in Fig. 7.35*b*. Device B1 has five bar-shaped emitter elements. Each element is a 3-μm $\times$11.8-μm rectangle which is surrounded by a self-aligned base finger that leads to the base pad. The base fingers are separated by 24 μm (edge to edge). The collector is interdigitated between the base fingers. Device D1 has 25 dot-shaped emitter elements. Each row of five dots is surrounded by a self-aligned base finger that leads to the base probe pad. Dots in each base finger are 3 μm in diameter and are separated by 6 μm (edge to edge). The D1 device has a wider thermal shunt than that of the B1 device, due to dot spacing requirements. The total emitter area of both devices is 177 μm^2. A summary of the geometries of the two devices and their thermal resistances at different power dissipations is given in Table 6.1.

Figure 7.36 shows the collector current versus the collector–emitter voltage of device D1 measured at room temperature. The negative differential resistance in the forward active mode is due to self-heating. Current collapse in a nonthermal shunt device occurs at a power density of about 1.33 mW/μm^2 [27], while the thermally-shunted devices have been taken up to a power density of 3.50 mW/μm^2 without showing signs of current collapse. The measured cutoff frequencies are between 50 and 60 GHz and $f_{\max}$ between 85 and 100 GHz. These results compare favorably with

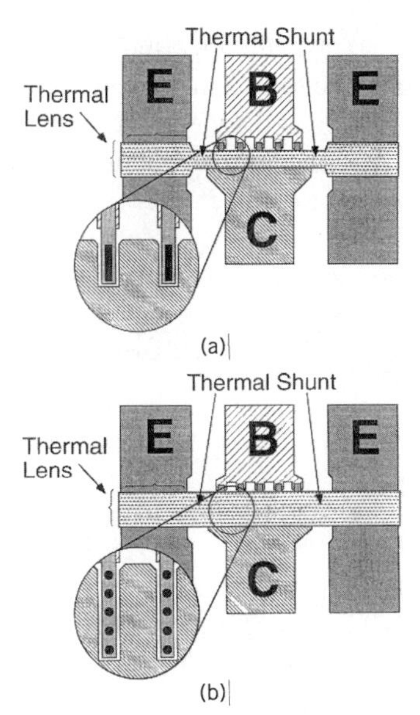

FIGURE 7.35 Layouts for (*a*) an emitter-bar device and (*b*) an emitter-dot device (after Sewell et al., Ref. 32 © IEEE).

those of the non-thermal-shunt device, suggesting that the extra parasitic reactance elements due to the thermal shunt are not significant.

Figure 7.37 shows the junction temperature and extrapolated thermal resistance versus power for bar-shaped emitter and dot-shaped emitter HBTs. The increase in thermal resistance at higher power densities is due to the temperature-dependent thermal conductivity of the GaAs substrate. The dot-shaped emitter devices have

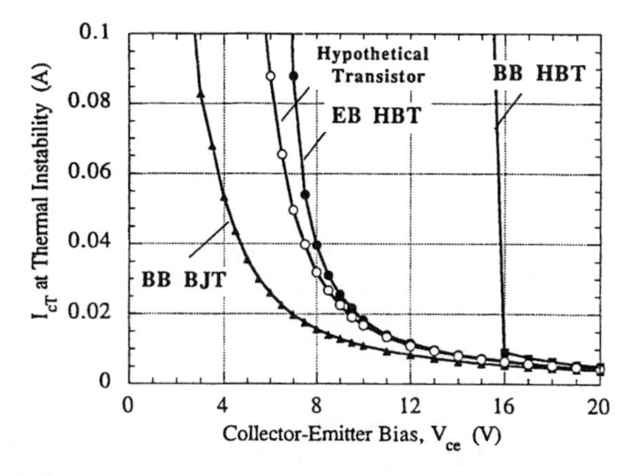

FIGURE 7.36 Collector current versus collector–emitter voltage (after Sewell et al., Ref. 32 © IEEE).

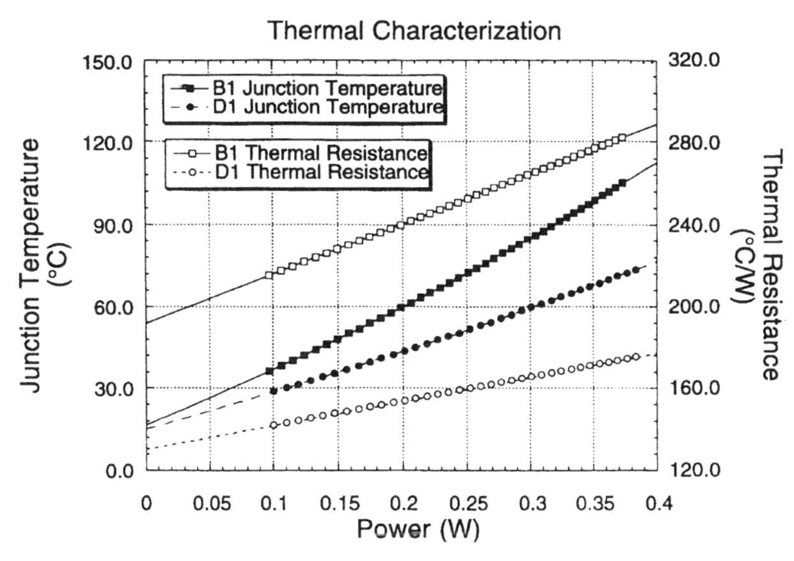

FIGURE 7.37 Junction temperature and thermal resistance versus power (after Sewell et al., Ref. 32 © IEEE).

smaller thermal resistances and junction temperatures than those of the bar-shaped emitter devices over a wide range of power. The lower thermal resistance for the D1 device is due primarily to the spreading of the heat sources and the larger thermal shunt width. At a power of 0.1 W, thermal resistance is 147 and 232°C/W for D1 and B1 devices, respectively. For a conventional HBT without thermal shunt, thermal resistances of 465°C/W (120 μm^2 emitter area) [27], 1000°C/W (60 μm^2 emitter area) [33], and 1500 °C/W (41 μm^2 emitter area) [34] have been reported at 0.1 W.

7.3.3 Base Ballasting Resistors

The use of emitter ballasting resistors is somewhat effective to prevent the collapse of current gain in AlGaAs/GaAs HBTs. The ballasting resistor is placed in the collector current path. The collector efficiency of the overall device is thus degraded. Therefore, the emitter resistance value chosen for a practical HBT is often compromised by performance issues, and absolute thermal stability in the entire operating region is not achieved. An alternative approach is to place the ballasting resistor in the base [35]. Because the base-ballasting (BB) resistance resides outside the collector current path, the trade-off between having absolute thermal stability and attaining respectable large-signal performance is possible. Since the base current is substantially less than the collector current, a large value base-ballasting resistor can be chosen before the dc power dissipation across the ballasted resistor becomes noticeable. With a base-ballasting resistance on the order of 100 Ω per finger, a capacitor with a practical area can then be connected in parallel to bypass the ballasted resistor at microwave frequencies. In this manner, minimization of the resistance's deleterious effects at RF and minimization of the power dissipation at dc is achieved simultaneously. The use of base-ballasting resistors therefore guarantees absolute thermal stability without adversely affecting the overall large-signal power performance.

Figure 7.38 shows the I–V characteristics of an unballasted (UB) two-finger HBT, displaying two distinctive regions separated by the collapse loci. Each finger has an area of 2 ×30 μm^2. The parasitic emitter contact resistance is measured to be 4 Ω per finger. As soon as V_{CE} increases beyond the collapse loci, only one finger remains active while the other ceases to function. Adding an emitter ballasting of 3 Ω per finger moves the collapse loci only slightly; the collapse of current gain still occurs at a low V_{CE} of about 6 V. If the ballasting resistance is increased further, the power dissipation in the resistor becomes a significant portion of the overall dc power dissipation. The ballasted transistor would then suffer from degraded collector efficiency. Figure 7.39 shows the I–V characteristics of a BB HBT grown on the same wafer as the UB HBT in Fig. 7.39. The I–V characteristics in Fig. 7.39 show that the collector current decreases gradually over the entire bias range and exhibits no sign of the current gain collapse. The apparent lack of collapse signifies that base ballasting is an effective alternative to emitter ballasting. Further evidence of the absolute thermal stability of the BB HBT is needed.

Figure 7.40 illustrates the regression characteristics of a BB HBT, obtained by applying a certain V_{CE} and measuring both collector current and base–emitter voltage as the base current increases. This measurement differs from a Gummel measurement in which collector and base currents are measured as V_{BE} increases. For comparison, the regression characteristics of both UB and EB HBTs is also shown in Fig. 7.40. It has been established that the transistor thermal instability, and consequently the collapse, occurs at the bias condition where $\partial I_C/\partial V_{BE} \rightarrow \infty$ (i.e., at the regression point). Compared to EB and UB HBTs, the BB HBT shows that V_{BE} continues to increase throughout the entire range of operation without ever reaching a maximum

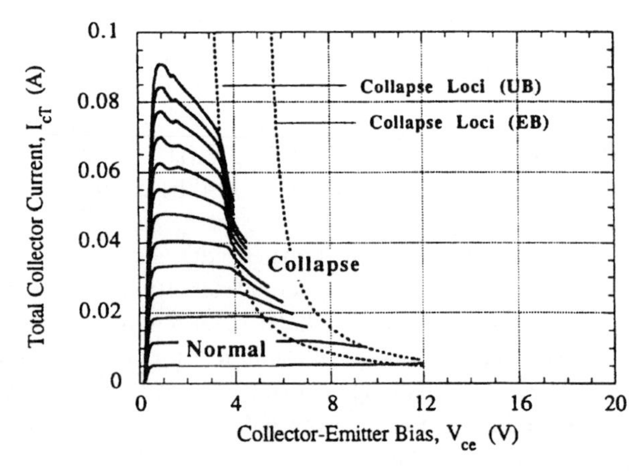

FIGURE 7.38 I–V characteristics for UB and EB HBTs (after Liu et al., Ref. 35 © IEEE).

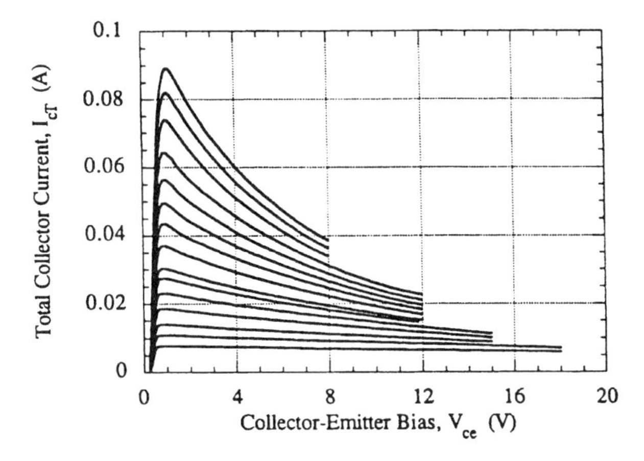

FIGURE 7.39 I–V characteristics for BB HBTs (after Liu et al., Ref. 35 © IEEE).

value and then regressing toward lower values. Since thermal instability never occurs in the BB HBT, Fig. 7.39 does not exhibit the collapse of current gain.

Figure 7.41 shows the large-signal output versus the input power for EB and BB HBTs measured at 9 GHz. Each HBT has an eight-finger unit cell, with a total emitter area of 800 μm^2. The two transistors are physically adjacent, making comparison meaningful. The UB HBTs are found to operate improperly under these operating conditions, possibly due to thermal instability occurring in part of the RF cycle. Therefore, only the results of the BB HBT and the EB HBT are compared. The collector bias was between 9 and 10 V to maximize the output power. It is clear from Fig. 7.41 that the BB HBT consistently has larger output power and power-added efficiency across the input power range. Figure 7.42 shows the frequency dependence

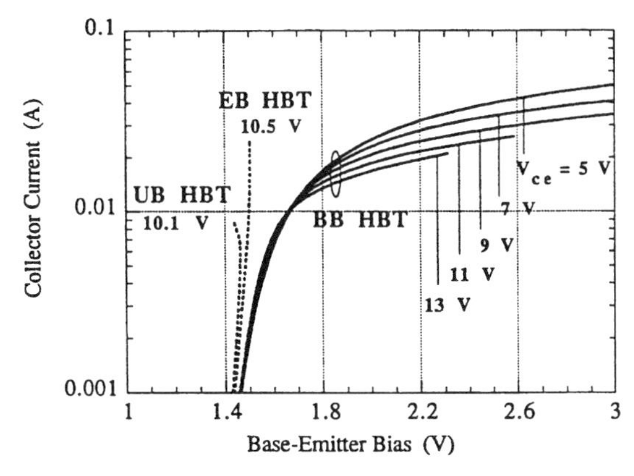

FIGURE 7.40 Regression characteristics of UB, EB, and BB HBTs (after Liu et al., Ref. 35 © IEEE).

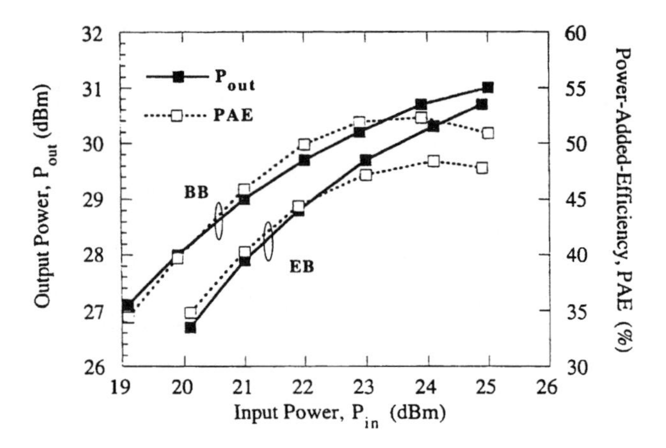

FIGURE 7.41 Output power versus input power (after Liu et al., Ref. 35 © IEEE).

of the large-signal performance. Each data point represents the average result of five devices. The input RF power is fixed at 24 dBm. As shown in Fig. 7.42, the collector efficiency of the BB HBT maintains at about 70% throughout the measured band between 8.5 and 10.5 GHz, whereas it decreases to about 55% for the EB HBT as frequency increases. The significant difference in the collector efficiency is attributed to the difference in the power dissipations across the ballasted resistors. Because the bypass capacitance effectively shunts the RF signal away from the ballasted resistor in the BB HBT, all of the RF signal goes through the ballasted resistor in the EB

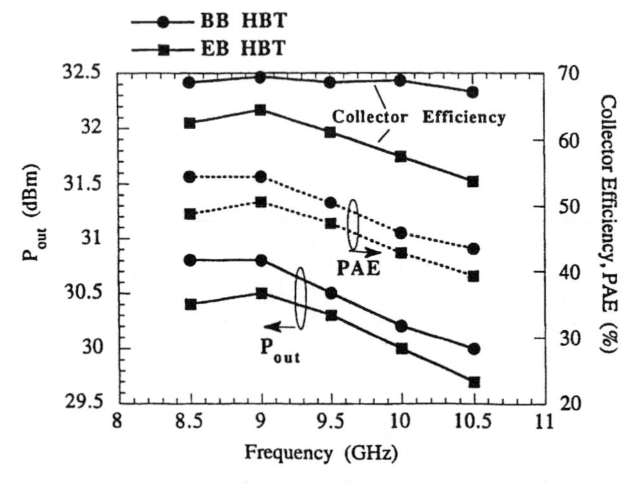

FIGURE 7.42 Output power, collector efficiency, and power-added efficiency versus frequency (after Liu et al., Ref. 35 © IEEE).

HBT. The large signal power is larger in the BB HBT by about 0.5 dB. This leads to the larger output power consistently measured from the BB HBT. The combined superiority in both the collector efficiency and the output power naturally results in the larger power added efficiency observed in the BB HBT. These results demonstrate that base ballasting is the preferred ballasting scheme, not only because it guarantees absolute thermal stability at dc, but also results in better large-signal performance at RF.

REFERENCES

1. S. Tiwari, *Compound Semiconductor Device Physics*, Academic Press, San Diego, CA (1992).
2. S. M. Sze, *Physics of Semiconductor Devices*, 2nd ed., Wiley, New York (1981).
3. J. J. Liou and J. S. Yuan, "Modeling the reverse base current phenomenon due to avalanche effect in advanced bipolar transistors," *IEEE Trans. Electron Devices*, **ED-37**, 2274 (1990).
4. E. Zanoni, E. F. Crabbé, J. M. C. Stork, P. Pavan, G. Verzellesi, L. Vendrame, and C. Canali, "Measurements and simulation of avalanche breakdown in advanced Si bipolar transistors," *International Electron Device Meeting*, 927 (1992).
5. T. M. Liu, T.-Z. Chiu, V. D. Archer, and H. H. Kim, "Characteristics of impact-ionization current in the advanced self-aligned polysilicon-emitter bipolar transistor," *IEEE Trans. Electron Devices*, **ED-38**, 1845 (1991).
6. K. Sakui, T. Hasegawa, T. Fuse, S. Watanabe, K. Ohuchi, and F. Masuoka, "A new static memory cell based on the reverse base current effect of bipolar transistors," *IEEE Trans. Electron Devices*, **ED-36**, 1215 (1989).
7. P.-T. Lu and T.-C Chen, "Collector–base junction avalanche effects in advanced double-poly self-aligned bipolar transistors," *IEEE Trans. Electron Devices*, **ED-36**, 1182 (1989).
8. E. Zanoni, R. Malik, P. Pavan, J. Nagle, A. Paccagnella, and C. Canali, "Negative base current and impact ionization phenomenon in AlGaAs/GaAs HBTs," *IEEE Electron Device Lett.*, **EDL-13**, 253 (1992).
9. T. K. Liu, T.-Y. Chiu, C. D. Archer III, and H. H. Kim, "Characteristics of impact-ionization current in the advanced self-aligned polysilicon-emitter bipolar transistor," *IEEE Trans. Electron Devices*, **ED-38**, 1845 (1991).
10. E. Zanoni and G. Zandler, "Experimental and Monte Carlo analysis of impact-ionization in AlGaAs/GaAs HBTs," *IEEE Trans. Electron Devices*, **ED-43**, 1769 (1996).
11. C. Canali, P. Pavan, A. D. Carlo, P. Lugli, R. Malik, M. Manfredi, A. Neviani, and L. Vendrame, "Dead-space effects under near-breakdown conditions in AlGaAs/GaAs HBTs," *IEEE Electron Device Lett.*, **EDL-14**, 103 (1993).
12. E. Zanoni, E. F. Crabbé, J. M. C. Stork, P. Pavan, G. Verzellesi, L. Vendrame, and C. Canali, "Extension of impact-ionization multiplication coefficient measurements to high electric fields in advanced Si BJTs," *IEEE Electron Device Lett.*, **EDL-14**, 69 (1993).
13. G.-B. Hong and J. G. Fossum, "Implementation of nonlocal model for impact ionization current in bipolar circuit simulation and application to SiGe HBT design optimization," *IEEE Trans. Electron Devices*, **ED-42**, 166 (1995),

14. Y. Apanovich, P. Blakey, R. Cottle, E. Lyumkis, B. Polsky, A. Shur, and A. Tcherniaev, "Numerical simulation of submicrometer devices including coupled nonlocal transport and nonisothermal effects," *IEEE Trans. Electron Devices*, **ED-42**, 890 (1995).

15. B. Jalali, Y.-K. Chen, R. N. Nottenburg, D. Sivco, D. A. Humpherey, and A. Y. Cho, "Influence of base thickness on collector breakdown in abrupt AlInAs/InGaAs heterojunction bipolar transistors," *IEEE Electron Device Lett.*, **EDL-11**, 400 (1990).

16. C. L. Anderson and C. R. Crowell, "Threshold energies for electron–hole pair production by impact ionization in semiconductors," *Phys. Rev. B*, **5**, 2267 (1972).

17. J. S. Yuan and J. Ning, "Effect of impact ionization on C_{JC} of heterojunction bipolar transistors," *Solid-State Electron.*, **3**, 742 (1995).

18. R. J. Malik, N. Chand, J. Nagle, R. W. Ryan, K. Alavi, and A. Y. Cho, "Temperature dependence of common-emitter I–V collector breakdown voltage characteristics in AlGaAs/GaAs and AlInAs/GaInAs HBTs grown by MBE," *IEEE Electron Device Lett.*, **EDL-13**, 557 (1992).

19. B. Jalali and S. J. Pearton, eds., *InP HBTs: Growth, Processing, and Applications*, Artech House, Norwood, MA (1995).

20. H. F. Chau, J. Hu, D. Pavlidis, and K. Tomizawa, "Breakdown-speed considerations in AlGaAs/GaAs heterojunction bipolar transistors with special collector design," *IEEE Trans. Electron Devices*, **ED-39**, 2711 (1992).

21. H.-F. Chau, D. Pavlidis, J. Hu, and K. Tomizawa, "Breakdown-speed considerations in InP/InGaAs single- and double-heterostructure bipolar transistors," *IEEE Trans. Electron Devices*, **ED-40**, 2 (1993).

22. X. Gui, G.-B. Gao, and H. Morkoç, "Simulation study of peak junction temperature and power limitation of AlGaAs/GaAs HBTs under pulse and CW operation," *IEEE Electron Device Lett.*, **EDL-13**, 411 (1992).

23. W. Liu, S. Nelson, D. Hill, and A. Khatibzadeh, "Current gain collapse in microwave multifinger heterojunction bipolar transistors operated at very high power density," *IEEE Trans. Electron Devices*, **ED-40**, 1917 (1993).

24. L. L. Liou and B. Bayraktaroglu, "Thermal stability analysis of AlGaAs/GaAs heterojunction bipolar transistors with multiple emitter fingers," *IEEE Trans. Electron Devices*, **ED-41**, 629 (1994).

25. W. Liu, "The interdependence between the collapse phenomenon and the avalanche breakdown in AlGaAs/GaAs power heterojunction bipolar transistors," *IEEE Trans. Electron Devices*, **ED-42**, 591 (1995).

26. W. Liu, H.-F. Chau, and E. Beam III, "Thermal properties and thermal instabilities of InP-based heterojunction bipolar transistors," *IEEE Trans. Electron Devices*, **ED-43**, 388 (1996).

27. W. Liu and A. Khatibzadeh, "The collapse of current gain in multi-finger heterojunction bipolar transistors: its substrate temperature dependence, instability criteria, and modeling," *IEEE Trans. Electron Devices*, **ED-41**, 1698 (1994).

28. I. E. Getreu, *Modeling the Bipolar Transistor*, Elsevier, New York (1978).

29. P. L. Hower and P. K. Govil, "Comparison of one- and two-dimensional models or transistor thermal instability," *IEEE Trans. Electron Devices*, **ED-40**, 1917 (1993).

30. G.-B. Gao, M. S. Unlu, H. Morkoç, and D. Blackburn, "Emitter ballasting resistor design for, and current handling capability of AlGaAs/GaAs power heterojunction bipolar transistors," *IEEE Trans. Electron Devices*, **ED-38**, 185 (1991).

31. B. Bayraktaroglu, J. Barrette, L. Kehias, C. I Hunag, R. Fitch, R. Neikhard, and R. Scherer, "Very high-power-density CW operation of GaAs/AlGaAs microwave heterojunction bipolar transistors," *IEEE Electron Device Lett.*, **EDL-40**, 493 (1993).

32. J. Sewell, L. L. Liou, D. Barlage, J. Barrette, C. Bozada, R. Dettmer, R. Fitch, T. Jenkins, R. Lee, M. Mack, G. Trombley, and P. Watson, "Thermal characterization of thermally-shunted heterojunction bipolar transistors," *IEEE Trans. Electron Devices*, **ED-17**, 19 (1996).

33. M. G. Adlerstein and M. P. Zaitlin, "Thermal resistance measurements for AlGaAs/GaAs heterojunction bipolar transistors," *IEEE Trans. Electron Devices*, **ED-38**, 1553 (1991).

34. J. R. Waldrop, K. C. Wang, and P. M. Asbeck, "Determination of junction temperature in AlGaAs/GaAs heterojunction bipolar transistors by electrical measurement," *IEEE Trans. Electron Devices*, **ED-39**, 1248 (1992).

35. W. Liu, A. Khatibzadeh, J. Sweder, and H.-F. Chau, "The use of base ballasting to prevent the collapse of current gain in AlGaAs/GaAs heterojunction bipolar transistors," *IEEE Trans. Electron Devices*, **ED-43**, 245 (1996).

PROBLEMS

7.1. Show that when the avalanche process is initiated by electrons, the avalanche integral of interest is related by the equation

$$1 - \frac{1}{M_n} = \int_0^W \alpha_n \exp\left[-\int_W^x (\alpha_n - \alpha_p)\,dy\right] dx$$

7.2. Derive the multiplication factor as a function of α_n and $E(x)$.

7.3. Compare the collector–base junction capacitance with and without avalanche effect for local and nonlocal models.

7.4. Derive the Early voltage, including avalanche effect.

7.5. Derive the free carrier capacitance using the expression

$$C_F = \frac{\partial}{\partial V}\left(q\int_0^W \Delta p\,dx\right)$$

where $\Delta p = p(0)(M - 1)$ and $M = 1/[1 - (V/BV_{CBO})^n]$.

7.6. A two-finger HBT has an area of $2 \times 20\ \mu\text{m}^2$ of each finger sitting on top of a 100-μm GaAs substrate. Assume that the mutual thermal coupling between the two fingers is negligible. Estimate the thermal resistance of each of the fingers. What is the overall thermal resistance of the two-finger device?

7.7. Explain qualitatively the thermal stability of multifinger Si BJTs and SiGe HBTs.

7.8. Discuss how to design the HBT to improve thermal stability.

CHAPTER EIGHT

Reliability

Heterojunction bipolar transistors have demonstrated excellent performance for high-frequency IC applications and for insertion of this technology into practical systems. The ultimate usefulness of this technology for system applications is determined by its reliability. The reliability-related parameters of interest, such as failure rates, mean time to failure (MTTF), and activation energies, are commonly determined by stressing devices and circuits under high forward biases and/or elevated temperatures. This procedure is known as high-current and high-temperature accelerated aging or life testing. The interpretation of the statistical life-test data is very important to ensure that the high-temperature and high-current stress tests induce realistic failures, which are expected under normal operating conditions. For bipolar transistors it is important to determine the actual device junction temperature or "hot spot" under biased-stress conditions. The junction temperature differential depends on the device thermal resistance, device size, and temperature.

8.1 ELECTRICAL AND THERMAL OVERSTRESS

Reliability requirements are dictated by the application. Reliability qualification work is aimed at establishing robust process and design rules for achieving high mean time between failure. General HBT reliability testing includes stabilization bake, accelerated life testing, and step-stress testing. An initial stabilization bake is performed to stabilize the device parameters prior to accelerated aging and step-stress tests, and to screen for infant mortality. Devices are typically subject to 200 to 240°C for 48 hours in a nitrogen ambient atmosphere to stabilize dc parameters.

Accelerated life tests include aging at 200 to 260°C under dc bias with step-stress testing to extract the activation energy and MTTF for degradation designed to explore accelerated lifetimes. This is more realistic because high-temperature aging induces a failure mechanism that may never be encountered in normal operation conditions. High-temperature testing is performed to reduce the time needed to achieve median lifetimes.

Reliability testing of GaAs HBTs has focused on accelerated aging of discrete devices under normal operating dc bias. The packaged devices are removed periodically and tested for dc current gain, base–emitter turn-on voltage, ohmic contact resistance, and junction leakage. Figures 8.1*a–d* show the experimental data of base–emitter leakage current, current gain, collector–emitter offset voltage, and base–collector leakage versus stress time. These key parameters enable test engineers to assess the stability and integrity of the base–emitter junction, the base layer, and the base–collector junction [1,2]. The base–emitter leakage current in Fig. 8.1*a* is seen to decrease steadily and uniformly with increasing stress time in the test population. The current gain in Fig. 8.1*b* has an initial increase of approximately 10%, which is attributed to a reduction in the surface recombination velocity as a result of exposure to high temperature. As the stress increases, this process reverses and the current gain decays gradually. The offset voltage increases with stress time as shown in Fig. 8.1*c*. Intially,

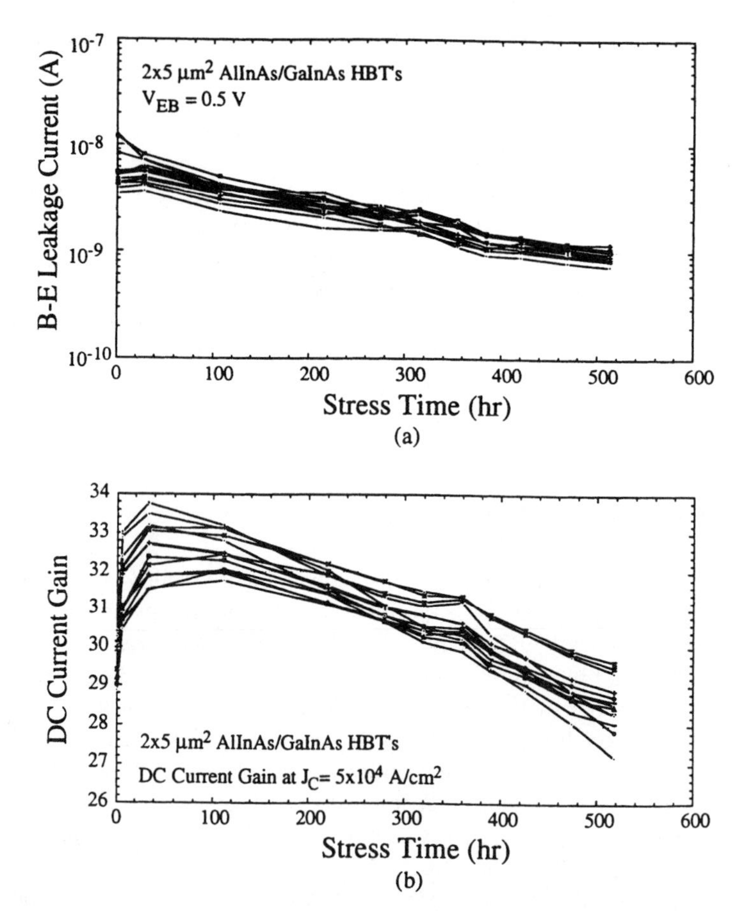

FIGURE 8.1 (*a*) Base–emitter leakage current and (*b*) current gain versus stress time (after Hafizi et al., Ref. 1 © IEEE).

the offset voltage is unchanged. It is then followed by a period of stress time and displays an identical trend to that of the base–collector leakage increase in Fig. 8.1*d*. It is clear that the leakage current decreases steadily in approximately the first 300 hours of stress. This is then followed by a rapid increase in the leakage current of the base–collector junction. The increase in the leakage current occurs very uniformly over a short interval of time, as opposed to the case where the degradation spreads out over the entire length of the stress period. This feature is highly desirable. The confinement and the uniformity of device degradation results in a failure distribution for the life-test population, which has a very low dispersion. Another apparent feature is the lack of early device failures or rapid degradation in the early hours of stress, known as *infant mortality*.

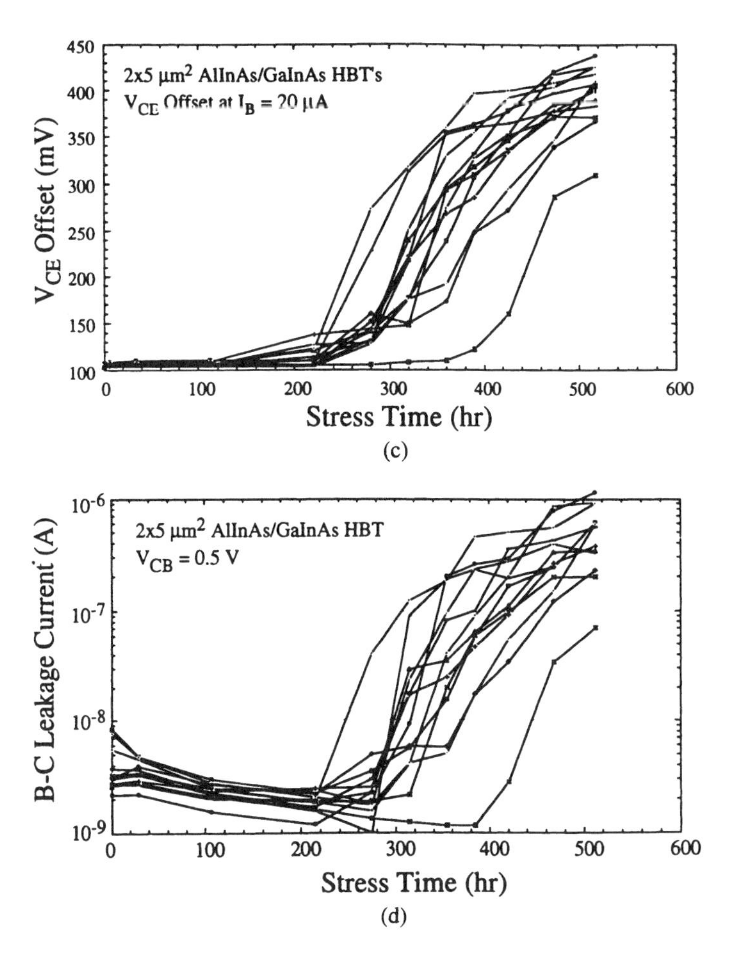

FIGURE 8.1 (*c*) Collector–emitter offset voltage and (*d*) base–collector leakage current versus stress time (after Hafizi et al., Ref. 1 © IEEE).

8.1.1 Forward- and Reverse-Bias Stress Effects

When bipolar transistors are forward-biased at high current densities, changes in the transistor performance characteristics, including leakage currents, parasitic resistances, and current gain can result. As bipolar technology is scaled to smaller dimensions, the operating current densities increase and the implications of forward-bias degradation for device design are of increasing concern.

Studies by Hemmert et al. [3,4] showed that forward bias stress induced gradual current gain degradation due to an increase in base current. They attributed the gain changes to electromigration. Other physical mechanisms to explain the gain degradation, such as Auger recombination, produced hot electrons were reported [5]. Sun et al. [6] examined forward-biased stress effects on silicon bipolar transistors in gain and noise characteristics. Test bipolar transistors from three different fabrication technologies were used. Type A devices were fabricated by conventional oxide isolation techniques with emitter dimensions of 2.5 μm $\times$ 4.3 μm. Type B devices were fabricated with polysilicon base contacts and sidewall oxide, which had emitter dimensions of 1.5μm $\times$ 2.0 μm. Metal contacts were via metal studs through oxide vias to reduce eletromigration-induced stress directly over the transistor contacts. Type C devices are from a test structure [2] designed particularly to investigate electromigration effects. Type C devices have emitter dimensions of 2.5 μm $\times$ 4.0 μm. Transistors were stressed with the base–collector junction short-circuited and the base–emitter junction forward biased. There was no reverse-biased junction to eliminate the possibility of degradation due to reverse-bias produced hot electrons. The temperature of the stressed devices was varied between -75 and 240°C. Stressing was interrupted at selected times to measure the Gummel plots and the noise spectra of the stressed devices. Control devices were used that received temperature stress but not electrical stress. Device measurements were performed at a temperature of 23 $\pm$ 1°C in an environmental chamber.

Stress effects on bipolar current gain have been studied. Figure 8.2 shows current gain versus base–emitter bias for technology A transistor. The forward stressing was done at 240°C with a current density of 2 mA/μm^2. These devices showed a pronounced and systematic gradual degradation of current gain with increasing stress time when subject to a combination of high forward-bias current and temperature. Normalized current gain (average current gain after stress/average current gain before stress) versus stress time at different current levels and ambient temperatures is shown in Fig. 8.3. In this plot solid squares represent 1 mA/μm^2 at -75°C, empty squares 1 mA/μm^2 at 175°C, solid circles 1 mA/μm^2 at 240°C, and triangles 2 mA/μm^2 at 240°C. The resulting gain degradation after 500-hour of stress at 240°C is approximately 20% when the stress current is 2 mA/μm^2 and approximately 10% after 750 hours when the stress current is 1 mA/μm^2. No degradation was observed during the 500-hour stress with ambient temperatures of -75 and 175°C. The gradual degradation and the temperature dependence is consistent both with electromigration-induced compressive stress and with Auger hot-electron-induced damage.

For type B devices, the degradation was quite different than for type A devices. Figure 8.4 shows the result of forward-bias stress with 5 mA/μm^2 at a temperature

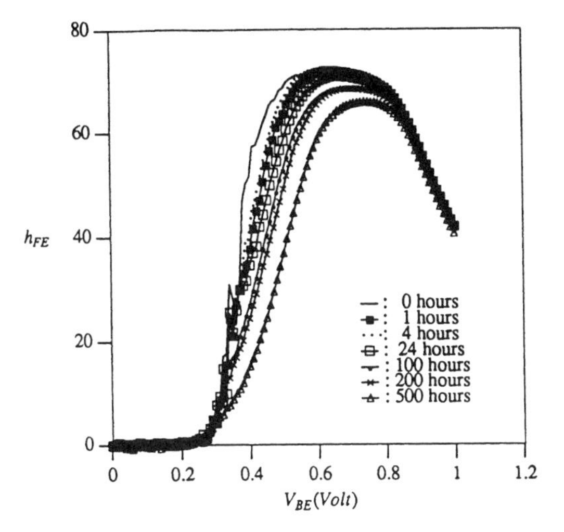

FIGURE 8.2 Current gain versus base–emitter voltage (after Sun et al., Ref. 6 © IEEE).

of 225°C. No systematic change was observed up to 100 hours, but significant degradation occurs later. In contrast to the result for type A technology, degradation for a given device was precipitous rather than gradual. Once degradation had begun, in about 30 to 40 hours the process culminated in an open-circuit transistor. The time required for 50% drop out of the population was approximately 400 hours. The nature of the failure is indicative of electromigration. Electromigration opens in the line initially, which produces an increase in series resistance and results in an open circuit when a line is opened completely. Scanning electron microscopy (SEM) confirmed this conclusion.

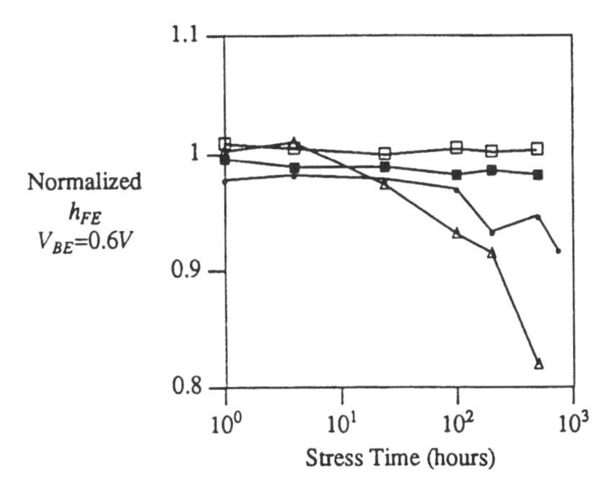

FIGURE 8.3 Normalized current gain versus stress time for technology A device (after Sun et al., Ref. 6 © IEEE).

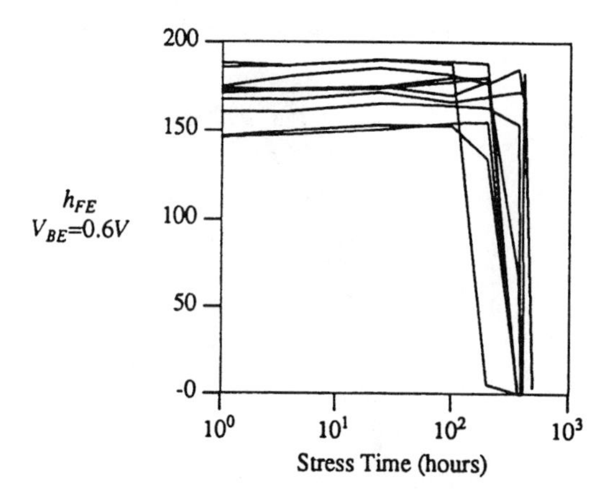

FIGURE 8.4 Current gain versus stress time for technology B device (after Sun et al., Ref. 6 © IEEE).

The gain degradation on type C devices is displayed in Fig. 8.5. The transistor that received the direct electrical stress shows the largest degradation in current gain after 500 hours of stressing with 5 mA/μm^2 at 200°C. For the other transistors, the gain degradation decreases with increasing distance from the first transistor, consistent with expectations for electromigration-induced compressive stress. For all type C transistors, the decrease in gain is due to an increase in base current.

The low-frequency noise characteristics were also measured for the three different bipolar technologies. In each case the noise measurement was done with a base–emitter bias in the range 0.77 to 0.82 V, corresponding to emitter currents below 0.5 mA. Under these bias conditions, the shot noise is much less than the $1/f$ noise

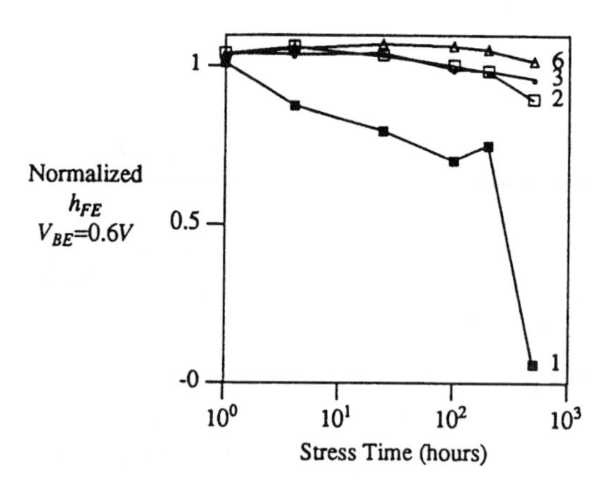

FIGURE 8.5 Normalized current gain versus stress time for technology C device (after Sun et al., Ref. 6 © IEEE).

in the frequency range 1 to 100 Hz. The technology C transistors were stressed with a current density of 5 mA/μm^2 at $T = 200°C$ for up to 500 hours. As shown in Fig. 8.6, the noise initially dropped for measurements taken after 1 hour of stress. Subsequent noise measurement of cumulative stress showed an increase in noise level with stress time. This noise decrease is in contrast to noise changes caused by hot-electron-induced stress since hot electrons produce an increase in noise [7]. Experiments performed on both technology A and B transistors show similar noise characteristic changes due to forward-bias stress. Technology A transistors showed an initial decrease in low-frequency noise when stressed at 240°C with a current density of 2 mA/μm^2. After a noise minimum is reached, the noise for technology A devices eventually increases with longer stress time (i.e., 100 hours of cumulative stress). This behavior is similar to that observed for technology C transistors. For both technology A and C transistors, the noise starts to increase, after its initial decrease, at about the same time that the gain begins to drop substantially. Technology B transistors also show a drop in the low-frequency noise characteristics as shown in Fig. 8.7. This technology, however, did not show the later increase in the low-frequency noise as did the other technologies. Note that technology B transistors did not show a gradual gain degradation as did technology A and C devices; rather, they showed an abrupt failure due to open circuits caused by electromigration.

Comparison of time to failure of SiGe and Si bipolar transistors has been made [8]. The effects of Ge in the epitaxial base on the reliability of SiGe HBTs were investigated. The 10-year time to failure under emitter–base junction reverse-bias stress was measured at the designed operation voltage by the current-acceleration method and compared to that of Si BJTs with no Ge content in the base. High-performance Si and SiGe bipolar transistors were fabricated epitaxially on Si wafers using rapid thermal chemical vapor deposition (RTCVD) technology [9]. Substrates were $\langle 100 \rangle$Si:Sb (2×10^{18} Sb/cm^{-3}). Following growth of 2.5-μm Si collector

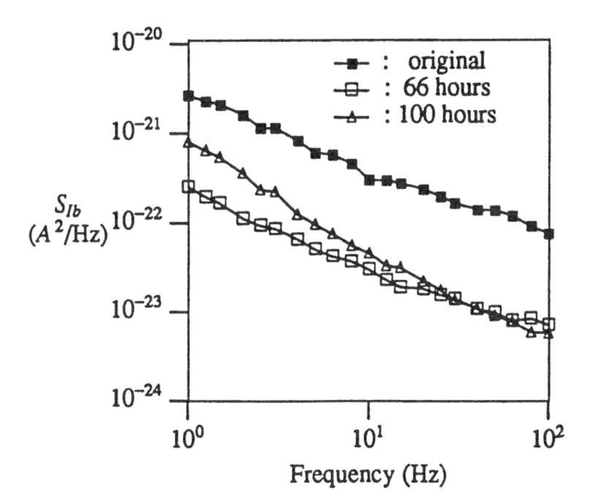

FIGURE 8.6 Noise versus frequency for technology A device (after Sun et al., Ref. 6 © IEEE).

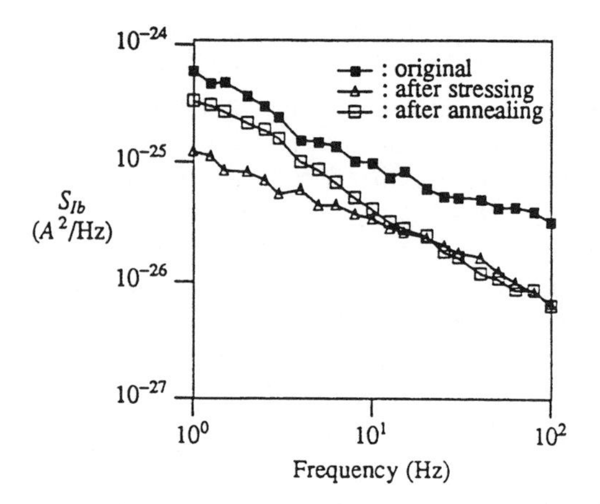

FIGURE 8.7 Noise versus frequency for technology B device (after Sun et al., Ref. 6 © IEEE).

(5×10^{15} As/cm^3), the active regions were isolated by a modified LOCOS process and RCA-cleaned prior to nonselective deposition of the SiGe base. Boron implantation in the extrinsic base was then performed, and an oxide/nitride stack was deposited and etched to open the emitter region followed by a polysilicon and As implant. All implants were activated by a single RTA. A 3000-Å undoped low-T oxide was then deposited. Contacts were then opened followed by TiW/AlCuSi metallization. The Ge concentration in the base and the dopant impurity concentrations were measured by SIMS. The Ge content is linearly graded from 0% at the emitter–base junction to a peak of 10% at the base–collector junction. The base dopant density is nearly constant (ca. 4×10^{18} cm^{-3}), and the base thickness is about 800 Å. The baseline Si BJT was fabricated with identical fabrication steps except that it has no Gc in thc quasi-neutral base. The TEM cross section does not show dislocations in the SiGe base indicating that the SiGe layer is not relaxed after the fabrication is completed.

Figure 8.8 shows the time to failure (TTF) versus reverse emitter–base stress voltage of the SiGe and Si bipolar transistors. The TTF was obtained from two stress and measure experiments: (1) the open-collector (OC) method, and (2) the forward-collector (FC) method, using V_{BC} in the range 0.75 to 0.9 V. First, the integrated terminal charge to failure, $Q_{TF} = Q_{EB\text{-stress}} =\equiv \int j_{E\text{-stress}}(t)\,dt$ required for $\Delta\beta_F = -10\%$ at $V_{BE} = 0.76$ V, is plotted against the stress voltage $V_{BE\text{-stress}}$ for both OC and FC measurements. Then the FC data are normalized twice with respect to the OC data to account for the different geometrical current distributions and the kinetic energy difference between the bond-breaking hot carriers at the two bias conditions. The TTF is then obtained by dividing the normalized Q_{TF} data by the open-collector emitter current $J_{E\text{-stress}}$. It is clear from Fig. 8.8 that the TTF is not affected by Ge. The slightly higher TTF in the SiGe HBTs may be due to a smaller reverse current. Note that the OC data (open circles and triangles) gave TTF only for $V_{EB\text{-stress}} > 2.75$ V because of the long stress time (> 105 s) required at lower stress

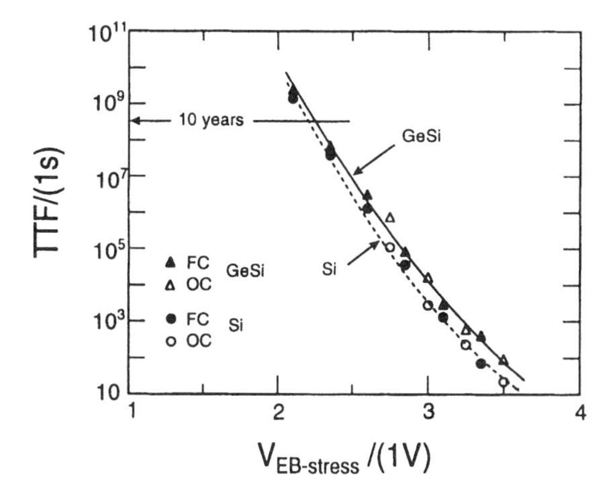

FIGURE 8.8 TTF as a function of $V_{EB\text{-stress}}$ (after Neugroschel et al., Ref. 8 © IEEE).

voltages. This OC TTF increases rapidly beyond reasonable time as the stress voltage is lowered below 2.75 V. The current-accelerated FC data (filled circles and triangles), however, gave TTF as high as 2.3×10^9 s (ca. 745 years). From Fig. 8.8 the 10-year TTF was attained when $V_{EB\text{-stress}} \leq 2.2$ V in the Si BJT and $V_{EB\text{-stress}} \leq 2.26$ V in the SiGe HBT.

8.1.2 Thermal Overstress

The thermal stress test is performed at given bias conditions and at elevated ambient temperatures, assuming that the failure mechanisms are thermally accelerated. Discrete devices with a substrate thickness of 100 μm were packaged to perform the accelerated life test [1]. The transistors were biased in the forward active mode with a constant collector current density of 7×10^4 A/cm. Over 2,500 hours of accelerated life test was carried out at 193, 208, and 228°C ambient temperatures. Extensive dc characterization was performed periodically during the life test. For devices with compositionally graded base–emitter junctions, an increase of approximately 1.8 mV in V_{BE} per angstrom of Be diffusion was determined. This indicates that the V_{BE} is a very sensitive parameter for monitoring the stability of the Be. Collector–emitter offset voltage and junction leakage currents were also recorded to monitor the stability of the base–emitter and base–collector junctions.

Figure 8.9 shows measured base–emitter and base–collector junction leakage currents before and after stress at 208°C. At low base–emitter biases, the base current decreases and leads to an increase of about 10% in the dc current gain. The base–emitter junction leakage currents, in both forward and reverse biases, decreased steadily under stress. The crystallographic orientation of the emitter mesa or a Si_3N_4

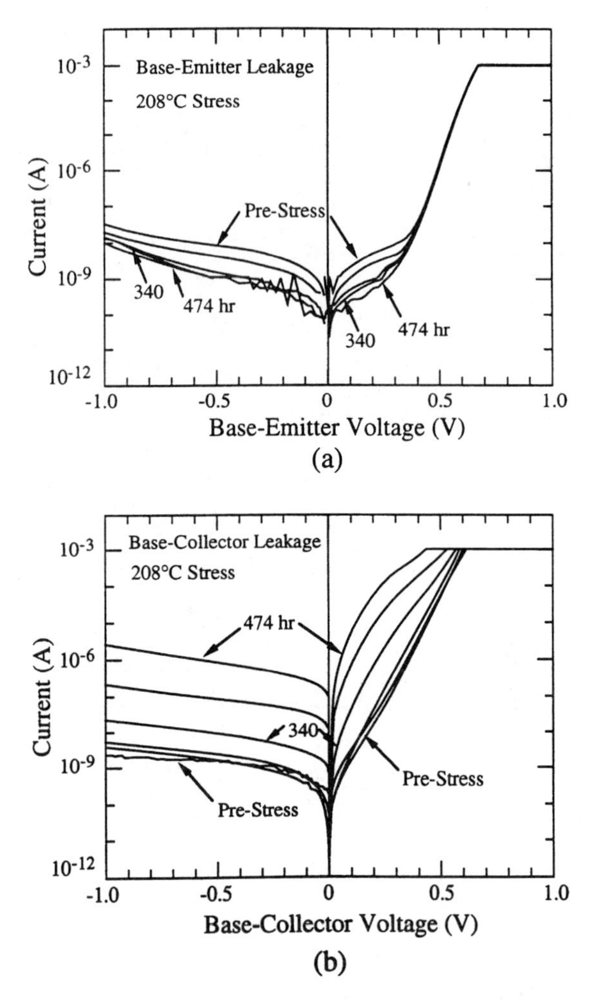

FIGURE 8.9 (*a*) Base emitter and (*b*) base–collector currents before and after stress at 208°C (after Jalali and Pearton, Ref. 10 © Artech House).

surface passivation did not have any effect on the stability of the B-E junction, due to the use of a nine-period superlattice inserted between the Be-doped GaInAs base and the AlInAs emitter. The superlattice effectively blocks the movement of any interstitial Be^+ diffusion from the base into the emitter. The base–collector leakage current in the reverse bias (see Fig. 8.9*b*) increases rapidly as the bias stress progresses. In the forward bias region, in addition to an emerging leakage current, the turn-on voltage of the junction also steadily decreases. The stress-increased leakage current has a voltage dependence of $\exp(qV_{BE}/nkT)$, where n increases from 1 to greater than 2 as the stress progresses. When the junction is under reverse bias, surface states are generated by hot electrons at or near the metallurgical junction where the electric field is greatest. The interface traps act as generation–recombination centers, causing an increase in the reverse leakage current and the nonideal current in the forward direction. The stress-induced leakage currents of the B-C junction correlate with the amount of reverse-bias voltage applied to the B-C junction.

Reliability of InP-based HBTs has been demonstrated for applications requiring MTTF in excess of 10^7 hours (at 125°C junction temperature) at high current drive levels. It appears that because of the excellent reliability performance, low-power operation, and relative ease of fabrication for this new technology, there are excellent prospects for insertion of InP-based integrated circuits into practical products in commercial and space applications. Figure 8.10 shows a failure plot for an InP-based HBT biased at $V_{CE} = 2$ and 3 V. In this plot the logarithm of time to failure is shown as a function of cumulative failure distribution. The 50% cumulative failure point in this graph yields the MTTF value for each of the life tests. The MTTF values obtained were 22, 122, and 390 hours at different I_C, V_{CE}, and temperatures. The dispersion (σ) of the log-normal distribution is approximately 0.3. The σ is estimated as log(time to 50% failure) minus log(time to 16% failure). There is no apparent change in the value of σ over the range of temperature of the tests, indicating the validity of the accelerated stress experiments. From an Arrhenius plot of the MTTF values (figure not shown here), an activation energy of 1.92 eV was obtained and a MTTF of 1.23×10^7 hours at 125°C junction temperature was projected. The MTTF at 50°C junction temperature is 5.7×10^{12} hours. The low dispersion of the failure distribution observed in the life test data is desirable for system applications because it leads to a very small failure rate.

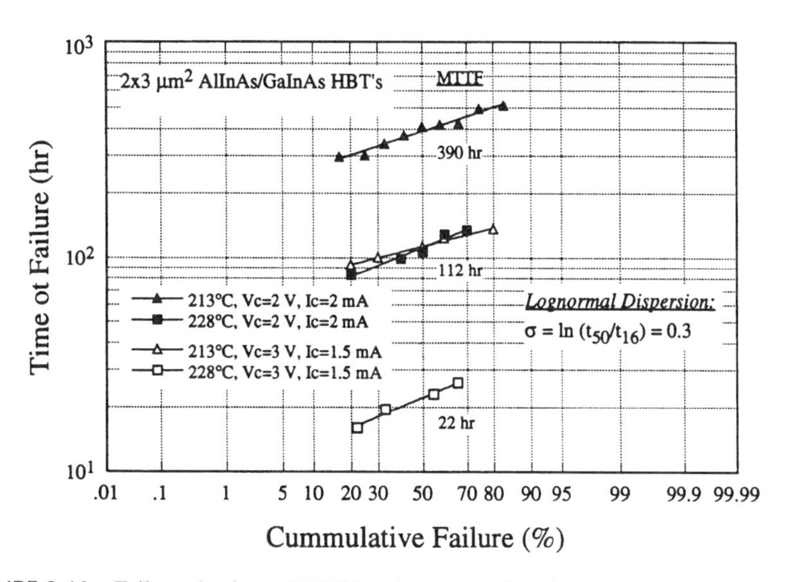

FIGURE 8.10 Failure plot for an HBT biased at $V_{CE} = 2$ and 3 V (after Jalali and Pearton, Ref. 10 © Artech House).

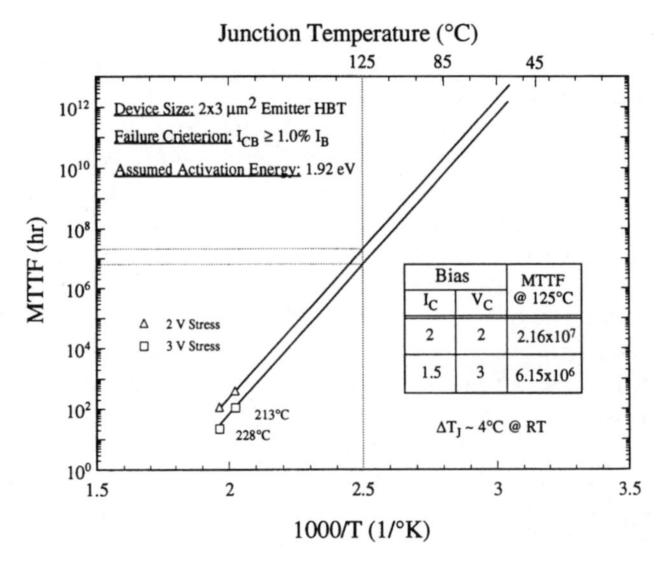

FIGURE 8.11 Arrhenius plot for data shown in Fig. 8.10 (after Jalali and Pearton, Ref. 10 © Artech House).

Figure 8.11 shows an Arrhenius plot of MTTF for the 2 V and 3 V life tests. The projected MTTF is 2.16×10^7 and 6.15×10^6 hours, respectively, at a junction temperature of 125°C. For these predictions, the failure criterion was set as the collector–base leakage current $I_{CB} > 1\%$ of I_B at a reverse bias of $V_{CE} = 2.5$ V. The activation energy was assumed to be 1.92 eV. A similar reliability performance has been predicted for double-heterojunction bipolar transistors incorporating an InP collector [11]. The failure mechanism and the performance degradation in InP-based DHBTs is identical to that of single-heterojunction devices.

8.1.3 Burn-in

Burn-in carried out in a high-temperature/current environment is used widely to assess the reliability of semiconductor devices. Such a test is particularly useful to determine the long-term performance of heterojunction bipolar transistors since semiconductor bulk and surface properties are susceptible to thermal and electrical stresses. Experimental results show that for a fixed bias condition, the burn-in could increase or decrease the base current of the HBT but does not change the collector current. An abnormal base current with an ideality factor of about 3 has been observed in the Gummel plot of an HBT [12] subject to a relatively long-hour burn-in test as shown in Fig. 8.12.

Two theories have been reported to explain the abnormal base current in the post-burn-in HBT [13,14]. The first theory [13] suggests that defects in a semiconductor can redistribute themselves and diffuse to dislocations due to enhanced recombination process and/or high thermal energy. Since the HBT undergoing the burn-in test is

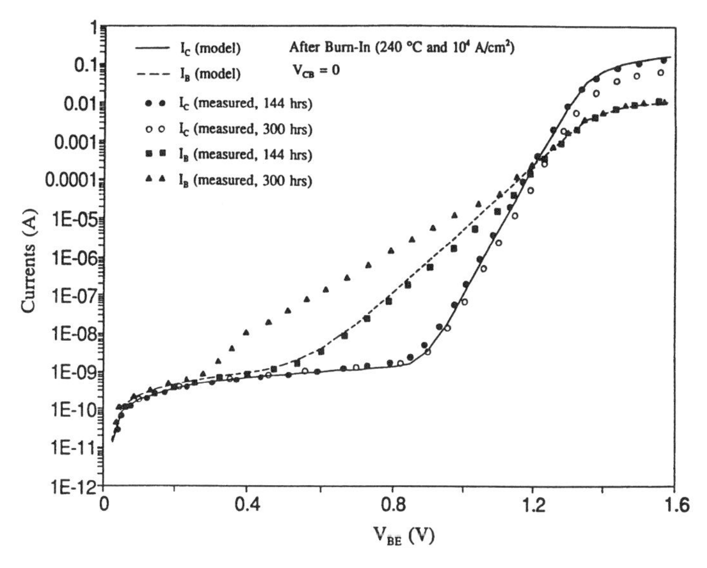

FIGURE 8.12 Base and collector currents after burn-in (after Liou and Huang, Ref. 12 © Elsevier Science).

subject to high temperature and high injection stresses, such a recombination/thermal-enhanced defect diffusion is a likely mechanism attributing to the unique post-burn-in behavior. The second theory [14] suggests that the abnormal current is attributed to a significant increase in the number of defects in the strained base. This is confirmed by deep-level transient spectroscopy, which indicates that the trapping states generated by the burn-in test are located in the base region.

Note that burn-in could also result in a decrease of base current. This burn-in effect is suggested related to the interface between the passivation layer and the semiconductor in the extrinsic base surface region [15]. Correlation between the burn-in effect and the extrinsic base surface quality in carbon-doped GaInP/GaAs HBTs was studied [16]. The wafers employed for the fabrication of the HBTs were grown by MOCVD with different passivation techniques. Figure 8.13 shows cross sections of the investigated HBTs. Sample A devices have SiN passivation, while sample B devices have a ledge passivation. After an initial characterization, the devices were stressed at room temperature in the forward active region. During the stress the applied collector–emitter voltage was 5.7 V and the stress collector current density was kept constant at 10^4 A/cm^2. The base current variation of these two passivated devices versus stress time is shown in Fig. 8.14. A decrease of base current for both A and B samples has been observed. The SiN passivated devices exhibit a larger burn-in effect associated with a reduction of the base current. This burn-in effect is a surface-related phenomenon. The defects involved in the burn-in effect are probably located at the extrinsic base surface region and possibly at the passivation/semiconductor

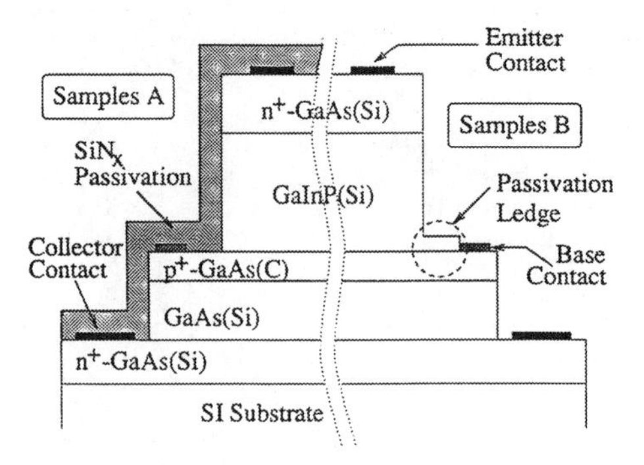

FIGURE 8.13 Two device structures for the investigation of burn-in effect (after Borgarino et al., Ref. 16 © IEE).

interface. From Fig. 8.14 it indicates that the ledge passivation may be a very promising technology to improve device stability and reliability.

8.2 PROCESS-RELATED RELIABILITY ISSUES

8.2.1 Base Dopant Out-diffusion

The effects of base dopant out-diffusion on SiGe [17] and GaAs [18] heterojunction bipolar transistors have been studied. Small amounts of base dopant out-diffusion from a heavily doped base into a more lightly doped emitter can seriously change the device performance, due to the formation of parasitic barriers. Once the barriers occur, the device performance cannot be recovered.

Consider an n-p-n SiGe heterojunction bipolar transistor with a flat Ge profile in the base. The interfaces between the Si–SiGe emitter–base junction and the SiGe–

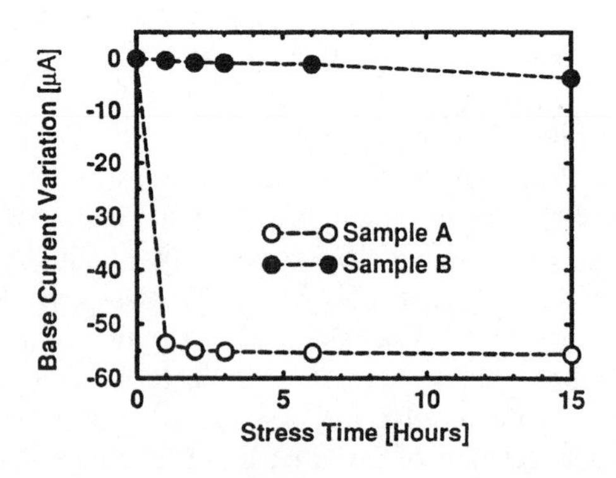

FIGURE 8.14 Base current variation versus stress time at $V_{CE} = 1.4$ V and $I_C = 4$ mA (after Borgarino et al., Ref. 16 © IEE).

Si collector–base junction have abrupt dopant transitions. The entire base layer is heavily doped with boron in a boxlike profile. The transistor structures were grown in situ on a ⟨100⟩ n-type silicon substrate using RTCVD. After chemical cleaning, the wafers were loaded into the RTCVD reactor and baked in hydrogen carrier gas. Heavily doped n^+ buffer layers and n-type collector layers doped about 5×10^{17} cm^{-3} were grown at 100°C. The p-type $Si_{0.8}Ge_{0.2}$ base has a doping of 10^{19} or 10^{20} cm^{-3} with 300-Å thickness and were grown at 850°C for 3 minutes. The base layers are strained because of a negligible number of misfit dislocations observed from TEM measurement. Base contact was established by boron implant, while emitter contact had a shallow arsenic implantation. The device was passivated with SiO_2 film deposited by plasma deposition at 350°C. The implants were then annealed for 10 minutes at 800°C. Titanium–aluminum metallization and patterning were used at the end of processing.

The valence-band discontinuity ΔE_V for a strained $Si_{1-x}G_x$ layer grown on ⟨100⟩ Si is close to the total bandgap difference between Si and SiGe [19]. Autodoping at interfaces during growth or boron out-diffusion into the adjacent silicon layers during growth can move the electrical p-n junction into the silicon emitter and collector regions [20]. Figure 8.15 shows that even small amounts of boron out-diffusion cause large parasitic barriers for electrons at both emitter–base and collector–base heterojunctions when the device is in forward-active mode. For example, for an out-diffusion length of 25 Å, a parasitic barrier of a height of 80 meV is formed at the

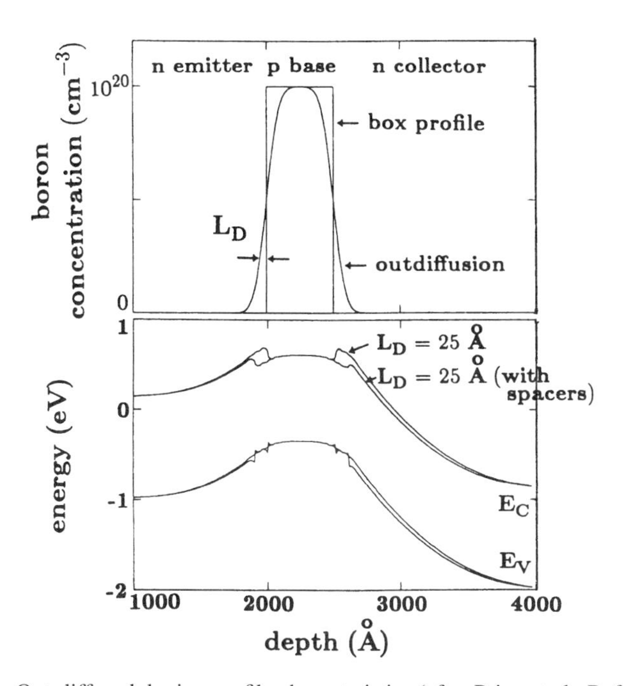

FIGURE 8.15 Out-diffused doping profile characteristics (after Prinz et al., Ref. 17 © IEEE).

emitter–base junction. This parasitic barrier impedes electrons injected from emitter to base and degrades the emitter efficiency and collector current. The overall barrier for holes traveling from base to emitter, however, is not changed. Device simulation results of normalized collector current versus inverse temperature for different out-diffusion lengths are shown in Fig. 8.16. In Fig. 8.16 curves (a), (b), (c), and (d) correspond to no out-diffusion and $L_D = 25$, 50, and 75 Å, respectively. Curve (e) corresponds to $L_D = 25$ Å with 100 Å spacer. The devices with out-diffusion have significant collector current degradation. The out-diffusion also reduces the slope of the collector current versus inverse temperature. The degradation in heavily doped devices indicates that the parasitic barrier effect is much larger than any beneficial effect of bandgap narrowing in the heavily doped bases. In addition, severe degradation is expected for more heavily doped bases because of the higher doping levels in the out-diffusion tails and because of the linear dependence of the boron diffusion coefficient on boron concentration [20].

The reliability characteristics of AlGaAs/GaAs n-p-n bipolar transistors with beryllium and carbon-doped base layers have been investigated [18]. Reliability assessment was based on a constant-stress life test conducted at three temperatures and using the dc current gain as the critical device parameter. For this life test, a 10% degradation in prestress current gain at a 1-mA collector current was established as the failure criteria. The experiment is a constant-stress lifetime at 240, 260, and 280°C ambient temperature. Over 215 discrete HBTs were studied. Postepitaxial growth processing is assumed to be identical over the various wafers. Electrical testing at 25°C was done periodically to monitor degradation in dc current gain at 1 mA. The life test with periodic electrical testing was continued until greater than 70% of a cell's population failed. During thermal stress, all HBTs, excluding control devices, were subject to a forward bias of $V_{CE} = 3$ V and $I_C = 2$ mA (6.67×10^3 A/cm). No RF bias was applied during thermal stress.

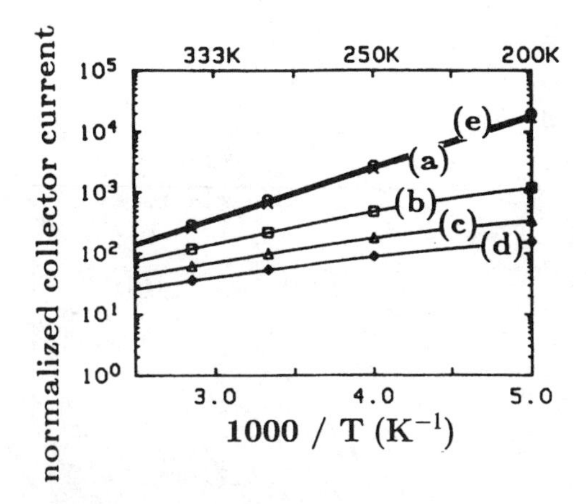

FIGURE 8.16 Normalized collector current versus inverse temperature (after Prinz et al., Ref. 17 © IEEE).

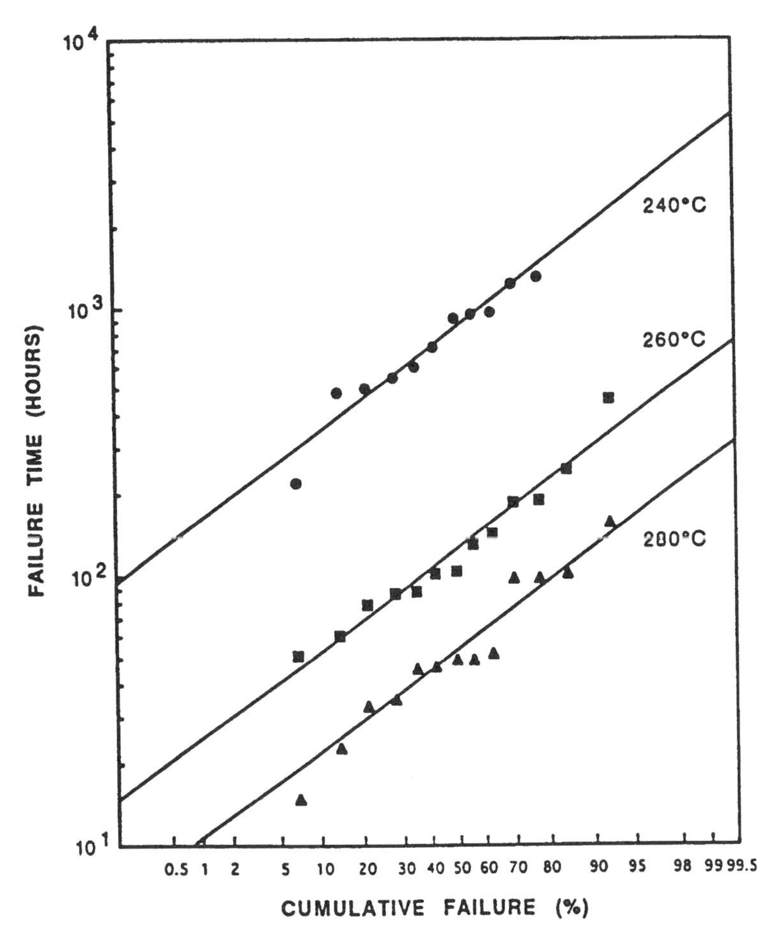

FIGURE 8.17 Failure time versus cumulative failure of Be-doped devices (after Yamada et al., Ref. 18 © IEEE).

Figure 8.17 shows the cumulative failure of the Be device EM^+ at 570°C. For this group of devices, the standard deviation $\sigma = 0.7$ was calculated by least-squares fitting of the log-normal distribution. The failure time decreases with increasing temperature, as expected. An Arrhenius plot illustrating median life versus inverse junction temperature is shown in Fig. 8.18. Compensation for an estimated junction temperature rise of 12°C over ambient temperature has been accounted for in Fig. 8.18. The calculated activation energy for Be/570, Be/570,EM^+, Be/580, and carbon devices are 1.52, 1.77, 1.19, and 0.86 eV, respectively. The difference between the Be/570 and the Be/570,EM^+ device in fabrication is that in the latter device the subsequent emitter layer growth was performed at an elevated substrate temperature. At 125°C the MTTF is 8.4×10^7, 1.87×10^8, 3.3×10^6, and 4.2×10^5 hours for Be/570, Be/570,EM^+, Be/580, and carbon devices, respectively.

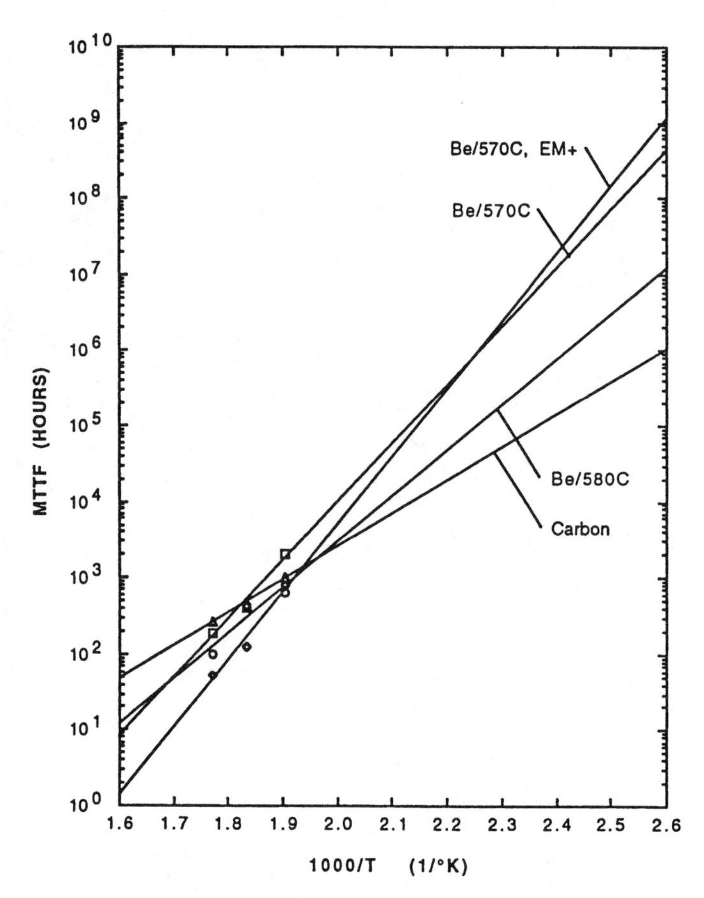

FIGURE 8.18 MTTF current versus inverse temperature (after Yamada et al., Ref. 18 © IEEE).

For the Be-doped HBTs, the dominant failure mechanism is the diffusion of interstitial beryllium (Be^+) from the base to graded base–emitter AlGaAs during forward bias operation. In devices with compositional grading at the base–emitter junction, the steady diffusion of the Be^+ displays the p-n junction by moving it gradually into the graded region (toward the emitter) as the bias stress progresses. As a result of Be^+ redistribution, a potential spike in the conduction-band at the emitter–base junction is formed. This increases the device turn-on voltage and causes a severe decrease in the dc current gain. The diffusion of interstitial Be^+ from the base into the emitter is activated only under applied bias and is accelerated by temperature. This mechanism differs from growth-related Be diffusion [21], which is thermally activated. The bias-dependent Be diffusion widely reported for AlGaAs/GaAs HBTs depends strongly on the Be doping level and the amount of bias applied during the stress. It is also affected by the crystallographic orientation of the emitter mesa, such as [011], [010], or $[0\bar{1}1]$ and by surface passivation films on the base–emitter junction.

Low substrate growth temperature and a high As/Ga flux ratio during MBE growth of beryllium could suppress Be^+ formation, resulting in reliable Be-doped HBTs. Lower Be-doping concentration improves device reliability also [22]. The failure mechanism is believed to be the same for all three Be groups. For carbon-doped HBTs the median lifetime is highest among all four groups at 260 and 280°C and second highest at 240°C. However, its low activation energy projects a MTTF at 125°C that is lowest among the four HBT groups in Fig. 8.18. In GaAs, the carbon has a high acceptor efficiency, and low diffusivity. This value of the diffusion coefficient is at least three orders of magnitude smaller than the value for beryllium. Thus a carbon-doped base improves the device reliability.

8.2.2 Sensitivity of Emitter–Base Junction Design

Base out-diffusion changes not only the collector current but also the ideality factor, Early voltage, and device switching speed. The simulations and experiments in [23] show that a strong correlation exists between the drop of collector saturation current, an increase of its ideality factor, and a rise in the switching time due to additional emitter delay from the parasitic barrier effects [24].

Examining Eq. (5.10), it is clear that the small part of the base dopant that lies outside the SiGe layer affects the integral significantly, due to the smaller n_i^2 in silicon. Although a very small fraction of the total base dopant is involved, the change in collector saturation current density may be significant. By the same token, a modulation of the space charge layer will cause an enormous variation in the effective base Gummel number, resulting in a low Early voltage or an ideality factor greater than 1, depending on whether the side of the out-diffusion is at the base–collector or base–emitter junction.

A SiGe HBT with base doping 5×10^{19} cm^{-3}, emitter doping 1×10^{18} cm^{-3}, and ideal abrupt doping profiles has been investigated using one-dimensional device simulation [23]. The valence-band discontinuity is about 200 meV at 25% Ge content [25]. Its normalized collector current density, collector current ideality factor, and emitter–base delay time as a function of the Si/SiGe interface position with respect to the p-n junction at $x = 0$ are shown in Fig. 8.19. Consider the ideal abrupt doping profile, which is represented by the curve a in Fig. 8.19. At $x = 0$ the Si/SiGe interface and the p-n junction coincide. The Si/SiGe interface lies within the base for $x > 0$ and within the emitter for $x < 0$. In the case of $x > 0$, a rapid decrease in collector saturation current density can be observed with increasing distance x because of the increase in the base Gummel number. At the same time the ideality factor rises due to formation of a parasitic conduction-band barrier. The ideality factor increases with interface position, reaches a maximum, and then drops when the barrier becomes too stable to be modulated by V_{BE}. The emitter–base delay increases rapidly for $x > 0$. This is due to a reduction in electron injection into the emitter which increases the emitter delay. The emitter delay is important for an HBT with a heavily doped base and a lightly doped emitter. The base delay remains almost constant. Moving the Si/SiGe interface into the emitter ($x < 0$), the collector saturation current density, ideality factor, and emitter–base delay are essentially unchanged.

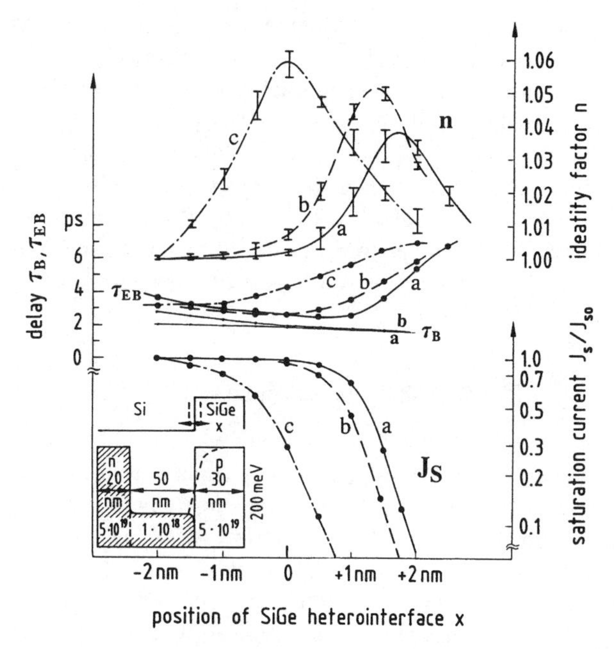

FIGURE 8.19 J_C, n, and E versus position (after Gruhle, Ref. 23 © IEEE).

When the rectangular base dopant profile has been broadened due to diffusion during growth, this leads to an error function distribution with $N_A/2$ at the original step interface position. Curve b in Fig. 8.19 corresponds to a diffusion length of $l = 2.5$ nm, and a curve of c corresponds to $l = 5$ nm. The new position of the p-n junction moves 3.7 nm (curve b) and 7.4 nm (curve c) to the left. The general behavior of the SiGe HBT remains similar. At $x > 0$, parasitic barriers form when out-diffusion takes place, and both the ideality factor and emitter–base junction increase.

The out-diffusion effect may be improved by using an undoped spacer layer at the emitter–base heterojunction. Zhang et al. [26] performed short-term current stress tests on $In_{0.53}Ga_{0.47}As/In_{0.52}Al_{0.48}As$ HBTs with lattice-matched and strained (tensile and compressive) spacer layers inserted between the base and emitter regions. No degradation of the dc and RF performance of the devices with a tensilely strained spacer layer was reported, while small changes in devices with lattice-matched and compressively strained spacer layers were recorded. The tensilely strained spacer layer is very effective in controlling the out-diffusion of Be dopant atoms from the base to the emitter region.

8.2.3 Influence of Dislocations on the Transistor Current Gain

The use of epitaxial layers grown on lattice-mismatched substrates can significantly expand the selection and combinations of material systems for various device applications. When the mismatched layer thickness becomes larger than the critical thickness, however, a high density of dislocations is generated at the mismatched interface, and these penetrate through the epitaxial layers grown above. The influence of dislocations on the minority-carrier lifetime in p^+-GaAs was investigated [27]. i-$Al_{0.3}Ga_{0.7}As/p^+$-GaAs/i-$Al_{0.3}Ga_{0.7}As$ (0.05/1.0/0.1 μm) DH samples were grown by MBE on (100)-oriented semi-insulating GaAs substrates. The GaAs layer was doped with Be to 1×10^{19} cm^{-3}. Three different dislocation density substrates were used: low-dislocation-density (LDD; 2×10^3 cm^{-2}) substrates, medium-dislocation-density (MDD) substrates, and a high-dislocation-density (HDD) substrate. They were created by inserting a lattice-mismatched InAs layer (0.1 μm) underneath an 1.1-μm undoped GaAs buffer layer. Each structure was grown simultaneously in a multiwafer MBE chamber. The substrate temperature during growth was 650°C. Minority electron lifetime was measured using a time-corrected single photon counting method [28].

Figure 8.20 shows the dependence of minority electron lifetime on dislocation density. In this figure the solid circles are the experimental data, the line τ_{df} is the experimentally reduced dislocation-free electron lifetime, and the line τ_d is the dislocation-limited electron lifetime calculated using

$$\tau_d = \frac{4q}{\mu_n k T \pi^3 N_d} \tag{8.1}$$

where N_d is the average dislocation density. The minority electron lifetime measured in the MDD samples is almost the same as that in the LDD sample (260 versus 250 ps). This indicates that the existence of 3×10^5 cm^{-2} dislocations does not significantly affect the minority electron lifetime in the epilayer. On the contrary, the minority electron lifetime in the HDD sample (40 ps) is significantly smaller than that in the

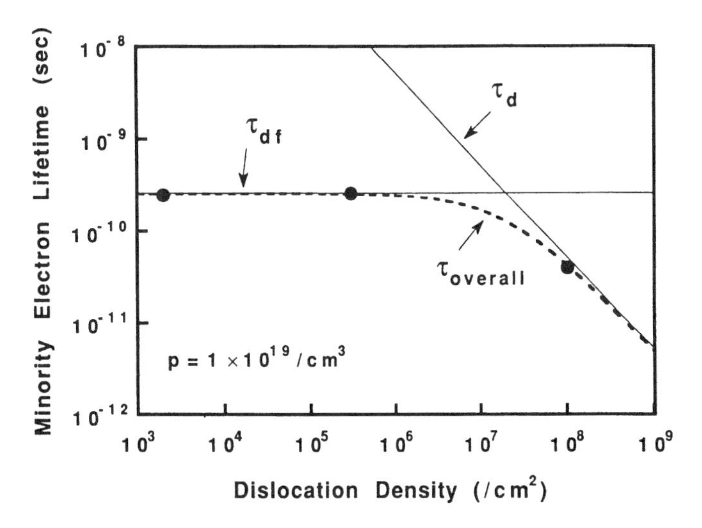

FIGURE 8.20 Lifetime versus dislocation (after Ito et al., Ref. 27 © IEEE).

MDD and LDD samples. A reduction in the minority electron lifetime in the HDD sample is attributed to excess nonradiative recombination through recombination centers in the HDD material. The overall lifetime ($1/\tau_{\text{overall}} = 1/\tau_{df} + 1/\tau_d$) is affected significantly by recombination at dislocations when the dislocation density is higher than 10^7 cm^{-2}.

A 3×10^{17} cm^{-3} doped $Al_{0.3}Ga_{0.7}As$ emitter layer with compositional grading from $x = 0.12$ to 0.3 over 300 Å, an 800-Å-thick 1×10^{19} cm^{-3} doped $Al_xGa_{1-x}As$ ($x = 0$ to 0.12) compositionally graded base layer, and a 3000-Å-thick undoped GaAs collector layer were fabricated. The transistors had an emitter area of 3.6×3.6 μm^2. Current gains at a current density of 2.5×10^4 A/cm^2 as a function of dislocation density are shown in Fig. 8.21. In this plot solid circles are the experimental data, and lines represent β_{df0}, β_{df}, β_{re}, β_{rb}, and β_{r0}. β_{df0} is calculated from dislocation-free current gain using the dislocation-free minority electron lifetime. β_{df} is the measured dislocation-free current gain. β_{re} is the dislocation-related emitter–base junction recombination-limited current gain. β_{rb} is the base recombination–lifetime limited current gain. β_{r0} is the overall dislocation-limited current gain. The overall current gain is determined by $1/\beta_{\text{overall}} = 1/\beta_{df} + 1/\beta_{r0}$. The results in Fig. 8.21 suggest that as long as the dislocation density is less than 10^7 cm^{-2}, there is relatively little current gain reduction for 1×10^{19} doped 800-Å base AlGaAs/GaAs HBTs. On the other hand, current gain reduction is prominent when the dislocation density is greater than 10^7 cm^{-2}.

8.2.4 Effect of Passivation on InAlAs/InGaAs HBTs

Current-induced degradation such as a shift in turn-on voltage of AlGaAs/GaAs HBTs due to Be diffusion has been understood. Stability of beryllium as a p-dopant in high-performance AlInAs/GaInAs HBTs with compositionally graded and abrupt

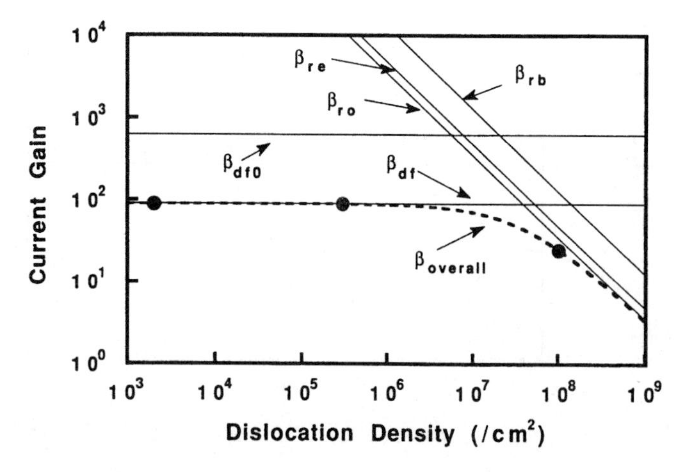

FIGURE 8.21 Current gain versus dislocation (after Ito et al., Ref. 27 © IEEE).

base–emitter junctions has been reported [1]. Compositionally abrupt AlInAs/GaInAs HBTs exhibited similar bias-dependent Be diffusion as observed in GaAs HBTs, but with a slower rate of degradation. Devices that were compositionally graded by inserting a nine-period AlInAs/GaInAs superlattice at the base–emitter junction were found to be extremely stable under a constant collector current stress of $J_C = 7 \times 10^4$ A/cm^2 [29].

In InGaAs HBTs polymide has been used widely for passivation and planarization. The use of polymide passivation results in low surface recombination velocity. This increases the dc current gain and decreases the $1/f$ noise of the HBT. Polyimide passivation also affects the stability and reliability of the heterojunction bipolar transistor [30]. The comparison of the Gummel plot for AlGaAs/GaAs and InAlAs/InGaAs HBTs before and after stress is displayed in Fig. 8.22. In the Gummel plots, the base current I_B rather than the collector current I_C presents a more striking contrast between the two HBTs. Whereas the base current for the GaAs-HBT is drastically reduced by the stressing current, it is essentially unchanged for the InGaAs-HBT. In the case of the GaAs-HBT, the base–emitter voltage shift for I_B is more drastic than that for I_C, and this may be different from the observation of SiO_2-passivated HBTs [22]. The stress current behavior of InAlAs/InGaAs HBTs is related to low emitter surface and emitter–base space-charge region recombination from the InGaAs material. The remarkably stable *I–V* characteristics of InAlAs/InGaAs HBTs suggests minimum stress-current–induced Be diffusion, up to a current density of 1.5×10^5 A/cm^2.

8.2.5 Effect of Hydrogen Out-diffusion in InGaP/GaAs HBTs

Reliability of InGaP/GaAs HBTs grown by MOCVD and GSMBE has been investigated [31]. Differences between MOCVD-grown and GSMBE HBTs have been found. The N-p-n heterostructures had a 500-Å-thick $In_{0.5}Ga_{0.5}P$ emitter doped with Si to 3×10^{17} cm^{-3}, a 800-Å base doped with C to 4×10^{19} cm^{-3}, and a 6000-Å GaAs collector doped with Si to 2×10^{16} cm^{-3}. N^+ $In_{0.5}Ga_{0.5}As$ was used in the nonalloyed

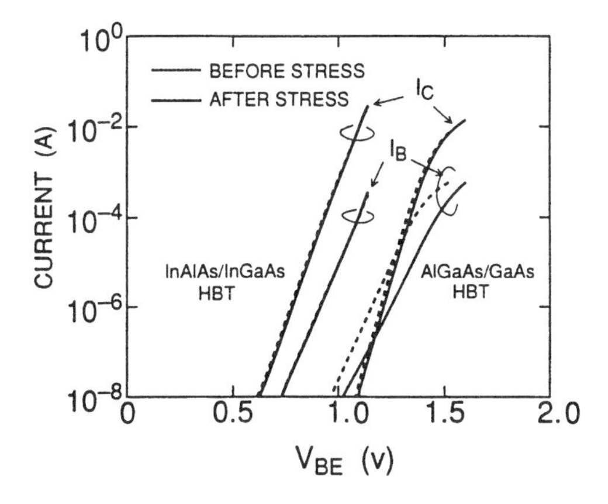

FIGURE 8.22 Gummel plot before and after stress (after Tanaka et al., Ref. 29 © IEEE).

emitter contact layer. The emitter area is $3 \times 3\ \mu m^2$ and has an emitter–base spacing $L_{EB} = 0.5$ and $2\ \mu m$. The stress was carried out at an ambient temperature of 200°C for 40 hours at $V_{CE} = 5$ V and $I_C = 5 \times 10^4$ A/cm^2. The junction temperature was estimated to be 320°C. The stress was interrupted every 30 min. The current gain at 1×10^4 A/cm^2, the turn-on voltage at $I_C = 1$ A/cm^2, and the base resistance were measured.

The MOCVD devices had unstressed 200°C values of current gain of 34.3 and 36.2 for emitter–base spacing $L_{EB} = 0.5$ and $2\ \mu m$, respectively. The GSMBE device had a current gain of 39.6. Unstressed turn-on voltage at 200°C was 0.806 V for the MOCVD devices and 0.834 V for the GSMBE device. Base sheet resistances were 189 and 218 Ω/sq, respectively. When stress is applied, the MOCVD devices show an initial burn-in during which current gain rises by about 5%, followed by an exponential-like decrease to about 85% of the starting value, regardless of emitter–base spacing. The GSMBE device, however, shows no significant degradation. After 40 hours of stress, the current gain at $I_C = 1 \times 10^4$ A/cm^2 changes by about 1% from the unstressed value. Similarly, the turn-on voltage increases about 9 mV for the MOCVD devices while remaining fairly stable for the GSMBE device. A larger emitter–base turn-on voltage could arise either from an increase in base doping, or the diffusion of the base dopant into the emitter. The latter effect, however, should not result in a decrease in base resistance, while the former effect should.

The presence of hydrogen in the epi is suspected when the base doping changes, since it is known that hydrogen passivates carbon acceptors [32]. The MOCVD epi has more in-grown hydrogen than the GSMBE epi, as shown by a SIM analysis result in Fig. 8.23. For the MOCVD epi, hydrogen concentrations of 1.25×10^{19} cm^{-3} and 1.07×10^{19} cm^{-3} were measured at sputter rates of 14 and 25.1 Å/s, respectively. In contrast, the GSMBE epi at a sputter rate of 14 Å/s showed a hydrogen concentration of 2.65×10^{18} cm^{-3} in the base; about five times lower than that obtained at the same sputter rate for the MOCVD-grown epi.

Since the MOCVD-grown epi has a high concentration of in-grown hydrogen, a substantial number of C-acceptors in the base are neutralized by the formation of C–H complexes. During stresses, the hydrogen is freed up and diffuses out of the intrinsic base region, thereby increasing the effective base doping. This hypothesis is supported by the experimental data in [32]. Since hydrogen moves out of the intrinsic base, the intrinsic base resistance will decrease, whereas the extrinsic base will remain largely unchanged. Therefore, for a given emitter area, the decrease in base resistance should be roughly independent of emitter–base spacing. The data in [32] show that the base resistance drops by 1.7 Ω for MOCVD-grown devices with both $L_{EB} = 0.5$ and $2\ \mu m$, in agreement with this expectation. The decrease in current gain and increase in turn-on voltage should be independent of emitter–base spacing also, assuming that the surface recombination in the extrinsic base is negligible. The data show that current gain for both device types degrades about 20% and that turn-on voltage increases about 7 mV after burn-in.

Using Fermi–Dirac statistics, Bahl et al. [31] estimated that an 8.6% increase in base doping without the effects of bandgap narrowing would correspond to a

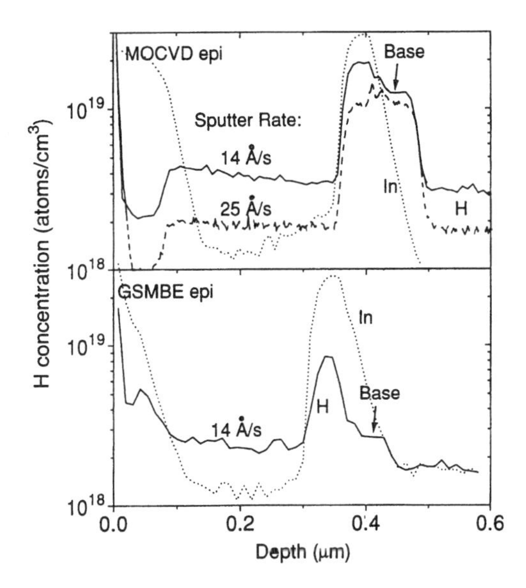

FIGURE 8.23 Hydroden concentration versus depth for MOCVD and GSMBE epi from SIMS analysis (after Bahl et al., Ref. 31 © IEEE).

7-mV increase in turn-on voltage. The base resistance of the MOCVD-grown device ($L_{EB} = 0.5\ \mu$m) dropped by 5.8%, which is not unrealistic for an 8.6% increase in base doping. In contrast, the GSMBE-grown device showed a 1.5% increase in base resistance and a 1-mV increase in turn-on voltage after burn-in. The results above are all consistent with dopant activation caused by hydrogen out-diffusion from the MOCVD-grown device during stress.

To summarize the analysis, C-doped base InGaP/GaAs HBTs grown by MOCVD had a hydrogen concentration in the base of 8.4×10^{18} cm^{-3}, about 20% of the base doping. During reliability testing, a medium-term degradation in current gain, a reduction in base resistance, and an increase in turn-on voltage were observed. The extrapolated activation energy of 0.64 eV is consistent with hydrogen movement in GaAs. The change in electrical characteristics is consistent with the base becoming more heavily doped as a result of C-acceptor activation due to hydrogen depassivation. In contrast, GSMBE-grown epi had five times less hydrogen concentration in the base, and HBT electrical characteristics were relatively stable under stress.

8.3 HOT CARRIER BEHAVIOR

As the dimensions of semiconductor devices shrink and the internal field rise, a large fraction of carriers in the active regions of the device are in states of high kinetic energy. At a given point in space and time, the velocity distribution of carriers may be narrowly peaked, in which case one speaks about "ballistic" electron packets. The term *hot electron* implies a nonequilibrium ensemble of high-energy carriers which have effective electron temperatures higher than the lattice temperature.

Ballistic motion of electrons can be launched by an electric field, thermionic emission from a wider-gap material, or tunneling. In the ballistic devices, electrons are injected into a narrow base layer at a high initial energy in the direction normal to the plane of the layer. Performance of these devices is limited by various energy-loss mechanisms in the base and by the finite probability of a reflection at the base–collector barrier. For example, electrons of HBTs are tunnel injected from the emitter to the conduction-band of the p-type quasi-neutral base. Initially, the injected electrons have average excess kinetic energy. At first the electron momentum is predominantly in the x direction (i.e., perpendicular to the plane of the injector). Later, while traversing the base, the injected electron can experience elastic (impurity) and inelastic collisions. This results in a redistribution of kinetic energy and directional motion. After transiting the base, electrons impinge on the collector barrier. The energy of this barrier may be controlled by the collector–base voltage. Derivation of the collector current with respect to the collector–base voltage for the AlGaAs/GaAs HBT is sketched in Fig. 8.24. The upper curve of this plot represents the initial energy of 220 meV and the lower curve represents the initial energy of 170 meV. The HBT has a p-type GaAs base with a concentration of 3×10^{18} cm^{-3} and a base thickness of 260 Å.

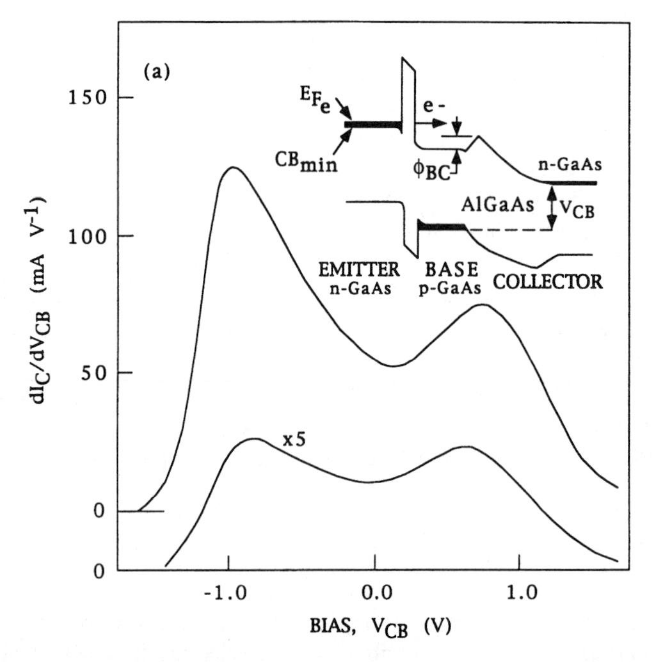

FIGURE 8.24 Derivation of the collector current dI_C/dV_{CB} as a function of V_{CB} (after Berthod et al., Ref. 33 © American Institute of Physics).

dI_C/dV_{CB} as a function of V_{CB} gives information related to a projection of the electron momentum distribution perpendicular to the plane of the collector barrier. Hot electron spectroscopy has been used to demonstrate the existence of nonequilibrium electron transport in the base of an AlGaAs/GaAs HBT in which the base width is comparable to the electron mean free path. In general, hot-electron HBTs show both high-gain and exceptionally high-speed operation.

The hot electrons, which introduce the current gain degradation, are those injected into the Si/SiO_2 interface of Si and SiGe bipolar transistors. Degradation of the bipolar current gain under emitter–base junction reverse-bias stress has been investigated extensively in the past decade [34–39]. It was believed that the current gain degradation was due to stress-generated interface traps at the oxide-covered emitter–base junction space-charge region. These interfacial traps were thought to be due to the dangling bonds from ruptured weak bonds during hot electron impact. Higher interface trap density gives higher electron–hole recombination rate, larger base current, and lower current gain via the SRH recombination current in the p-n junction space-charge region.

Figure 8.25 shows an n^+-p junction and its energy band diagram under reverse-bias stress. Under the oxide there are interface traps D_{IT} (open triangles) at weak bonds

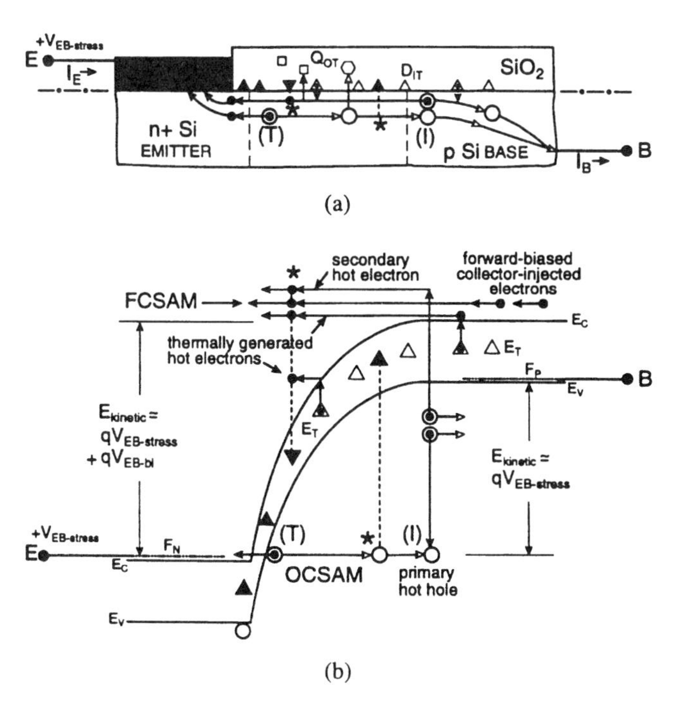

FIGURE 8.25 Thermally generated hot electrons, primary hot holes, and secondary hot electrons in (*a*) the cross sectional view and (*b*) the energy-band diagram along the SiO_2/Si interface (after Neugroschel et al., Ref. 40 © IEEE).

(filled triangles) and charging of oxide electron (open squares) and hole (open hexagons) traps Q_{OT} by hot electrons and hot holes. Since the only direct source of hot electrons in n-p-n bipolar transistors under emitter–base reverse-bias stress is the thermal emission of trapped electrons, these thermal electrons are accelerated to high kinetic energies by the applied emitter–base voltage. The thermal emission rate increases rapidly with increasing temperature. Examine the measured base currents taken at 77 and 298 K shown in Fig. 8.26. The stress-induced base current increments are not a strong function of temperature. This indicates that thermally activated hot electrons are not the dominant hot carriers in a reverse-bias emitter–base junction. Because of the very high base and emitter doping concentrations, the emitter reverse current should be dominated by tunneling, such as tunneling of the valence-band electrons in the p-base into the unoccupied conduction-band states. This tunneling occurs predominately at the highest electric field location near the n^+/p dopant-compensation boundary of the emitter–base junction. The tunneling is consistent with the small temperature dependence observed in Fig. 8.26. The tunneled valence-band electron labeled by the circled dots next to (T) leaves a thermal hole in the valence band of the emitter space-charge region. This thermal hole (the circles with the dot removed or electron dot tunneling out to the n^+ emitter's conduction-band) is then accelerated by the applied $V_{EB\text{-stress}}$, moving in parallel with and adjacent to the SiO_2/Si interface toward the p-base boundary of the emitter–base SCR at the SiO_2/Si interface. This accelerated and hot primary hole can then generate secondary hot electrons by intervened impact generation.

Kizilyalli and Bude [41] used Monte Carlo simulations to determine the hot-electron and hot-hole distributions in the EB junction at high kinetic energies or stress voltages. The current gain degradation due to both the primary hot holes and secondary hot electrons was suggested. Neugroschel et al. [42] demonstrated experimental evidence of the degradation of current gain during reverse emitter–base

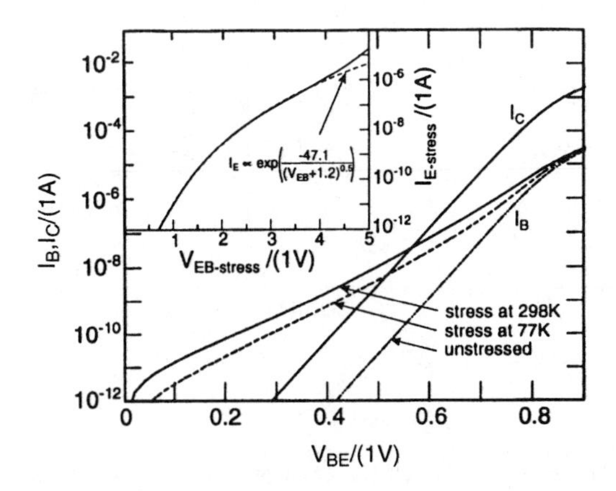

FIGURE 8.26 Collector and base currents versus base–emitter voltage (after Neugroschel et al., Ref. 40 © IEEE).

bias stress. Experimental data favor the primary hot-hole pathways over secondary hot-electron pathways. The experimental stress fluence, charge to failure (Q_{TF}), was computed by $Q_{EB\text{-stress}} = \int j_{E\text{-stress}}(t)\, dt$, for a 10% drop in current gain measured at $V_{BE} = 0.8$ V. Figure 8.27*a* shows Q_{TF} versus $V_{EB\text{-stress}}$ for the two stress-and-measure (SAM) experiments [42] that delineated the hot-electron and hot-hole contributions. In Fig. 8.27*a* the circles represent the data from the open-collector or reverse-collector bias SAM bias configurations encountered in normal BiCMOS circuit operations in which both the primary hot-hole and secondary hot-electron degradation pathways could contribute. The solid lines represent the data from forward-biased collector–base junction SAM measurements (FCSAM) in which the hot-electron current injected by the forward-biased collector–base junction.

The fundamental parameter of D_{IT} generation is the hot carrier kinetic energy, not $V_{EB\text{-stress}}$; thus the kinetic energy of the primary holes in the OCSAM experiment is about $E_G + E_{Fn_+}$ smaller than that in the FCSAM experimental, which

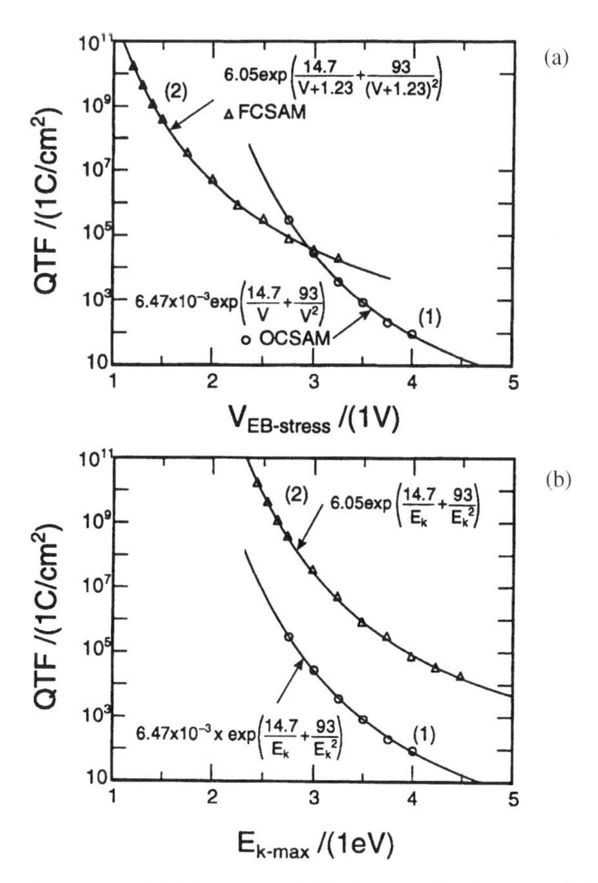

FIGURE 8.27 Q_{TF} versus (*a*) $V_{EB\text{-stress}}$ and (*b*) $E_{k\text{-max}}$ (after Neugroschel et al., Ref. 42 © IEEE).

accounts for the 1.23-V negative lateral shift of the FCSAM data from the OCSAM in Fig. 8.27*a*. Because only a small fraction of the electrons injected by the forward-biased collector–base junction in the FCSAM experiment will flow in the electron surface channel next to the SiO_2–Si interface at the EB junction perimeter to break the weak bonds and increase D_{IT}, the collector–base forward-bias voltage must be large, which gives high I_C and I_E values in the bulk region of the base layer. This large bulk component shifted the Q_{TF}(FCSAM) data upward from the OCSAM data by about $6.05/6.47 \times 10^{-3} \approx 935$, as indicated in the Q_{TF}–$E_{k-\max}$ plot in Fig. 8.27*b*. The nearly constant Q_{TF} ratio of the OCSAM and FCSAM data indicates that the primary hot holes dominate over secondary hot electrons under the open- and reverse-bias collector-circuit-operation bias configurations and that the nearly identical kinetic energy dependence of the generation efficiency of interface traps is initiated by tunnel-injected hot holes and substrate-injected hot electrons. Using the normalized FCSAM and OCSAM experimental data, time to failure (TTF $= Q_{TF}/J_{E\text{-stress}}$) versus $V_{EB\text{-stress}}$ is displayed in Fig. 8.28. The curve gives 2.35 V for the 10-year-TTF maximum operation voltage.

The base current relaxation transient following reverse emitter–base bias stress was measured in silicon bipolar transistors [43]. Figure 8.29 shows I_B versus time measured at $V_{BE\text{-mess}} = 0.56$ V during relaxation for a self-aligned bipolar transistor after reverse-bias stress at various stress voltages. The device was previously stressed at $V_{EB\text{-stress}} = 4.5$ V for 10^4 s. The inset in Fig. 8.29 shows I_B versus stress time. The base current increases with an increase in the stress time. In the I_B relaxation transient measurement, the base current decreases with time. The poststress I_B relaxation measured in the forward-biased EB junction results from the discharging of oxide traps. The oxide trap charging and discharging rates are determined by the trap distribution in the oxide rather than by the EB-junction perimeter electric field that controls the I_B degradation due to N_{IT} generation by hot carriers during reverse-bias stress. Device

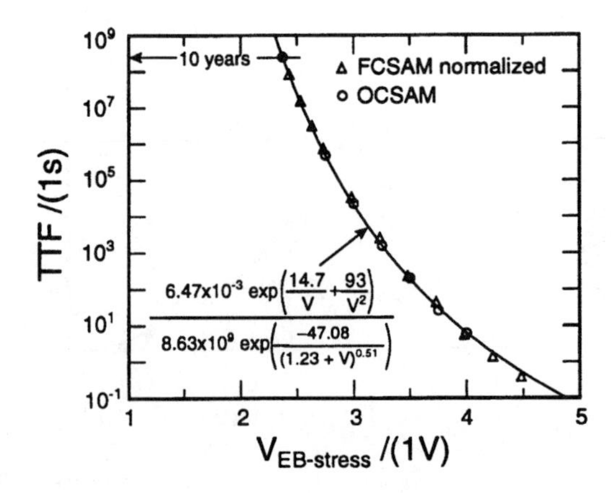

FIGURE 8.28 TTF versus $V_{EB\text{-stress}}$ (after Neugroschel et al., Ref. 42 © IEEE).

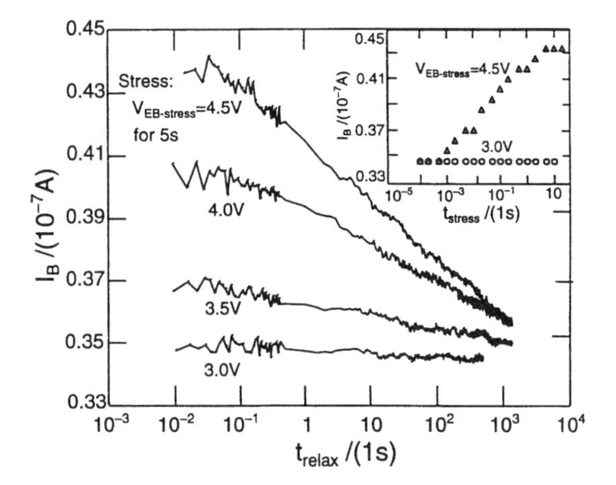

FIGURE 8.29 I_B versus time (after Neugroschel et al., Ref. 43 © IEEE).

temperature as a factor affecting the I_B relaxation characteristics is ruled out because the collector current remained constant during I_B relaxation transient measurements.

The analysis in [43] shows that the I_B relaxation is due to trapped holes in the oxide tunneling to the silicon valence band. The measured dependence of the I_B relaxation on time, temperature, stress voltage, and forward bias during relaxation all support the hole tunneling mechanism. Since the delay time of $\approx$ 1 ms for I_B relaxation to begin after termination of reverse-bias stress was observed, the bipolar reliability in high-speed circuits can be determined by measuring I_B within $\approx$ 1 ms after stress.

8.4 RADIATION EFFECTS

Semiconductor devices in space radiation and nuclear explosion environments can suffer crystalline lattice displacement (defects) and ionization (generation of electron–hole pairs) effects. Radiation effects are generally divided into four categories according to their sources: (1) total ionizing dose, (2) neutron, (3) dose rate (low and high), and (4) single-event upset (SEU). The displacement alters the device performance by reduction of minority-carrier lifetime, majority-carrier removal (trapping and compensation), and mobility degradation. The ionization interaction in a low-dose-rate environment causes gradual charging of dielectric regions, while a high-dose-rate environment causes transient photocurrents. Dose-rate effects could upset circuit functions and cause burnout of junctions or interconnections associated with high or multiplied photocurrents. High-energy galactic particles, protons, and solar flare ions are the main sources of SEU in space, while alpha particles from minute traces of radiotopes in ceramic chip packages can cause the same effect in non-space-based environments. Excess carriers generated by these heavily ionizing particles incident on sensitive junctions are collected to create subnanosecond current pulses. Circuit upsets can result from a single particle confined to a single node. In some cases, single-event currents are regeneratively amplified in a device to cause burnout. Radiation-intensive applications include satellite systems, high-energy particle accelerators, instrumentation in nuclear reactors, and many low-noise high-speed military systems.

Radiation introduces point defects such as vacancies, interstitial, and antisites in semiconductors. Various materials respond differently to the same radiation environment because of variations in the formation of radiation-induced lifetime-limiting defects. In Si, radiation-induced point defects interact with dopants to form lifetime-limiting traps, resulting in higher lifetime degradation with increased base doping concentration [44]. In InP some of the radiation-induced point defects interact with the dopants to form complexes that do not affect the lifetime [45]. This results in a lower net concentration of radiation-induced lifetime-limiting point defects in InP. Consequently, an increase in doping concentration results in a decrease in the lifetime degradation in InP. Unlike Si and InP, the radiation-induced point defects in GaAs do not interact appreciably with the dopants, resulting in only a weak dependence of lifetime degradation on base doping. In GaAs, lifetime-limiting defects are attributed directly to point defects, such as antisites or sublattice atom vacancies [46].

Figure 8.30 shows the calculated lifetime as a function of doping concentration for Si, GaAs, and InP after a 1-MeV electron irradiation at a fluence of 1×10^{15} cm^{-3}. Based on limited experimental data, radiation effects on SiGe HBTs are known as similar to that of Si bipolar transistors. The radiation hardness of GaAs HBTs are significantly greater than Si bipolar transistors, and InP-based HBTs are more radiation hard than are GaAs/AlGaAs HBTs.

8.4.1 Si-Based Bipolar Transistors

The detrimental effects of ionizing radiation on conventional double-diffused silicon bipolar transistors are well known. There has been renewed interest in the effects of ionizing radiation on modern double polysilicon self-aligned BJTs. Bipolar transistors are susceptible to particle bombardment in radiation environments, which create trapping centers in the emitter–base space-charge region and the base region and therefore increase electron–hole recombination in these regions [48]. For SiGe heterojunction bipolar transistors, however, very few radiation experimental data have been reported. Babcock et al. [49] reported in 1995 the effect of ionizing radiation on both the electrical and $1/f$ noise characteristics of advanced UHV/CVD SiGe HBTs for the first time. High-performance SiGe HBTs and Si BJTs manufactured by IBM were used in their investigation [50]. To ensure unambiguous comparison, the Si BJTs were processed in the same wafer lot, under processing conditions identical to those for SiGe HBTs. The only difference was the growth of Ge in the epitaxial base. Transistors were fabricated using a planar structure with a conventionally doped arsenic polysilicon emitter and a deep-trench isolation process, which is believed to give these transistors an inherent tolerance to ionizing radiation.

Irradiation was performed in a Gamma-Cell with a ^{60}Co source providing a dose rate of about 48 rad(Si)/s. Pre- and postradiation electrical measurements were made

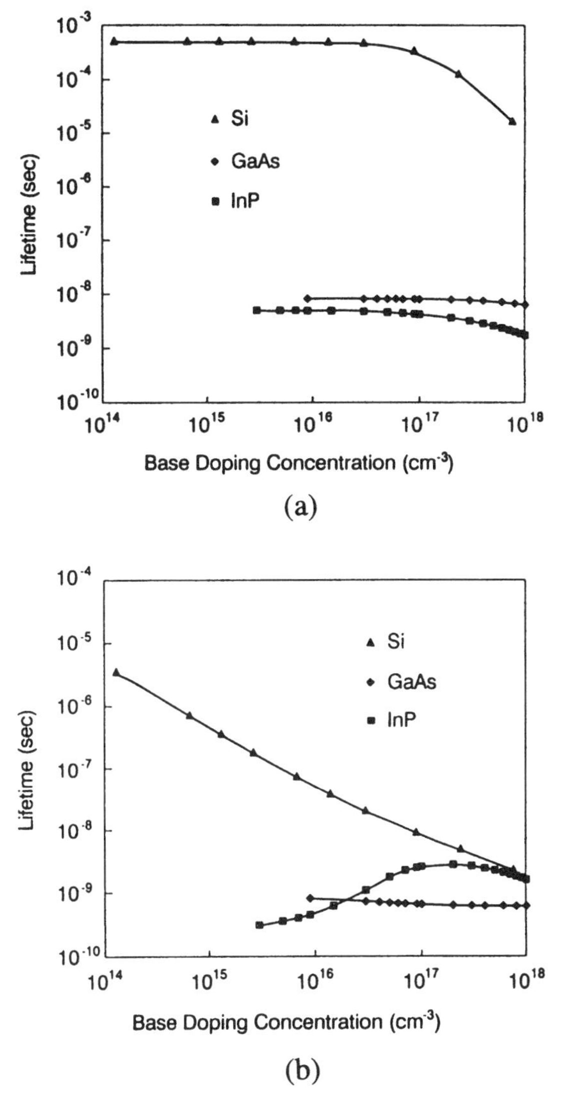

FIGURE 8.30 Calculated lifetime of Si, GaAs, and InP versus base doping concentration (*a*) before and (*b*) after irradiation (after Augustine et al., Ref. 47 © IEEE).

after devices were removed from the source at 300 K using a computer-controlled Keithley measurement system. Delay time between irradiation and remote testing was approximately 15 minutes. All device terminals were grounded during irradiation and the dose rate during irradiation was confirmed using TLD dosimeters. Pre- and postradiation low-frequency $1/f$ noise measurements were made using an HP 3561 dynamic signal analyzer.

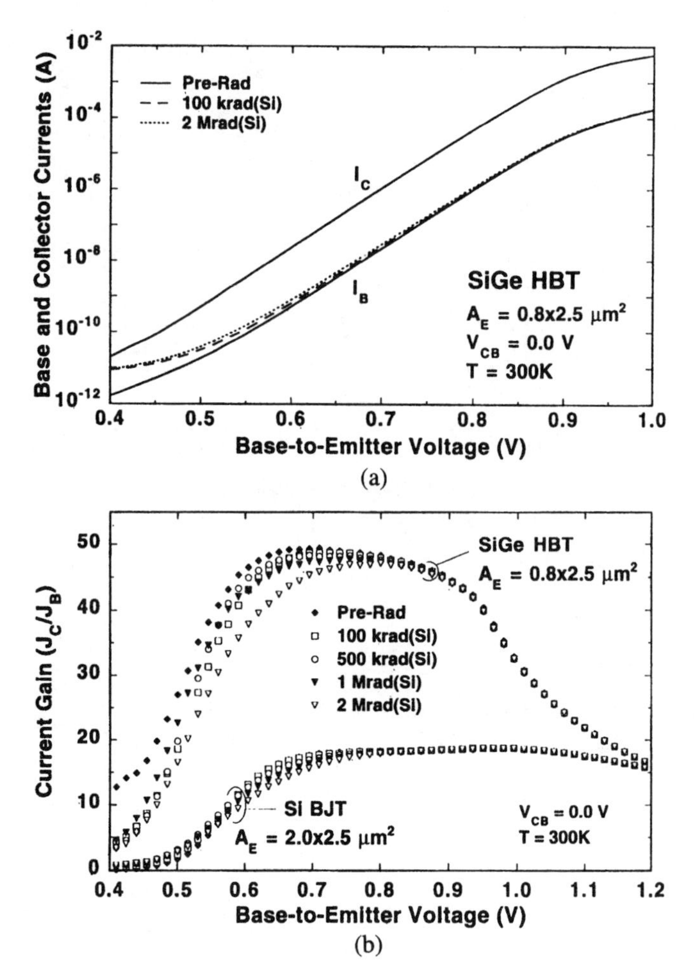

FIGURE 8.31 (*a*) I_C and I_B and (*b*) current gain versus V_{BE} (after Babcock et al., Ref. 49 © IEEE).

The typical response of a SiGe HBT to ionizing radiation is displayed in the Gummel and β plots in Fig. 8.31*a* and *b*, respectively. As shown in Fig. 8.31*a*, ionizing radiation causes an increase in base current at low base–emitter voltages but does not affect the collector current of the SiGe HBT. Pre- and postradiation current gain in Fig. 8.31*b* indicate the reduction of current gain at low base–emitter voltages. A similar response to ionizing radiation was also observed in the electrical characteristics of Si BJTs. Radiation exposure to 2.0 Mrad(Si) results in virtually no current gain degradation for $V_{BE} \geq 0.8$ V ($J_C \approx 24\ \mu$A/μm^2 in the SiGe HBT). For $V_{BE} < 0.8$ V, degradation in the current gain of both the SiGe HBT and the Si BJT becomes significant, especially for low base–emitter voltages. Less than 8% shift in the peak current gain ($V_{BE} = 0.7$ V) after exposure to 2 Mrad(Si) has been observed. These results are remarkable compared to modern Si bipolar transistors, which typically have a 50% or greater reduction in the peak current gain after exposure to 500 krad(Si) [51]. Enhanced tolerance of the UHV/CVD SiGe HBTs and the Si BJTs used in the investigation is believed to be due to the result of the thin nitride/oxide spacer and the high doping at the surface of the epi base region. Utilization of a thin nitride/oxide spacer region over the intrinsic base results in significant reduction in the radiation-induced oxide trapped charge and interface trapped charge over conventional SiO_2 spacers. UHV/CVD growth of the epitaxial base region allows for the incorporation of dopant atoms during the growth phase and thus eliminates implant damage in the screen oxide, which is typically found in more conventional bipolar processes [52]. The high doping at the surface of the epi base region requires a proportionally larger amount of radiation-induced trapped charge to cause a parasitic inversion region to form when compared to a lower-doped base.

Figure 8.32 shows the percent change in normalized base current difference $(\Delta I_B/I_B)$ versus total radiation dose at constant V_{BE}. ΔI_B is the difference between the post- and preradiation base current. In Fig. 8.32, almost identical shifts in base current with total radiation dose for both the SiGe HBT and the Si BJT have been observed. These results suggest that the introduction of SiGe strained layers into the epitaxial base of the SiGe HBT do not impose any additional reliability risk for ionizing radiation tolerance compared to Si BJTs.

The flicker noise of SiGe HBTs before and after ionizing radiation was measured [49]. Exposure to ionizing radiation resulted in an observed increase in the $1/f$ noise in several of the SiGe HBTs and Si BJTs investigated. Figure 8.33 shows the measured pre- and postradiation equivalent base noise power spectral densities $(S_I B)$ in a SiGe HBT after exposure to 2 Mrad(Si). An increase in the $1/f$ noise is apparent after irradiation. The multiplication of S_{IB} by the frequency reveals a Lorentzian-shaped

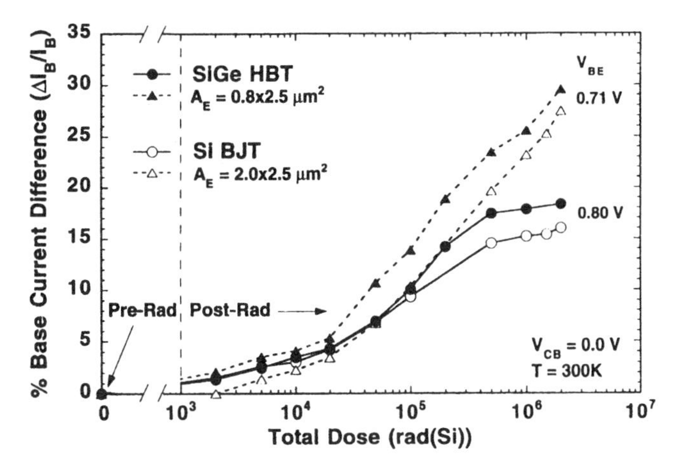

FIGURE 8.32 Percent change in base current versus total dose (after Babcock et al., Ref. 49 © IEEE).

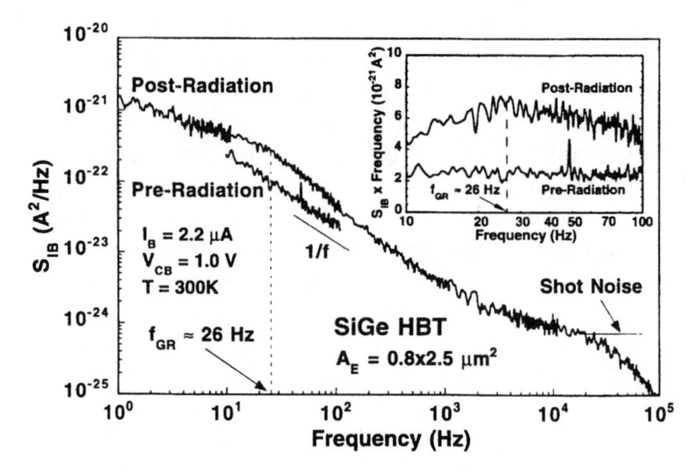

FIGURE 8.33 Equivalent base noise power spectral density (S_{IB}) versus frequency before and after ionizing radiation (after Babcock et al., Ref. 49 © IEEE).

hump superimposed on the $1/f$ component which is associated with generation–recombination (G/R) centers [53]. The time constant associated with G/R centers contributing to the postradiation noise is found by taking $1/\omega_{G/R} \approx 6.1$ ms, where $\omega_{G/R} = 2\pi f_{G/R}$ and $f_{G/R}$ is the 3-dB break frequency associated with the G/R noise (26 Hz in this case).

8.4.1.1 Oxide Trapped Charge and Excess Base Current

Exposure to ionizing radiation degrades the current gain of bipolar transistors by increasing the base current while leaving the collector current approximately unchanged, as evidenced by the Gummel plot in Fig. 8.33 [54]. The role of the net positive oxide trapped charge and surface recombination velocity on excess base current is identified. The oxide charge changes the surface potential along the base surface, causing depletion in the n-p-n bipolar transistor. The interface states near midgap cause an increase in the surface recombination velocity. Because of the surface potential behavior, the effects of ΔN_{ot} and ΔN_{it} on ΔI_B are interactive and not simply additive. In addition, the defects of interest in BJTs are not precisely the same as those in MOS transistors. For example, ΔN_{it} in MOSFETs is a measure of the number of charged interface states at threshold. In BJTs, on the other hand, the excess base current depends on the number of interface states (recombination centers) near midgap, not the total number between midgap and threshold. The excess base currents due to changes in surface potential depend on the total radiation-induced oxide charge at the bias condition and the lateral position of interest.

Consider the I_B curves in Fig. 8.34, which exhibit two distinct slopes. The slope corresponding to the ideality factor $1 < n < 2$ is a clear indication that surface recombination near the junction is dominant. The transition to $n = 2$ at higher values of V_{BE} is a signature of predominately bulk rather than surface recombination. The

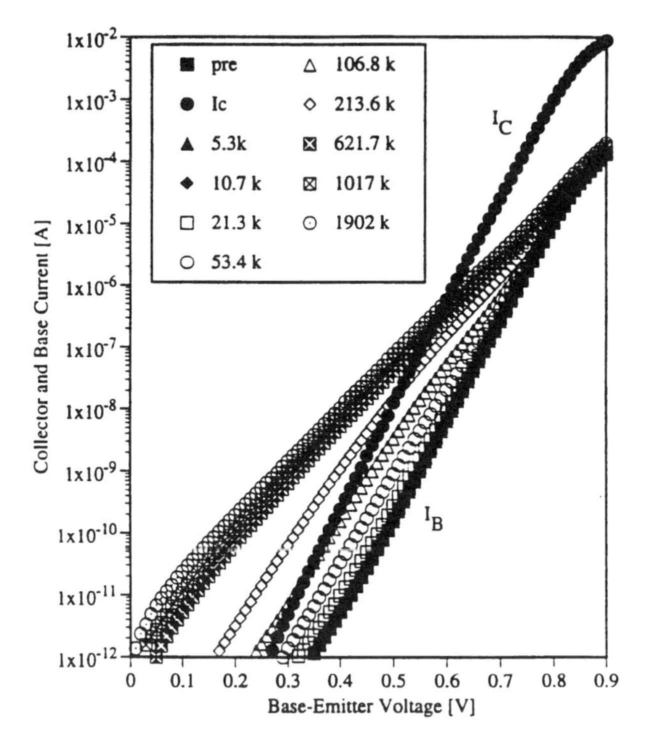

FIGURE 8.34 Base and collector currents versus base–emitter voltage at various radiation levels (after Kosier et al., Ref. 54 © IEEE).

voltage at which this transition occurs is an indication of the amount of radiation-induced charge in the oxide covering the emitter–base junction and the intrinsic base.

8.4.1.2 Low-Dose-Rate Radiation

Investigations of the total dose response of modern n-p-n bipolar transistors have shown that the gain degradation is much greater at a dose rate below 10 rad/s than at a dose rate above 100 rad/s. Recent analysis of low-dose-rate response shows that excess base current, oxide charge, and effective surface recombination velocity are highest at a lower dose rate ($<$ 10 rad/s) and decrease with increasing dose rate until they become independent of dose rate at higher dose rates ($>$ 100 rad/s). The dose-rate response of the BJT is very different from the dose-rate response of MOSFETs [55], where oxide charge tends to be reduced at low dose rates due to trapped-hole neutralization and/or interface trap buildup. The enhanced degradation at low dose rate is a result of a larger net trapped positive charge N_{OX} in base oxides at low electric field [52]. This larger N_{OX} (20 to 30% higher) can cause a much larger (5 to 10 times) value of excess base current ΔI_B when the surface potential at the intrinsic base surface is near the crossover point.

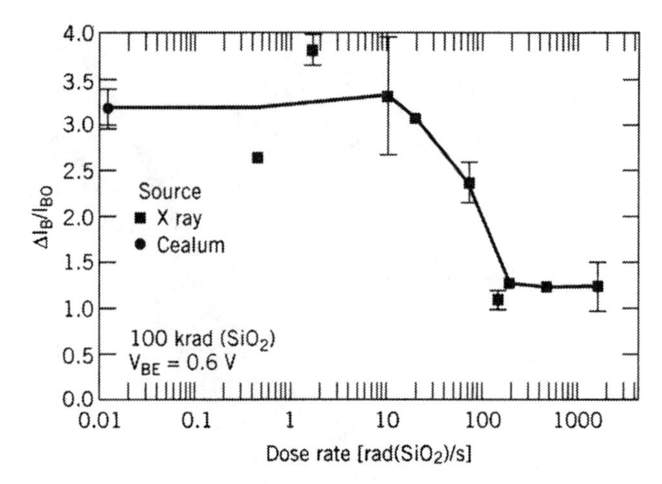

FIGURE 8.35 Normalized excess base current versus dose rate (after Nowlin et al., Ref. 55 © IEEE).

Figure 8.35 shows dose-rate dependence of the normalized excess base current for devices irradiated at room temperature. The magnitude of the total-dose-induced excess base current is independent of dose rate above 150 rad(SiO_2)/s, and it increases below 150 rad(SiO_2)/s. The degradation at low dose rates is about three to four times greater than that at the high dose rates. Moreover, these results are relatively independent of irradiation source.

The implications for radiation hardness assurance are discussed as follows. Since the dose-rate response saturates at low dose rates, it is not necessary to test at even lower dose rates to determine the device response for space applications because of the cost and time required for low-dose-rate experiments. One may have to test at dose rates about three- to fourfold worse than the high-dose-rate response when these devices are exposed to 100 krad(SiO_2)/s. An overtest in dose by a factor of 3 can allow one to conservatively estimate device suitability for space applications if other circuit elements do not interfere by causing earlier failure.

8.4.1.3 Implications for Circuit Behavior

The total dose-rate radiation affects the input offset voltage, input offset current, and open-loop gain of linear amplifiers [56]. At a low dose rate, the average input offset voltage increases to −11 mV at 100 krad, whereas at high dose rate, the average showed essentially little change in Fig. 8.36*a*. Following a 168-hour 25°C anneal, the high-dose-rate average increased to −1.5 mV, then decreased slightly after an additional 300-hour 100°C anneal. Thus the change in input offset voltage at low dose rate is much greater than the change at high dose rate plus anneal. The input offset current increases significantly with total dose for both low and high dose rates, as shown in Fig. 8.36*b*. In Fig. 8.36*c* the open-loop gain decreases with total dose. The degradation is enhanced for low dose rate.

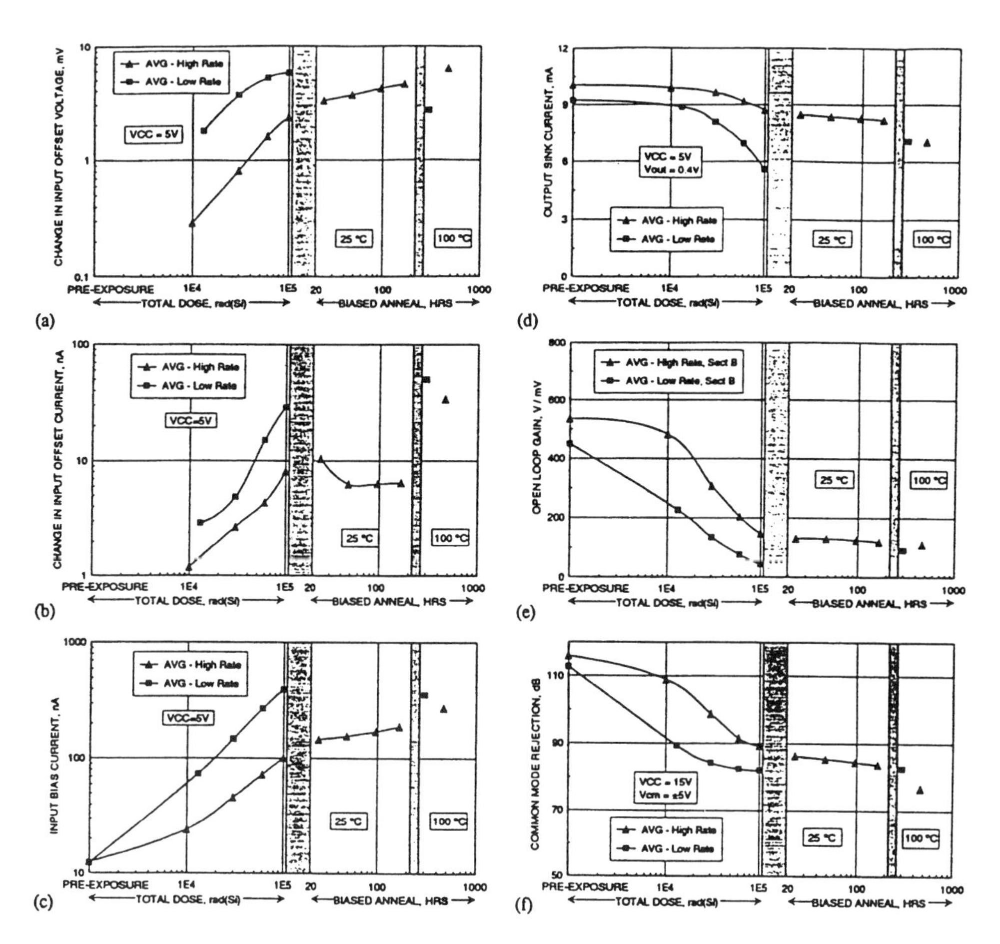

FIGURE 8.36 Change in (*a*) input offset voltage, (*b*) input offset current, (*c*) input bias current, (*d*) output sink current, (*e*) open-loop gain, and (*f*) common-mode rejection ratio versus total dose and anneal hours of a BJT linear amplifier (after McClure et al., Ref. 56 © IEEE)

Analog-to-digital converters (ADCs) are sensitive to the total ionizing dose. The most sensitive parameters were transfer curve nonlinearities, reference voltage, and tristate leakage current. The converters failed functionally at lower dose level with the low-dose-rate testing. After total dose irradiation, the internal reference voltage degraded significantly. The internal track-and-hold amplifier on the analog input allows the converter to accurately convert an input sine wave of 6 V peak-to-peak amplitude accuracy. Noise in this critical path along with internal reference noise will add significant distortions to the input sine wave and will eventually degrade the dynamic performance of the converter. The harmonic distortion of the converter increases substantially as the total dose radiation level increases. The signal-to-noise ratio decreases significantly as the noise level and number of harmonics increases. Similar noise results for a low-noise amplifier after total dose irradiation were observed. The voltage noise of this device was essentially unchanged by radiation, but the current noise degraded severely.

8.4.2 GaAs- and InP-Based Bipolar Transistors

To date, there has been little work on radiation effects on III/V compound HBTs. Schrantz et al. [57] reported that AlGaAs/GaAs HBTs irradiated with 10^{15} neutrons/cm^{-2} demonstrated much higher levels of immunity than did Si bipolar transistors. Devices fabricated from III/V materials generally display more resistance to total dose effects from all forms of radiation than does a comparable Si-based structure. The greater radiation hardness of III/V compound HBTs arises from shielding of the active junctions from the semi-insulating substrate by the highly doped subcollector layer. Also, there is no n-p isolation junction effects or Fermi-level pinning at the surface. GaAs/AlGaAs heterojunction inverted transistor integrated logic gate arrays proved to be relatively insensitive to energetic protons and heavy ions between 37 and 66 meV.

Yamaguchi and co-workers [58–61] studied p-type InP with a report of 1.0-MeV electron irradiation-induced electron traps at activation energies of $E_A = 0.15, 0.22,$ 0.2, 0.37, and 0.52 eV and an electron trap at $E_A = 0.19$ eV. They attribute the radiation resistance of InP to the migration energy of In- and P-displaced atoms in InP, which is favorable compared to GaAs. Sibille et al. [62,63] reported on 1.0-MeV electron irradiation of p-InP using a Ti–Au Schottky contact. They reported hole traps with $E_A = 0.17, 0.33, 0.37,$ and 0.54 eV, in good agreement with Yamaguchi, although capture cross section differed in some cases by an order of magnitude. Sibille observed a significant reduction in radiation damage by annealing at 470 K for 10 min. They also identified the 0.37-eV level as a point defect and the 0.52-eV level as a point defect-impurity complex.

8.4.2.1 Total Dose Effects

In total dose testing a ^{60}Co source provides gamma-ray radiation, which generates Compton electrons, accumulated over a long period. The damage is considered permanent. The total-dose environment appears to be less harmful to HBTs than to Si devices. InP-based devices are expected to be even more radiation-hard than their GaAs-based counterparts.

Song et al. [64] examined the radiation hardness of GaAs/AlGaAs HBTs in neutron, total γ-ray dose, and dose-rate environments. The HBTs were $3 \times 10\ \mu m^2$ devices, with Be base doping of 1×10^{19} cm^{-3}, grown by MBE. Fission neutrons with energy greater than 10 keV were used to irradiate samples up to doses of 1.3×10^{14} neutrons/cm^2. The dc gain of the devices decreased from 60 to 48 after a neutron flux of $\approx 10^{14}$ cm^{-2}, with the onset of degradation at $\approx 3 \times 10^{13}$ cm^{-2}. It was postulated that much of the gain decrease resulted from an increase in space-charge recombination at the emitter–base heterojunction. The RF performance was unchanged by a neutron dose of 10^{14} cm^{-2}. Similarly, only minor changes in dc gain

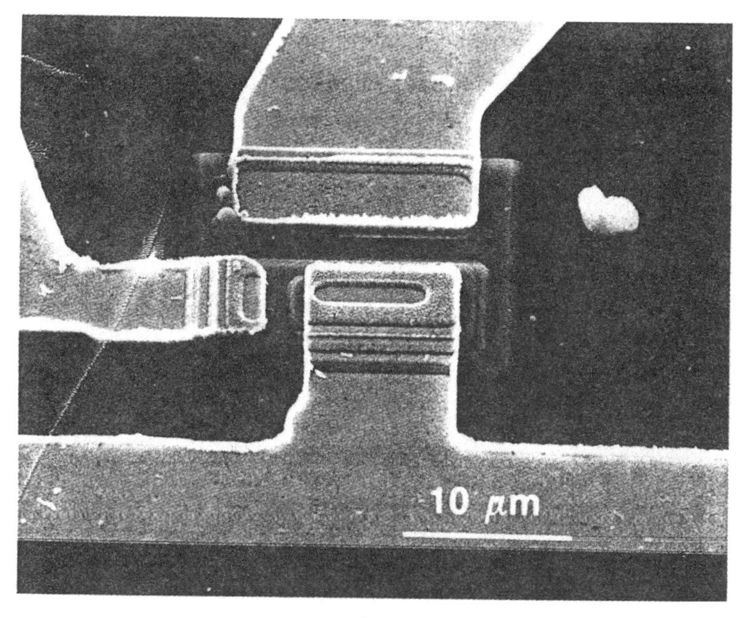

(*a*)

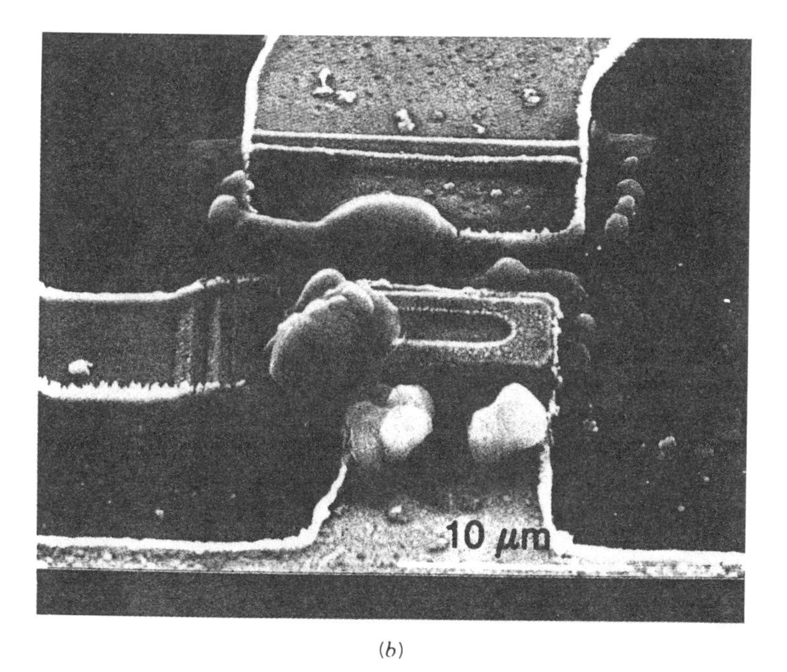

(*b*)

FIGURE 8.37 SEM micrographs of AlInAs/InGaAs before (*a*) and after (*b*) catastrophic failure (after Jalali and Pearton, Ref. 10 © Artech House).

of these devices was observed after a total γ-ray dose of 467 Mrad (7% reduction in current gain).

Minimal changes in the dc characteristic of the InP-based HBT were observed for doses up to 20 Mrad, while at 40 Mrad some of the devices started to fail catastrophically. The failure-mode characteristic, an open circuit from base to emitter, was the same in all failed HBTs. The base-to-collector junction was not affected. The failed devices showed the presence of large quantities of oxygen and carbon-rich debris around the contact as displayed in Fig. 8.37. Parametric changes, such as decreased current gain, transconductance, and increased output resistance, were observed in the devices at high doses. Figure 8.38 shows a Gummel plot of a $2 \times 4 \mu m^2$ AlInAs/InGaAs HBT before and after 40-MRad radiation. Significant degradation of collector current occurs for $V_{BE} > 1.0$ V, while the base current is increased at high base–emitter biases.

8.4.2.2 Transient Radiation Effects

Dose-rate tests have been performed by using flash x-ray pulses, 22 ns wide. The HBT transient collector current recovery time is on the order of a few microseconds, as shown in Fig. 8.39. The lack of long-term transients in HBTs is due to the vertical structure, where the active junctions are isolated from surfaces and substrate interface. Another transient radiation effect is the single-event upset (SEU). High-energy galactic particles, protons, and solar flare ions are the main sources of SEU in space, while alpha particles (helium nuclei) from minute traces of radioisotopes in ceramic chip packages can cause the same effect in non-space-based environments. Excess carriers generated by these heavily ionizing particles incident on sensitive junctions

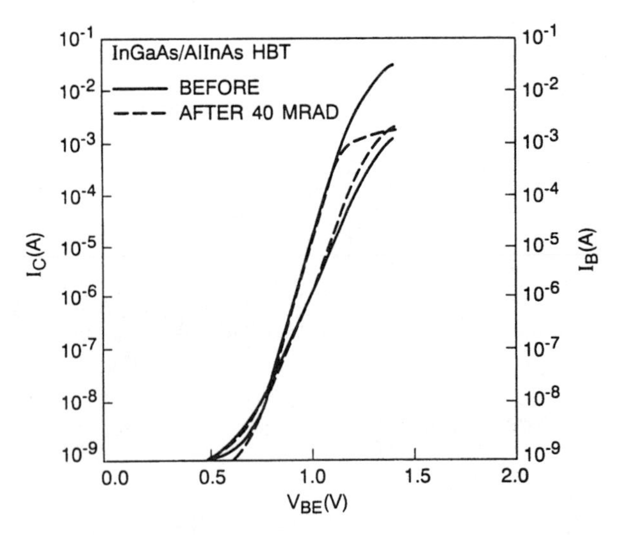

FIGURE 8.38 I_C and I_B of an AlInAs/InGaAs HBT before and after a 40-Mrad dose of ^{60}Co γ-ray radiation (after Jalali and Pearton, Ref. 10 © Artech House).

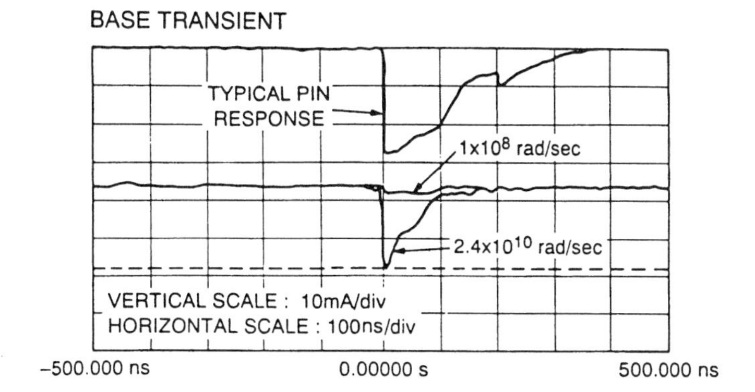

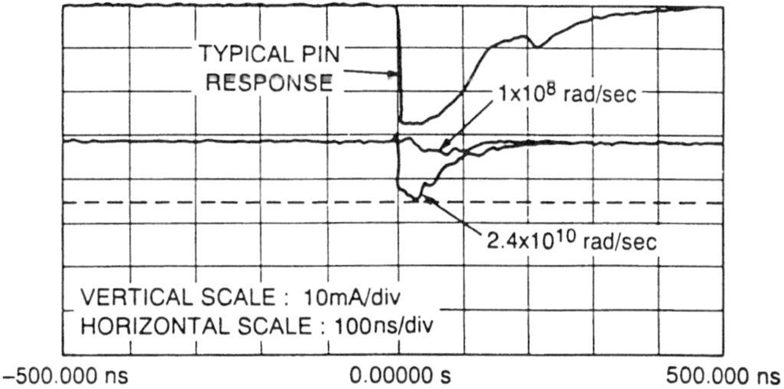

FIGURE 8.39 Transient collector and base currents of an AlGaAs/GaAs HBT during a 120-ns ionizing radiation pulse (after Jalali and Pearton, Ref. 10 © Artech House).

are collected to create a subnanosecond current pulse. In some cases, single-event currents are amplified regeneratively in a device to cause burnout. An LSI RAM or shift register is attractive for SEU monitoring.

REFERENCES

1. M. Hafizi, W. W. Stanchina, R. A. Metzger, J. F. Jensen, and F. Williams, "Reliability of AlInAs/GaInAs heterojunction bipolar transistors," *IEEE Trans. Electron Devices*, **ED-40**, 2178 (1993).
2. M. Hafizi, W. E. Stanchina, F. Williams, Jr., and J. F. Jensen, "Reliability of InP-based HBT IC technology for high-speed, low-power applications," *IEEE Trans. Microwave Theory Tech.*, **MTT-43**, 3048 (1995).
3. R. S. Hemmert, G. S. Prokop, J. R. Lloyd, P. M. Smith, and B. M. Calabrese, "The relationship among electromigration, passivation thickness, and common-emitter current

gain degradation within shallow junction npn bipolar transistors," *J. Appl. Phys.*, **53**, 4456 (1982).

4. R. S. Hemmert and M. Costa, "Electromigration-induced compressive stresses in encapsulated thin-film conductors," *Proceedings of the 29th IEEE International Reliability Physics Symposium*, 64 (1991).
5. R. A. Wachnik, T. J. Bucelot, and G. P. Li, "Degradation of bipolar transistors under high current stress at 300 K," *J. Appl. Phys.*, **63**, 4734 (1988).
6. C. J. Sun, T. A. Grotjohn, C.-J. Huang, D. K. Reihard, and C.-C. W. Yu, "Forward-bias stress effects on BJT gain and noise characteristics," *IEEE Trans. Electron Devices*, **ED-41**, 787 (1994).
7. C. J. Sun, D. K. Reinhard, C.-J. Huang, T. A. Grotjohn, and C.-C. W. Yu, "Hot-carrier-induced degradation and post-stress recovery of bipolar transistor gain and noise characteristics," *IEEE Trans. Electron Devices*, **ED-39**, 2178 (1992).
8. A. Neugroschel, C.-T. Sah, J. M. Ford, J. Steele, R. Tang, and C. Stein, "Comparison of time-to-failure of GeSi and Si bipolar transistors," *IEEE Electron Device Lett.*, **EDL-17**, 211 (1996).
9. M. Hong, E. de Frésart, J. Streele, A. Zlotnicka, C. Stein, G. Tam, M. Racanelli, L. Knoch, Y.-C. See, and K. Evans, "High-performance SiGe epitaxial base bipolar transistors produced by a reduced-pressure CVD reactor," *IEEE Electron Device Lett.*, **EDL-14**, 450 (1993).
10. B. Jalali and S. J. Pearton, eds., *InP HBTs: Growth, Processing, and Applications*, Artech House, Norwood, MA (1995).
11. M. Hafizi, T. Liu, A. E. Schmitx, P. A. McDonald, M. Lui, and F. Williams, "Power performance and reliability of AlInAs/GaInAs/InP double heterojunction bipolar transistors," *Proceedings of the 6th International Conference on InP and Related Materials*, Santa Barbara, CA, 527 (1994).
12. J. J. Liou and C. I. Huang, "Base and collector currents of pre- and post-burn-in AlGaAs/GaAs heterojunction bipolar transistors," *Solid-State Electron.*, **37**, 1349 (1994).
13. J. L. Benton, M. Levinson, A. T. Macrander, H. Temkin, and L. C. Kimberling, "Recombination enhanced defect annealing in n-InP," *Appl. Phys. Lett.*, **45**, 566 (1984).
14. H. Sugahara, J. Nagano, R. Nittono, and K. Ogawa, "Improved reliability of AlGaAs/GaAs heterojunction bipolar transistors with a strain-relaxed base," *IEEE GaAs IC Symposium*, 115 (1993).
15. M. Borgarino, R. Plana, J. G. Tartarin, S. Delage, H. Blanck, F. Fantini, and J. Graffeuil, "On the stability of the DC and RF gain of GaInP/GaAs HBTs," *1996 Workshop on High Performance Electron Devices for Microwave and Optolectronic Applications*, 37 (1996).
16. M. Borgarino, J. G. Tartarin, S. Delage, R. Plana, F. Fantini, and J. Graffeuil, "Correlation between the burn-in effect and the extrinsic base surface quality in C-doped GaInP/GaAs HBTs," *1997 Workshop on High Performance Electron Devices for Microwave and Optolectronic Applications*, 43 (1997).
17. E. J. Prinz, P. M. Garone, P. V. Schwartz, X. Xiao, and J. C. Sturm, "The effects of base dopant out-diffusion and undoped $Si_{1-x}Ge_x$ junction spacer layers in $Si/Si_{1-x}Ge_x/Si$ heterojunction bipolar transistors," *IEEE Electron Device Lett.*, **EDL-2**, 42 (1991).
18. F. M. Yamada, A. K. Oki, D. C. Steet, Y. Saito, D. K. Umemoto, L. T. Tran, S. Bui, J. R. Velebir, and G. W. Mclver, "Reliability analysis of microwave GaAs/AlGaAs HBTs with beryllium and carbon doped base," *Dig. IEEE MTT*, 739 (1992).

19. R. People and J. C. Bean, "Band alignment of coherently strained Ge_xSi_{1x}-/Si heterostructure on ⟨00⟩ Ge_ySi_{1-y} substrates," *Appl. Phys. Lett.*, **48**, 538 (1986).
20. R. Fair, "Boron diffusion in silicon: concentration and orientation dependence, background effects and profile estimation," *J. Electrochem. Soc.*, **122**, 800 (1975).
21. D. L. Miller and P. M. Asbeck, "Be redistribution during growth of GaAs and AlGaAs by molecular beam epitaxy," *J. Appl. Phys.*, **57**, 1816 (1985).
22. O. Nakajima, H. Ito, T. Nittono, and K. Nagata, "Current induced degradation of Be-doped AlGaAs/GaAs HBTs and its suppression by Zn diffusion into extrinsic base layer," *IEDM Tech. Dig.*, 673 (1990).
23. A. Gruhle, "The influence of emitter–base junction design on collector saturation current, ideality factor, Early voltage, and device switching speed of Si/SiGe HBTs," *IEEE Trans. Electron Devices*, **ED-41**, 198 (1994).
24. J. W. Slotboom, G. Streutker, A. Pruijmboom, and D. Gravesteijn, "Parasitic energy barriers in SiGe HBTs," *IEEE Electron Device Lett.*, **EDL-12**, 486 (1991).
25. R. People and J. C. Beam, "Band alignments of coherently strained GeSi/Si heterostructures," *Appl. Phys. Lett.*, **48**, 538 (1986).
26. K. Zhang, X. Zhang, P. Bhattacharya, and J. Singh, "Influence of psedomorphic base–emitter spacer layers on current-induced degradation of beryllium-doped InGaAs/InAlAs heterojunction bipolar transistors," *IEEE Trans. Electron Devices*, **ED-43**, 8 (1996).
27. H. Ito, O. Nakajima, T. Furuta, and J. S. Harris, Jr., "Influence of dislocations on the dc characteristics of AlGaAs/GaAs heterojunction bipolar transistors," *IEEE Electron Device Lett.*, **EDL-13**, 232 (1992).
28. H. Ito, T. Furuta, and T. Ishibashi, "Minority electron lifetimes in heavily doped p-type GaAs grown by molecular beam epitaxy," *Appl. Phys. Lett.*, **58**, 2936 (1991).
29. S. Tanaka, K. Kasahara, H. Shimawaki, and K. Honjo, "Stress current behavior of InAlAs/InGaAs and AlGaAs/GaAs HBTs with polyimide passivation," *IEEE Electron Device Lett.*, **EDL-13**, 560 (1992).
30. M. Hafizi, R. A. Metzger, and W. E. Stanchina, "Stability of beryllium-doped compositionally graded and abrupt AlInAs/GaInAs HBTs," *Appl. Phys. Lett.*, **63**, 93 (1993).
31. S. R. Bahl, L. H. Camnitz, D. Houng, and M. Mierzwinski, "Reliability investigation of InGaP/GaAs heterojunction bipolar transistors," *IEEE Electron Device Lett.*, **EDL-17**, 446 (1996).
32. S. J. Pearton, C. R. Abernathy, and J. W. Lee, "Comparison of H^+ and He^+ implant isolation of GaAs-based heterojunction bipolar transistors," *J. Vac. Sci. Technol.*, **B13**, 15 (1995).
33. K. Berthod, A. F. J. Levi, J. Walker, and R. J. Malik, "Extreme nonequilibrium electron transport in heterojunction bipolar transistors," *Appl. Phys. Lett.*, **52** (26), 2247 (1988).
34. D. R. Collins, "Excess current generation due to reverse bias P-N junction stress," *Appl. Phys. Lett.*, **14**, 264 (1968).
35. D. D. Tang and E. Hackbarth, "Junction degradation n bipolar transistors and the reliability imposed constraints to scaling and design," *IEEE Trans. Electron Devices*, **ED-35**, 2101 (1988).
36. J. D. Burnett and C. Hu, "Modeling hot carrier effects in polysilicon emitter bipolar transistors," *IEEE Trans. Electron Devices*, **ED-35**, 2238 (1988).

37. D. Quon, P. K. Gopi, G. J. Sonet, and G. P. Li, "Hot carrier induced bipolar transistor degradation due to base dopant compensation by hydrogen: theory and experiment," *IEEE Trans. Electron Devices*, **ED-41**, 1824 (1994).

38. H. S. Momose and H. Iwai, "Analysis of the temperature dependence of hot-carrier-induced degradation in bipolar transistors for BiCMOS," *IEEE Trans. Electron Devices*, **ED-41**, 978 (1994).

39. S. L. Kosier, A. Wei, R. D. Schrimpf, D. M. Fleetwood, M. D. DeLaus, R. L. Pease, and W. E. Combs, "Physically based comparison of hot-carrier-induced and ionizing-radiation-induced degradation in BJTs," *IEEE Trans. Electron Devices*, **ED-42**, 436 (1995).

40. A. Neugroschel, C.-T. Sah, and M. S. Carrol, "Degradation of bipolar transistor current gain by hot holes during reverse emitter–base bias stress," *IEEE Trans. Electron Devices*, **ED-43**, 1286 (1996).

41. I. C. Kizilyalli and J. D. Bude, "Degradation of gain in bipolar transistors ," *IEEE Trans. Electron Devices*, **ED-41**, 1083 (1994).

42. A. Neugroschel, C. T. Sah, and M. S. Carrol, "Accelerated reverse emitter–base stress methodologies and time-to-failure applications," *IEEE Electron Device Lett.*, **EDL-17**, 112 (1996).

43. A. Neugroschel, C.-T. Sah, M. S. Carrol, and K. G. Pfaff, "Base current relaxation transient in reverse emitter–base stressed silicon bipolar junction transistors," *IEEE Trans. Electron Devices*, **ED-44**, 792 (1997).

44. A. Rohatgi, "Radiation tolerance of boron doped dendritic web silicon solar cells," *4th Solar Cell High Efficiency and Radiation Damage Conference*, Cleveland, OH, 281 (1980).

45. M. Yamaguchi and K. Ando, "Effects of impurities on radiation damage in InP," *J. Appl. Phys.*, **60**, 935 (1986).

46. D. Pons, A. Mircea, and J. Bourgoin, "An annealing study of electron irradiation-induced defects in GaAs," *J. Appl. Phys.*, **51**, 4150 (1980).

47. G. Augustine, A. Rohatgi, and N. M. Jokerst, "Base doping optimization for radiation-hard Si, GaAs, and InP solar cells," *IEEE Trans. Electron Devices*, **ED-39**, 2395 (1992).

48. J. J. Liou, J. S. Yuan, and H. Shakouri, "Modulating the bipolar junction transistor subjected to neutron irradiation for integrated circuit simulation," *IEEE Trans. Electron Devices*, **ED-39**, 593 (1992).

49. J. A. Babcock, J. D. Cressler, L. S. Vempati, S. D Clark, R. C. Jaeger, and D. L. Harame, "Ionizing radiation tolerance and low-frequency noise degradation in UHV/CVD SiGe HBTs," *IEEE Electron Device Lett.*, **EDL-16**, 351 (1995).

50. D. L. Harame, J. M. C. Stork, B. S. Meyerson, K. Y. Hsu, J. Cotte, K. B. W. Scharf, and J. A. Yasaitis, "Optimization of SiGe HBT technology for high speed analog and mixed-signal applications," *IEDM Tech. Dig.*, 874 (1993).

51. R. N. Enlow, R. L. Pease, W. E. Combs, R. D. Schrimpf, and R. N. Nowlin, "Response of advanced bipolar process to ionizing radiation," *IEEE Trans. Nucl. Sci.*, **NS-39**, 2206 (1992).

52. D. M. Fleetwood et al., "Physical mechanisms contributing to enhanced bipolar gain degradation at low dose rates," *IEEE Trans. Nucl. Sci.*, **NS-41**, 1871 (1994).

53. Y. Dai, "Deep-level impurity analysis for p-n junctions of a bipolar transistor from low-frequency G-R noise measurements," *Solid-State Electron.*, **32**, 439 (1989).

54. S. L. Kosier, R. D. Schrimpf, R. N. Nowlin, D. M. Fleetwood, M. DeLaus, R. L. Pease, W. E. Combs, A. Wei, and F. Chai, "Charge separation for bipolar transistors," *IEEE Trans. Nucl. Sci.*, **NS-40**, 1276 (1993).

55. R. N. Nowlin, D. M. Fleetwood, and R. D. Schrimpf, "Saturation of the dose-rate response of single-poly BJTs below 10 rad(SiO_2)/s: implication for hardness assurance," *IEEE Trans. Nucl. Sci.*, **NS-41**, 2637 (1994).

56. S. McClure, R. L. Pease, W. Will, and G. Perry, "Dependence of total dose response of bipolar linear microcircuits on applied dose rate," *IEEE Trans. Nucl. Sci.*, **NS-41**, 2544 (1994).

57. G. A. Schrantz, N. W. van Vonno, W. A. Krull, M. A. Rao, S. I. Long, and H. Kromer, "Neutron irradiation effects on AlGaAs/GaAs HBTs," *IEEE Trans. Nucl. Sci.*, **NS-35**, 1657 (1988).

58. M. Yamaguci, C. Uemura, and A. Yamamoto, "Radiation damage in InP single crystals and solar cells," *J. Appl. Phys.*, **55**, 1429 (1984).

59. M. Yamaguci, K. Ando, A. Yamamoto, and C. Uemura,, "Injection-enhanced annealing of InP solar-cell radiation damage" *J. Appl. Phys.*, **58**, 568 (1985).

60. M. Yamaguchi and K. Ando, "Effects of impurities on radiation damage in InP," *J. Appl. Phys.*, **60**, 935 (1986).

61. M. Yamaguchi and K. Ando, "Mechanism for radiation resistance in InP solar cells," *J. Appl. Phys.*, **63**, 5555 (1988).

62. A. Sibille and J. C. Bourgoin, "Electron irradiation induced deep levels in p-InP," *Appl. Phys. Lett.*, **41**, 956 (1982).

63. A. Sibille, "Origin of the main deep electron trap in electron irradiated InP," *Appl. Phys. Lett.*, **48**, 593 (1986).

64. Y. Song, M. E. Kim, A. K. Oki, M. E. Hafizi, J. B. Camou, and K. W. Kobayashi, "Radiation hardness characteristics of GaAs/AlGaAs HBTs," *11th GaAs IC Symp. Tech. Dig.*, 155 (1989).

PROBLEMS

8.1. A reliability experiment was performed at 100, 200, and 300°C. In each case the junction temperature rise due to self-heating is 50°C. The MTTFs are 600, 700, and 800 hours, respectively. What should be the best estimate of the relibility lifetime at a junction temperature of 120°C?

8.2. Explain why the insertion of a space layer in the emitter–base junction reduces the out-diffusion effect.

8.3. Explain the radiation effect on the transconductance, mobility, and collector–emitter offset voltage of the bipolar transistor.

8.4. Use the device cross section to explain why III/V HBTs are more radiation hard than Si BJTs.

8.5. Analyze low-dose-rate radiation effect on the base current of the bipolar transistor.

8.6. Consider a semiconductor exposed to intense radiation that creates a high injection level; that is,

$$\Delta n \approx \Delta p \gg n_0, p_0.$$

The radiation is removed instantaneously, and the excess carrier densities begin to decrease via predominately band–band Auger recombination.

(a) What carrier lifetime τ value describes the rate of this recombination, $\Delta n/\tau = \Delta p/\tau$?

(b) Is the decay of Δn and Δp exponential in time? Explain.

8.7. A one-dimensional bipolar transistor is shown in Fig. P8.1. The current gain of the bipolar transistor is controlled predominately by recombination in the quasi-neutral emitter, which is characterized by the saturation current density J_{E0}. Ionizing radiation produces G electron–hole pairs per unit area per unit time in the collector–base space-charge region. Neglect thermal generation and assume low-level injection.

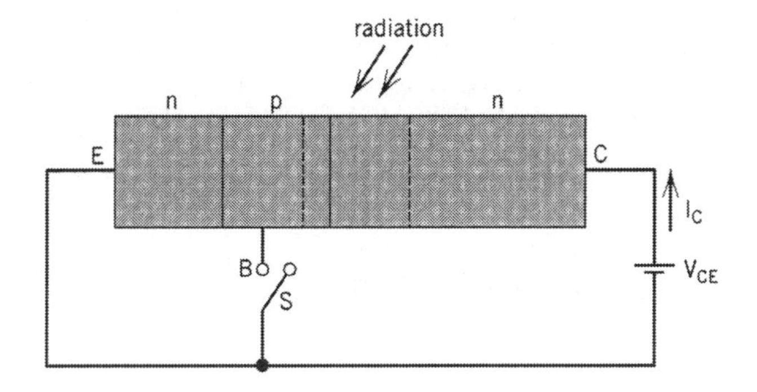

(a) When the switch is closed, what is the collector current?

(b) When the switch is open, what is the collector current?

(c) What is the common-emitter breakdown voltage when the switch is open? The carrier multiplication factor in the collector–base space-charge region is written as

$$M = \frac{1}{1 - \left(V_{CB}/BV_{CB0}\right)^n}$$

CHAPTER NINE

RF and Digital Circuits for Low-Voltage Applications

The drive toward HBT technology maturation has been motivated largely by the prospect of improving military electronic systems. The high f_T and $f_{\max}$ values, as well as other inherent device advantages, make the HBT ideal for front-end function from dc to microwave and millimeter-wave frequencies. The high transconductance values at low collector current make possible microwave amplifiers with very low power dissipation. The HBT's high linearity, associated with its low output conductance, is attractive for low-harmonic-distortion applications such as a low-dc-power output-state amplifiers for use with HEMT low-noise input amplifiers. The high intrinsic gain (g_m/g_o) combined with high speed and freedom from trapping effects make HBTs ideal for wideband amplifiers. The exponential output current/input voltage relation combined with high-frequency performance and high substrate isolation make the HBT attractive for monolithic wide-dynamic-range logarithm IF amplifiers and analog multiplier-mixers. The low $1/f$ noise makes the HBT attractive for low-phase-noise oscillators. The high current density and high breakdown voltage are attractive for high-efficiency power amplifiers. For digital functions the HBT's higher intrinsic $f_{\max}$ value, low base resistance, and lower substrate capacitance will permit higher speeds than Si bipolar transistors. Application-specific ICs (ASICs) that are fully customized will probably be the rule for HBTs, due to its high performance and limited integration complexity. The robust high-speed nature of the HBT is attractive for many digital circuits, such as multiplexers and demultiplexers as well as basic logic gates. The HBT's high-speed capabilities also make it ideal for frequency dividers. The low phase noise and fast rise time make the digital-phase detection function attractive for phase-locked-loop frequency synthesizers. The excellent device matching, high intrinsic gain, high linearity, and high speed together make the HBT ideally suited for A/D conversion applications. Furthermore, the high-current-gain, low-noise, and high-speed characteristics of the HBT make it a good candidate for phototransistor applications.

9.1 LOW-VOLTAGE APPLICATIONS

One of the most important advantages of SiGe- and InP-based HBTs in circuit design is a low base–emitter turn-on voltage. At a given current level, the best way to reduce power consumption is by reducing the supply voltage. The device base–emitter voltage has a large impact on the minimum supply voltage. Figure 9.1 shows the significantly lower turn-on voltage offered by InP-based HBT technology with a GaInAs base layer. The turn-on voltage of the graded InGaAs HBT is about 780 mV smaller than that of the graded GaAs HBTs and is about 220 mV smaller than that of the Si bipolar transistor. The turn-on voltage of the SiGe HBT is comparable to that of the graded InGaAs HBT. The higher speed of InP-based HBTs combined with the lower power supply voltage produces a substantial improvement in speed–power product.

Comparison of f_T for a typical InP heterojunction bipolar transistor and a scaled Si bipolar transistor as a function of emitter current density is given in Fig. 9.2. At a given emitter current density, the InP HBT has a much higher cutoff frequency. For a given cutoff frequency, the emitter current density of the InP HBT is much smaller. For example, to obtain $f_T = 10$ GHz, the emitter current density is required about 10^3 A/cm^2, while it is about 10^4 A/cm^2 for the scaled Si BJT. A low emitter current density means low power dissipation in circuit operation.

9.2 WIDEBAND AMPLIFIERS

GaAs MESFET technology is currently the dominant technology for monolithic microwave integrated-circuit (MMIC) applications and is growing steadily in use for

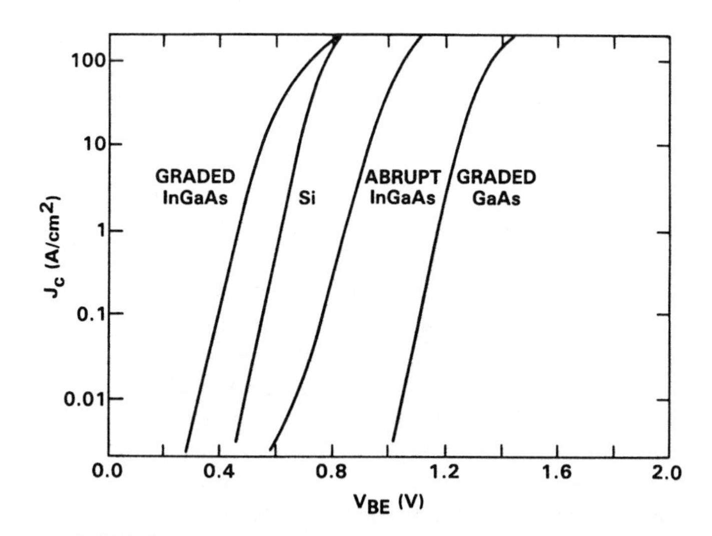

FIGURE 9.1 Collector current versus base–emitter voltage.

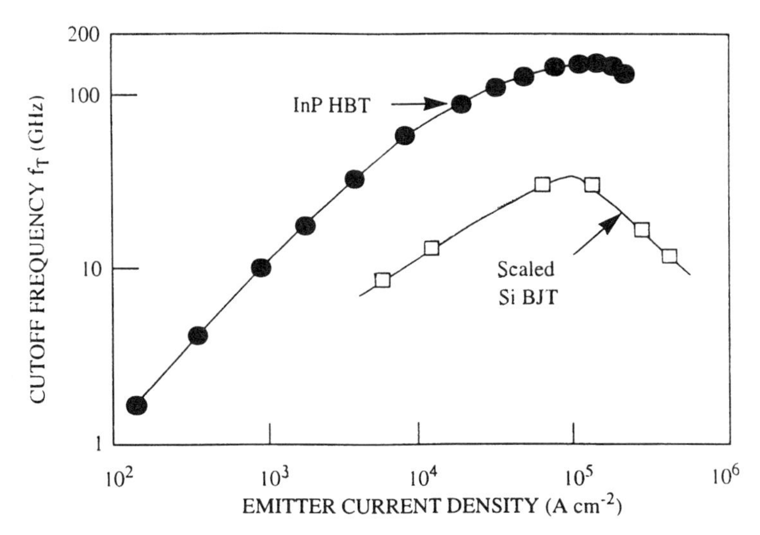

FIGURE 9.2 Cutoff frequency versus collector current density (after Jalali and Pearton, Ref. 1© Artech House).

high-speed digital applications. However, relatively little progress has been made in the area of analog applications. This is due to the MESFET's poor dc and low-frequency characteristics. In contrast, the HBT technology is much better suited to wideband analog amplifiers for communication and instrumentation applications because of its higher intrinsic gain, higher transconductance, better device matching, and the lack of hysteresis or backgating trapping effects. For monolithic fabrication in microwave ICs, cost reduction due to mass manufacturability is crucial. Greater circuit complexities can be achieved with minimal circuit size and weight. The success of monolithic techniques, however, relies heavily on reproducible device characteristics, both on the same wafer and from wafer to wafer. HBTs comfortably satisfy reproducibility requirements for monolithic circuits because the critical device parameters are set by the vertical device structure, which can be controlled with great accuracy by MBE and MOCVD. Also, device performance is not critically dependent on emitter width. The uniformity of device performance is therefore not limited by lithography. The device's active area is almost totally shielded from the surface effects, due to its vertical device structure.

The wideband amplifier is a critical component in the front end of almost all analog-to-digital converters. This amplifier must take relatively low-level input signals over a wide bandwidth and amplify them with minimal addition of distortion, offset, and noise while maintaining flat-gain versus frequency characteristics. As with most components, low power is an additional constraint. The HBT broadband amplifier performance from dc to 20 GHz can be realized by a very simple monolithic Darlington-coupled feedback amplifier.

Silicon bipolar technology has traditionally dominated wideband amplifier implementations. Compared to their respective bipolar transistors, FET technologies suffer from higher distortion due to lower transconductance and larger offsets due to mismatch. For bandwidth in the tens of megahertz, the lowest power implementation of a wideband amplifier will be in silicon bipolar technology. The dc performance of wideband amplifiers implemented in SiGe, GaAs, and InP HBT technologies is similar, due to the identical intrinsic transconductance characteristics. For wider bandwidth (e.g., $>$ 100 MHz) applications, GaAs and InP HBT technology offer advantages over silicon and SiGe technology. Capacitances in the signal path not only lower circuit bandwidth but can cause ac distortion as well. Here, ac distortion is caused by a capacitive load on a nonlinear impedance (e.g., transistor transconductance) or by a nonlinear capacitor. The dominate source of high-frequency nonlinearity in wideband amplifiers is typically the nonlinear collector capacitances. Due to the semi-insulating substrate of GaAs and InP, the HBT transistor's substrate capacitance is over an order of magnitude lower than for silicon transistors. Thus the reduced overall parasitic capacitances in GaAs and InP technology not only provide wider bandwidth amplifiers, but result in better ac linearity as well.

As discussed in Chapter 3, the self-heating effect is pronounced for III/V compound HBTs compared to their silicon bipolar counterparts. Although InP HBTs are generally lower power and have better thermal conductivity than the equivalent GaAs counterpart, thermal-related effects are still a greater concern than for Si. If a transistor is connected in an emitter-follower configuration, a thermally induced gain loss would be evident at low frequencies but not at higher frequencies (due to the slow thermal time constant). In differential pair connections, self-heating causes V_{BE} modulation, which leads to distortion. To mitigate thermal effects, the designer should minimize current and power in circuit areas where self-heating effects are important. If that cannot be avoided, due to high-speed biasing requirements, emitter-followers should be bootstrapped so that the collector–emitter voltage remains constant with input signal, and amplifiers should employ the V_{BE} compensation methods shown in Fig. 9.3. This compensation method is based on replicating and canceling the V_{BE} error. The inner differential pair Q_5–Q_6 has a transconductance identical to that of the Q_1–Q_2 pair. The cascodes Q_3–Q_4 replicate the V_{BE} error in Q_1–Q_2, and the transconductance stage subtracts this error from the output nodes. Proper device size selection and adequate spacing of adjacent devices are also required to minimize potential thermal problems [1].

A wideband low-power-consumption monolithic amplifier using SiGe heterojunction bipolar transistors has been reported [2]. This amplifier provides 9.5 dB of gain from dc through 18 GHz, with a 5.3-dB noise figure at 50 mW power consumption. Low-supply-voltage operation is possible with slight bandwidth reduction and less power consumption. The circuit topology of the SiGe wideband amplifier is shown in Fig. 9.4. This circuit follows the parallel-feedback Darlington pair and requires no bias terminals other than the RF ports. A similar direct-coupled Darlington feedback amplifier with a 7.8-dB gain and a -3-dB bandwidth of 30 GHz using AlGaAs/GaAs HBTs has also been reported [3]. In Fig. 9.4 the feedback resistor, $R_1 = 300\ \Omega$, was chosen for good input and output match at low frequencies. Additional series

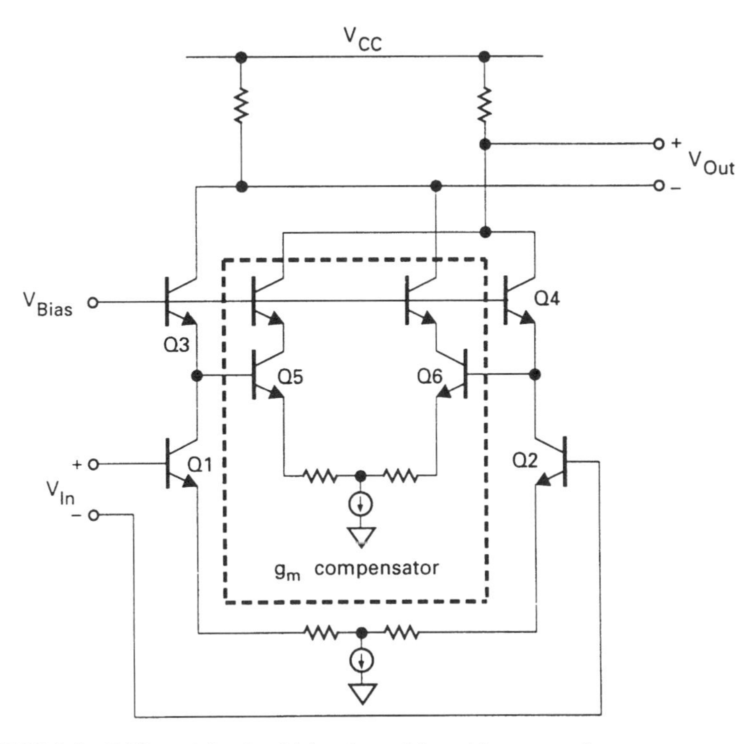

FIGURE 9.3 Differential pair wideband amplifier with transconductance compensation.

feedback is provided through resistor $R_4 = 3\ \Omega$ in the emitter of transistor T_2 to improve the output match. The base-biasing resistor R_2 is 412 Ω; R_3 is 174 Ω. With the resistor values chosen, the amplifier will operate from a 3-V supply voltage at the output port with collector currents of 5 mA for T_1 and 10 mA for T_2. The inductance

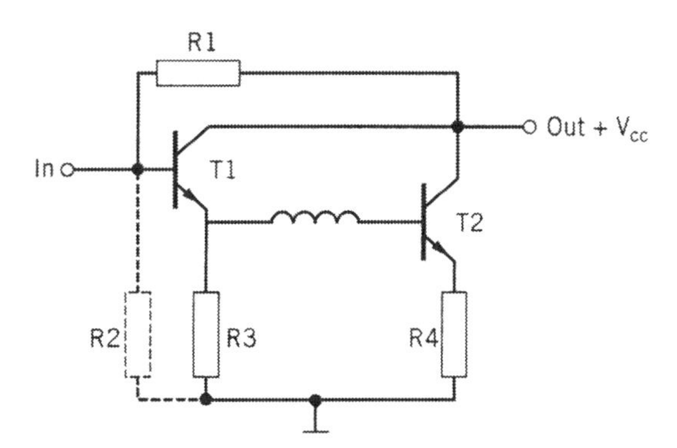

FIGURE 9.4 Parallel-feedback Darlington pair.

between T_1 and T_2 is realized as a short line and enhances the bandwidth by partly canceling the input capacitance of T_2.

The fabricated amplifiers were tested on-wafer using a HP 8510C network analyzer, an ATN NP5B parameter test set, Picoprobe wideband wafer probes, and a Cascade MicroTech Summit 9000 probe station. The transducer gain G_{50} with 50-Ω source and load impedances is shown in Fig. 9.5. At a 3-V supply voltage and 17-mA total current, the circuit shows a gain of 9.5 dB and a −3-dB bandwidth of 18 GHz. When the supply voltage is reduced to 1.6 V and the total current to 11 mA, the flat-band gain remains essentially unchanged and the bandwidth degrades by 17%, to 15 GHz. The slightly increased low-frequency gain at the lower supply voltage is believed to result from a higher output resistance of T_2 at the low collector–emitter voltage.

The noise figure is shown in Fig. 9.6. F_{50} and F_{min} are almost independent of frequency and average 5.3 dB over the bandwidth of the amplifier. The noise-optimized source reflection coefficient, Γ_{opt}, is close to zero. Gain and noise figures for the 50-Ω source and load are therefore virtually identical to the optimum values. This very low-noise figure over a wide frequency range is a direct result of the high base doping concentration, resulting in a low base resistance. Also, the relatively large first-stage transistor is of benefit here in further reducing base resistance.

9.3 RF POWER AMPLIFIERS

Power amplifiers are used in many military and commercial microwave systems, ranging from radar to communication, where a transmitter is needed. Although each application has its unique set of requirements, power amplifiers are usually needed to amplify signals, requiring the smallest device, highest reliability, lowest signal distortion, and highest power efficiency. For communication system applications, power

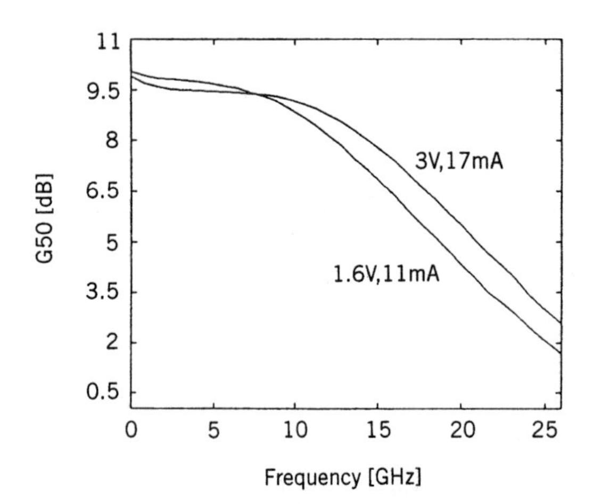

FIGURE 9.5 Gain versus frequency of a SiGe wideband amplifier (after Schmacher et al., Ref. 2 © IEEE).

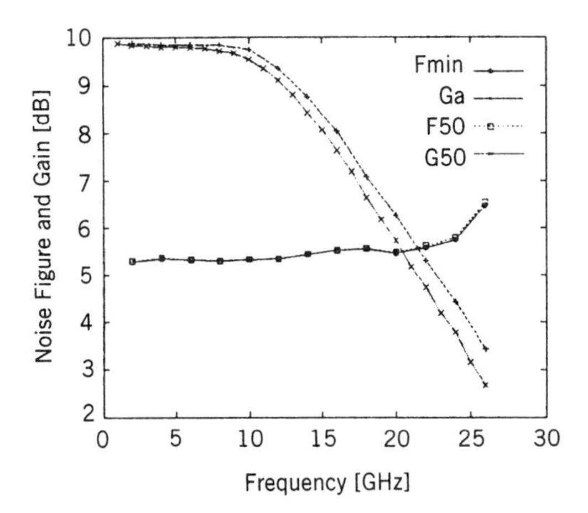

FIGURE 9.6 Noise figure and gain versus frequency (after Schmatcher et al., Ref. 2 © IEEE).

amplifiers are operated in continuous-wave (CW) mode. In some other applications, such as high-resolution radar systems, power amplifiers are operated under very short pulses. The output requirement for power amplifiers in these applications is usually more demanding because the total energy in each short pulse must be sufficient to guarantee adequate return from the target.

As a power amplifier, the HBT can be regarded as an extension of the Si bipolar technology into the microwave frequency range. For a GaAs material system, a CW output power of 2 W with 72% power-added efficiency (PAE) at L-band (1.5 GHz) was reported [4]. At S-band (3 GHz), a CW output power of 1.1 W with a 12.3 power gain and 61% PAE was obtained [5]. A high power gain of 11.6 dB and a high power-added efficiency of 67.8% were achieved at 10 GHz [6]. At 18 GHz, 358-mW output power with 11.4 dB power gain and 43% power added efficiency were demonstrated [7]. In the area of high-power microwave amplifiers, a 5.3-W output power at X-band (8 GHz) CW operation with a power gain of 4.3 dB and 33% PAE was obtained [8]. At Ku-band (12 GHz) a 1-W output power (5.0 W/mm) with 72% PAE was achieved [9]. In pulse operation at 10 GHz and at a peak output power of 560 mW, a power density of 18.7 W/mm was achieved for a 300-ns pulse duration with a 33% duty cycle [10]. For an InP material system, at S-band a CW output of 1.51 W and a PAE of 52% using a GaInP/GaAs/GaInP double-heterojunction bipolar transistor was reported [11]. At X-band (9 GHz) an output power of 2 W (5.6 W/mm) with 70% PAE using an AlInAs/GaInAs/InP HBT was achieved [12].

9.3.1 Power-Added Efficiency

For solid-state microwave amplifiers used for mobile communication systems, high power-added efficiency is desirable. Compared to GaAs power MESFETs or power AlGaAs/GaAs MODFETs, AlGaAs/GaAs power HBTs have a higher PAE, ranging from about 40 to 70%. The power-added efficiency is defined as

$$\eta \equiv \frac{P_{\text{out}} - P_{\text{in}}}{P_{\text{dc}}} = \eta_C \left(1 - \frac{1}{G_p}\right) \tag{9.1}$$

where η_C is the collector efficiency, P_{dc} the dc power from the power supply, P_{in} the ac input power from the high-frequency power generator, and P_{out} the ac output power at the load. If a constant transconductance is assumed, η_C is given by

$$\eta_C = \frac{1}{2} E(\theta) \left(1 - \frac{RF v_{CES}}{V_{CC}}\right) \tag{9.2}$$

where $E(\theta)$ is a factor related to the current flow angle θ. It is clear from (9.1) and (9.2) that the higher the power gain and the lower the RF saturation voltage v_{CES}, the higher the power-added efficiency.

In single-ended class B and class C power amplifiers, the flow angle θ of the collector current is 90° or less, respectively (i.e., the transistors conduct during only one-half of the period or less). The high-frequency (HF) output characteristics are quite different from the dc output characteristics. In HF power amplifiers (class B or C), the transistor is loaded by a selective impedance having a real value of load resistance at the fundamental frequency (operating frequency) but a short circuit at harmonics. The half-cosinusoidal current pulses of peak value i_{CM} generate a fundamental frequency voltage that has a peak value approximately equal to the supply voltage. From Fourier analysis and a simple calculation, the output power P_{out} at the fundamental frequency is obtained:

$$P_{\text{out}} = \tfrac{1}{4} \alpha_1(\theta) i_{CM} (V_{CE\,\text{max}} - RF v_{CES}) \tag{9.3}$$

The power gain is the ratio of the power delivered to the load and the input power to the transistor. When input and output are conjugately matched simultaneously, the maximum available gain (MAG) G_{max} is obtained. The power gain for common-emitter class B and C operations is less than that of class A and is given by [13]

$$G_p = \frac{G(\theta) f_T}{8\pi (r_B + r_E + \pi f_T L_e) C_C f_0^2} \tag{9.4}$$

where $G(\theta) = \alpha_1(\theta)(1 - \cos\theta)$. The $G(\theta)$ is a factor related to the current flow angle θ, as shown in Fig. 9.7. For class B with a current flow angle $\theta = 90°$ and $G_p = 0.5$, its power gain is 50% of that of class A. Note that the feedback elements, base resistance, emitter resistance, and emitter lead inductance affect the power gain. For HBTs the intrinsic base resistance is so small that the base contact resistance has a notable influence on the power gain. The emitter ballasting resistor is effective for

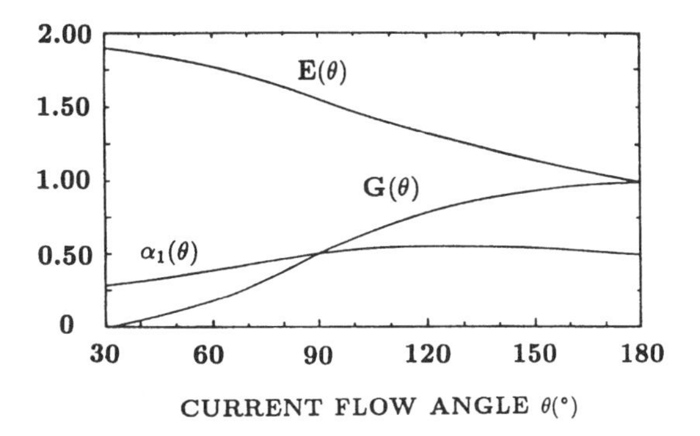

FIGURE 9.7 $G(\theta)$, $\alpha_1(\theta)$, and $E(\theta)$ versus current flow angle θ (after Gao et al., Ref. 13 © IEEE).

improving the saturation output power and the power gain at high input power, but it degrades the power gain at low power levels. The base ballasting resistor could be used to replace the emitter ballasting resistor for the improvement of thermal stability while achieving higher output power and power-added efficiency [14]. The emitter lead inductance has a dramatic impact on the power gain and should be kept as small as possible.

Another important consideration in the design of high-power HBT amplifiers is the choice of operating mode. Although common-emitter devices produce the most favorable impedance conditions for this application, common-base amplifiers are also used at some microwave frequencies. These two modes of operation can be compared by examining the measured S-parameters for the same size of device. In the common-emitter mode, the device's output impedance is larger than the input impedance ($S_{22} > S_{11}$). In the common-base mode, the device input impedance is larger than the output impedance. The HBT breakdown voltage is substantially larger under common-base-mode than under common-emitter-mode conditions. Common-base-device structures can be optimized to provide relatively high impedance levels and substantially larger breakdown voltages at a given frequency. The increased impedance level and power density bring the maximum available power from common-base and common-emitter devices close to each other. Nevertheless, the common-emitter mode of operation is chosen by most RF designers whenever possible because device stability is more easily ensured under varying load conditions.

9.3.2 Impact of Device Parameters on the HBT Large-Signal Gain

The ongoing developments in RF technology and the increased integration of analog and digital circuits on the same substrate provide new opportunities and challenges for communication circuit design. For wireless applications, the device modeling and circuit design are further complicated by the high-frequency operation. To meet these challenges, circuit designers must often draw knowledge from different areas of expertises. For example, the impact of device parameters on RF circuit performance can be examined using statistical analysis. The numerical experiments evaluate device parameters according to their statistical significance. Most important parameters for given circuits are then identified. The significant parameters identified can be used in the design of device doping profiles and dimensions to achieve the best RF circuit performance.

The common-emitter BJT and HBT large-signal amplifiers under class A operation are simulated using the microwave circuit simulator Libra [15]. The design starts with device models, dc biasing, RF chokes, and blocking capacitors, as shown in Fig. 9.8. From a stability analysis of the devices it is inferred that the bipolar transistor is unconditionally stable for all frequencies of interest, while the HBT is potentially unstable below 2.1 GHz. To make a fair comparison, it is desirable to make the operating conditions of the two devices as similar as possible. The stability differences are overcome by stabilizing the heterojunction bipolar transistor with resistive load in the base circuit. The RF performance of the HBT is somewhat degraded, but the maximum available gain (G_{max}) of 19 dB at 0.9 GHz has been obtained. This G_{max} value is more than twice that of the BJT. Simultaneous conjugate match reflection coefficients are provided using single stub matching for both circuits. Power sources are added to the circuit and the power is swept from −40 to +40 dBm. These data are used to determine the input power levels corresponding to a 1-dB compression point. After that the power levels are fixed at −6 and −17 dBm for the BJT and

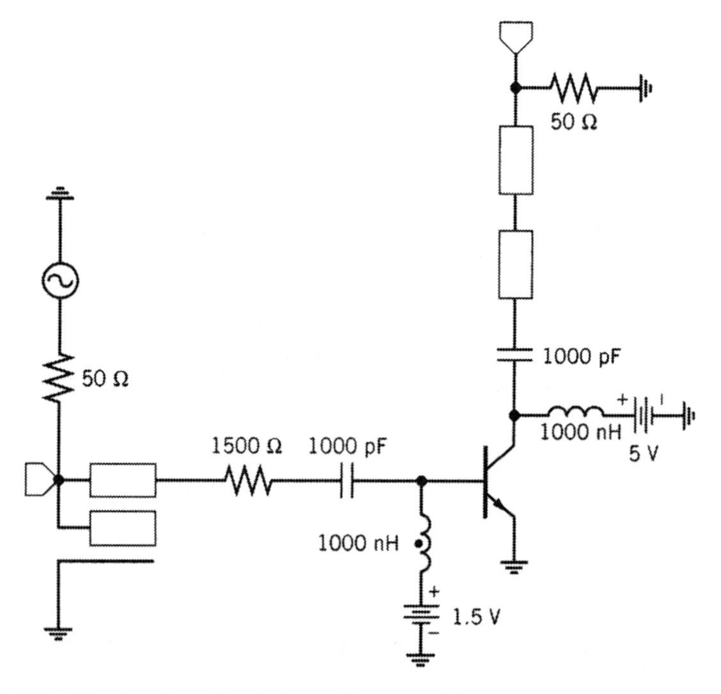

FIGURE 9.8 Circuit schematic of a large-signal amplifier used in the HP Libra simulation.

HBT devices, respectively, and the large-signal power-dependent S-parameters are determined. Simultaneous complex-conjugate match amplifiers are designed with these S-parameters.

Figure 9.9 shows the maximum available gain, S_{11}, S_{22}, and S_{21} for (a) BJT and (b) HBT large-signal amplifiers. The large-signal amplifiers are designed at 900-MHz operation, as evidenced by the peak values of S_{21} at that frequency. The maximum available gain of the HBT amplifier is larger than that of the BJT amplifier, as expected. Figure 9.10 shows the output power versus input power for BJT and HBT amplifiers. These curves provide information on the power-handling capabilities of the devices and the levels of input power at which compression occurs. At $P_{\text{in}} = 0$ dBm, P_{out} is 11 dBm for the HBT amplifier, while it is 2 dBm for the BJT amplifier.

Statistical analysis is used to identify the most significant transistor parameters on circuit performance. The model parameters are assigned low and high values. The low value is the nominal value of the parameter, while the high value is 20% larger than the nominal value. Using different combinations of low and high parameters, the output power at 1 dB compression point is recorded. The numerical experiments indicate that the most significant parameters for the BJT large-signal amplifier are the forward transit time τ_F, base resistance R_b, and collector–base junction capacitance C_{JC}. The most significant parameters for the HBT large-signal amplifier are the collector current ideality factor n_f, collector–base leakage current ideality factor n_e, and collector–base junction C_{JC}. For the HBT, the forward transit time and base resistance are very small compared to those of the BJT. This may be why τ_F and R_b are not identified as the most significant parameters by statistical analysis. The n_f and n_e values for the HBT, however, are larger than those for the BJT, due to the band discontinuity at the emitter–base heterojunction.

9.3.3 Heterojunction Bipolar Transistor Design for Power Applications

The output power P_{out} at a given operating frequency f_0 with an appropriate power gain G_p is an important parameter for power HBTs. Generally, for an HBT, the higher the f_0 value, the lower the G_p and P_{out} values for a given input power. The maximum available output power is determined by the maximum peak value of the high-frequency current i_{cm} and the breakdown voltage BV_{CB0}, while the operating frequency f_0 depends on the maximum oscillation frequency $f_{\max}$.

The output available from any device is directly proportional to the voltage swing that the device can sustain and the peak current that it can carry. Therefore, high-power devices should operate under large bias conditions. If the cross section of the current-conducting device has been maximized, its output power can be further enhanced by increasing the operating bias voltage. Transistor operating voltages are also restricted by considerations such as maximization of gain and surface breakdown voltage. Therefore, a device design trade-off is always needed.

The output power capability of an HBT is strongly influenced by the design of the collector layer. Thicker and lightly doped collector layers are always desired to sustain a large breakdown voltage at the base–collector junction. However, the requirement for a thick collector layer must be balanced by the need for a short transit

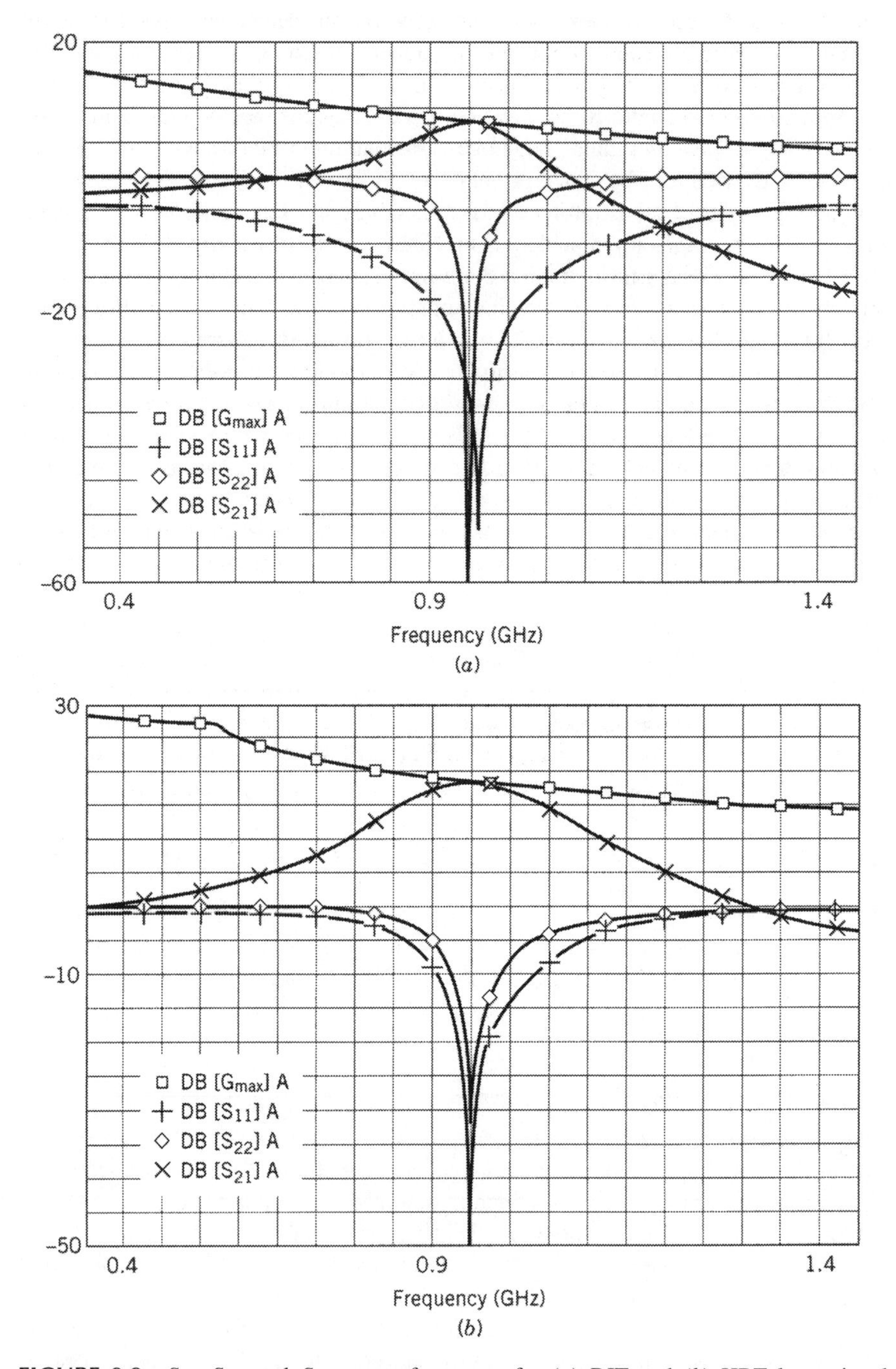

FIGURE 9.9 S_{11}, S_{21}, and S_{22} versus frequency for (*a*) BJT and (*b*) HBT large-signal amplifiers.

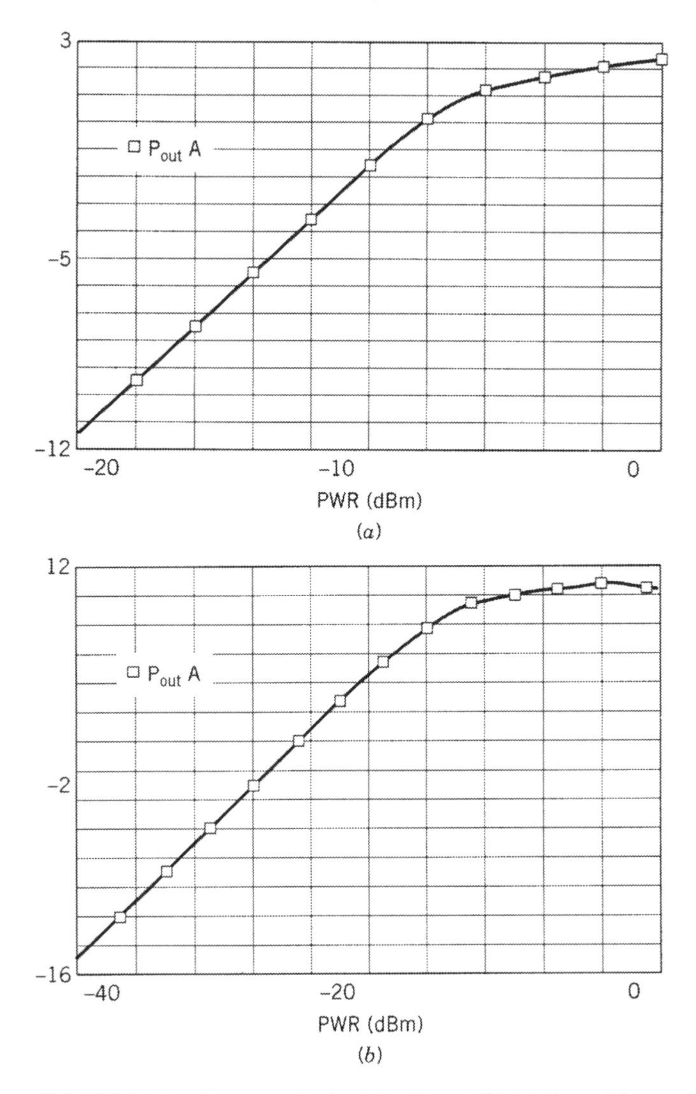

FIGURE 9.10 P_0 versus P_{in} for (*a*) BJT and (*b*) HBT amplifiers.

time through the collector layer. A rule of thumb in the design of power HBTs is to select a collector thickness that produces about 50% of the emitter–collector transit time delay. The collector structure of power heterojunction bipolar transistors also has an enormous influence on the maximum oscillation frequency, the output power, the power gain, and the efficiency. The high-field drift velocity determines not only the collector space-charge transit time but also the current-handling capability through the Kirk effect. For the collector layer it is shown that inserting an i-layer between the heavily doped base and the collector improves breakdown voltage. The effect of reduced field caused by the added i-layer also has the advantage of extending the region of velocity overshoot in the collector.

The speed advantages of InP/GaInAs and AlInAs/GaInAs HBTs compared to AlGaAs/GaAs HBTs are explained in Chapter 7. However, InP-based single-heterojunction bipolar transistors with a GaInAs collector are not suitable for microwave power applications. The high rates of impact ionization in GaInAs severely limit the range of collector voltage for safe operation. Therefore, the majority of microwave power HBTs today use AlGaAs/GaAs and GaInP/GaAs heterojunctions.

AlInAs/GaInAs/InP DHBTs, in which the collector is made of InP, are very promising for high-performance power amplifiers due to the low impact ionization rate, high peak and saturated velocities, and high thermal conductivity of InP. The conduction-band edge discontinuity at the base–collector heterointerface, however, often results in a current-blocking potential which increases the saturation voltage and causes gain compression to occur at a lower current density. This is a major problem for all DHBTs and has to be eliminated to exploit the advantages of the wide-gap collector. Second, since the thickness of the collector of an X-band power HBT is on the order of 1 μm to withstand collector voltages larger than 20 V, the speed of the device is generally limited by the transit time of electrons through the collector, and the high-frequency performance of the device can be enhanced noticeably by reducing the collector transit time.

A novel transistor design of AlInAs/InGaAs/InP DHBT in which the base–collector junction is bandgap engineered to eliminate the current blocking barrier completely and to allow hot electrons to be injected into the collector to enhance the speed of the device has been reported [16]. The base–collector design involves a linear bandgap variation and two delta doping layers at the ends of the grade, acceptor at the base end and donor at the collector end, to form a dipole. The quasi-electric field created by the linear bandgap variation is canceled by the electrostatic field arising from the ionized impurities of the doping sheets. As a result, the entire bandgap difference is transferred into the valence band, and the conduction-band becomes barrier-free, similar to that of a SHBT, as long as the delta doping sheets are depleted. Therefore, the need for a spacer layer, which would increase the transit time, is also eliminated.

A composite, linear-delta design using two wide-bandgap semiconductors to launch hot electrons into the collector is shown in Fig. 9.11. The device structure consists of an n^+ 700-nm GaInAs subcollector, a composite collector involving a 1.0-μm-thick InP layer doped at $n = 1.5 \times 10^{16}$ cm^{-3}, and an AlInAs/GaINAs chirped superlattice (CSL), followed by the GaInAs base and AlInAs emitter ($n = 8 \times 10^{17}$ cm^{-3}) with an n^+ GaInAs contact layer. The base consisted of a 60-nm p^+ GaInAs layer doped at 3×10^{19} cm^{-3} sandwiched between two GaInAs layers doped at $p = 2 \times 10^{18}$ cm^{-3}. The latter layers are intended to accommodate the diffusion of Be from the base during growth. The base–emitter junction was compositionally graded using a chirped superlattice. The linearly graded region of the base–collector junction was approximated by a 33-period chirp superlattice consisting of alternating GaInAs and AlInAs layers of varying thicknesses. The period of the CSL was kept constant at 1.5 nm. The p-type half of the CSL was doped at 1×10^{17} cm^{-3} and

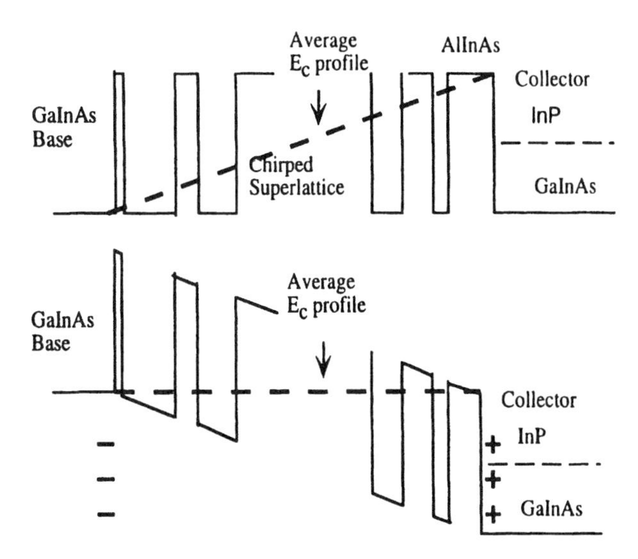

FIGURE 9.11 Schematic conduction-band-edge diagrams of a flat band and the effect of the doping dipole to approximate a barrier-free conduction-band (after Nguyen et al., Ref. 12 © IEEE).

the n-type half at 1.5×10^{16} cm^{-3}, so that the CSL was depleted completely at zero base–collector bias. The dipole was implemented with a 2.0-nm AlInAs layer doped at n = 2.0×10^{18} cm^{-3} at the collector end of the CSL. At the base end of the CSL, the heavily doped base naturally provided the necessary charge for the dipole.

Figure 9.12 shows the output power and power-added efficiency of a 12-finger DHBT power cell ($12 \times 2 \times 30\ \mu$m^2) operating in class B and measured on-wafer at 9 GHz using an active load-pull system. The solid lines were obtained with a load tuned for high output power. The output power was 2.5 W, corresponding to a power density of 6.9 W/mm, at 61% power-added efficiency. The associated gain was 7.6 dB. When the load was tuned for high PAE at 2 W output power, the PAE was 75% with an associated gain of 10 dB. The base–collector breakdown voltage was 38 V, and the emitter–collector breakdown voltage was approximately 32 V with the base terminal open. The excellent power performance of the power cell clearly demonstrated the advantage of the linear-delta InP-HBT for power applications.

A figure of merit (FOM) based on different material parameters has been studied for various heterojunction power transistors [17]. The figure of merit is defined as the product of the operating frequency and the output power with 3-dB power gain, $f_0 P_{\text{out}}$. The analysis concluded that normalized to AlGaAs/GaAs HBTs, the FOM of a Si BJT is lower by 42% if the thermal effects are neglected. The AlGaAs/GaAs/AlGaAs double HBT has a 54% improvement in the FOM. The AlGaAs/Ge/AlGaAs HBT offers an 81% improvement, whereas the improvement for the InP/InGaAs system is 71%. When thermal effects are taken into account, the junction temperature of the

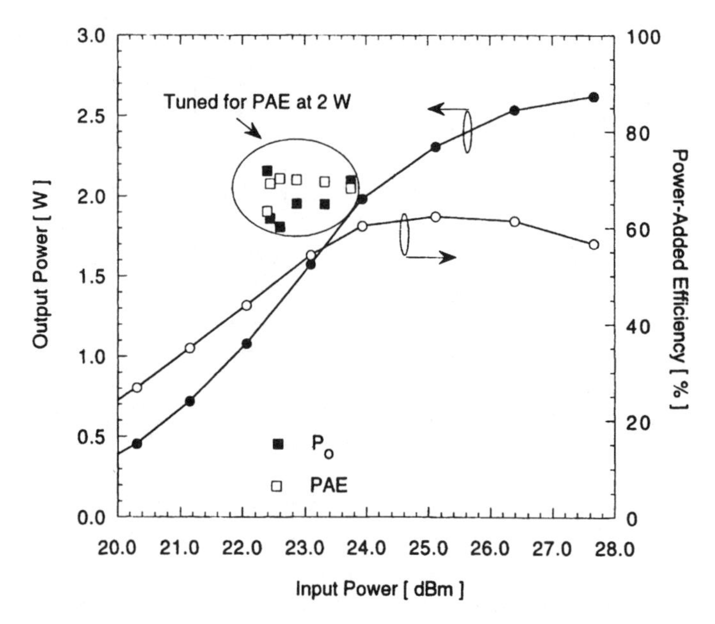

FIGURE 9.12 Output power and power-added efficiency versus input power (after Nguyen et al., Ref. 12 © IEEE).

power HBT increases and both the electron and hole mobilities in the base decrease, as does the electron drift velocity. This results in a corresponding decrease in the FOM.

Wide-bandgap materials such as SiC, GaN, and diamond are good choices for collectors, due to their high electron drift velocities at high field, high critical field, and high thermal conductivity. For wide-bandgap collector materials, the improvement in FOM is impressive. Awaiting technological breakthroughs, improvements of about 5, 17, and 20 are possible in SiC, GaN, and diamond collectors.

9.3.4 Class E Power Amplifiers

Class A, B, and C power amplifiers use the active transistor as a current source. Class D and E power amplifiers use the transistor as a switch. The ideal power efficiency of class D and E PAs is 100%. Class E switching-mode tuned power amplifiers and dc/dc power converters have become increasingly valuable building blocks in radio transmitters and switching-mode dc power-supply applications. These circuits offer extremely high power conversion efficiency and reduce the size and weight of the equipment. In 1975, Sokal and Sokal [18] published a pioneering work on the high-efficiency class E switching power amplifier.

The circuit schematic of a class E tuned power amplifier is shown in Fig. 9.13. The power amplifier has an active device used as a switch, a tuned network, and a load.

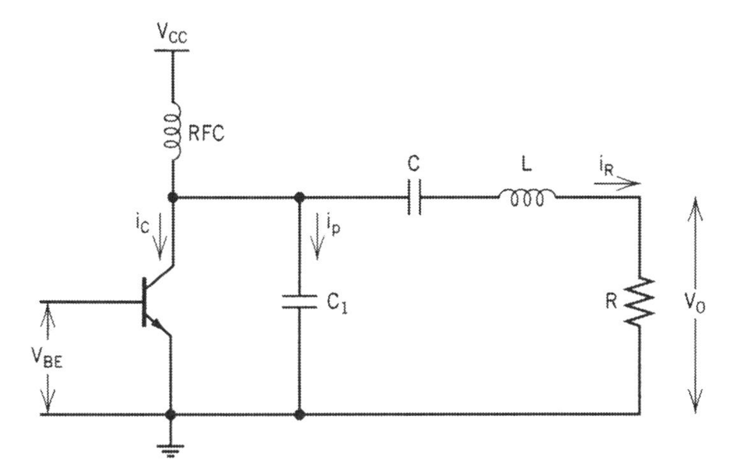

FIGURE 9.13 Class E power amplifier.

The following assumptions are usually made in analyzing the class E tuned power amplifier:

(1) The device acts as an ideal switch with zero saturating voltage, zero saturation resistance, infinite off resistance, and zero switching times.

(2) The quality factor Q of the series-tuned circuit is high enough so that the harmonic components are suppressed and the output current is essentially a sinusoid at the carrier frequency.

(3) The RF choke resistance is zero and its inductance is very high. Therefore, the current flowing through the RF choke is constant.

(4) The load network components are linear and lossless.

The ideal class E voltage and current waveforms are shown in Fig. 9.14. The collector current goes to zero when the collector–emitter voltage rises. Since the power dissipation of the active transistor $\int i_C(t)v_{CE}(t)\,dt$ is 0, it produces a power efficiency of 100%. For high-frequency operation, zero switching time is virtually impossible. An analysis of the effect of the collector current fall time on power efficiency was derived [19]. The collector current waveform during the turn-off transient is considered as an exponential decay.

The ac current in Fig. 9.13 is written as

$$I_{CC} - i_R(\omega t) = i_C(\omega t) + i_p(\omega t) \tag{9.5}$$

where I_{CC} is the current flowing through the RF choke, i_C the collector current, i_p the capacitive current flowing through C_1, and i_R is the load current. For an infinite quality factor Q, the load current can be modeled as a sinusoidal waveform with the

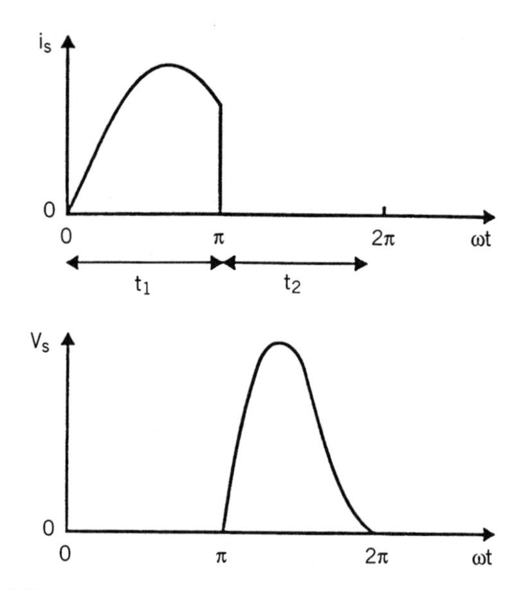

FIGURE 9.14 Ideal class E voltage and current waveforms.

amplitude I_R and an initial phase ϕ:

$$i_R(\omega t) = I_R \sin(\omega t + \phi) \tag{9.6}$$

For the time period $0 < \omega t \leq \pi$, the bipolar transistor is conducting. The first assumption above dictates that the collector–emitter voltage in this period be zero

$$v_{CE}(\omega t) = 0 \qquad \text{for } 0 < \omega t \leq \pi \tag{9.7}$$

The ac current i_p is thus

$$i_p(\omega t) = 0 \qquad \text{for } 0 < \omega t \leq \pi \tag{9.8}$$

From (9.5)–(9.8), the collector current in this period is given as

$$i_C(\omega t) = I_{CC} - I_R \sin(\omega t + \phi) \qquad \text{for } 0 < \omega t \leq \pi \tag{9.9}$$

For the time period $\pi < \omega t \leq \pi + \theta_f$, where θ_f is the collector fall time, the collector current can be described as

$$i_C(\omega t) = i_C(\pi) e^{-(\omega t - \pi)/\tau} \qquad \text{for } \pi < \omega t \leq \pi + \theta_f \tag{9.10}$$

where τ is the decay lifetime and

$$i_C(\pi) = I_{CC} + I_R \sin \phi \tag{9.11}$$

Substituting (9.6) and (9.10) into (9.5) gives the current flowing through the capacitor C_1 during the fall time:

$$i_p(\omega t) = I_{CC} - I_R \sin(\omega t + \phi) - (I_{CC} + I_R \sin\phi)e^{-(\omega t - \pi)/\tau} \quad \text{for } \pi < \omega t \leq \pi + \theta_f \tag{9.12}$$

The collector–emitter voltage is then

$$\begin{aligned} v_{CE}(\omega t) &= \frac{1}{\omega C_1} \int i_p(\omega t)\, d(\omega t) \\ &= \frac{1}{\omega C_1} \Big\{ I_{CC} \left[\omega t - \pi + \tau(e^{-(\omega t - \pi)/\tau} - 1)\right] \\ &\quad + I_R \left[\cos(\omega t + \phi) + \cos\phi + \tau \sin\phi (e^{-(\omega t - \pi)/\tau} - 1)\right]\Big\} \\ &\qquad \text{for } \pi < \omega t \leq \pi + \theta_f \end{aligned} \tag{9.13}$$

For the time period $\pi + \theta_f < \omega t \leq 2\pi$, the bipolar transistor is turning off, the collector current is approaching zero, and the capacitive current i_p is

$$i_p(\omega t) = I_{CC} - I_R \sin(\omega t + \phi) \qquad \text{for } \pi + \theta_f < \omega t \leq 2\pi \tag{9.14}$$

The collector–emitter voltage is

$$\begin{aligned} v_{CE}(\omega t) &= \frac{1}{\omega C_1} \int i_p(\omega t)\, d(\omega t) + v_{CE}(\pi + \theta_f) \\ &= \frac{1}{\omega C_1} \Big\{ I_{CC} \left[\omega t - \pi + \tau(e^{-\theta_f/\tau} - 1)\right] \\ &\quad + I_R \left[\cos(\omega t + \phi) + \cos\phi + \tau \sin\phi (e^{-\theta_f/\tau} - 1)\right]\Big\} \\ &\qquad \text{for } \pi + \theta_f < \omega t \leq 2\pi \end{aligned} \tag{9.15}$$

To design a high-efficiency class E power amplifier, the collector–emitter voltage and its slope when the collector current begins to rise should be set to zero. The boundary conditions for the class E power amplifier are

$$v_{CE}(2\pi) = 0 \tag{9.16a}$$

$$\left.\frac{dv_{CE}(\omega t)}{d(\omega t)}\right|_{\omega t = 2\pi} = 0 \tag{9.16b}$$

Using (9.16a) and (9.16b), one obtains

$$I_R = \frac{-I_{CC}[\pi + \tau(e^{-\theta_f/\tau} - 1)]}{2\cos\phi + \tau \sin\phi (e^{-\theta_f/\tau} - 1)} \tag{9.17}$$

Substituting (9.17) into (9.15) and using (9.16b) yields

$$\tan\phi = \frac{-2}{\pi + 2\tau(e^{-\theta_f/\tau} - 1)} \tag{9.18}$$

Now, once ϕ is known from (9.18), $i_R(\omega t)$, $i_C(\omega t)$, $i_p(\omega t)$, and $v_{CE}(\omega t)$ are determined. Combining (9.6), (9.10)–(9.12), and (9.18), the collector current waveform for a complete cycle of the operating frequency is

$$\begin{aligned}
\frac{i_C(\omega t)}{I_{CC}} &= \frac{\pi - \theta}{2}\sin\omega t - \cos\omega t + 1 && \text{for } 0 < \omega t \le \pi \\
&= \left[1 - \frac{\sin\phi[\pi + \tau(e^{-\theta_f/\tau} - 1)]}{2\cos\phi + \tau\sin\phi(e^{-\theta_f/\tau} - 1)}\right] e^{-(\omega t - \pi)/\tau} && \text{for } \pi < \omega t \le \pi + \theta_f \\
&\to 0 \quad \text{for } \pi + \theta_f < \omega t \le 2\pi
\end{aligned} \tag{9.19}$$

Combining (9.7), (9.13), (9.15), and (9.17) and using $V_{CC} = (1/2\pi)\int_0^{2\pi} v_{CE}(\omega t)d(\omega t)$, the collector–emitter voltage for a complete cycle is given by

$$\begin{aligned}
\frac{v_{CE}(\omega t)}{V_{CC}} &= 0 \quad \text{for } 0 < \omega t \le \pi \\
&= \frac{1}{\Delta}\Bigg[\omega t - \pi + \tau(e^{-(\omega t + \pi)/\tau} - 1) \\
&\quad - \frac{\pi + \tau(e^{-\theta_f/\tau} - 1)\cos(\omega t + \phi) + \cos\phi + \tau\sin\phi(e^{(-\omega t + \pi)/\tau} - 1)}{2\cos\phi + \tau\sin\phi(e^{-\theta_f/\tau} - 1)}\Bigg] \\
&\qquad \text{for } \pi < \omega t \le \pi + \theta_f \\
&= \frac{1}{\Delta}\Bigg[\omega t - \pi + \tau(e^{-\theta_f/\tau} - 1) \\
&\quad - \frac{\pi + \tau(e^{-\theta_f/\tau} - 1)\cos(\omega t + \phi) + \cos\phi + \tau\sin\phi(e^{-\theta_f/\tau} - 1)}{2\cos\phi + \tau\sin\phi(e^{-\theta_f/\tau} - 1)}\Bigg] \\
&\qquad \text{for } \pi + \theta_f < \omega t \le 2\pi
\end{aligned} \tag{9.20}$$

where

$$\Delta = \frac{1}{2\pi}\Bigg\{\frac{\pi^2}{2} + \tau(e^{-\theta_f/\tau} - 1)(\pi - \theta_f) - \tau^2(e^{-\theta_f/\tau} - 1) - \tau\theta_f$$

$$- \frac{\pi + \tau(e^{-\theta_f/\tau} - 1)}{2\cos\phi + \tau\sin\phi(e^{-\theta_f/\tau} - 1)}$$

$$\times \left[\theta_f \cos\phi - \sin\phi\tau^2(e^{-\theta_f/\tau} - 1) - \tau\theta_f \sin\phi + \cos\phi(\pi - \theta_f)\right.$$
$$\left.\left. + \tau \sin\phi(e^{-\theta_f/\tau} - 1)(\pi - \theta_f) + 2\sin\phi\right]\right\}$$

For a full cycle, the power consumption of the bipolar transistor is

$$P_{dtf} = \frac{1}{2\pi} \int_0^{2\pi} i_C(\omega t) v_{CE}(\omega t)\, d(\omega t) \tag{9.21}$$

The efficiency η is then given by

$$\eta = 1 - \frac{P_{dtf}}{I_{CC} V_{CC}} \tag{9.22}$$

The voltage and current waveforms derived are computed to demonstrate the utility of the analytical equations. Figure 9.15 shows the collector–emitter voltage waveform normalized to V_{CC} versus time at $\theta_f = 0$, 30, and 60°. The corresponding collector current waveforms normalized to I_{CC} are shown in Fig. 9.11*b*. For a higher θ_f value, the collector current has a longer exponential decay when the transistor is off and the collector–emitter voltage is shifted to higher degrees. The power efficiency as a function of θ_f is depicted in Fig. 9.16. In this figure the solid line represents the analytical predictions using an exponential collector falling waveform, and the squares represent the theoretical result using a linear collector falling waveform [20]. Using

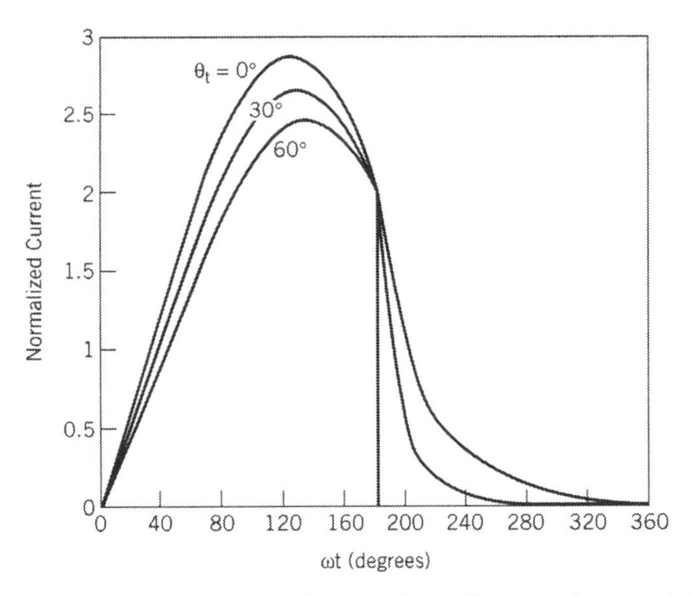

FIGURE 9.15 Collector current and collector–emitter voltage as a function of time (after Blanchard and Yuan, Ref. 19 © IEEE).

$\tau = 0.5\theta_f$ in the exponential collector falling waveform, good agreement between the present analytical results and the result calculated from [20] is obtained. For $\theta_f = 0, 30$, and 60°, the collector efficiency predicted by the analytical model is 100, 96.8, and 86.6%, and as calculated, from [20] it is 100, 97.7, and 90.8%. The power efficiency decreases with an increase in θ_f. This is due to a larger transistor power loss when θ_f increases. To improve power efficiency of the class E power amplifier at high frequencies, high-speed devices such as heterojunction bipolar transistors are needed. The high cutoff frequency and maximum oscillation frequency of the HBTs improve the collector current fall time and power amplifier power efficiency.

The main requirements for power amplifiers used in mobile transceivers are high power efficiency and the ability to work at low power supplies. A class E power amplifier for low-voltage mobile communications was evaluated [21]. A fully integrated class E power amplifier module operating at 835 MHz is designed, fabricated, and tested. The amplifier delivers 24 dBm of power to the 50-Ω load with a power-added efficiency greater than 50% at a supply voltage of 2.5 V. The power dissipation in the integrated matching networks is 1.5 times the power dissipated in the transistors.

Figure 9.17 shows the circuit diagram of a class F driver and a class E power amplifier. An interdigitated depletion-mode MESFET is used as the switch. Similar design ideas may be used for HBT class E power amplifiers. In Fig. 9.13 the MESFET transistor can generate the current harmonics of the input voltage applied to its gate. When this transistor is used as an amplifier with two parallel-tuned LC circuits in series with the drain, one tuned to the first harmonic and the other tuned to the third harmonic of the input frequency, it will generate a drain voltage with the necessary first

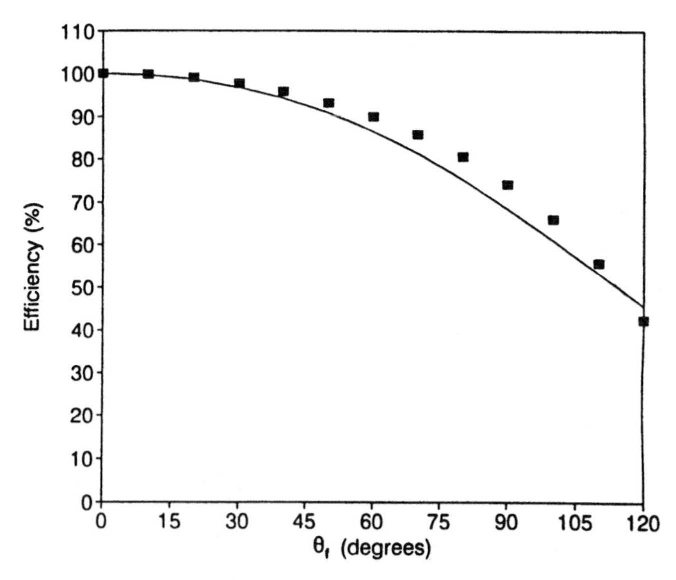

FIGURE 9.16 Power efficiency versus switching time delay θ_f (after Blanchard and Yuan, Ref. 19 © IEEE).

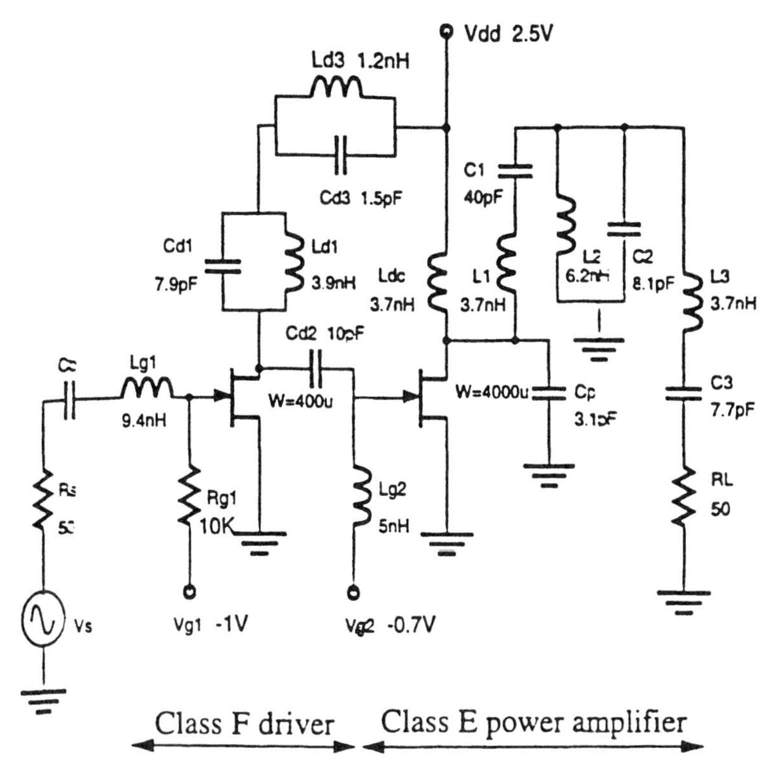

FIGURE 9.17 Circuit diagram of class F driver and class E power amplifier (after Sowlati et al., Ref. 21 © IEEE).

and third harmonic contents which are a good approximation of a square waveform. This operation is called a class F amplifier [22]. The class F driver is designed to provide the suitable switching waveform for the class E power amplifier. The class E power amplifier in Fig. 9.17 has two parallel-tuned *LC* circuits resonating at the first and third harmonics of the input frequency and an input matching network. A micrograph of the power amplifier module is shown in Fig. 9.18. The circuit, which occupies an area of 8.4 mm^2, operates at 835 MHz and delivers 24 dBm to the load at a supply voltage of 2.5 V. The power gain is 20 dB, and the second and third harmonics are 20 and 45 dB below the carrier. The measured and simulated output power and power-added efficiency versus the frequency at the supply voltage $V_{dd} = 2.5$ V are shown in Fig. 9.19. The output power and PAE are almost constant in the frequency range of interest (824 to 849 MHz).

9.3.5 Third-Order Intermodulation

Communication systems require linear power amplifiers with high-efficiency and very low intermodulation distortion. Intermodulation (IM) distortion often defines the

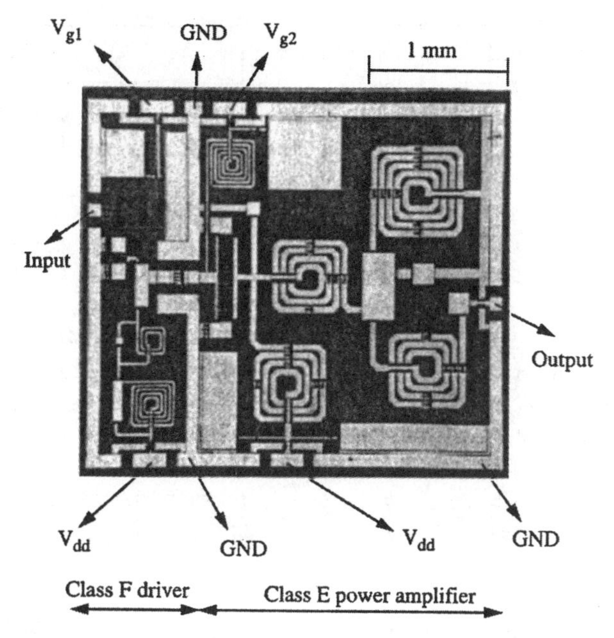

FIGURE 9.18 Micrograph of the power amplifier module (after Sowlati et al., Ref. 21 © IEEE).

upper limit of the signal-handling capability of a microwave receiver. It is particularly serious in broadband receivers designed for communications or spectral surveillance.

Consider a stage having a transfer function given by the power series expansion

$$f(x) = a_0(x)^0 + a_1(x)^1 + a_2(x)^2 + a_3(x)^3 + a_4(x)^4 + \cdots \tag{9.23}$$

Two input signals at frequencies Δf and $2\Delta f$ from the on-channel input frequency are involved in producing a third-order intermodulation response in an amplifier. Let these two signals be designated by A and B, where $A = f + \Delta f$, $B = f + 2\Delta f$, and $x = A + B$.

The power series expansion becomes

$$f(x) = a_0(A + B)^0 + a_1(A + B)^1 + a_2(A + B)^2 + a_3(A + B)^3 + \cdots \tag{9.24}$$

Evaluating the subexpansion of the third-order term in (10.24) gives

$$a_3A^3 + \underline{3a_3A^2B} + 3a_3AB^2 + a_3B^3$$

The underlined term in the expression above produces the IM response. To determine the slope m of this response, express this term in decibels:.

$$\text{IM(dB)} = 20 \log 3a_3 + 20 \log A^2 + 20 \log B \tag{9.25}$$

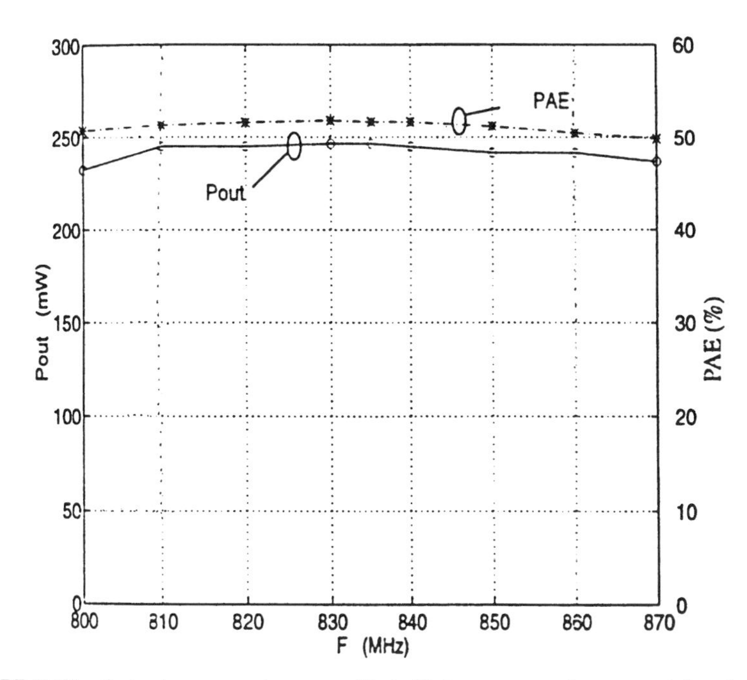

FIGURE 9.19 Output power and power-added efficiency versus frequency (after Sowlati et al., Ref. 21 © IEEE).

Since signals A and B are both allowed to vary, a 1-dB increase in A and B results in a 3-dB increase in undesired IM response. Hence for a third-order IM in an amplifier stage, $m = 3$. Knowing the slope is useful when evaluating the intermodulation distortion because for every 1-dB change in the two input signals, the IM response seen on the spectrum analyzer should change by 3 dB.

Figure 9.20 shows a plot of output power versus input power. The curve with slope $m = 1$ is the fundamental response. The curve with slope $m = 3$ is the third-order response. The third-order intermodulation intercept point (IP3) is the extrapolated intersection of the third-order intermodulation product and the fundamental output power versus the input power. The high output impedance of the HBT associated with the high Early voltage, high transconductance, and slow variation of current gain with collector current can be used to achieve high linearity capabilities in device and circuit applications. For comparable devices and biases, the HBT is observed to have an IP3/P_{dc} power ratio 5 to 10 times higher than that of typical MESFET and HEMT devices.

The unusually high linearity of the heterojunction bipolar transistor at relatively low levels of dc bias power is one of its best features. This high linearity is most remarkable in view of the exponential dependence of the HBT emitter current on the base–emitter voltage, which is characterized by having an extremely pronounced nonlinearity. The IM distortion of the HBT devices is low because of cancellation of

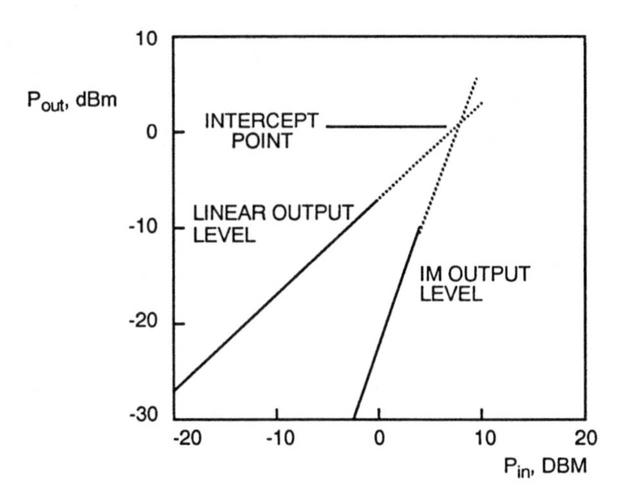

FIGURE 9.20 Plot of input power versus output power for illustration of IP3.

IM components arising in the nonlinear base-to-emitter junction resistance and capacitance [23]. A study using Microwave Harmonica [24] simulation was analyzed to determine the device parameters of the HBT that affect the third-order intermodulation distortion in a mixer. The physical parameters assumed to affect IP3 were the emitter junction capacitance, Early voltage, base resistance, collector junction capacitance, forward transit time, and current gain. From the harmonic balance simulation results, it shows that IP3 is not sensitive to the forward transit time, base resistance, and current gain. However, IP3 is very sensitive to C_{JE}. The dependence of intermodulation on the emitter junction capacitance decreases with increasing emitter junction capacitance, and after a minimum, increases with increasing emitter junction capacitance. This is explained as follows: The intermodulation caused by the resistive nonlinearity of the emitter junction is nearly 180° out of phase with respect to the intermodulation caused by the capacitive nonlinearity of the junction. Thus these nonlinearities partially cancel each other at a given range of junction capacitance. Outside this range, the intermodulation distortions caused by the resistive and capacitive nonlinearities of the emitter junction are no longer near the 180° phase difference, and hence the overall intermodulation increases with increased emitter junction capacitance.

It is known that despite strong nonlinear (exponential) dependence of junction current on voltage, HBTs exhibit excellent linearity. The elements with the strongest nonlinearity are the emitter–base junction conductance g_e and the diffusion capacitance C_d, both of which are linear functions of emitter current, forcing a cancellation of the nonlinearity in the C_d/g_e product. However, this is not a complete cancellation because the junction capacitance consists of the diffusion capacitance and the depletion capacitance. The depletion capacitance has a much weaker nonlinearity. In general, the depletion capacitance dominates at low bias, whereas the diffusion capacitance is dominant at high forward-bias conditions. Thus efforts aimed at reducing the depletion capacitance of the emitter–base junction will lead to improved linearity.

Additional insight into the HBT nonlinearities was gained by using the Volterra series expansion method [25]. Calculation of the nonlinear transfer functions is performed sequentially, from lowest to highest order, by solving a linear system of equations. The first-order transfer functions determine the response of the linear circuit, while second- and higher-order functions account for the device's nonlinear behavior. The second (or third)-order current is proportional to the second (or third) derivative of the charge of a particular element with respect to the base–emitter voltage and collector–emitter voltage. The analysis is carried out at the frequency of harmonics. It is concluded that the nonlinear current generated by the emitter–base junction capacitance turns out to be the strongest of all elements (C_{JE}, C_{JC}, g_{ie}, and g_m) followed by the transconductance. At different bias or frequency conditions, emitter–base conductance g_{ie} and emitter–base capacitance C_{JE} show a more pronounced canceling effect. Compared to C_{JE} and g_m, collector–base junction capacitance, C_{JC} generates a weaker current. However, the interaction of C_{JC} with other elements results in more pronounced nonlinear characteristics. Furthermore, the presence of nonlinearities due to C_{JE}, g_{je} and g_m together improved the third-order intermodulation.

9.3.6 Self-Linearizing Technique for the L-Band HBT Power Amplifier

Power amplifiers operating with high-efficiency and high linearity at a single low supply voltage are in strong demand for mobile communication systems. InGaP/GaAs HBTs are the most promising devices for mobile communication handsets, due to their superior power characteristics at low bias voltage and their higher power-handling capability with smaller chip size.

The effects of load and source impedance on phase distortion and adjacent channel leakage power (ACP) have been studied [26]. ACP originates from the nonlinearity of a device. Power gain compression and phase distortion result in an ACP increase as the output power increases. The nonlinear effect of device parameters on phase distortion was analyzed using the simplified equivalent circuit of a common-emitter HBT in Fig. 9.21. In the equivalent circuit, the collector–base capacitance and conductance are neglected to simplify the analysis.

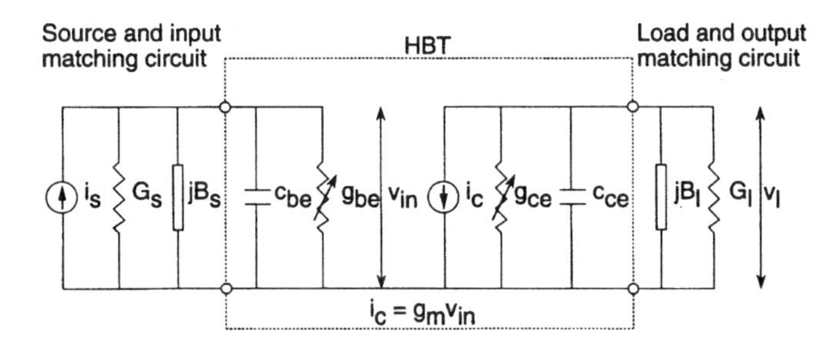

FIGURE 9.21 Small-signal equivalent circuit for calculating output phase distortion.

The output voltage in Fig. 9.21 is given by

$$v_l = \frac{g_m i_s e^{j(\pi-\theta_l-\theta_s)}}{\left|g_{ce} + G_l + j(\omega C_{ce} + B_l)\right| \left|g_{be} + G_s + j(\omega C_{bc} + B_S)\right|} \tag{9.26}$$

$$\tan\,\theta_l = \frac{\omega C_{ce} + B_l}{g_{ce} + G_l} \tag{9.27}$$

$$\tan\theta_S = \frac{\omega C_{bc} + B_S}{g_{be} + G_S} \tag{9.28}$$

where G_s is the conductance of the input matching circuit, B_s the susceptance of the input matching circuit, G_l the matching load conductance, and B_l the matching load susceptance. As the output power increases, g_{ce} increases and g_m decreases because of the current clipping at the saturation region in the common-emitter *I–V* curves. This results in compression of the output voltage swing v_l.

The output phase distortion $\Delta\theta_l$ and the input phase distortion $\Delta\theta_s$ can be calculated from the differentials of the θ_l and θ_s

$$\Delta\theta_l = -\frac{\omega C_{ce} + B_l}{(g_{ce} + G_l)^2 + (\omega C_{ce} + B_l)^2}\,\Delta g_{ce} \tag{9.29}$$

$$\begin{aligned}\Delta\theta_s = &- \frac{\omega C_{bc} + B_s}{(g_{be} + G_s)^2 + (\omega C_{bc} + B_s)^2}\,\Delta g_{be} \\ &+ \frac{g_{be} + G_s}{(g_{be} + G_s)^2 + (\omega C_{be} + B_s)^2}\,\omega\,\Delta C_{be}\end{aligned} \tag{9.30}$$

The total phase distortion, $\Delta\theta$, is given by

$$\begin{aligned}\Delta\phi &= -\Delta\theta_l - \Delta\theta_s \\ &= \frac{\omega C_{ce} + B_l}{(g_{be} + G_l)^2 + (\omega C_{ce} + B_l)^2}\,\Delta g_{ce} + \frac{\omega C_{be} + B_s}{(g_{be} + G_s)^2 + (\omega C_{be} + B_s)^2}\,\Delta g_{be} \\ &\quad - \frac{g_{be} + G_s}{(g_{be} + G_s)^2 + (\omega C_{be} + B_s)^2}\,\omega \Delta C_{be}\end{aligned} \tag{9.31}$$

When an input-matching circuit is conjugate matched for maximum gain, $\omega C_{be} + B_s$ is zero. When an output matching circuit is tuned for maximum output power or efficiency, $\omega C_{ce} + B_l$ is not zero because the load impedance deviates from the conjugate-matched impedance. The phase distortion $\Delta\theta$ depends mainly on Δg_{ce} and ΔC_{be} in the foregoing matching condition, as seen in (9.31). The phase distortion, however, can be canceled with the nonlinearity of input conductance Δg_{be}. In HBTs, nonlinearity of the input conductance is greatest when the input base–emitter circuit is forward biased. By optimizing an input-matching circuit impedance, the value of the sign of the second term in (9.31) can be adjusted to cancel the other terms. Then phase distortion can be self-compensated by the nonlinear input conductance of the device itself and the susceptive component of the input-matching circuit impedance. Therefore, the phase distortion is reduced to achieve superior ACP performance with this self-linearizing technique.

Figure 9.22 shows the contour lines for several values of power gain and ACP at a 50-kHz offset frequency from the source-pull measurement results. The load impedance is set to a minimum ACP with a PAE of over 50% when the source impedance is gain matched. The output power is 31 dBm at $V_{CE} = 3.5$ V under class AB operation. In Fig. 9.23, point A is the gain-matched impedance, points B and C are the impedance changed by subtracting the susceptance value from point A while keeping the conductance at the same value. The impedances for points A, B, and C are $0.182 + j0.092$, $0.199 - j0.041$, and $0.192 - j0.0116$ S, respectively. In Fig. 9.23*b* a constant-conductance circle is also plotted. From point A to point C the ACP

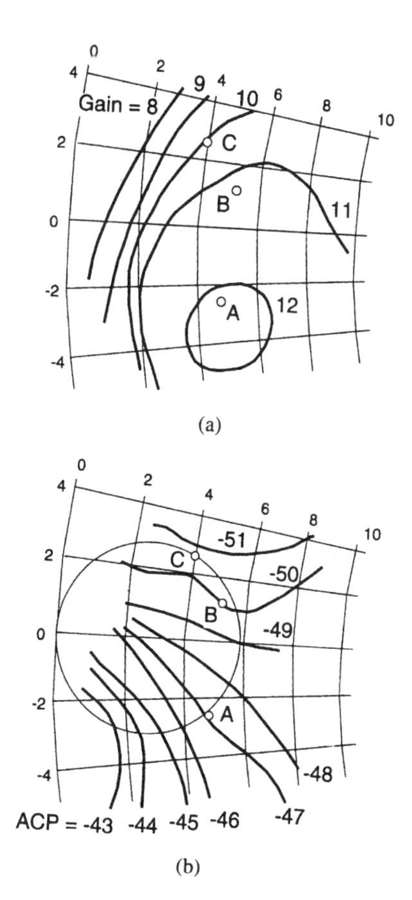

FIGURE 9.22 Contour lines for (*a*) power gain and (*b*) for ACP for three different impedance points (after Yamada et al., Ref. 26 © IEEE).

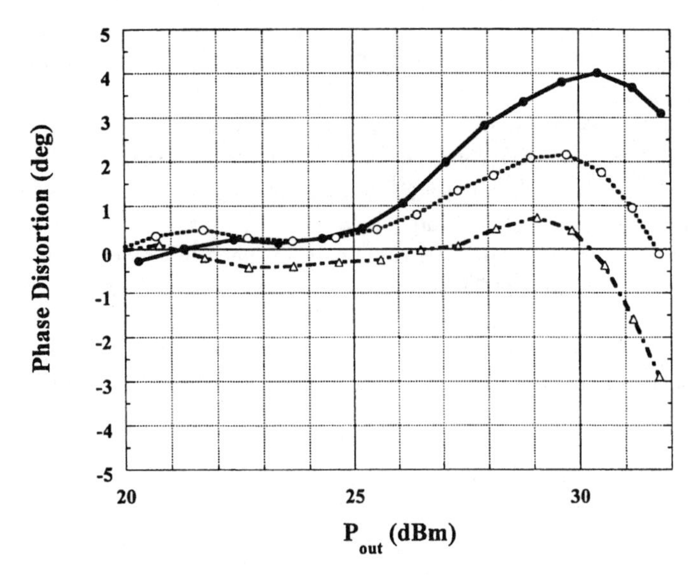

FIGURE 9.23 Phase distortion versus output power (after Yamada et al., Ref. 26 © IEEE).

improves while the gain decreases. The constant ACP contour lines are perpendicular to the constant conductance circuit near point A.

Figures 9.23 and 9.24 show phase distortion and gain compression as a function of output power. When the output power is over 25 dBm, phase distortion is greatly reduced as susceptance decreases below the gain-matched impedance. However, there is only a very slight corresponding degradation in gain compression. At the gain-matched impedance A, the phase distortion increases with output power. The phase distortion is improved at impedance points B and C.

9.4 LOW-NOISE AMPLIFIERS

In a microwave amplifier, even when there is no input signal, a small output voltage can be measured. This small output power is referred to as the *amplifier noise power*. The total noise output power is composed of the amplified noise input power plus the noise output power produced by the amplifier.

The noise figure NF describes quantitatively the performance of a noisy microwave amplifier. The *noise figure* of a microwave amplifier is defined as the ratio of the total available noise power at the output of the amplifier to the available noise power at the output due to thermal noise. The noise figure is expressed as

$$\mathrm{NF} = \frac{P_{No}}{P_{Ni} G_A} = \frac{P_{Si}/P_{Ni}}{P_{So}/P_{No}} \tag{9.32}$$

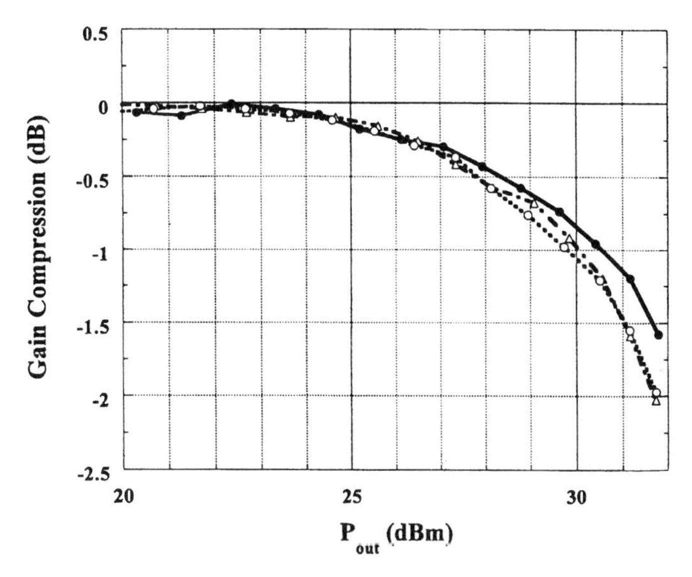

FIGURE 9.24 Gain compression versus output power (after Yamada et al., Ref. 26 © IEEE).

where P_{No} is the total available noise power at the output of the amplifier, $P_{Ni} = kT\ \Delta f$ is the available noise power in a bandwidth Δf, G_A the available power gain, P_{So} the available signal power at the output, and P_{Si} the available signal power at the input.

A model [27] for the calculation of the noise figure of a two-stage amplifier is shown in Fig. 9.25. Here the total available power at the output is given by

$$P_{No,\text{total}} = G_{A2}(G_{A1}P_{Ni} + P_{n1}) + P_{n2} \tag{9.33}$$

The noise figure of the two-stage amplifier is given by

$$\text{NF} = \frac{P_{No,\text{total}}}{P_{Ni}G_{A1}G_{A2}} = 1 + \frac{P_{n1}}{P_{Ni}G_{A1}} + \frac{P_{n2}}{P_{Ni}G_{A1}G_{A2}} \tag{9.34}$$

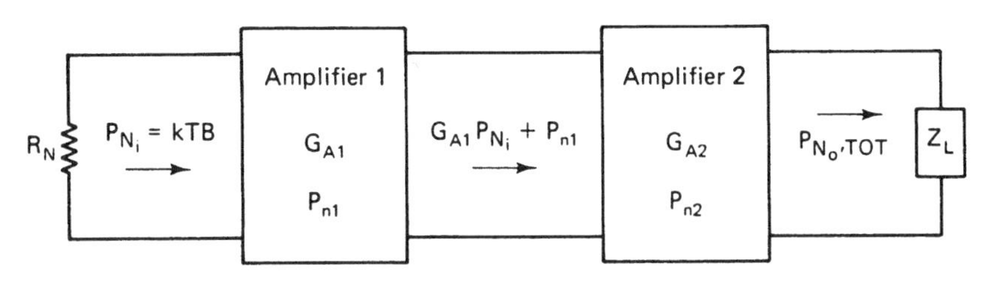

FIGURE 9.25 Noise-figure model of a two-stage amplifier (after Gonzalez, Ref. 27 © IEEE).

or

$$NF = NF_1 + \frac{NF_2 - 1}{G_{A1}} \tag{9.35}$$

where

$$NF_1 = 1 + \frac{P_{n1}}{P_{Ni} G_{A1}}$$

and

$$NF_2 = 1 + \frac{P_{n2}}{P_{Ni} G_{A2}}$$

Equation (9.35) shows that the noise figure of the second stage is reduced by the power gain of the first stage, G_{A1}. Therefore, the noise contribution from the second stage is smaller than that from the first stage.

Since a minimum noise figure and maximum power gain cannot be obtained simultaneously, constant-noise-figure circles, together with constant-available-power-gain circles, can be drawn on the Smith chart, and reflection coefficients can be selected that compromise between noise figure and gain performance. Figure 9.26 is a plot of the noise figure and power gain circles.

The noise figure of a two-port amplifier is given by

$$NF = NF_{\min} + \frac{r_n}{g_s} |Y_s - Y_o|^2 \tag{9.36}$$

where r_n is the equivalent normalized noise resistance of the two-port, $Y_s = g_s + jb_s$ represents the source admittance, and $Y_o = g_o + jb_o$ represents the source admittance that results in the minimum noise figure $NF_{\min}$.

Using the equivalent noise and network parameters of the HBT, the minimum noise figure as a function of device parameters is obtained [28]

$$NF_{\min} = 1 + \frac{1}{\beta} + \eta g_\pi R_b + \eta g_m R \left(\frac{f}{f_T}\right)^2$$
$$+ \sqrt{2 g_m R \left(\frac{f}{f_T}\right)^2 + (2 g_m R_b)^2 \left(\frac{f}{f_T}\right)^4 + \eta g_\pi R + \frac{\eta}{2\beta}} \tag{9.37}$$

where $R = R_b + R_e$.

Equation (9.37) indicates that as the device is operated with a high collector current, $NF_{\min}$ monotonically increases due to a decrease in β and an increase in g_π. However, reduction of I_C to a low level will reduce f_T and increase $NF_{\min}$. For a given I_C, when

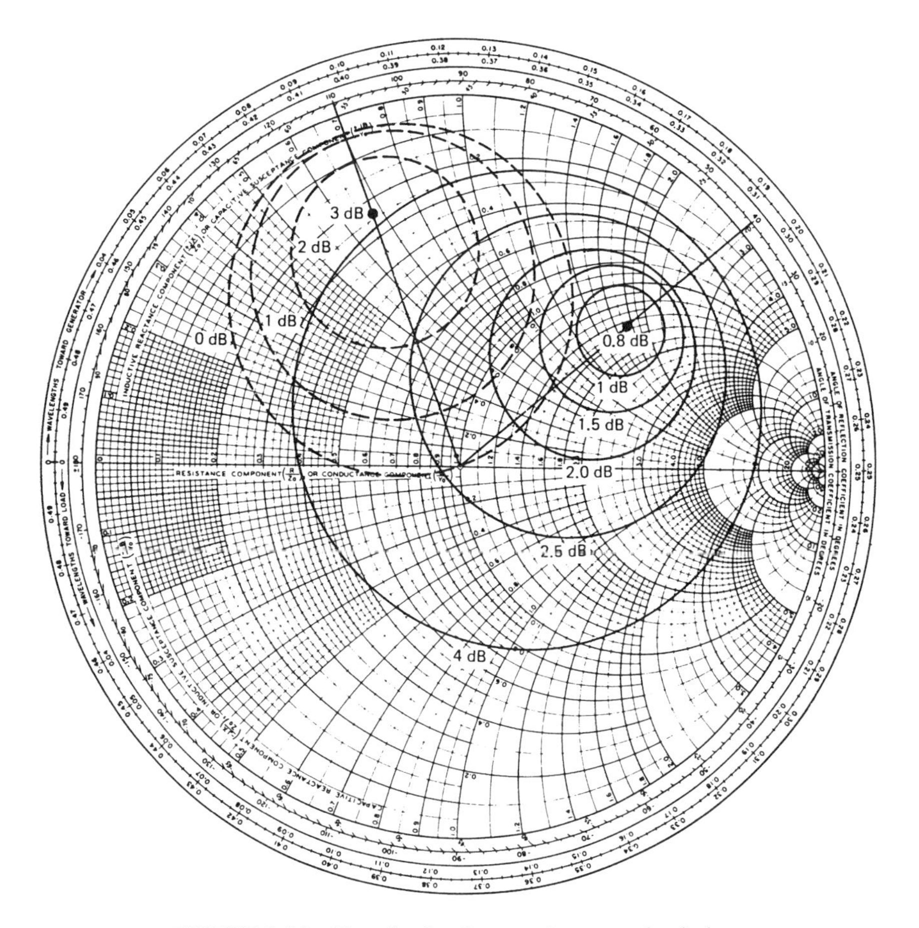

FIGURE 9.26 Plot of noise-figure and power-gain circles.

$f \ll f_T$, the noise figure will increase with frequency and can be approximated as

$$\mathrm{NF}_{\min} \approx 1 + \sqrt{2}\,\frac{f}{f_T}\sqrt{g_m(R_e + R_b)} \tag{9.38}$$

In portable consumer applications where conserving battery life affects cost directly, obtaining high gain and a low-noise figure under low-dc bias operation is highly desirable. HBTs are more attractive than FETs for these applications because of their high device transconductance and linearity under low-dc-bias operation, small size, and low-device-noise-figure capability at L-, S-, and C-band frequencies. Figure 9.27 shows a plot of the gain-to-dc power ratio versus noise figure for several L- and S-band low-noise amplifiers. The HBT low-noise amplifiers (LNAs) showed gain/P_{dc} ratios

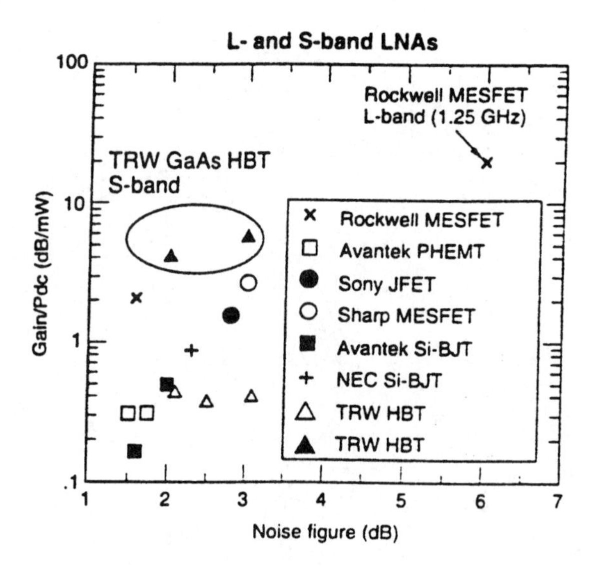

FIGURE 9.27 Comparison of gain/P_{dc} versus noise figure for several low-noise amplifiers (after Kobayashi et al., Ref. 29 © IEEE).

of 4.2 and 5.65 and corresponding noise figures of 2.0 and 3.01 dB at 2 GHz [29]. The HBT LNAs were fabricated with TRW's AlGaAs/GaAs SABM technology.

For HBT devices, lower current generally means lower minimum noise. For low-noise-amplifier design, the source impedance required to achieve minimum noise needs to be considered over bias. Figure 9.28 shows a plot of the optimum noise source impedance (gamma optimum) from 300 MHz to 3 GHz of a $3 \times 10\ \mu\text{m}^2$ quad-emitter HBT as a function of collector current. The impedance plot shows that at lower currents the optimum impedance has a large real part, greater than 50 Ω, with a significant inductive reactance. For higher collector currents, gamma optimum is mostly real and decreases below 50 Ω. At 16 mA, the impedance is close to 50 Ω over the frequency range. At this bias point, gamma optimum is nearly coincident with 50 Ω, which makes it easier to design both good input-return-loss and low-noise figures over a broad band. However, the minimum achievable noise figure at this bias will be higher than at lower bias currents. For low-noise and low-dc-bias operation, a design employing series inductive matching at the input of the HBT device will result in optimum noise match over a narrow band.

A schematic of a 2-GHz low-dc-power, low-noise HBT amplifier is shown in Fig. 9.29*a*. The LNA is a one-stage narrowband design that is matched for a center frequency of 1.9 to 2.0 GHz. A 3.8-nH input series inductor, L_b, is used to match the input of a $3 \times 10\ \mu\text{m}^2$ quad-emitter HBT for minimum noise. A collector bias current of 1 mA and a $V_{CE} = 2.0$ V were chosen to realize a gain greater than 8 dB and a minimum noise figure less than 2.0 dB with a total power consumption of 2 mW at

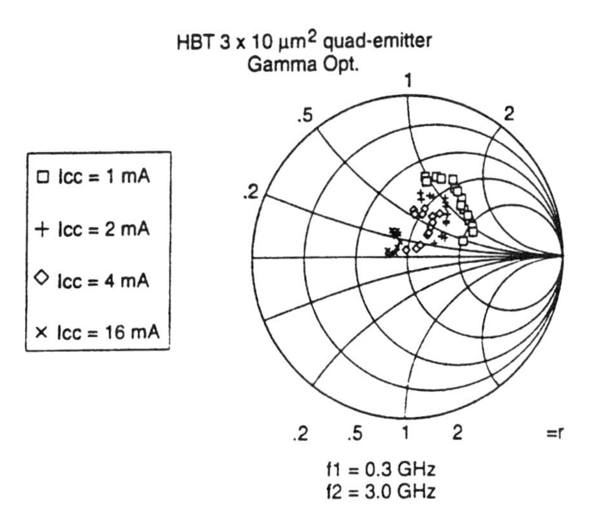

FIGURE 9.28 Optimum noise source impedance (gamma optimal) loci from 300 MHz to 3 GHz (after Kobayashi et al., Ref. 29 © IEEE).

2 GHz. A 0.5-nF spiral inductor L_e in series with the emitter of the HBT is used to tune gamma optimum so that it coincides more closely with the 50-Ω source impedance to achieve optimum noise and input-return-loss match. The output of the amplifier is matched using a series LC matching network comprised of C_{out} and TLIN_o. An inductive choke L_e, in series with a small load resistance R_L, provide both a high-pass ac load and a means for biasing the collector of the HBT.

A schematic of a two-stage C-band low-noise amplifier using the same narrow-band noise-matched topology is shown in Fig. 9.29*b*. Two stages are used, to obtain reasonable gain at this C-band frequency. In addition, resistive self-biasing comprised of resistors R_1, R_2, and $R_{L\text{dc}}$, is used to simplify the biasing of the circuit, which is convenient for on-wafer evaluation. The dc current flowing through the resistive bias network is on the order of 0.2 mA per stage to ensure proper operation over variations in dc current gain from wafer to wafer. The HBTs of each stage of the amplifier are biased at 10 mA and $V_{CE} = 2$ V. At this bias, a nominal gain greater than 15 dB and a minimum noise figure of 2.52 dB are obtained at 5.7 GHz. The losses of the spiral inductors in series with the base of the HBTs explain the $\approx$ 0.4 dB noise figure.

Noise figures and gain were measured for the S-band low-noise amplifier at various V_{CC} values to find the optimum noise bias of the amplifier. Figure 9.30*a* shows gain and noise-figure performance at 2 GHz as a function of biases at $I_{CC} = 5$ mA. This plot shows that at a V_{CC} value $\geq$ 2.0 V, the amplifier achieves a minimum noise figure and a maximum gain performance. This may be explained by the fact that the collector–base capacitance is fully depleted under reverse biases greater than about 0.6 V, which corresponds to $V_{CC} \approx V_{CE} = 2.0$ V. At this optimum collector voltage, the gain and noise figures were measured as a function of I_{CC} at 2 GHz as shown

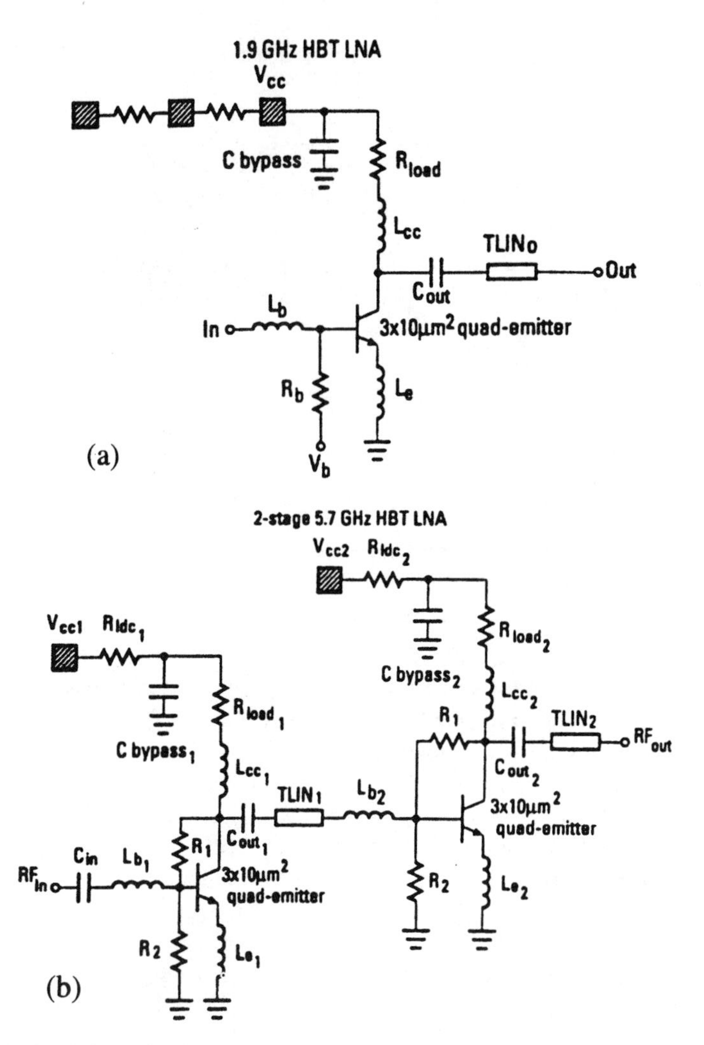

FIGURE 9.29 Schematic of (*a*) one-stage and (*b*) two-stage low-power, low-noise amplifiers (after Kobayashi et al., Ref. 29 © IEEE).

in Fig. 9.30*a*. This figure shows that the optimum low-noise bias current is between 2 and 4 mA for the LNA. At currents of less than 2 mA, the noise figure increases quickly, due to a rapid change in the gamma optimum over bias.

Figure 9.31*a* shows the measured gain and return loss at $I_{CC} = 2$ mA and $V_{CC} =$ 2.0 V. The nominal gain is 11.1 dB at 2 GHz. The input and output return loss at this frequency are -13.7 and -14.7 dB, respectively. The low-noise amplifier has a 40% 1-dB bandwidth from 1.5 to 2.3 GHz. Figure 9.30*b* shows noise figures versus frequency. The noise figures range from a minimum of 1.9 dB to a maximum of 2.2 dB from 1.5 to 2.8 GHz.

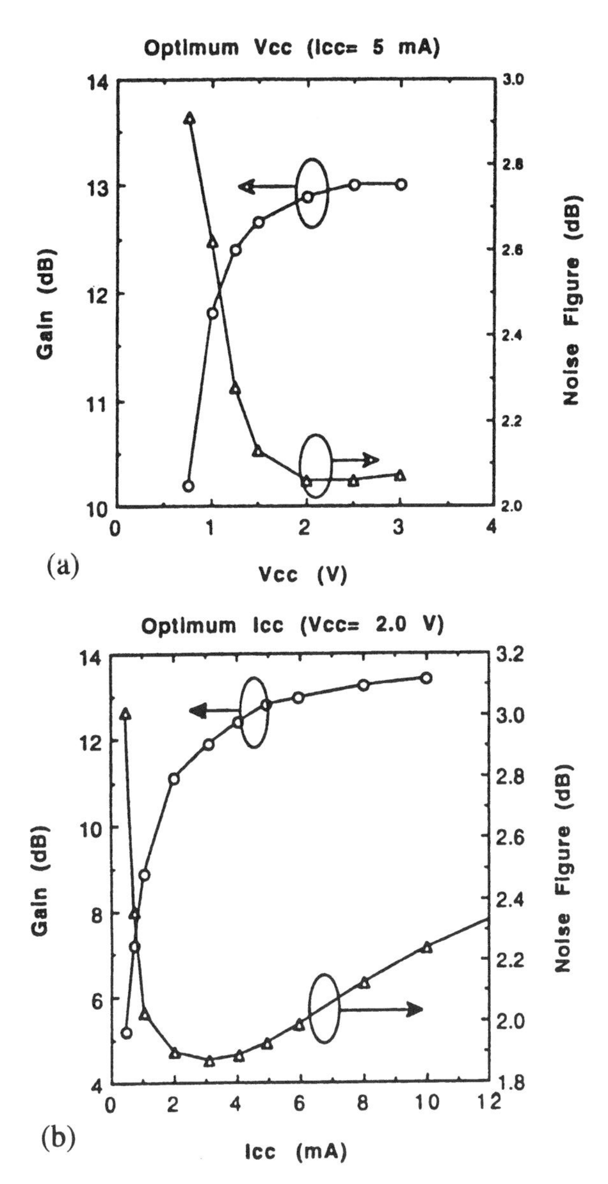

FIGURE 9.30 Measured gain and noise figure as a function of (*a*) bias and (*b*) collector current at 2 GHz (after Kobayashi et al., Ref. 29 © IEEE).

9.5 HBT OSCILLATORS

Many microwave systems require the use of oscillators to generate a reference signal. A voltage-controlled oscillator (VCO) is central to the operation of many important

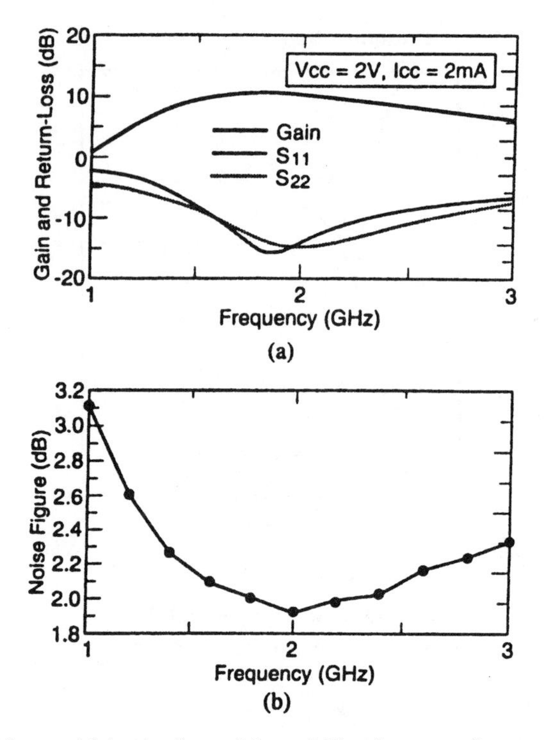

FIGURE 9.31 Measured (*a*) gain, S_{11}, and S_{22} and (*b*) noise versus frequency at $I_{CC} = 2$ mA and $V_{CC} = 2.0$ V(after Kobayashi et al., Ref. 29 © IEEE).

electronic systems. It is used in a phase-locked loop (PLL), for demodulation of FM signals, and in frequency synthesizers, which are particularly important for wireless applications. Furthermore, VCOs are the critical components of FMCW radars used for range and velocity measurements. As such, VCOs must have very low phase noise and be highly linear. For example, for an accurate FMCW radar, the VCO must have a very linear sweep, typically about 0.5% for most applications as well as have ultralow phase noise. In oscillators, the low-frequency flicker noise is mixed with the high-frequency oscillation and results in phase noise around the carrier frequency, which degrades system performance. In the FMCW radar, phase noise in the VCO broadens the spectrum of the detected signal and makes targets that are physically close together difficult to distinguish, and small Doppler shifts indiscernible.

A number of solid-state devices, such as Gunn diodes, Si bipolar transistors, GaAs FETs, and HBTs, are used in the design of microwave oscillators. The choice of device type is dictated primarily by the performance requirement of each application and the cost and availability of devices. The low-noise in the HBT compared to the MESFET and the HEMT results in oscillators with superior phase noise, especially for InP-based HBT oscillators.

A high-power, high-efficiency, and small-size monolithic coplanar waveguide oscillator incorporating a single-stage buffer amplifier on the same chip has been reported [30]. The equivalent circuit for the K-band monolithic oscillator using a heterojunction FET is shown in Fig. 9.32. The oscillator design employed the device-circuit interaction design concept. Similar design concepts may be applied to the monolithic oscillator using the HBT. The oscillator design is explained as follows:

(1) Inject RF power from an external source at a desired oscillation frequency causing RF current I_{RF}, and measure the negative impedance Z_{NEG} for various I_{RF} values by changing the power level of the external source. Initial lengths of T_S and T_G are selected so that -Re(Z_{ENG}) is maximum at small-signal operation.

(2) Choose I_{RF} values at large-signal operation. Three load lines at different RF currents are identified when matching loads satisfy the oscillation condition.

(3) Calculate the output power using $P_{out} = -\text{Re}(Z_{ENG})|I_{RF}|^2$. For the I_{RF} conditions selected, optimum lengths of T_S and T_G can be found to obtain the maximum P_{out} value at the frequency desired.

(4) Design each matching circuit so that its impedance, Z_{LOAD}, is equal to -Z_{ENG} for the optimum lengths of T_S and T_G. The matching circuit is designed to transform Z_{LOAD} into a 50-Ω load.

(5) Calculate the inverse oscillator reflection coefficients (Γ_{OSC}^{-1}) and plot them versus RF, as shown in Fig. 9.33. It is important to note that the locus of Γ_{OSC}^{-1} can be changed by varying an intercept angle (θ) by adjusting the load line. In Fig. 9.33, the buffer amplifier S_{11} value is traced versus frequency. In the case of line A, the loci of Γ_{OSC}^{-1} and S_{11} cross nearly orthogonally, as is desired ($\theta \approx 90°$). Output power variation among the three types is within 1 dB and is acceptable.

(6) Design the buffer amplifier to be driven by the oscillator to achieve an overall high-power, high-efficiency operation. The overall dc-RF efficiency is

$$\eta_{ALL} = \frac{G}{1 + (P_{dc,AMP}/P_{dc,OSC})} \eta_{OSC} \tag{9.39}$$

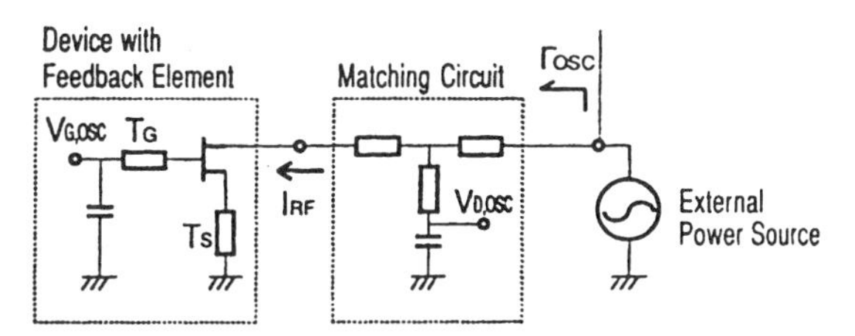

FIGURE 9.32 Setup of a K-band FET oscillator for calculating oscillator reflection coefficients (after Maruhashi et al., Ref. 30 © IEEE).

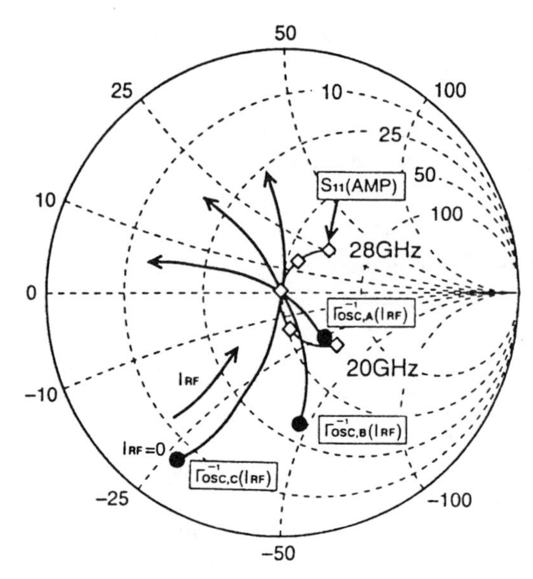

FIGURE 9.33 Γ_{OSC}^{-1} and S_{11} versus frequency (after Maruhashi et al., Ref. 30 © IEEE).

where G is the gain under operation, $P_{\text{dc,OSC}}$ and $P_{\text{dc,AMP}}$ are the oscillator and amplifier dc power dissipation, respectively, and η_{OSC} is the oscillator dc-RF efficiency.

The requirements for the buffer amplifier design are (1) operation at a center frequency, (2) incorporation of an inductive feedback line in the input to ensure the stability of the amplifier, and (3) determination of the length of the feedback line to make the amplifier operate near the output saturation point with a high gain. Applying a large-signal harmonic balance method, the circuit parameters were optimized to satisfy the foregoing requirements. Both the input and output ports of the amplifiers were designed to be widely matched to a 50-Ω impedance.

Figure 9.34 shows the maximum output power and overall dc-RF efficiency for each oscillator at supply voltages of 2.0 and 3.0 V as a function of oscillation frequency. When the bias voltage is increased from 2 V to 3 V, the output power is increased by about 4 dB. For all oscillators, the output power is higher than 16 dBm. The 21-GHz oscillator output power is 17 dBm with an overall dc-RF efficiency of 22% at 3-V supply voltage.

In a microwave oscillator, two types of noise are concerned. The first is the background noise at the oscillation frequency band caused by thermally generated current and voltage in the semiconductor device. This noise is "white" in character and sets the noise floor for the oscillators. The magnitude of this noise is usually very small. The other noise source is the $1/f$ noise, which has phase and amplitude components. When the noise of an oscillator is added to the carrier and fed back through the

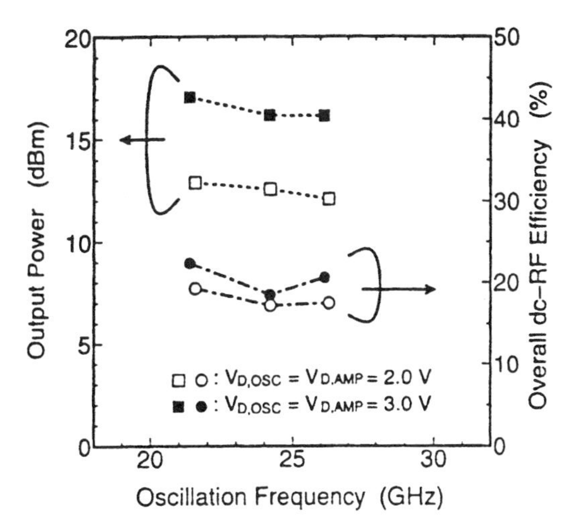

FIGURE 9.34 Maximum output power and overall dc-RF conversion efficiency versus frequency (after Maruhashi et al., Ref. 30 © IEEE).

resonator, the resulting signal is both amplitude and phase modulated. The limiting amplifier will remove the amplitude-modulated component, but does not affect the phase modulation. Phase noise dominates the near-carrier spectrum of the oscillator and often determines one endpoint of the dynamic range in a system. The maximum undistorted signal depends on system linearity, while the minimum is set by noise.

Phase noise is crucial to the performance of many wireless communication systems. In cellular communication systems, for example, the phase noise of the local oscillator determines the minimum spacing between adjacent channels to avoid interference. In Doppler radar systems, phase noise can disguise weak, reflected incoming signals. The phase-noise spectrum of bipolar oscillators is dominated by their $1/f$ noise [31,32]. Recent experimental data show that both phase noise and low-frequency $1/f$ noise originate from diffusion coefficient fluctuation in the base [33]. The intrinsic baseband $1/f$ noise is up-converted to the oscillation frequency through nonlinearities in the device [34,35]. Although the power in the sidebands is not very high, its presence degrades the performance of microwave systems. For example, in a coherent radar system, the return signal is mixed with a portion of the transmitter signal for detection. If the noise in the sidebands is too high, the return signal magnitude falls below the noise level and the detection sensitivity of the radar is reduced accordingly. For the case where up-conversion of ideal $1/f$ noise dominates, within the resonator bandwidth, the phase noise varies as $1/f_m^3$, where f_m is the offset frequency.

The phase-noise spectral density of an 11-GHz HBT dielectric resonator oscillator has been investigated [36]. The schematic diagram of the dielectric resonator oscillator is shown in Fig. 9.35. Parallel feedback is used, with the dielectric resonator providing the necessary inductive feedback between the collector and the base. Series feedback

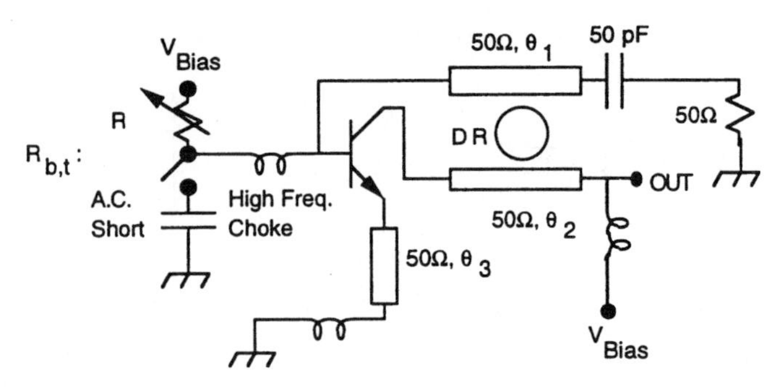

FIGURE 9.35 Schematic diagram of a dielectric resonator oscillator (after Tutt et al., Ref. 36 © IEEE).

in the emitter is used to improve output matching. The collector and base are biased from a single-variable dc voltage source which is connected to the circuit through high-frequency chokes. The base current is controlled via a 10-Ω wirewound potentiometer. The low-frequency base termination $R_{b,t}$ can be switched from a value R set by the potentiometer to approximately 0 Ω. The high-frequency choke allows signal transmission below 100 kHz.

The spectrum measured for the low-frequency short-circuit base termination case, with $V_{CE} = 5$ V and $I_C = 50$ mA is shown in Fig. 9.36. The oscillation frequency is 11.02 GHz, and the output power is 5.6 dBm. No tuning of the emitter stub was attempted to maximize the output power. The external measured quality factor Q ranges from 1200 to 2200 as the collector current decreases from 50 mA to 20 mA.

The phase noise of the oscillator was measured from 10 to 100 kHz using an HP3048A phase-noise measurement system. A plot of $Ł(f_m)$ versus frequency for $I_C = 50$ mA and $V_{CE} = 5.0$ V for different base terminations is shown in Fig. 9.37. This figure clearly shows the impact of the base termination on $Ł(f_m)$. The noise was reduced by about 4 to 7 dB over the entire measurement band when the base was ac short circuited at baseband frequencies. At 10 kHz $Ł(f_m) = -101$ dBc/Hz for $R_{b,t} = 0$ and -95 dBc/Hz for $R_{b,t} = R$.

9.5.1 Modeling the Bipolar Phase Noise

Phase noise corresponds to instability in the phase or frequency of a signal. It is measured as the ratio of the power in the noise to that in the carrier. Noise power is measured at a specified offset frequency in a specified bandwidth. Phase noise may be extracted from the small-signal equivalent circuit. In a voltage-controlled oscillator, since the device is operated under large-signal conditions, the low-frequency noise is up-converted to sideband noise around the carrier due to device nonlinearity. The nonlinear circuit elements for a bipolar oscillator are the base–emitter capacitance, transconductance, and current gain.

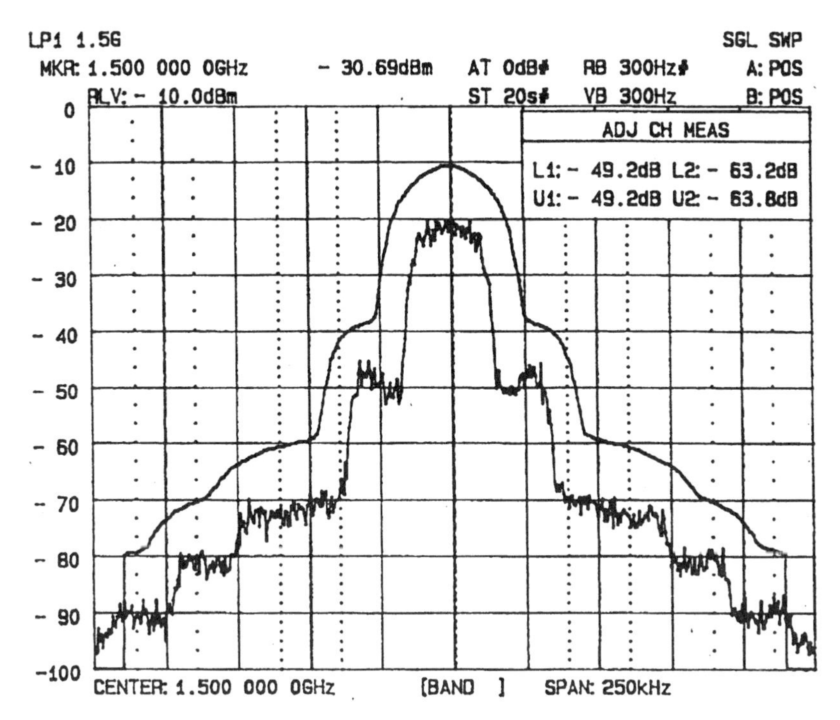

FIGURE 9.36 Measured spectrum of the dielectric resonator oscillator (after Tutt et al., Ref. 36 © IEEE).

To analyze phase noise, the equivalent circuit in Fig. 9.38 is used. In this figure L_b is the base packaging inductance and L_e is the emitter packaging inductance. Small-signal components g_m, r_π, C_π, and C_μ have their conventional meaning in the hybrid-π equivalent circuit.

The voltage between the base and emitter terminals in Fig. 10.38 is expressed as [37]

$$V_1 \approx I_{b0}(R_b + j\omega L_b) + i_B Z_{be} + g_m i_B Z_{be}(R_e + j\omega L_e) \tag{9.40}$$

where ω is the radian frequency and $Z_{be} = r_\pi/(1 + j\omega r_\pi C_\pi)$.

Since $i_B = i_C/g_m Z_{be}$ and $r_\pi = \beta/g_m$, (9.40) is rewritten as

$$\frac{i_C}{V_1} = \frac{\beta}{R_b + r_\pi + \beta R_e - \omega^2 L_b r_\pi C_\pi + j\omega(L_b + \beta L_e + R_b r_\pi C_\pi)} \tag{9.41}$$

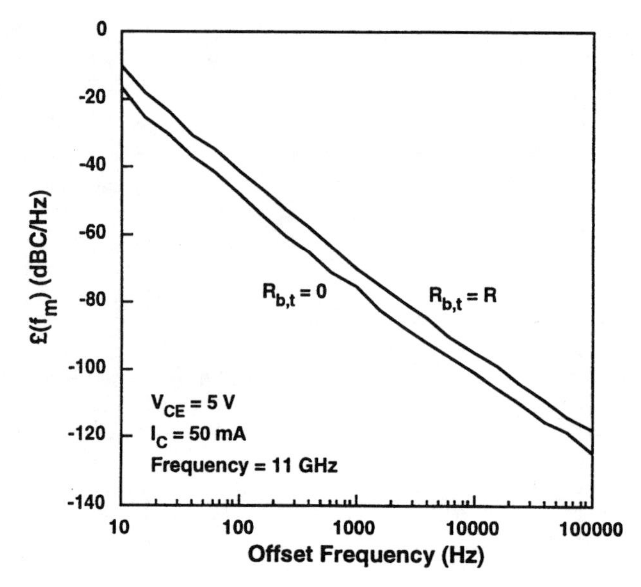

FIGURE 9.37 Phase noise versus offset frequency (after Tutt et al., Ref. 36 © IEEE).

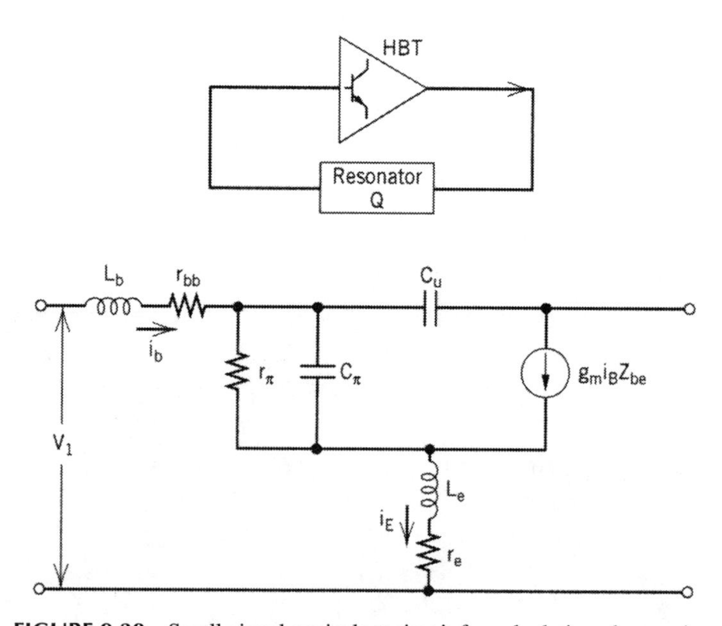

FIGURE 9.38 Small-signal equivalent circuit for calculating phase noise.

From (9.41) one finds the phase change as

$$\frac{d\phi}{di_C} = \frac{d}{di_C}\left(\tan^{-1}\frac{i_C}{V_1}\right)$$
$$= \frac{\omega\beta V_T\{[L_b + \beta L_e + \beta R_b\tau_F - (R_b + \beta R_e)R_b C_{JE}] - \omega^2(L_b + \beta L_e)L_b C_{JE}\}}{D} \tag{9.42}$$

where

$$D = \left[(R_b + \beta R_e - \omega^2 L_b\beta\tau_F)I_C + \beta V_T - \omega^2 L_b\beta V_T C_{JE}\right]^2 + \left[\omega(L_b + \beta L_e + \beta R_b\tau_F)I_C + \omega\beta V_T R_b C_{JE}\right]^2$$

The sideband noise (SBN) due to noise up-conversion is expressed as

$$\text{SBN} = 20\log\left(\frac{d\phi}{di_C}\frac{di_C f_0}{2\sqrt{2}Q_L f_m}\right) \tag{9.43}$$

where f_0 is the oscillator frequency, Q_L the oscillator loaded quality factor, and di_C the collector current variation due to thermal noise, shot noise, and $1/f$ noise, given by

$$di_C = \sqrt{\frac{4kT\,\Delta f}{R_e}} + \sqrt{4kTR_b g_m^2\,\Delta f} + \sqrt{\frac{k_f I_B^{af}\,\Delta f}{f}} \tag{9.44}$$

The first term in (9.44) represents the thermal noise due to the resistance R_e, the second term represents the thermal noise due to R_b, and the third term represents the flicker noise. It is clear from (9.43) and (9.44) that limiting noise up-conversion through a high-Q resonator and reducing the low-frequency noise are two effective ways to reduce phase noise.

Using (9.43), phase noise as a function of device and circuit parameters can be evaluated. Figure 9.39 shows the phase noise of an oscillator versus the collector current. In this plot the line represents the model predictions and the triangles represent the experimental data. The device and circuit parameters include $R_b = 30\ \Omega$, $R_e = 68\ \Omega$, $L_b = 2$ nH, $L_e = 2$ nH, $\tau_F = 15$ ps, $f_0 = 900$ MHz, $f_{\max} = 25$ kHz, $a_f = 1$, $k_f = 9 \times 10^{-10}$, and $Q_L = 95$. The oscillator phase noise decreases with the collector current at low I_C and then increases with the collector current at high I_C. The decrease in phase noise with increasing I_C stems from an increase in transistor power. The increase of phase noise at high I_C is due to the transistor saturation effect. Figure 9.40 shows the collector–base voltage versus time for the bipolar transistor biased at $I_C = 6$ mA. It is clear from this plot that the collector–base junction becomes forward biased during some period of time. The collector–base junction becomes more forward biased when the collector current is sufficiently high.

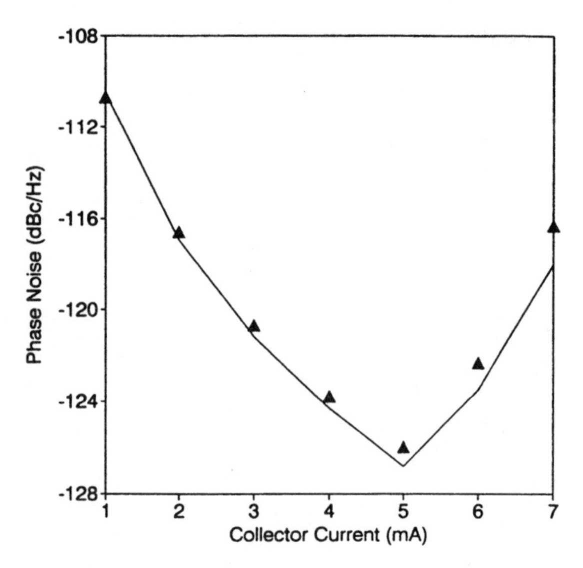

FIGURE 9.39 Phase noise versus collector current (after Yuan, Ref. 37 © Elsevier Science).

A forward-biased collector–base junction induces the hole injection from the base into the collector and increases the base current and flicker noise. This is why the phase noise increases significantly at high collector currents. Phase noise versus forward transit time and parasitic inductance is shown in Fig. 9.41*a* and *b*, respectively. Phase noise increases with forward transit time and parasitic inductance.

From the results above, the pros and cons of the use of silicon bipolar transistors and heterojunction bipolar transistors in an oscillator will now be discussed. Since

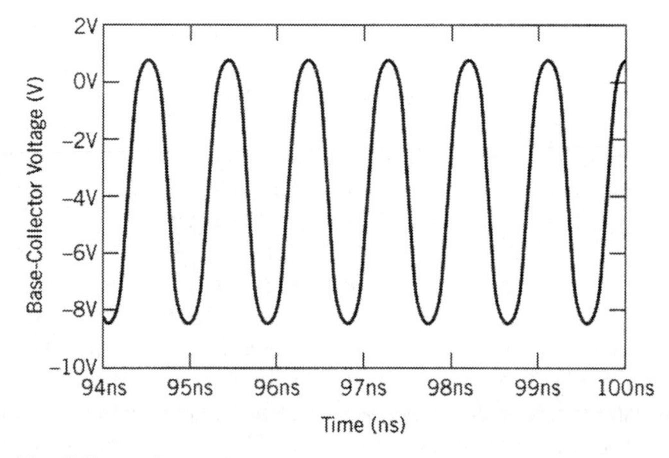

FIGURE 9.40 Collector–base voltage versus time (after Yuan, Ref. 37 © Elsevier Science).

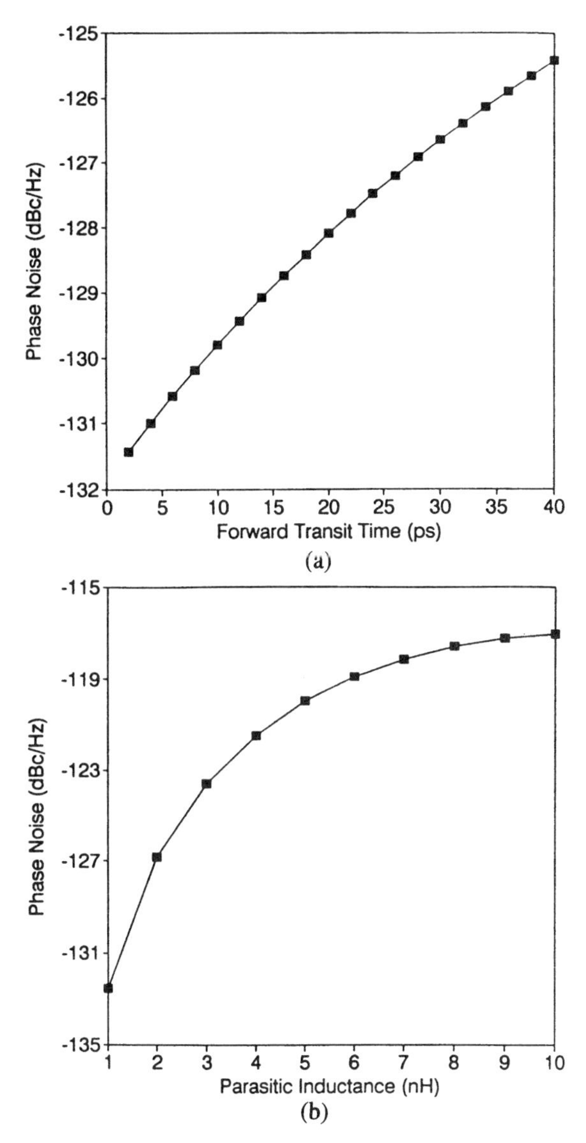

FIGURE 9.41 Phase noise versus (*a*) forward transit time and (*b*) parasitic inductance (after Yuan, Ref. 37 © Elsevier Science).

phase noise exhibits $1/f^2$ slope when the offset frequency is below half of the resonator bandwidth ($f_0/2Q_L$) or $1/f^3$ slope when the offset frequency is below $1/f$ noise corner frequency f_c [38], a lower corner frequency of $1/f$ noise will result in lower phase noise. Because the GaAs surface has a higher surface state density than that of silicon semiconductors, silicon bipolar transistors may have advantages over GaAs HBTs for an oscillator designed at relatively low frequencies. The surface state density and recombination of the GaAs semiconductor, however, can be reduced by using surface passivation on top of the extrinsic base of the HBT [39]. Furthermore, InGaAs and InP semiconductors have lower surface state densities and surface recombination velocities [40] and lower corner frequencies. The InGaAs HBT also has a smaller forward transit time, base resistance, and junction capacitance than its silicon counterparts. These features suggest that the InGaAs/InP HBTs will be good candidates for oscillators operated at very high frequencies and low voltages. Use of the double-heterojunction bipolar transistor suppresses the saturation effect, which causes an increase in phase noise of the bipolar transistor biased at high collector current. Phase noise at high collector current of the DHBT can be reduced.

9.6 ANALOG MULTIPLIERS

From dc to RF frequency, Gillbert cell analog multipliers can be used for double-balanced active mixers and up-converters in microwave applications and for highly sensitive detection mixers in coherent optical heterodyne receivers. Compared with conventional diode double-balanced mixers, the active double-balanced mixer has the advantages of low local oscillator (LO) driver power with high linearity, the ability to provide positive conversion gain, and elimination of the need for bulky hybrid balun circuitry.

The circuit configuration on an analog multiplier is shown in Fig. 9.42. The Gilbert cell configuration is selected for its double-balanced implementation, which offers high conversion gain and improved spur performance in a very compact size. The

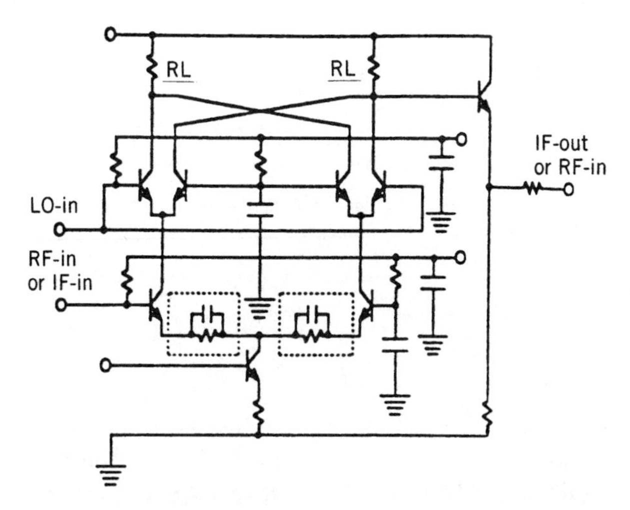

FIGURE 9.42 Circuit configuration of an analog multiplier (after Osafune and Yamauchi, Ref. 41 © IEEE).

RF and intermediate frequency (IF) signal enters a lower amplifier formed by the emitter-coupled pair. The LO signal enters upper cross-coupled devices. Wideband impedance matching is set by 50-Ω resistors in shunt with the high input impedances of the RF or IF and LO devices, and by a 10-Ω resistor in series with the low output impedance of the emitter-follower. Several MIM capacitors are adopted to enable single-ended operation of the RF or IF and LO parts without baluns. To obtain a double-balanced mixer and an up-converter in microwave applications, the series negative-feedback emitter resistors and peaking capacitors enclosed in dotted lines in Fig. 9.42 were removed from the circuit. Using this configuration, a very high conversion gain from dc to ultrahigh-frequency can be obtained by fully exploiting the HBT's high performance [41].

Figure 9.43*a* shows the conversion gain of the mixer versus the IF frequency. The LO input frequency is at 20 GHz, the input power is at 0 dBm, and $R_L = 400\ \Omega$. The 3-dB gain-reduction bandwidth is 3.5 GHz. The RF/LO input return losses and the IF output return loss are kept below 10 dB from dc to above 20 GHz. The circuit achieves an IF output power of −1.3 dBm at 1-dB gain compression. Typical spurious output product suppression ratios are about 30 dB for the zero-order LO and the first-order RF, and about 40 dB for the second-order LO and the second-order RF. SSB noise figures for LO inputs from 2 to 20 GHz were measured to be from 16.5 to 17.8 dB. The circuit operates at 9-V supply voltage with a power dissipation of 90 mW. Figure 9.43*b* shows the conversion gain versus RF frequency. The RF frequency is the LO frequency plus the IF frequency of 1 GHz. The conversion gains are 5 dB for 5-GHz RF output, and 0 dB for 8.5-GHz RF output with 23-dB LO-RF isolation at 0 dBm LO input power. The 3-dB gain-reduction IF bandwidth is 2.3 GHz. The measured conversion gain of the mixer versus LO frequency under a −7.5-dBm LO input condition is shown in Fig. 9.43*c*. The RF frequency is the LO frequency plus the IF frequency of 0.5 GHz. The conversion loss is a very small 1 dB and the 3-dB gain-reduction bandwidth is 15 GHz.

9.7 A/D CONVERTERS

Analog-to-digital conversion is critical in digital signal processing and digital control of analog functions. GaAs HBT technological development was motivated to a large degree by these A/D conversion applications. In addition, the trend toward wider real-time bandwidth in oscilloscopes and transient waveform recorders creates an immediate niche market for III/V HBT ADCs. The function of an ADC is to convert a continuous-time analog-input signal into a series of discrete values (quantization) at discrete time intervals (sampling). A typical wideband n-bit ADC can be divided into three primary components: sample-and-hold (S/H), quantizer, and wideband amplifier. The S/H circuit provides nonslewing samples uniformly spaced in time. The quantizer quantizes the held voltage into 1 of 2^n binary codes. The wideband amplifier amplifies the analog input signal to a level compatible with the full-scale voltage range of the quantizer. In addition, an integrated wideband amplifier enables the ADC to be driven directly with low-level signals from the front-end RF circuits for reduced overall system power dissipation.

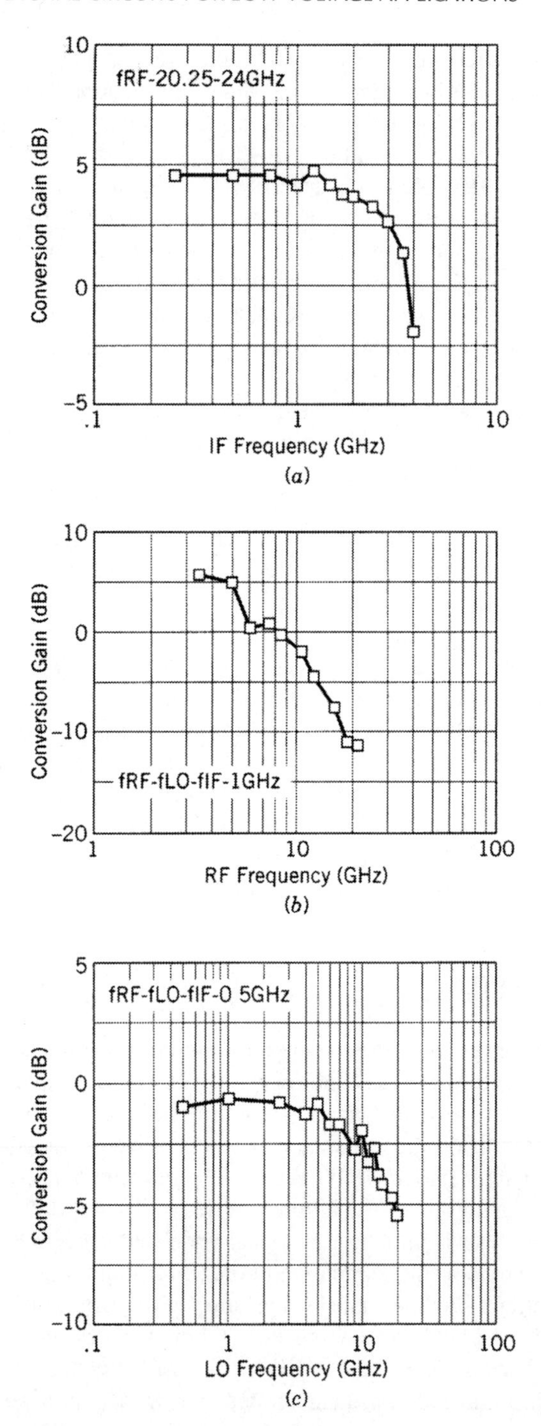

FIGURE 9.43 Conversion gain versus (*a*) IF frequency, (*b*) RF frequency, and (*c*) LO frequency (after Osafune and Yamauchi, Ref. 41 © IEEE).

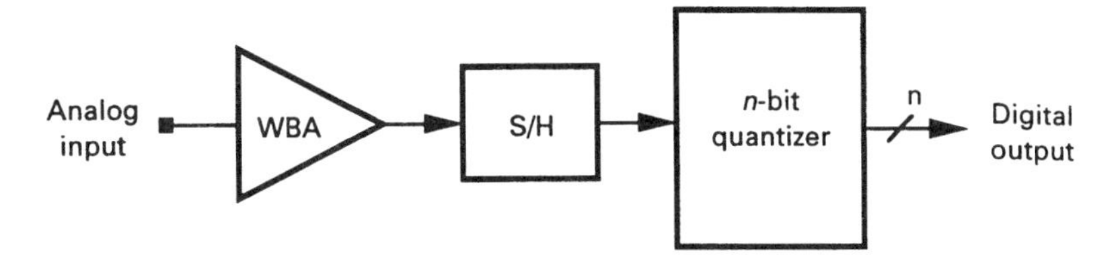

FIGURE 9.44 Full-parallel (flash) ADC architecture.

The fastest ADC architecture employs the full-parallel or flash approach, described in Fig. 9.44. This ADC is often used without S/H, letting the comparator strobing provide the sampling function. The quantizer uses $2^n - 1$ comparators and is characterized by a relatively high input capacitance. Its accuracy is limited by the offset and hysteresis of the comparators, thus favoring the bipolar device over FETs. For higher-resolution ADCs or higher analog input frequencies, the comparator used as a sampler limits performance. Adding an S/H in front of the quantizer can be difficult due to the high input capacitance of the quantizer. The flash ADC is an excellent candidate in the range 2 to 6 bits and for some applications it may be useful up to 8 bits. A four-chip GaAs HBT 8-bit ADC module is shown in Fig. 9.45.

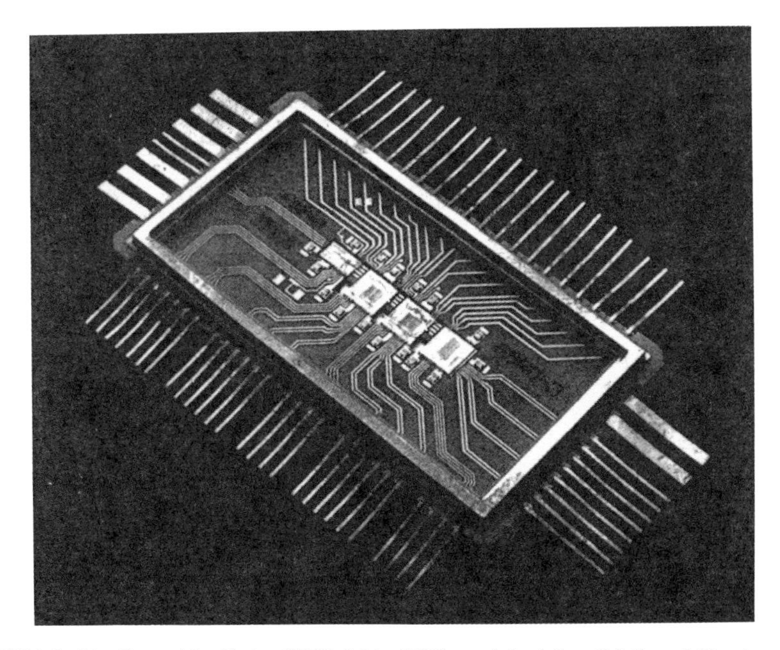

FIGURE 9.45 Four-chip GaAs HBT 8-bit ADC module (after Jalali and Pearton, Ref. 1© Artech House).

The voltage comparator, which has the function of outputting a digital decision based on the value of two analog inputs, is the heart of every quantizer design. The maximum sample rate of the quantizer is not the same as the digital switching rate; it depends on the ability of the comparator to operate with an input overdrive that is a small fraction of the least significant bit. High-performance comparator operation must have transistors with excellent matching, low parasitic capacitances for high speed, and high gain for good sensitivity.

One important parameter that affects the accuracy of a voltage comparator is the hysteresis. For a given reference voltage for a comparator, the hysteresis is the difference in the input voltage thresholds that lead to output transitions from logic low to high and from high to low. There are many possible causes of hysteresis, including the effects of carrier traps, parasitic feedback of output to input, and temperature differences of the transistors at the inputs.

The circuit and timing diagrams of the HBT voltage comparators shown in Fig. 9.46 have been studied by Wang et al. [42]. The circuits consist of two clocked current-mode logic bilevel latches in a master–slave configuration implemented with AlGaAs/GaAs HBTs. The transistors were grown by MBE with emitter dimensions of 1.6 μm $\times$ 4 μm. A schematic diagram of the test setup is shown in Fig. 9.47. The comparator input V_{in} was switched periodically between a soak voltage and a ground. V_{in} was held at +0.5 V or −0.5 V for 95% of the time and at 0 V for the remaining 5% of the time. The reference voltage V_{ref} was an adjustable dc voltage. The clock frequency that governed the comparator sampling rate was varied from 1 to 50 MHz. Supply voltages were $V_{CC} = 1$ V and $V_{EE} = -4.5$ V. The total emitter current including both master and slave latches was biased at 2 mA.

The timing diagram of the positive soaking test is shown in Fig. 9.48. Threshold voltage V_{th} values were obtained at a series of clock frequencies at the second sampling pulse and at the fifth sampling pulse after both the positive and negative soak periods. The unsoak time is defined as the time delay from the time when soak was turned off to the time when the balanced V_{th} was measured. Due to the uncertainty in the timing of the unsoak pulse relative to the clock, the unsoak time for measurements during the second pulse was estimated to be 1.5 times the clock period, and for those during the fifth pulse, 4.5 times the clock period. Absolute values of the corrected V_{th} were plotted as a function of the unsoak time, as shown in Fig. 9.49. The sign of V_{th} was positive for positive soak voltages and negative for negative soak voltages. The sum of the absolute values of V_{th} for positive and negative soak voltages is termed the *dynamic hysteresis voltage* of the comparator. The results in Fig. 9.49 are consistent with the effect of heating due to current flowing through input transistors Q_1 and Q_2 and through latching transistors Q_3 and Q_4.

When the input voltage was switched rapidly after prolonged soaking at positive and negative voltages, heating effects related to the temperature differences of the input and latching transistors were identified as the major source of comparator inaccuracy [42]. No contribution due to carrier trapping effects was found. To reduce the thermal hysteresis, the input circuits need to be operated with minimum power dissipation. Substrates with good heat conductivity are preferred.

The InP-based HBTs offer several advantages over GaAs HBTs for A/D converters:

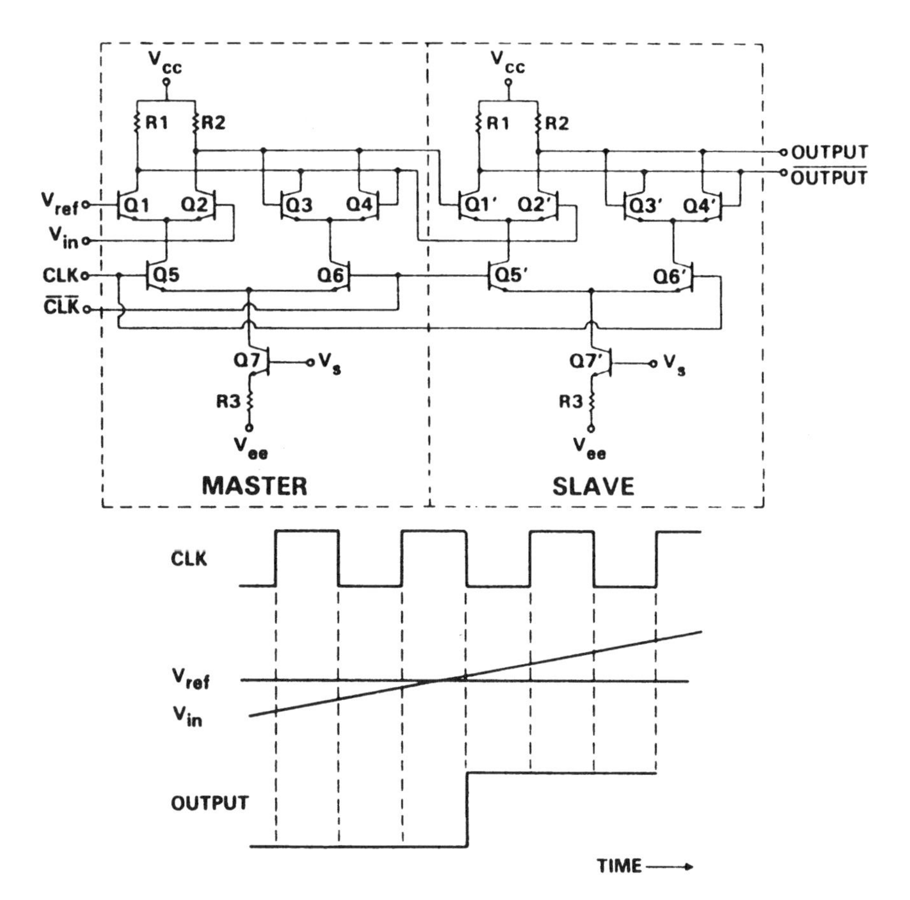

FIGURE 9.46 HBT voltage comparator (after Wang et al., Ref. 42 © IEEE).

- Lower turn-on voltage V_{BE} due to narrow-bandgap base material, resulting in 30 to 50% lower supply voltages and corresponding reductions in power dissipation.
- Higher circuit speeds due to the higher intrinsic mobility and reduced forward transit time.
- Less device self-heating, due to the higher thermal conductivity, allowing greater circuit design and layout flexibility.
- Lower-power and higher-density circuits, due to lower surface recombination velocity and therefore extended smaller emitter areas.

The primary disadvantage of the InP HBT compared to the GaAs HBT is its lower breakdown voltage. By using a wide-bandgap collector and/or bandgap engineering at the collector–base junction, however, adequate to high breakdown voltages can be obtained.

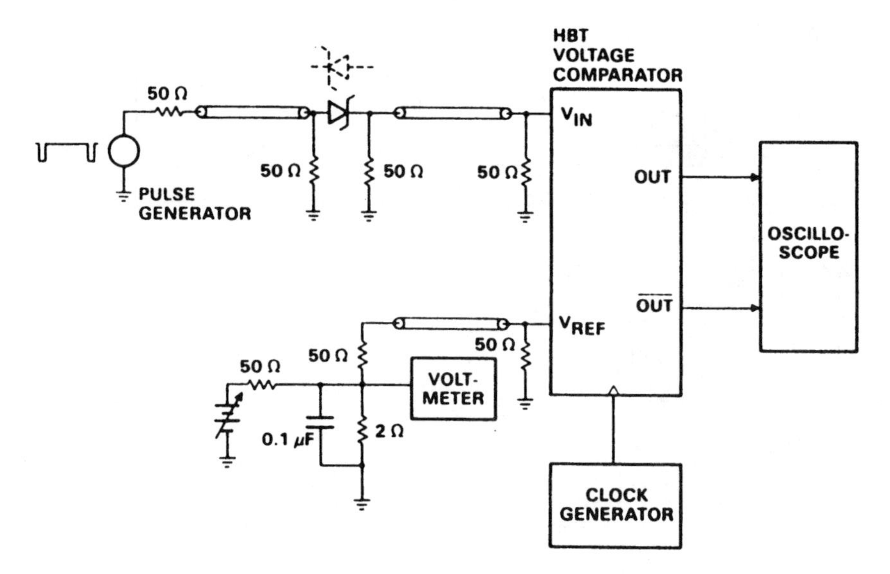

FIGURE 9.47 Schematic diagram of the test setup for the HBT voltage comparators (after Wang et al., Ref. 42 © IEEE).

9.8 DIODE-HBT LOGIC WITH ECL/CML CIRCUITS

In circuit applications, the highest-speed performance is achieved by using nonsaturating transistor operation. For digital and A/D applications, emitter-coupled logic (ECL) and its variant, current-mode logic (CML), provide the highest-speed performance. This is opposed to the slower-saturating integrated-injection logic (I^2L). In the ECL type of logic, differential pair transistors steer current from a current source through load resistors based on the logic state of the inputs. In CML, the emitter-follower output drive transistor of ECL is eliminated; the differential pair transistors serve as both the logic switch and the output driver. Many designers argue that ECL is faster than CML because the emitter-follower isolates the load from the switching time constant. For an SSI family in the transistor-limited speed range, this may be true. For LSI and VLSI circuits operated at current levels that produce RC time-constant-limited gate speeds, CML without emitter-follower buffers produce lower speed–power products. A CML gate consumes 0.7 the power of an ECL gate at the same speed at RC time-constant-limited current levels. A gate driving large fan-out or high capacitive loads will still use the emitter-follower buffer. Because of the very short transistor transit times, InP-based HBT CML circuits are usually RC time-constant limited even at high current densities. ECL and CML typically have high power dissipation due to the use of constant-current-source and vertical logic structure. However, power dissipation can be traded for speed. Because of the HBT's high intrinsic speed, low power–delay products can be achieved. Furthermore, the

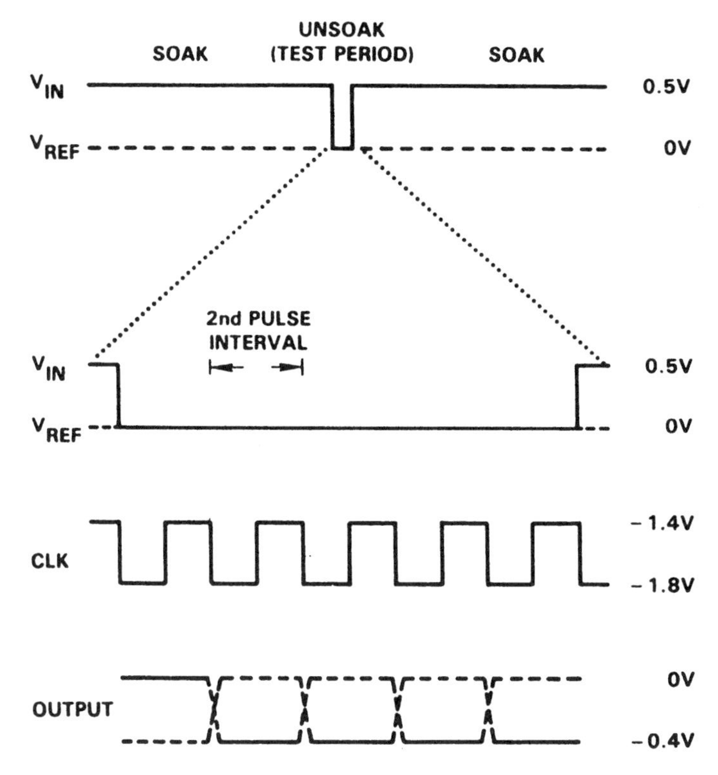

FIGURE 9.48 Timing diagram of the positive soaking test (after Wang et al., Ref. 42 © IEEE).

HBT's high-performance switching and driving capabilities make the simpler CML approach attractive for many circuit applications.

A novel logic approach, diode-HBT logic, which is implemented with AlGaAs/GaAs HBTs and Schottky diodes to provide high-density, low-power digital circuit operation was proposed [43]. The schematic diagram of a basic diode-HBT-logic gate, configured as an AND–OR–INVERT gate is shown in Fig. 9.50. A single transistor is used per bilevel gate while switching is performed with Schottky diodes. The HBT is used in saturation; Schottky clamping has been demonstrated with HBTs to reduce the associated charge storage. A comparison of different HBT diode-transistor logic gates is given in Fig. 9.51. These gates include heterojunction-diode-transistor logic (HDTL), heterojunction-Schottky logic (HSTL), heterojunction-integrated-injection logic (HI^2L), Schottky-transistor logic (STL), and diode-HBT logic (DHL). Approaches based on diode logic are more conveniently implemented in GaAs than in Si, because of the simplicity and quality of GaAs Schottky diodes. The HI^2L operates with the HBTs in deep saturation, so the speed is limited. The STL used a clamping diode to avoid the deep saturation, but it

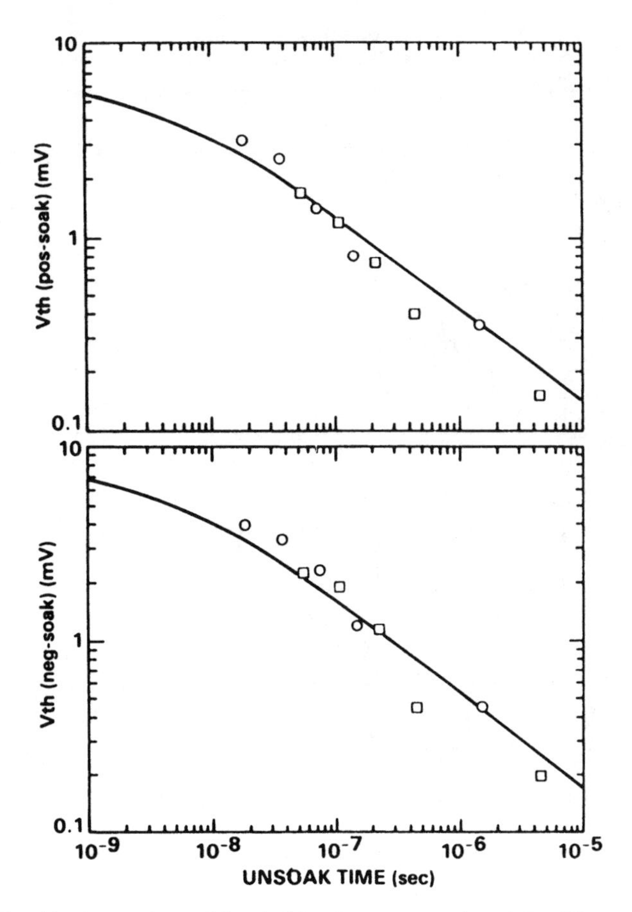

FIGURE 9.49 Absolute values of threshold voltage versus time (after Wang et al., Ref. 42 © IEEE).

requires that the switching diodes and clamping diode have different Schottky barrier heights. Also, the voltage swing of the logic is limited to the difference of the Schottky barrier heights. The DHL uses level-shifting diodes to avoid the deep saturation and the need for using diodes of different types. The logic OR function can be implemented easily with the level-shifting diodes. Therefore, this approach provides wired AND, diode-implemented OR, and INVERT. This combination is particularly worthwhile for efficient implementation of latches that require two gates of this type.

The DHL gate, ring oscillators, and frequency dividers were fabricated with a manufacturable baseline HBT process at Rockwell International. The diodes were realized using the same epitaxial structure and process as the HBTs. A ring oscillator of 19 stages of a bilevel DHL gate was designed as shown in Fig. 9.52. The propagation delay per gate measured was 160 ps. The associated power–delay product was 180 fJ.

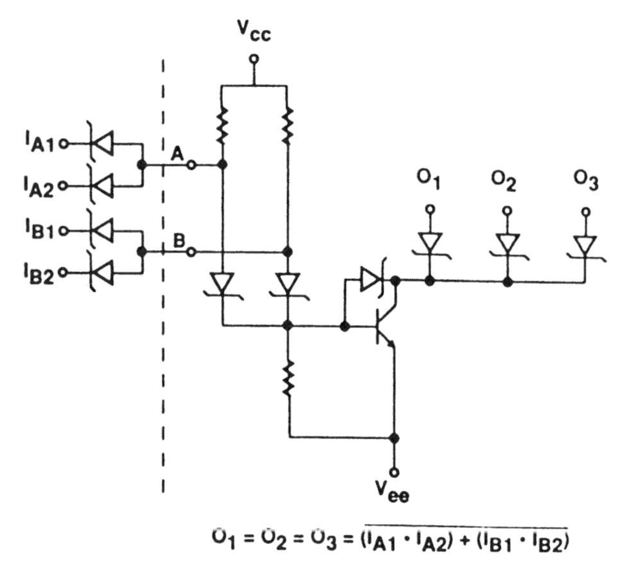

FIGURE 9.50 Diode-HBT logic (after Wang et al., Ref. 43 © IEEE).

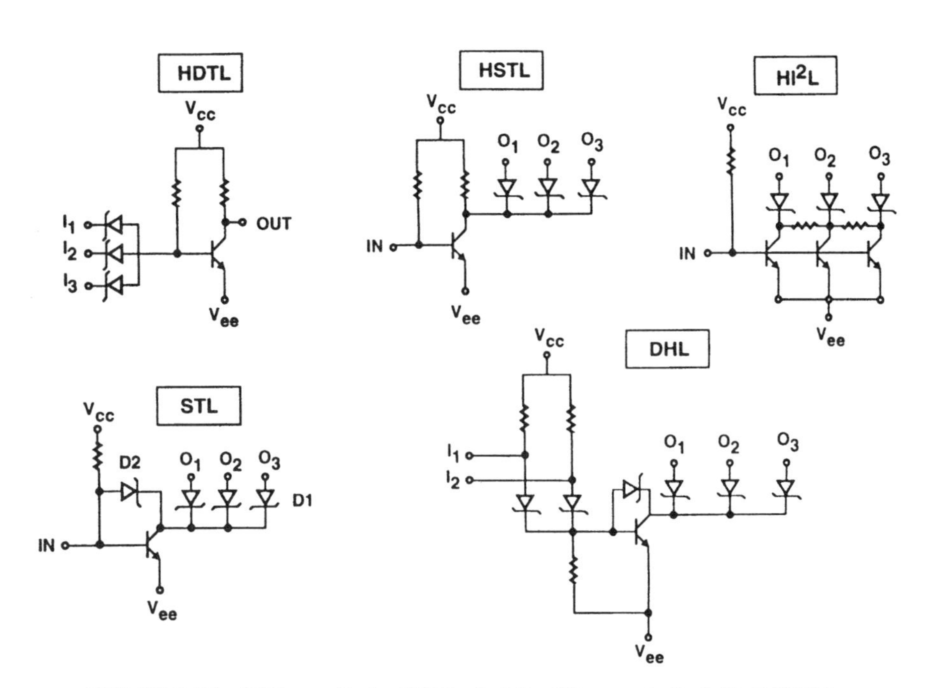

FIGURE 9.51 Different diode-HBT logic (after Wang et al., Ref. 43 © IEEE).

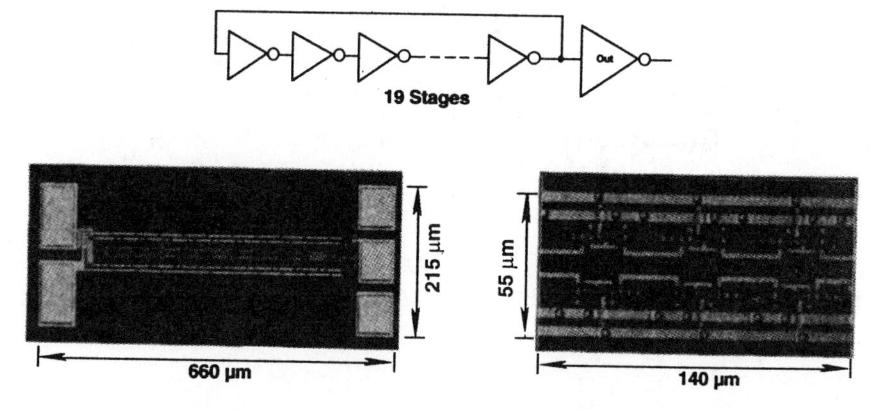

FIGURE 9.52 Fabricated DHL ring oscillator.

Measured and simulated gate delay versus the power of the DHL ring oscillator is illustrated in Fig. 9.53.

To further benchmark the speed, density, and power dissipation of DHL versus alternative logic families, a DHL static frequency divider was designed. The divide-by-4 circuit was based on cascaded master/slave flip-flop divide-by-2 stages, together with an input receiver, internal buffers, and an output driver, as shown in Fig. 9.54. The divider operated well with input frequencies up to 6 GHz. Power per divide-by-2 stage was 8 mW at 2 GHz, 10 mW at 4 GHz, and 15 mW at 6 GHz. Operation of the divider at 4 GHz is shown in Fig. 9.55.

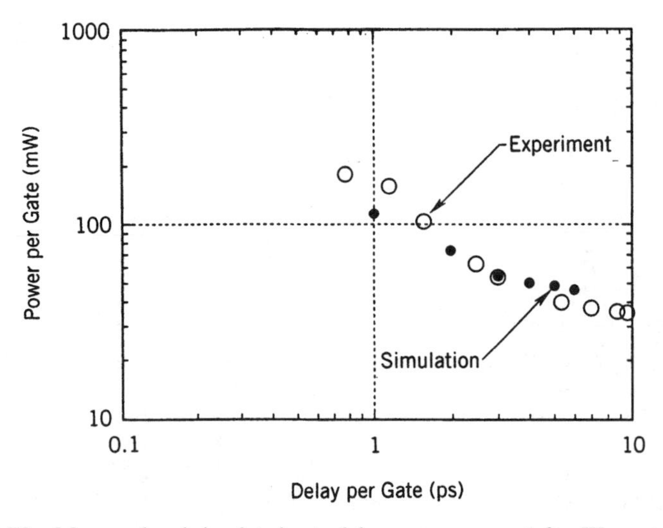

FIGURE 9.53 Measured and simulated gate delay versus power (after Wang et al., Ref. 43 © IEEE).

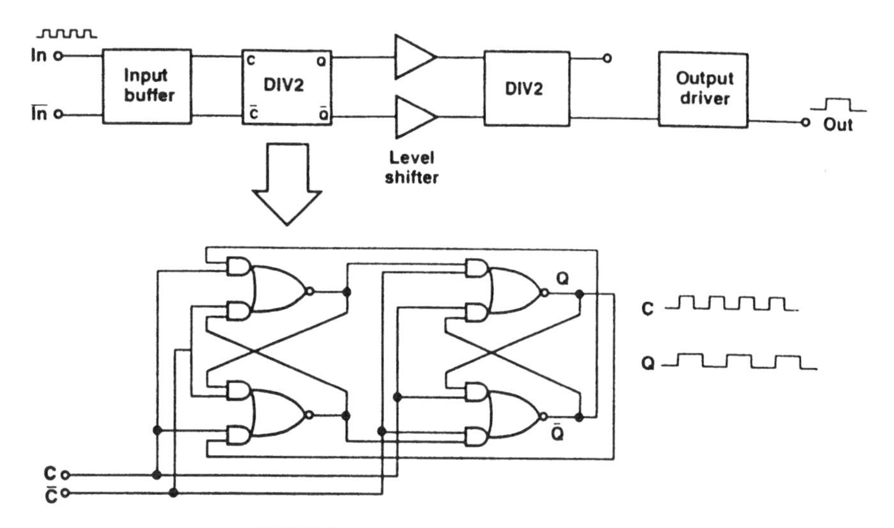

FIGURE 9.54 DHL frequency divider.

The monolithic integration of AlGaAs/GaAs N-p-n and P-n-p HBTs has been demonstrated [44]. A schematic of the complementary HBT I^2L gate consisting of an N-p-n switching transistor with Schottky clamp and P-n-p current injector with Schottky-level shifter is shown in Fig. 9.56. Its scanning electron micrograph is displayed in Fig. 9.57. A 65-ps delay per gate with power dissipation of 13 mW per gate for a 850-fJ power–delay product was obtained.

9.8.1 Gate Delay Versus Power

Minimization of propagation delay at an acceptable power level is of prime concern in the design of high-speed bipolar logic circuits. If the transistors are not driven into saturation (such as with CML and ECL), the propagation delay is determined by device

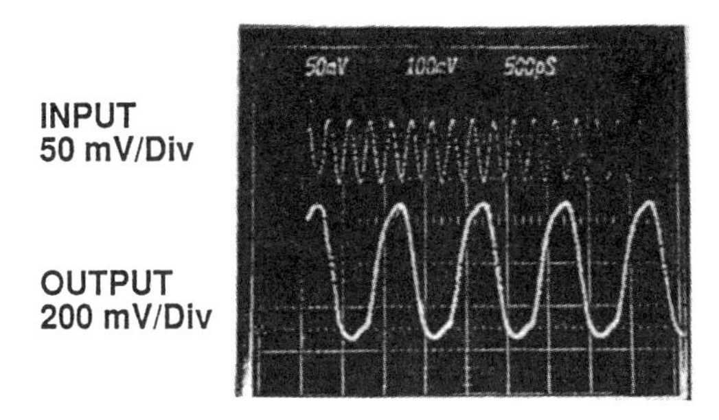

FIGURE 9.55 Measured input and output waveforms of a DHL frequency divider operated at 4 GHz (after Wang et al., Ref. 43 © IEEE).

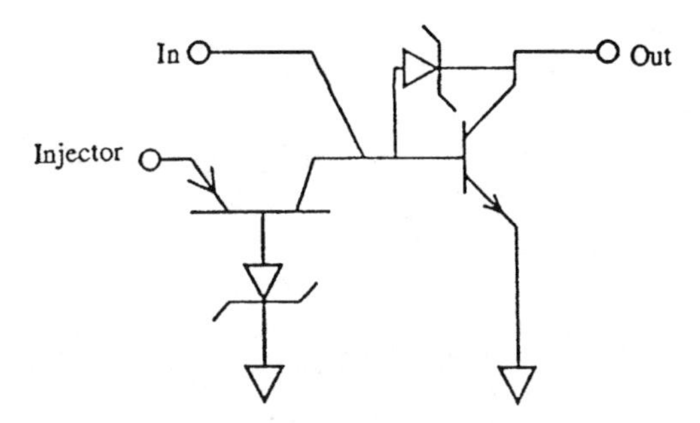

FIGURE 9.56 Complementary HBT I^2L logic (after Enquist et al., Ref. 44 © IEEE).

transit time and the rise and fall times associated with charging and discharging various device and circuit capacitances. The propagation delay per gate varies inversely with the power per gate over a substantial range of operating current choices (constant power–delay product). For VLSI circuits, the maximum power that can be dissipated in the chip package or that can be accommodated in the system power budget frequently limits the power per gate, and therefore the delay per gate. Consideration of the gate power–delay product is therefore of fundamental importance. Unlike CMOS logic, the power dissipated in ECL is virtually independent of the rate of switching of the gate. The gate delay decreases with the collector current at low to moderate I_C and then increases with the collector current at high I_C. Optimizing the transistor current to obtain the minimum gate delay is important in digital circuit design.

FIGURE 9.57 Scanning electron micrograph of the I^2L logic in Fig. 9.56 (after Enquist et al., Ref. 44 © IEEE).

9.8.2 Figure of Merit for CML

The figures of merit f_T and f_{max} for RF circuits do not correspond closely to the maximum possible frequency of operation of logic circuits and are not necessarily the best guides for transistor design of digital circuits. Simple expressions to quantify the behavior of transistors in logic circuits are needed.

The charging and discharging effects are important for CML digital switching. At low collector current the delay is dominated by device and circuit capacitance. As the gate current is increased, these capacitances are charged and discharged at a faster rate, leading to a reduction in propagation delay. At high current the combined effects of device transit time and series resistance lead to an increase in delay with current. Since the logic swing and gate current are related through the collector load resistance, the minimum delay point is related to an optimum load resistance as well. The cutoff frequency is not an appropriate figure of merit for digital circuits because it does not include the effects of base resistance and collector load resistance. The maximum frequency of oscillation includes base resistance, but it does not include the effect of collector load resistance.

A simple CML inverter is shown in Fig. 9.58. It is a differential pair of transistors with equal collector-load resistors, a current source, and differential inputs. Routing all high-speed signals differentially to prevent unwanted coupling, to increase noise margin, and to eliminate any common-mode signal distortion is recommended. For single-end operation, a reference voltage centered in the middle of the logic swing replaces the complementary input. An analytical expression for the delay of the CML gate has been derived by Greenwich [45]. In the analysis the following assumptions are made: (1) the base–emitter and base–collector junction capacitances are represented by effective average value capacitances, independent of bias; (2) the base transit time is a constant, independent of collector current; and (3) base resistance is a constant.

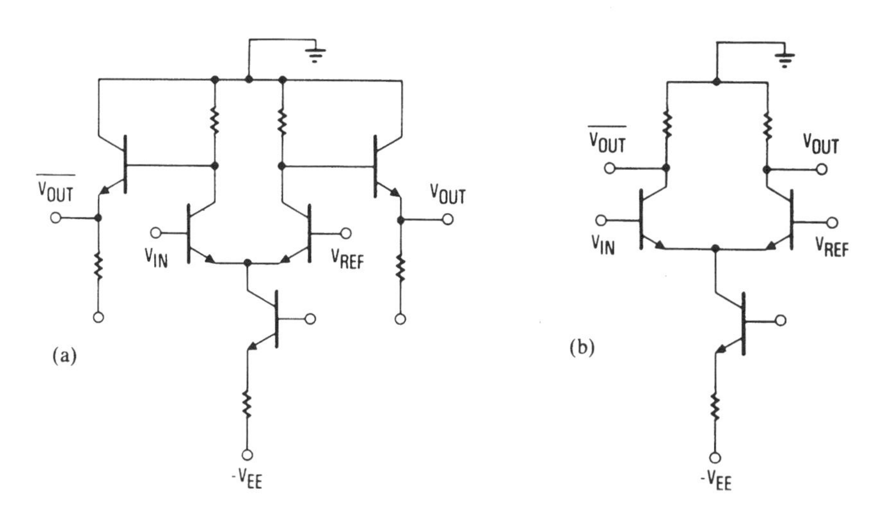

FIGURE 9.58 (*a*) ECL and (*b*) CML inverter.

The propagation delay of the CML gate shown in Fig. 9.59*a* is taken to be the average of the 50% rise and fall response of the gate to a step input voltage. The delay is estimated by calculating the large-signal 3-dB cutoff frequency of the differential mode half-circuit in Fig. 9.59*b*. The circuit elements are linearized by calculating their effective average values. If there are no dominant zeros and there is a single dominate pole, the −3-dB frequency is approximately

$$\omega_{-3\text{dB}} = \frac{1}{\sum_{i=1}^{n} \left(-1/p_i\right)} \tag{9.45}$$

where p_i are the poles of the circuit.

The sum of the zero-value time constants is equal to the sum of the reciprocal poles and thus may be used directly to calculate the −3-dB frequency. In the analysis of frequency response using zero-value time constants, all capacitances in the circuit are set to zero and the resulting resistance seen across each capacitor is calculated. The value of resistance times the capacitance is then the zero-value time constant for that capacitor. For the equivalent circuit of Fig. 9.59*b*, the time constants are

$$T_0 = \left(R_B \| r_\pi\right) C_\pi + \left[\left(R_B \| r_\pi\right)\left(1 + g_m R_L\right) + R_L\right] C_\mu \tag{9.46}$$

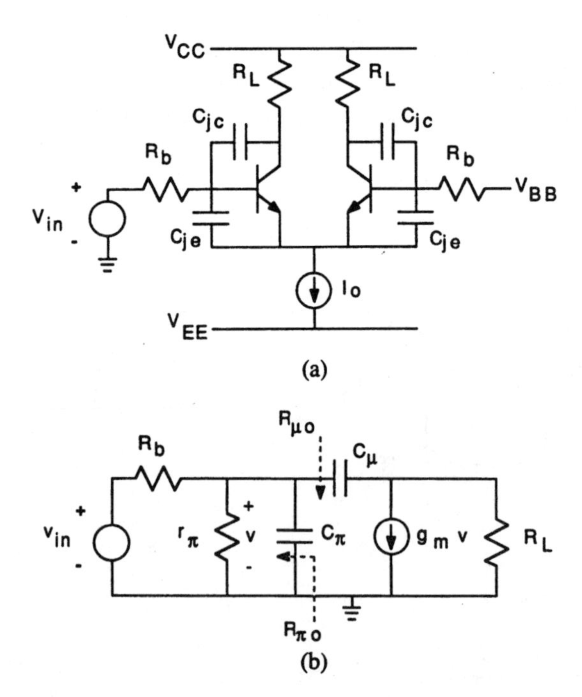

FIGURE 9.59 (*a*) CML gate with base resistance and junction capacitances and (*b*) differential mode half-circuit.

Analysis of the circuit in Fig. 9.59*b* gives two poles:

$$p_1 = -\frac{1}{T_0}$$

$$p_2 = -\left[\frac{1}{R_L C_\mu} + \frac{1}{(R_B||r_\mu)C_\pi} + \frac{1}{R_L C_\pi} + \frac{g_m}{C_\pi}\right]$$

Note that the last term in p_2 is approximately $2\pi f_T$. It is clear that p_2 is not the dominant pole, validating the use of (9.46) to approximate the -3-dB frequency.

Ashar [46] presents a method for estimating the delay of switching circuits based on the response to a delta function input:

$$\lim_{s=0} \frac{-\partial F(s)/\partial s}{F(s)}$$

where $F(s)$ is the transfer function of the circuit. Applying this to a CML gate and eliminating negligible terms yields an expression identical to (9.46).

The effective average values of the time-constant parameters are calculated:

$$\begin{aligned}\overline{g_m} &= \frac{qI_S}{kT\ \Delta V_{BE}} \int_{v_L/2}^{v_L} e^{qV_{BE}/kT}\, dV_{BE} \\ &= \frac{2I_o}{v_L} = \frac{2}{R_L}\end{aligned} \tag{9.47}$$

where v_L is the logic swing. Note that for the CML gate, $\Delta V_{BE} = v_L/2$, due to the constant reference potential applied to one side of the gate.

$$\overline{C_D} = \overline{g_m \tau_F} = \frac{2\tau_F}{R_L} \tag{9.48}$$

$$\overline{r_\pi} = \frac{\beta_0}{\overline{g_m}} = \frac{\beta_0 R_L}{2} \tag{9.49}$$

Since r_π is typically much larger than R_B, $R_B||r_\pi \approx R_B$. From (9.46) the effective time constant sum is

$$\overline{T_0} = R_B(C_{JE} + 3C_{JC}) + \frac{2R_B\tau_F}{R_L} + R_L C_{JC} \tag{9.50}$$

The average 50% delay is the $\overline{\tau_d} = \overline{T_0}\ \ln\ 2$.

The calculated gate delay versus load resistance is plotted in Fig. 9.60. The SPICE simulation results are compared with the prediction using equation (9.50). The agreement between the theoretical calculations and SPICE predictions is very good. The

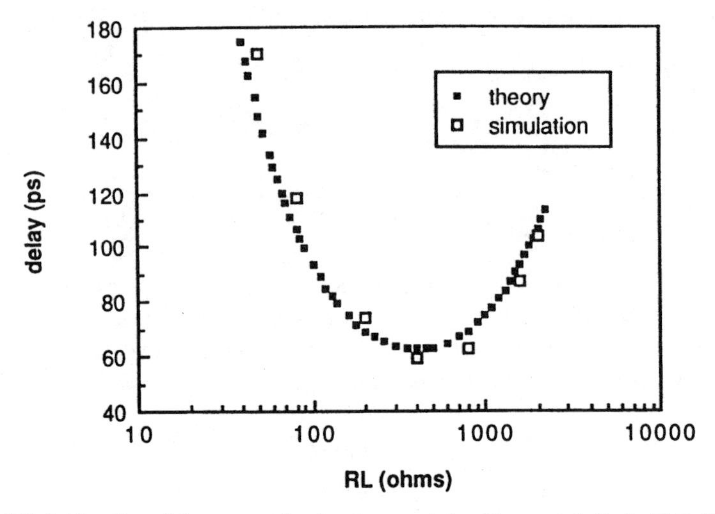

FIGURE 9.60 Gate delay versus load resistance (after Greenwich, Ref. 45 © IEEE).

minimum delay as determined by (9.50) is given at an optimum load resistance of

$$R_L(\text{opt}) = \sqrt{\frac{2R_B\tau_F}{C_{JC}}} \tag{9.51}$$

Inserting R_L(opt) into (9.50) gives a minimum delay:

$$T_0(\text{opt}) = R_B(C_{JE} + 3C_{JC}) + 2\sqrt{2R_B\tau_F C_{JC}} \tag{9.52}$$

The optimum delay using (9.52), normalized to $R_B C_{JC}$, as a function of $\tau_F/(R_B C_{JC})$ is displayed in Fig. 9.61. The open circles in Fig. 9.61 represent SPICE simulation and the solid lines represent the theoretical predictions using (9.52). The agreement between the model predictions and the SPICE simulation is very good. The parameters used for this simulation are $I_o = 400\ \mu$A to 16 mA, $R_L = 50$ to 2000 Ω, $\tau_F = 2$ to 20 ps, $R_B = 50$ to 1000 Ω, and $C_{JE,C} = 40$ to 400 fF. Using the optimum delay as a figure of merit shows explicitly the relative effect of various device parameters on gate delay. From this, the advantage of using heterojunction bipolar transistors in CML gates that have smaller base resistance and forward transit time than those of Si bipolar transistors is clear.

Figure 9.62 shows the gate delay versus power dissipation per gate in a variety of HBT technologies. The scaled AlInAs/GaInAs HBT obviously has the best performance of gate delay over a wide range of power dissipation [1]. At a given low power dissipation, the AlInAs/GaInAs HBT achieves the lowest gate propagation delay. This demonstrates the superiority of scaled InP-based HBTs in low-power applications.

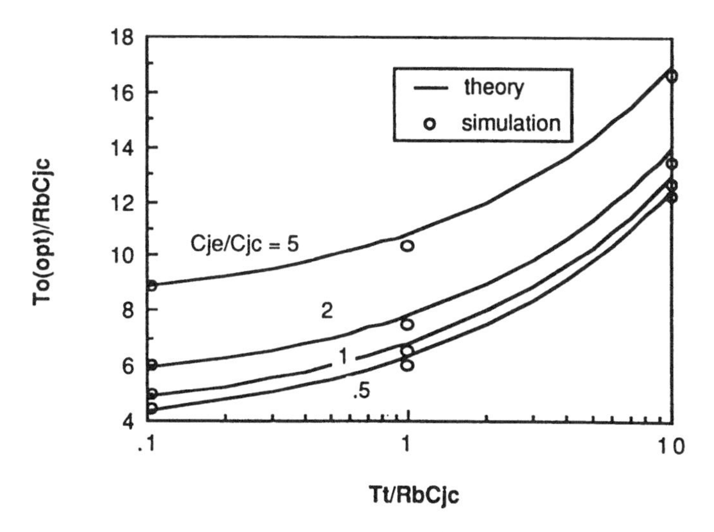

FIGURE 9.61 Optimum gate delay normalized to $R_B C_{JC}$ (after Greewich, Ref. 45 © IEEE).

9.8.3 Figure of Merit for ECL

The logic circuit gate delay can be computed either with numerical methods or analytically. The analytical delay equation is basically obtained from the differential equations that describe the node voltage or branch current of the circuit. The gate delay can be expressed analytically as the sum of the *RC* time constants, each weighted by a constant determined by the circuit topology. The resistance includes load resistors and the resistance in the transistor equivalent circuit, and the capacitance includes the

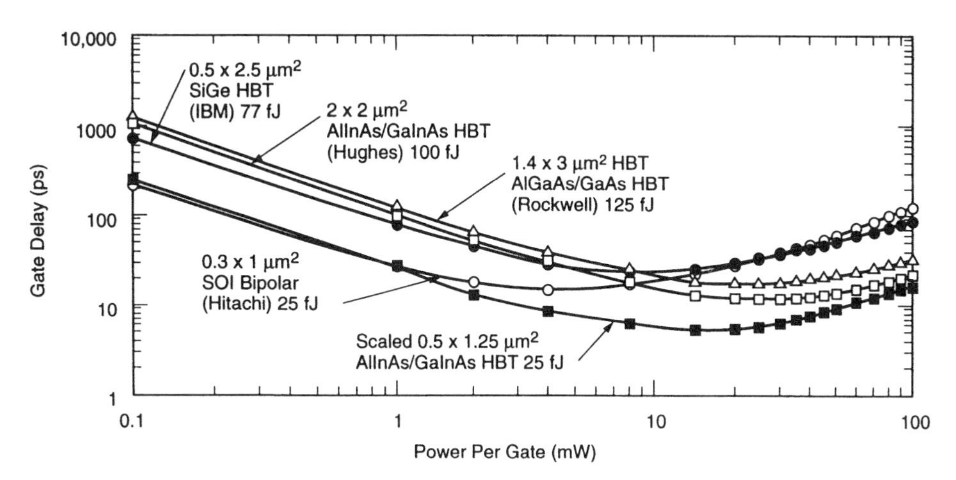

FIGURE 9.62 Speed relationship for CML gate for different state-of-the-art HBT technologies (after Jalali and Pearton, Ref. 1© Artech House).

load capacitors and the capacitors in the transistor equivalent circuits. The analytical equation derived usually has a limited accuracy. The delay equation derived from the sensitivity of the gate delay to the device parameters is more accurate over a wide range of parameters. The delay equation of the ECL gate obtained from the delay sensitivity of the transistor parameters is given as [47]

$$\tau_d = k_1 R_1 C_{JC} + k_2 R_1 C_{JE} + k_3 R_1 C_{JS} + k_4 R_1 C_1 + k_5 \tau_F + k_6 R_{BI} C_{JC} + k_7 R_{BI} C_d + k_8 R_{BX} C_d \tag{9.53}$$

where $C_d = 2I_C/V_1$, V_1 is the logic swing of the ECL gate, and $R_1 = V_1/I_C$. The coefficient k's are given as follows: $k_1 = 2.5 + 0.413 \times$ fan-out, $k_2 = 0.24 \times$ fan-out, $k_3 = 1.10$, $k_4 = 0.233$, $k_5 = 2.2 + 0.39 \times$ fan-out, $k_6 = 1.27$, $k_7 = 5.19$, and $k_8 = 2.21$. In (9.53) all units of capacitance are in femtofarads, time is in picoseconds, resistance is in kilohms, voltage is in volts, and current is in milliamperes. A more complicated delay equation suitable for heterojunction bipolar transistors was also derived [48].

The asymptotic line of each term in the delay equation in (9.53) is shown as a function of the collector current in Fig. 9.63. The terms related to R_1 decrease as current increases, since R_1 decreases as the current increases. This term also decreases as the area of the transistor decreases as a result of the reduction of the lithographic line width. The terms $R_B C_d$ increase with current. These terms also decrease as the base width of the transistor decreases. At low injection, the terms $R_{BI} C_{JC}$ and τ_F vary very slowly with current. At high injection τ_F increases and R_B decreases. The $R_{BI} C_{JC}$ term rapidly becomes insignificant as the transistor size shrinks. Under most circumstances, the $R_1 C_{JC}$, τ_F and $R_{BI} C_d$ terms in (9.53) are dominant. Equation (9.53) predicts an optimum operating point either when $k_1 R_1 C_{JC} = k_7 R_{BI} C_{JC}$ for structures with large $R_{BI} C_d$ or when $k_1 R_1 C_{JC} = k_5 \tau_F$ otherwise. The first case gives

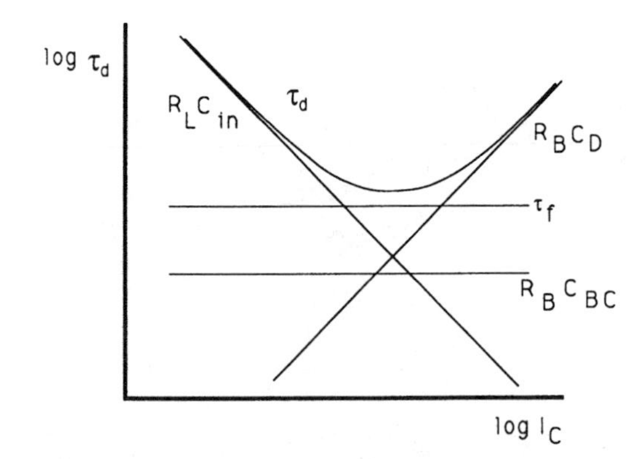

FIGURE 9.63 Different delay components versus collector current (after Tang and Soloman, Ref. 47 © IEEE).

a sharp minimum delay versus current, while the second gives a flat-bottomed delay versus current.

9.8.4 SiGe Digital Circuit Performance

Power consumption of electronic equipment is an important limitation in many applications. In the future low power consumption will be even more important as technology develops toward higher computation speed, increased complexity, and increased portability.

The design of Si/SiGe heterojunction bipolar transistors for low power consumption has been investigated [49]. The transistor is assumed to be placed in an I^2L-like circuit. Only the first transistor in a chain of inverters was taken into account, as shown in Fig. 9.64. The base current I_B to the transistor is provided by a current generator. By opening the external switch, the current will be forced to go through the base–emitter junction and open the transistor. Another current generator provides the collector current I_C that due to the base current is allowed to flow into the transistor. This second current has exactly the same magnitude as the base current. Under given conditions, such as the sizes and the Ge fraction in the base, it is possible to find the combination of doping values in the emitter, base, and collector of a transistor that gives the shortest total switching time. The total switching time is defined as the sum of the turn-on (or rise) time T_1, the storage time T_2, and the decay (or fall) time T_3. By using Laplace transformation on the Ebers–Moll model [50], the following equations for the switching times are derived:

$$T_1 = \frac{\ln \dfrac{I_B}{I_B - 9(1 - \alpha_N)I_C/10\alpha_N}}{(1 - \alpha_N)\omega_N} \tag{9.54}$$

$$T_2 = \frac{\ln [I_B\alpha_N/(1 - \alpha_N)I_C](\omega_N + \omega_I)}{(1 - \alpha_N\alpha_I)\omega_N\omega_I} \tag{9.55}$$

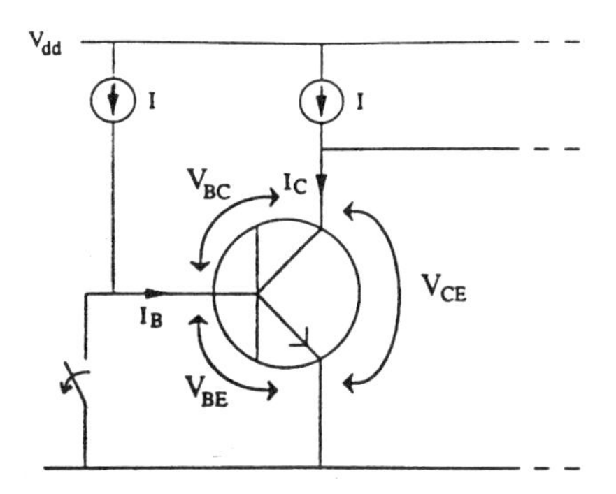

FIGURE 9.64 Switching transistor in the I^2L inverter.

$$T_3 = \frac{\ln 10}{(1 - \alpha_N)\omega_N} \tag{9.56}$$

where α_N and α_R are the common-base current gain in the forward and inverse active modes, respectively, and ω_N and ω_R are the forward and reverse frequencies. $\omega_{N,R} = 1/(\tau_E + \tau_B + \tau_C)$.

The total switching time versus Ge fraction at the currents 0.1, 1.0, and 10 μA is shown in Fig. 9.65. A Ge fraction ranging from 0 to 20% is used. Higher current levels give shorter total switching times. The total switch time for a given current is improved by the introduction of Ge in the base. Depending on the current, the shortest total switch time is found at a Ge fraction of around 12 to 18%.

Using (9.54)–(9.56), Shafi et al. [51] presented a comparison between the predicted propagation delays of ECL circuits composed of Si/SiGe HBTs and silicon BJTs. Important transistor parameters, such as the current gain, base transit time, base resistance, and emitter delay, are calculated for the heterojunction bipolar transistor as a function of Ge concentration in the SiGe base. The propagation delay of SiGe ECL circuits depends strongly on the base width and the base doping concentration of the transistor. The emitter doping concentration and emitter width have been optimized throughout to give a minimum propagation delay for any given base width (the emitter concentration was varied between 5×10^{17} and 10^{20} cm^{-3} and the emitter width between 0.1 and 0.3 μm). A Ge concentration of 12% was chosen for these calculations. This germanium concentration is sufficient to meet the $\beta \geq 50$ criterion for the majority of the base widths and doping concentrations. A minimum gate delay is reached for a given base doping, representing a compromise between high base resistance at narrow base widths and high base transit time at wider base widths.

Figure 9.66 shows a plot of the predicted propagation delay versus collector current density for heterojunction and homojunction circuits. The heterojunction bipolar tran-

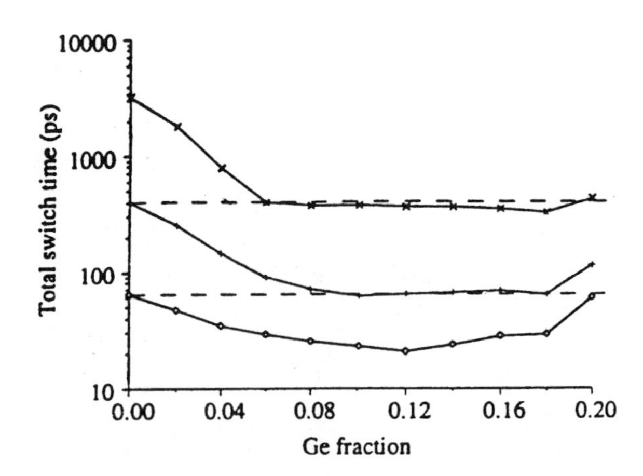

FIGURE 9.65 Total switching time versus Ge content (after Karlsteen and Willander, Ref. 49 © Elsevier Science).

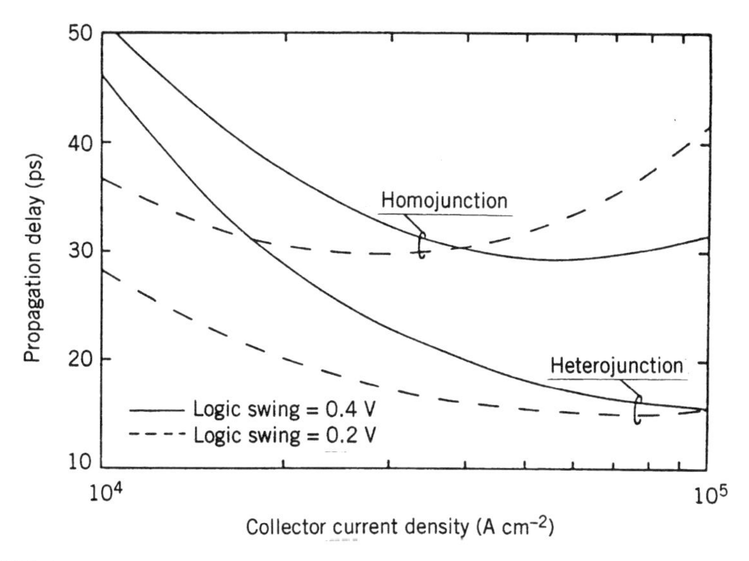

FIGURE 9.66 Predicted propagation delay versus collector current density (after Shafi et al., Ref. 51 © Elsevier Science).

sistors have an emitter doping of 2×10^{18} cm^{-3}, a base doping of 1×10^{20} cm^{-3}, an emitter width of 0.15 μm, and a base width of 0.02 μm. The homojunction transistors have an emitter doping of 1×10^{20} cm^{-3}, a base doping of 5×10^{18} cm^{-3}, an emitter width of 0.15 μm, and a base width of 0.05 μm. For the propagation delay of the homojunction circuit utilizing a 0.4-V logic swing, the delay decreases as the collector current density is increased and then reaches a minimum of 29.2 ps at $J_C = 5.5 \times 10^4$ A/cm^2 and starts increasing beyond this value. A similar trend is shown for the heterojunction propagation delay, but in this case the minimum delay of 15.5 ps occurs at much higher current densities. Similar results are also obtained for a logic swing of 0.2 V.

Average gate delay versus average switch current of an ECL using SiGe bipolar transistors is shown in Fig. 9.67. In this case the SiGe HBT was actually integrated with advanced 0.25-μm CMOS. In addition to reduction of the extrinsic device parasitic through improved layout, the extrinsic base of the device was fully silicided, yielding a device with low parasitic R_B and C_{BC}. A 50-GHz peak f_T and record 60-GHz peak $f_{\max}$ were obtained. ECL ring oscillators switched at a record 18.9 ps at 1.6-mA switch current in the 0.5×2.5 μm^2 SiGe ground-rule transistors, and 20 ps at 0.85 mA for transistors that were 0.5×1.0 μm^2, as shown in Fig. 9.67.

A primary motivation for a SiGe-based HBT technology is the ability to merge the high-performance bipolar device with conventional Si technologies, primarily CMOS technologies. The epitaxial base must be compatible with the existing CMOS tool sets and process technology to replace an ion-implanted base step and take full advantage of the existing silicon technology base. To establish a high performance SiGe-base BiCMOS process without compromising the performance of the n-p-n

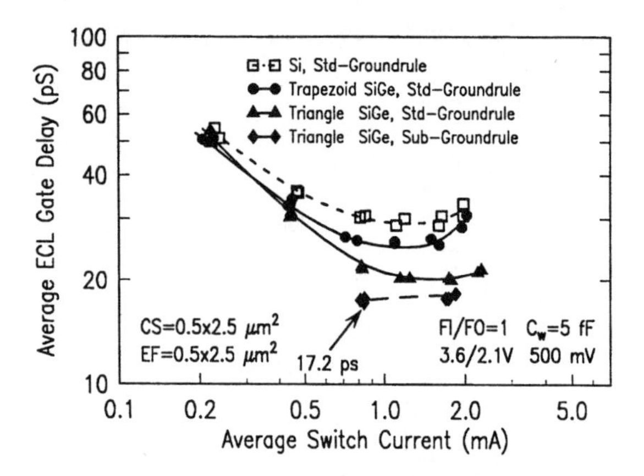

FIGURE 9.67 Delay versus collector current for SiGe ECL (after Harame et al., Ref. 52 © IEEE).

or CMOS technology, three key issues must be addressed: (1) how to ensure yield for the epi-base and CMOS, (2) how to share layers and processes to simplify the integration, and (3) how to combine the bipolar and CMOS without compromising the performance of either one due to incompatible thermal budget requirement.

SiGe-base heterojunction bipolar transistors have been used in a BiCMOS circuit to enhance its speed performance [52]. The influence of the HBT germanium profile in a BiCMOS inverter has been studied using an analytical model [53]. The BiCMOS inverter with an output load of 0.5 pF is displayed in Fig. 9.68. Figure 9.69 shows the rise time of the BiCMOS inverter versus the germanium concentration for a triangular germanium profile. The three peak germanium locations (0.8 W_B, 0.9 W_B, W_B) are close to the base-to-collector junction. As the germanium concentration is

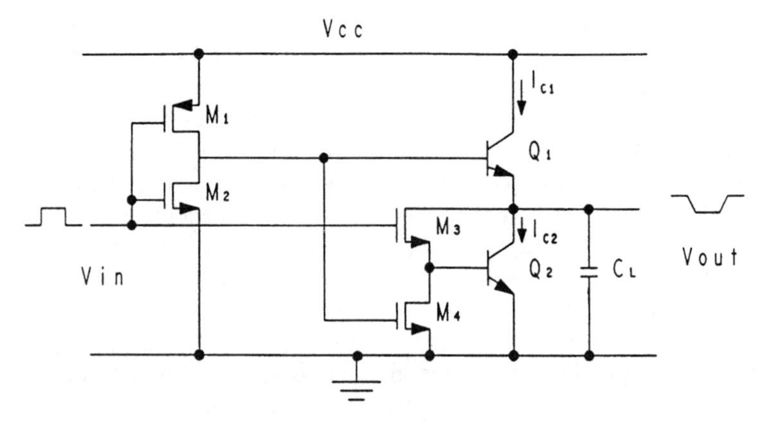

FIGURE 9.68 BiCMOS inverter.

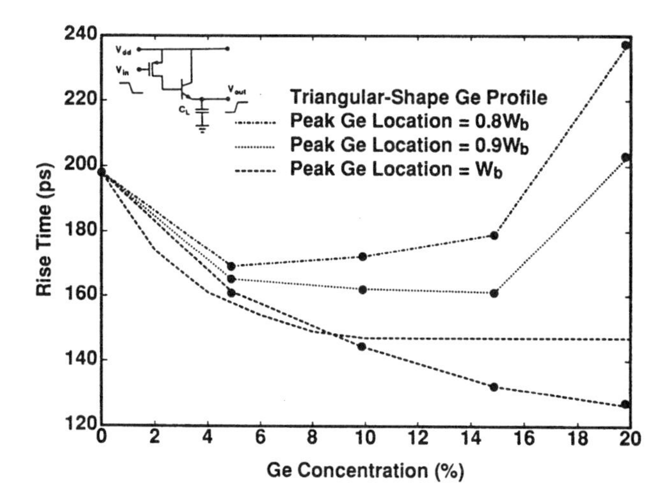

FIGURE 9.69 Rise time versus Ge concentration (after Lu and Kuo, Ref. 53 © IEEE).

less than 0.1, the difference in the rise time among the three cases is small. As the germanium concentration is high, however, the difference in rise time is substantial.

As indicated in [54], the rise time of the BiCMOS inverter is determined by the time constant:

$$2\left(\frac{1}{R_d C_{12}} + \frac{q_b}{\beta \tau_F} + \frac{C_p}{\tau_F C_{12}}\right)^{-1} \tag{9.57}$$

where R_d is the equivalent resistance accounting for the average channel resistance of the PMOS device and the base resistance of the bipolar device. C_{12} is the sum of the load capacitance and the parasitic capacitance at the output. C_p represents the sum of the total parasitic capacitance at the output. From (9.57) the rise time of the BiCMOS inverter is determined by the smallest of the three time constants: the load-related time constant ($R_d C_{12}$), the transit-time-related time constant ($\tau_F C_{12}/C_p$), and the high-injection-related time constant ($\beta \tau_F / q_b$). Depending on the output load condition, the importance of the high injection condition varies. A larger output load makes the high injection effect more influential in determining the speed of the BiCMOS transient response.

9.9 PHOTOTRANSISTORS

High-speed optical fiber transmission has been investigated extensively. Using an all-optical multiplexer/demultiplexer, speeds have reached 100 Gbits. Although optical devices are much faster than electronic devices, electronic devices have some advantages over optical devices, such as high functionality, small size, low cost, and high reliability. The interest in the development of optoelectronic integrated circuits

(OEICs) for fiber sensing and communication is increasing [55]. In OEIC, optical devices and electronic devices are integrated monolithically onto a semiconductor substrate, and they are connected with electrically conducting elements.

Heterojunction bipolar phototransistors (HPTs) have been found to be extremely fast detectors of optical signals. In impulse measurements, their response has been measured to be on the order of picoseconds. This suggests that they can be used in systems where millimeter waves are carried on optical signals and then converted into propagating signals using these heterojunction devices with appropriate antenna structures. InP-based heterojunction bipolar phototransistors have shown high-speed response and the promise of significant optical gain [56].

A phototransistor converts an input light signal to an electric signal at the output. The energy-band diagram in Fig. 9.70 illustrates the operation of an n-p-n HPT. The incident photocarriers (photons) pass through the wide-bandgap emitter, which functions as a transparent window [57]. The photons are absorbed in the base and collector regions, creating free electrons and holes. The major component of the photocurrent is due to the electron–hole pairs generated in the base, in the base–collector depletion region, and within a diffusion length of the depleted edge in the bulk collector. The electrons that are generated in these regions are collected by the field of the base–collector junction, leading to a current flow in the external circuit. The holes that are generated in these regions are swept into the base, thereby increasing the base potential. This in turn increases the base–emitter forward bias. A large number of electrons are then injected from the emitter into the base and collected in the collector. If the lifetime of the injected electrons in the base is longer than the base transit time, the current gain is achieved by normal transistor action.

Figure 9.71 shows the one-dimensional representation of a phototransistor biased in the floating base configuration. The shaded areas represent the space-charge regions in the phototransistor. Since there is no base contact, the emitter current is equal to the collector current. The hole back injection into the emitter J_{pE} is determined by the hole diffusion equation. The electron current is collected in the collector from the electron diffusion current in the base.

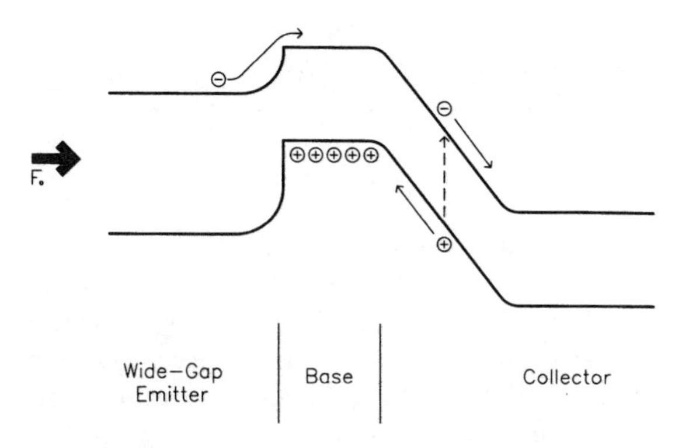

FIGURE 9.70 Energy-band diagram of a phototransistor.

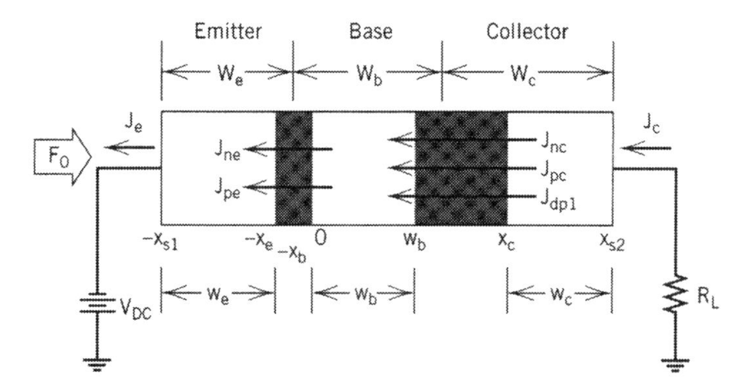

FIGURE 9.71 Phototransistor biased in the floating base configuration.

HPTs have an appreciable advantage over p-i-n photodetectors in optical gain. The optical gain is defined as

$$G = \frac{J_{C,\mathrm{opt}}}{qF_0} \tag{9.58}$$

where $J_{C,\mathrm{opt}}$ is the optical component of the collector current and F_0 is the incident radiant flux density at the surface of the emitter. The optical gain describes the relationship between the optically generated collector current and the number of incident photons.

The absorption of light in the emitter, base, collector, and collector–base depletion layer are important in determining the optical gain. The optical component of the collector current is found to be [1]

$$\begin{aligned} J_{C,\mathrm{opt}} = {} & J_{n0}\left(e^{qV_{BE}/kT} - 1\right)_{\mathrm{opt}} \\ & + F_0\left\{g_c + g_b f_{nc} + g\eta_c^{-(\alpha_e W_E + \alpha_b W_B)}\left[1 - e^{-\alpha_c(X_C - W_B)}\right]\right\} \end{aligned} \tag{9.59}$$

where the first term in (9.59) is the response of the emitter potential to the accumulation of photogenerated holes in the base region. The second, third, and fourth terms are the primary photocurrent, which results from absorption in the neutral region of the collector, the base, and the base–collector space-charge layer. In (9.59) η is the quantum efficiency, α the absorption efficient, and g the generation rate.

The relationship between the optical gain of the HPT and the common-emitter current gain of the HBT is given by [1]

$$G = \frac{\beta}{q}\left\{g_e + g_b\left(f_{nc}\cosh\frac{W_B}{L_{nB}} + f_{ne}\right) + g_c\cosh\frac{W_B}{L_{nb}}\right.$$

$$+q\eta_c \cosh \frac{W_B}{L_{nb}} e^{-(\alpha_e W_E + \alpha_b W_B)} \left[1 - e^{-\alpha_c (X_C - W_B)}\right]\Bigg\}$$

$$+ \frac{1}{q}\frac{\beta}{\Gamma}\left\{g_c + g_b f_{nc} + q\eta_c^{-(\alpha_e W_E + \alpha_b W_B)}\left[1 - e^{-\alpha_c (X_C - W_B)}\right]\right\} \quad (9.60)$$

The optical gain depends on the common-emitter current gain multiplied by the absorption of light in the emitter, base, collector, and base–collector depletion layer. In the emitter most of the photogenerated electron and hole pairs essentially recombine with each other. Only a fraction of the electron and hole pairs that are generated near the base–emitter depletion region can cause any excess current to flow. The absorption in the base contributes marginally to the total optical gain due to a very narrow base width. The primary contribution to the optical gain is from light that is absorbed in the neutral collector and base–collector depletion layer. Figure 9.72 shows the total optical gain versus absorption coefficient in the neutral collector and base–collector depletion regions. The optical gain has its strongest dependency on absorption in the collector and base–collector depletion regions. The optical gain is also a function of wavelength. Figure 9.73 shows the optical gain versus the wavelength, where the short-wavelength edge of the HPT is 520 nm and the long-wavelength edge is 780 nm.

The phototransistor transient response is determined by the charge-up and charge-down times of the base–emitter junction. The charge-up time to a step-function change in the optical signal depends primarily on the charging time of the junction capacitances through the primary photocurrent and the dark current. The charge-up time is

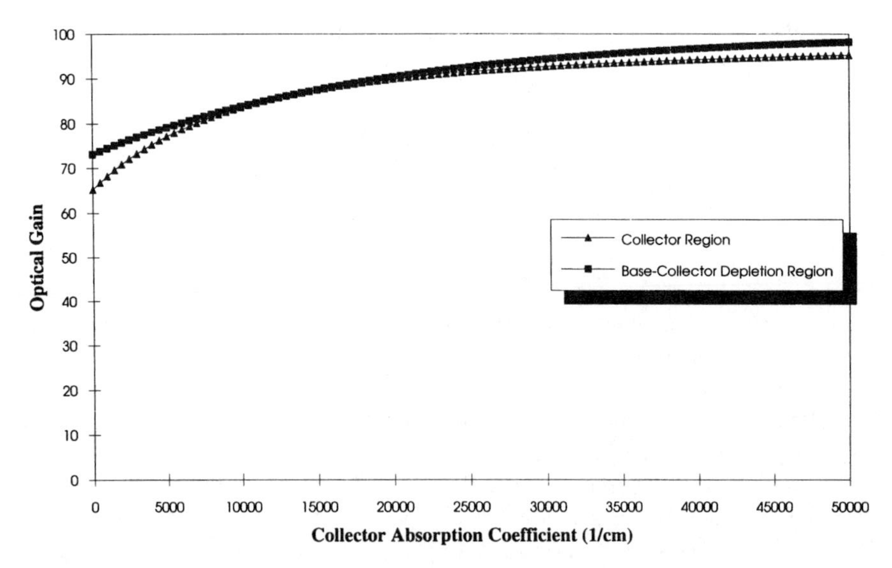

FIGURE 9.72 Total optical gain versus collector absorption coefficient (after Jalali and Pearton, Ref. 1© Artech House).

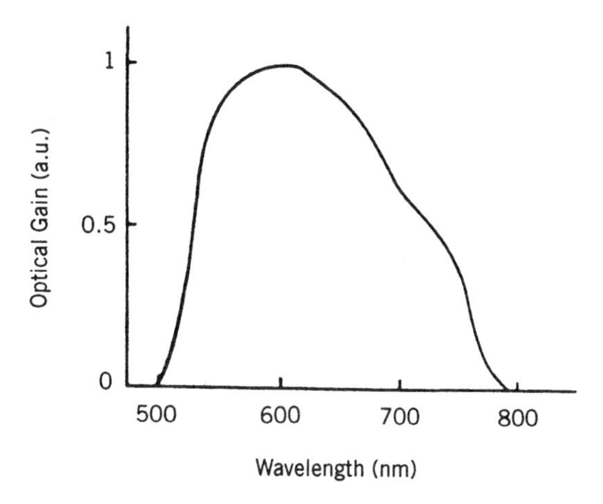

FIGURE 9.73 Spectral response of the HPT.

written as

$$\tau_{\mathrm{up}} = \frac{kT}{q} \frac{C_{BE} + C_{BC}}{I_{\mathrm{ph}} + 2I_{n0}\left[\cosh\left(W_B/L_{nB}\right) - 1\right] + I_{pE0} + I_{pC0}} \tag{9.61}$$

When the light of the HBT is turned off, $J_{\mathrm{ph}} = 0$. The charge-down time is given by

$$\tau_{\mathrm{down}} = \frac{kT}{q} \frac{C_{BE} + C_{BC}}{2I_{n0}\left[\cosh\left(W_B/L_{nB}\right) - 1\right] + I_{pE0} + I_{pC0}} \tag{9.62}$$

Comparing (9.61) and (9.62) one finds that the charge-down time is much larger than the charge-up time because only the dark current can charge-down the base–emitter junction. Thus the limiting mechanism for how fast the emitter potential (equivalently, the optical gain) can respond to optical modulation is determined by how fast the stored charge in the base can be removed.

The charge-down time can be reduced significantly by applying an electrical dc bias to the base terminal of the HBT [56,58]. With the additional bias, (9.62) becomes

$$\tau_{\mathrm{down}} = \frac{kT}{q} \frac{C_{BE} + C_{BC}}{2I_{n0}\left[\cosh\left(W_B/L_{nB}\right) - 1\right] + I_{pE0} + I_{pC0} + I_{B,\mathrm{dc}}} \tag{9.63}$$

Chandrasekhar et al. [58] showed that by applying the appropriate base bias, the 3-dB optical gain–bandwidth product is enhanced 15-fold over the same HBT with no base bias.

9.10 PHOTORECEIVERS

A light-to-light transducer converts an input light signal to an electric signal and then back to the output light signal [59,60]. The transducer has the function of light amplification, which can be extended to other functions, such as optical bistability, light-activated switching, light coherency conversion, and wavelength conversion. A device with the function of light amplification is required for information processing in an optical signal form in optoelectronics. A light amplifier would become a key device for detecting, sensing, and processing optical signals. The amplification principle is described here. An energy-band diagram of the device is shown in Fig. 9.74.

An input signal is incident onto the top of the device. Photons generate electrons and holes in the phototransistor. The photogenerated holes are swept into the base, increasing the forward bias of the base–emitter junction, which causes a large electron current to flow from emitter to collector. The heterojunction phototransistor (HPT) provides large photo-current gain without the high bias voltages. It also provides a larger signal-to-noise ratio than that of the avalanche photodiode. The emitter–base injection efficiency of this structure is very close to unity because the valence-band heterojunction discontinuity effectively eliminates hole injection from base to emitter when the emitter–base heterojunction is forward biased. Photogenerated electrons are absorbed in the n-type InGaAsP absorption layer. These electrons drive the light-emitting diode and produce an output light signal through electron–photon interaction. Light amplification can be expected when the following condition is satisfied. The

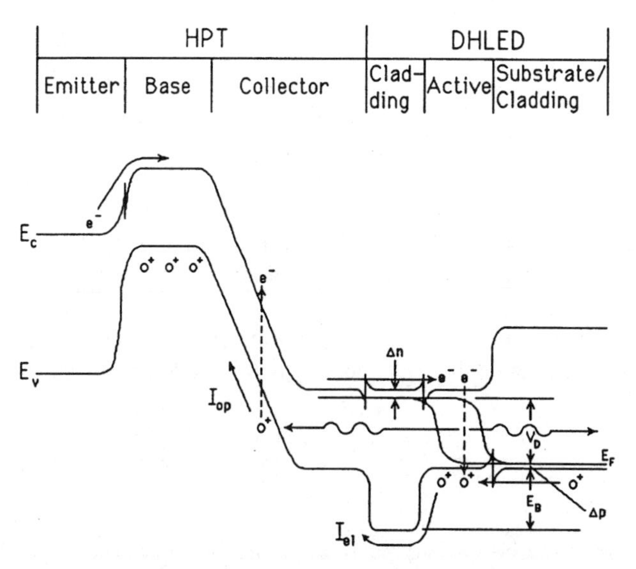

FIGURE 9.74 Band diagram of the LOAS (after Feld et al., Ref. 59 © IEEE).

light amplification gain is written as

$$G_o = \frac{i/e}{P_i/hv}\,\frac{P_o/hv}{i/e} = G\eta \tag{9.64}$$

where P_i and P_o denote the input and output light power, respectively, $G = (i/e)/(P_i/hv)$ is the optical conversion gain of the phototransistor, and $\eta = (P_o/hv)/(i/e)$ is the external quantum efficiency. We should have $G > 1/\eta$ to achieve a net positive gain $G_o > 1$.

Figure 9.75 shows amplification characteristics of the transducer. The amplification is shown in Fig. 9.77*a–c*, where the bias voltage was varied as $V_b = 3.75$, 4.0, and 4.25 V, respectively. The light gain is increased with increasing bias voltage and decreasing load resistance. Good linearity between the input and the output is observed. It is very desirable to obtain an output intensity linearly proportional to the input intensity.

The light-amplifying optical switch (LAOS) changes state from a low-current to a high-current condition via a negative differential resistance that is caused by the positive feedback from the output LED to the HBT input. The electrical and optical feedback mechanisms are interlocked. The optical feedback is due to generation of electron–hole pairs in the collector–base region of the HBT caused by the absorption of light radiated back from the LED. Thermal-assisted injection of holes over the barrier formed by the InP cladding layer of the LED provides the electrical feedback. These feedback mechanisms can be modeled as current sources or current-carrying elements, as shown in the equivalent circuit in Fig. 9.76. I_{in} is the equivalent base current produced by the input light. I_{el} and I_{op} are the equivalent electrical and optical feedback currents, respectively, which connect directly to the base and add to the input current. The Early effect and leakage are modeled as circuit elements that provide a shunt current path from collector to emitter. The magnitude of the optical feedback is proportional to the intensity of light generated by the LED, which is directly proportional to the diode current I.

Consider both optical and electrical feedback along with Early voltage, leakage, nonlinear gain, and series resistance R_S. An analytical expression for V_S as a function of the current was obtained [59]:

$$V_S = \frac{I - \beta_0^{n_1}\left(I_{\text{in}} + \gamma I + \chi I^{n_2}\right)^{n_1}}{\delta\left(I_{\text{in}} + \gamma I + \chi I^{n_2}\right) + 1/R_L} + \frac{n_2 kT}{q}\ln\left(\frac{I}{I_{01}} + 1\right) + R_S I \tag{9.65}$$

where I_{02} is the preexponential current of the LED diode current, n_2 the ideality factor of the diode current, I_{01} the preexponential current of the electrical feedback current, χ the electrical feedback coefficient, and γ the optical current coefficient. $\chi = I_{01}/I_{02}^{n_2}$; $\gamma = I_C/I_{\text{op}}$. Equation (9.65) can be solved numerically to find the breakover voltage as a function of γ, χ, and δ.

Figure 9.77 shows the calculated current through a LOAS as a function of supply voltage for various values of input light (or base current). In this plot V_{BO} is the

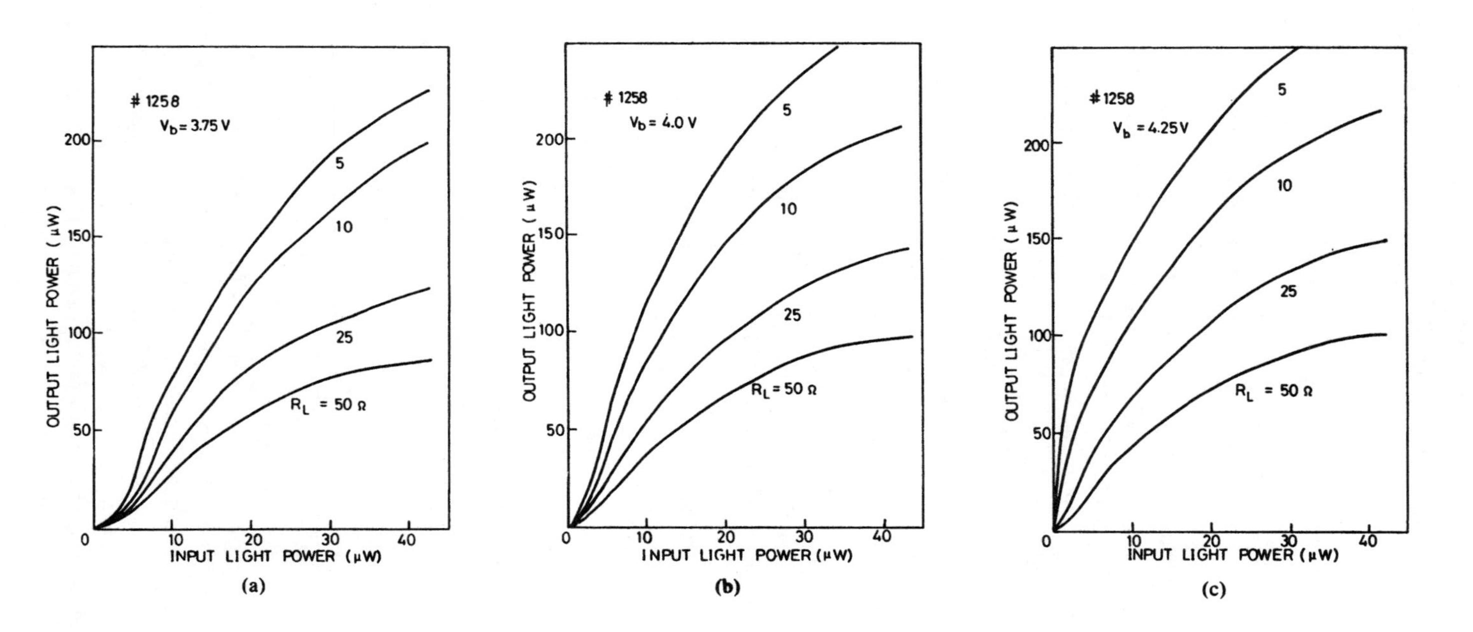

FIGURE 9.75 Amplification characteristics of a transducer (after Sasaki et al., Ref. 60 © IEEE).

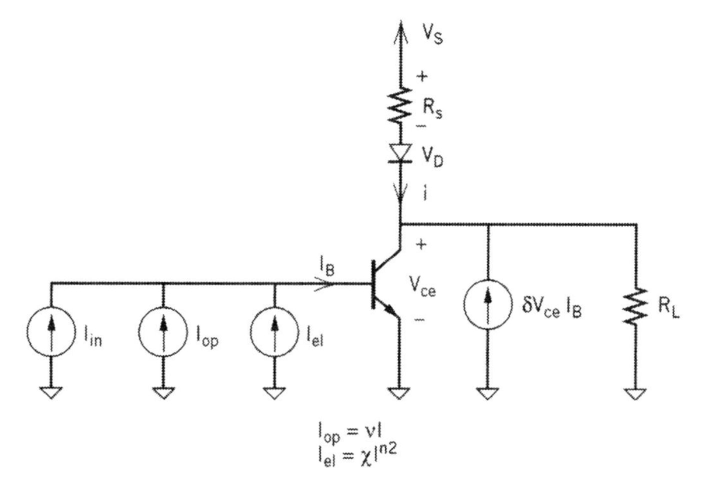

FIGURE 9.76 Equivalent circuit of a LAOS with electrical and optical feedback (after Feld et al., Ref. 59 © IEEE).

breakover voltage at which switching occurs from a low-current to a high-current state through an NDR. Comparison of the calculated and measured breakdown voltage as a function of the input light power is shown in Fig. 9.78. In this plot, the open circles represent the experimental data and the solid line represents the theoretical results. The experimental data were obtained by shining the output of a $\lambda = 1300$-nm laser on the LAOS in calibrated intensity steps and measuring the *I*–*V* characteristics. The theoretical results agree with the experimental data at moderate and high laser power

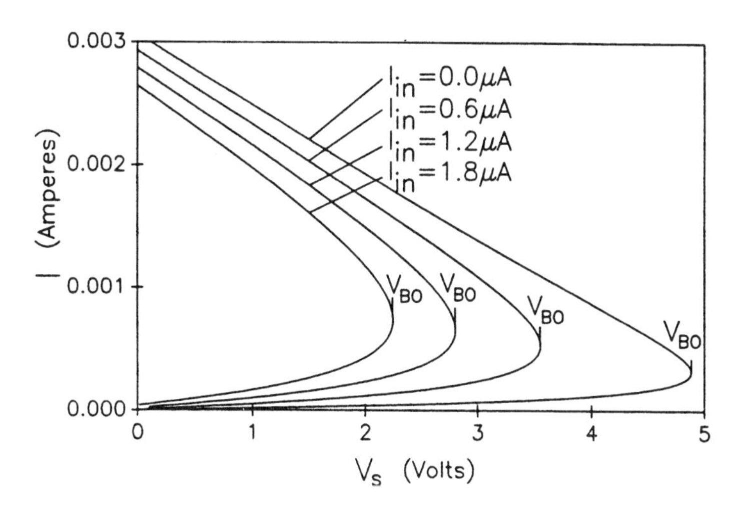

FIGURE 9.77 Calculated current versus supply voltage (after Feld et al., Ref. 59 © IEEE).

densities. At low input laser power, the model predicts higher breakover voltages than the experimental data. The discrepancy is caused by the avalanche breakdown, which is not included in the model equations. Avalanche occurs before the feedback currents and is strong enough to initiate the NDR. Therefore, the breakover voltage is reduced.

Very high-speed optical fiber transmission has been investigated extensively experimentally, and speeds have reached 100 Gbits/s as a result of using an all-optical multiplexer/demultiplexer. Nevertheless, electronic devices have some advantages over optical devices in small size, low cost, and high reliability. IC chip sets with an operating speed of 10 Gbits/s have been reported for Si bipolar transistors, GaAs MESFETs, and AlGaAs/GaAs HBTs. A lot of work has been conducted to achieve over-20-Gbit/s operation using different device technologies. Among them, InP-based HEMTs and HBTs are very attractive because of their excellent performance and compatibility with optoelectronic devices.

The requirement for the sensitivity and gain of optical receivers has been relaxed with the introduction of Er-doped fiber amplifiers (EDFAs). Hagimoto et al. [61] reported an optical transmission experiment using an hybrid receiver composed of an EDFA, pin-PD, and decision IC. It is desirable to integrate pin-PDs, amplifiers, and D flip-flops. Consider three types of phototransceivers in Fig. 9.79. The time constants of the transimpedances for the photoreceivers are [62]

$$\tau_{pa} = \frac{\beta[C_{BE} + (1 + g_m R_L)C_{BC}]}{g_m} \tag{9.66}$$

$$\tau_{pb} = \tau_F + \frac{(R_F + R_B)[C_{BE} + (1 + g_m R_L)C_{BC}]}{g_m R_L} \tag{9.67}$$

$$\tau_{pc} = \tau_F + \frac{(R_F + R_B)[C_{BE} + (1 + g_m R_L)C_{BC}] + R_F C_{PD}}{g_m R_L} \tag{9.68}$$

where R_L is the load resistance, R_F the feedback resistance, R_B the base resistance,

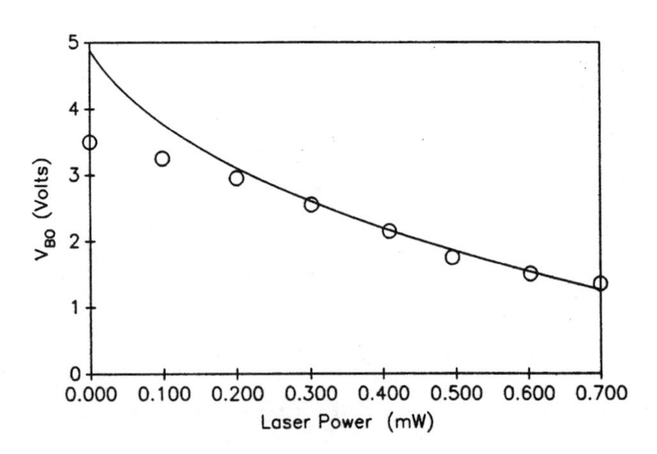

FIGURE 9.78 Breakover voltage versus laser power (after Feld et al., Ref. 59 © IEEE).

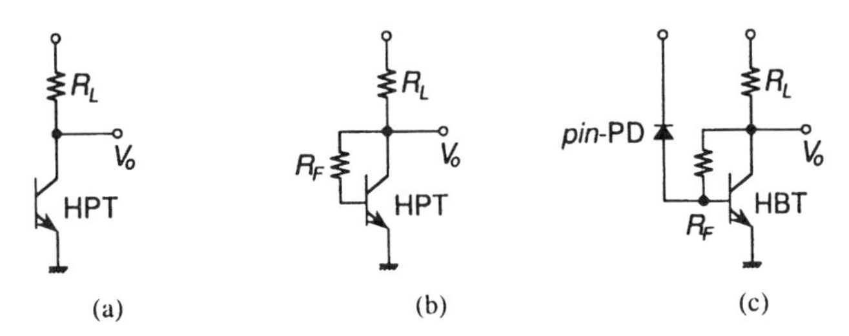

FIGURE 9.79 Three types of phototransceivers (after Sano et al., Ref. 62 © IEEE).

and C_{PD} the pin-PD junction capacitance. Since commonly used HPTs are much larger than HBTs, the capacitances of the HPT are usually larger than those of the HBT. Therefore, the combination of the pin-PD and HBT gives the highest bandwidth of the three structures. To enhance the phototransceiver response, the HBT device structure needs to be optimized for fiber optic applications.

The figures of merit for evaluation of the bandwidth of the HBT are the cutoff frequency f_T and maximum frequency of oscillation $f_{\max}$. The analytical expressions of the cutoff frequency and the maximum oscillation frequency are given in (4.15) and (4.16), respectively. Analyzing the figures of merit and the time constant given above, one can predict the HBT device structures and doping profiles to improve the speed and bandwidth of the photoreceivers. In addition, in optical fiber communication, the optical signal diminishes with distance. The photocurrent gain and sensitivity of a typical HPT drops significantly at low optical power, thereby limiting the sensitivity of the device. This behavior is attributed to different sources of recombination currents. Suppression of major recombination sources are the key to improvement of the sensitivity of HPTs.

Figure 9.80 shows the circuit configuration of the monolithic photoreceiver. The first stage is a conventional transimpedance type. The second stage is based on parallel feedback with a transistor (Q_4) and resistor (R_{F2}) to enlarge the bandwidth for a 50-Ω load drive. The capacitor C_s is the shot capacitance and the resistor R_D is the damping resistance. All the transistors had an effective emitter area of $1.6 \times 9.6\ \mu\text{m}^2$ and were biased at collector currents ranging from 7 to 15 mA. Nominal values of the resistors R_{F1} and R_{F2} were 500 and 100 Ω. The photoreceiver had an input pad for electrical scattering parameter measurements performed using a 50-GHz network analyzer.

Measured gain S_{21} and transimpedance Z_t characteristics [62] of the photoreceivers fabricated on the wafers with collector thicknesses of 430 nm and 830 nm are shown in Fig. 9.81. The dissipation current was 53 mA at $V_{CC} = 5$ V. The photoreceiver had a Z_t of 48.9 dB·Ω with a 3-dB bandwidth of 26.7 GHz and an S_{21} value of 17.1 dB with a 3-dB bandwidth of 26.3 GHz. These bandwidths include the influence of the *RC* time constant of the pin PD. The optical bandwidth of the photoreceiver is determined by the electrical bandwidth of 26.7 GHz and the carrier-transit-time-related bandwidth. Using 75 GHz for the carrier-transit-time-related bandwidth, the optical bandwidth of the photoreceiver is estimated to be 25.1 GHz. In another example, Fig. 9.81*b* shows the measured gain S_{21} and the transimpedance Z_t characteristics for the photoreceiver

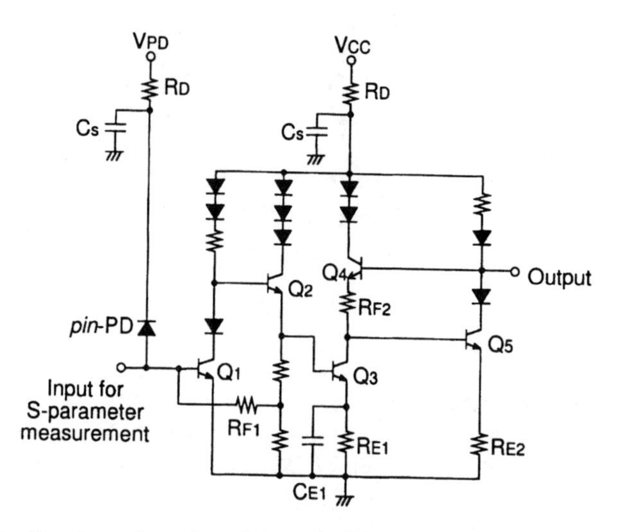

FIGURE 9.80 Circuit configuration of the monolithic photoreceiver (after Sano et al., Ref. 62 © IEEE).

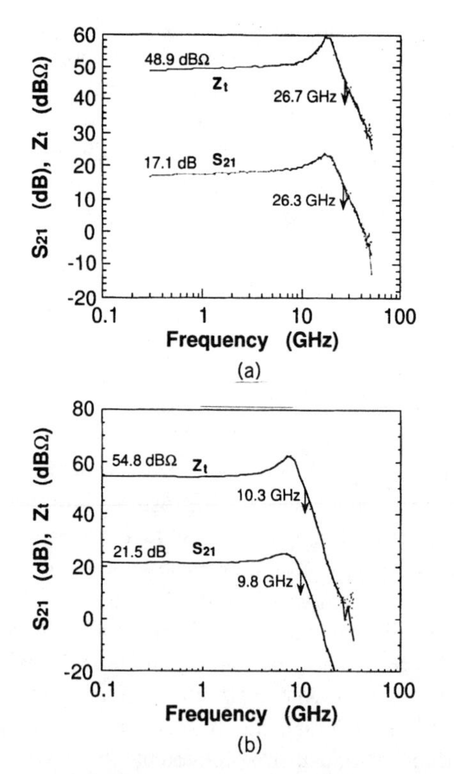

FIGURE 9.81 S_{21} and transimpedance Z_t versus frequency for collector thicknesses of (*a*) 430 nm and (*b*) 830 nm (after Sano et al., Ref. 62 © IEEE).

with a collector thickness of 830 nm. The photoreceiver had a Z_t value of 54.8 dB·Ω with a 3-dB bandwidth of 10.3 GHz and S_{21} of 21.5 dB with a 3-dB bandwidth of 9.8 GHz.

REFERENCES

1. B. Jalali and S. J. Pearton, *InP HBTs: Growth, Processing, and Applications*, Artech House, Norwood, MA (1995).

2. H. Schmacher, A. Gruhle, U. Erben, H. Kibbel, and U. König, "A 3V supply voltage, dc-18 GHz SiGe HBT wideband amplifier," *BCTM Tech. Dig.*, 190 (1995).

3. N. H. Sheng, W. J. Ho, N. L. Wang, R. L. Pierson, P. M. Asbeck, and W. L. Edwards, "A 30 GHz bandwidth AlGaAs-GaAs HBT direct-coupled feedback amplifier," *IEEE Microwave Guided Wave Lett.*, **8**, 208 (1991).

4. M. F. Chang and P. M. Asbeck, "III-V heterojunction bipolar transistors for high-speed applications," *Int. J. High Speed Electron.*, **1**, 245 (1990).

5. M. A. Khatibzadeh and B. Bayraktaroglu, "High-efficiency, class-B, S-band HBT power amplifiers," *IEEE MTT-S International Microwave Symposium*, 101 (1989).

6. N. L. Wang, N. H. Sheng, M. F. Chang, W. J. Ho, G. J. Sullivan, E. A. Sovero, J. A. Higgins, and P. M. Asbeck, "Ultrahigh power efficiency operation of common-emitter and common-base HBTs at 10 GHz," *IEEE Trans. Microwave Theory Tech.*, **MTT-38**, 1381 (1990).

7. N. L. Wang, N. H. Sheng, M. F. Chang, W. J. Ho, G. J. Sullivan, E. Sovero, J. A. Higgins, and P. M. Asbeck, "18 GHz high gain, high-efficiency power operation of AlGaAs/GaAs HBTs," *IEEE MTT-S International Symposium*, 997 (1990).

8. B. Bayraktaroglu, M. A. Khatibzadeh, and R. D. Hudgens, "Monolithic X-band heterojunction bipolar transistor power amplifiers," *IEEE GaAs IC Symposium*, 271 (1989).

9. T. Shimura, M. Sakai, S. Izumi, H. Nakano, H. Matsuoka, A. Inoue, J. Udomoto, K. Kosaki, T. Kuragaki, H. Takano, T. Sonoda, and S. Takamiya, "1 W Ku-band AlGaAs/GaAs power HBTs with 72% peak power-added efficiency," *IEEE Trans. Electron Devices*, **ED-42**, 1890 (1995).

10. M. G. Adlerstein, M. P. Zaitlin, G. Flynn, W. Hoke, J. Huang, G. Jackson, P. Lemonias, R. Maajarone, and E. Tong, "High power density pulsed X-band heterojunction bipolar transistors," *Electron. Lett.*, **27**, 148 (1991).

11. W. Liu, E. Beam, and A. Khatibxadeh, "1.5-W CW S-band GaInP/GaAs/GaInP double heterojunction bipolar transistor," *IEEE Electron Device Lett.*, **EDL-15**, 215 (1994).

12. C. Nguyen, T. Liu, M. Chen, H.-C Sun, and D. Rensch, "AlInAs/GaInAs/InP double heterojunction bipolar transistor with a novel base–collector design for power applications," *IEEE Electron Device Lett.*, **EDL-17**, 133 (1996).

13. C.-B. Gao, H. Morkoç, and M.-C. F. Chang, "Heterojunction bipolar transistor design for power applications," *IEEE Trans. Electron Devices*, **ED-39**, 1987 (1992).

14. W. Liu, A. Khatibzadeh, J. Sweder, and H.-F. Chau, "The use of base ballasting to prevent the collapse of current gain in AlGaAs/GaAs heterojunction bipolar transistors," *IEEE Trans. Electron Devices*, **ED-43**, 245 (1996).

15. T. Ivanov and J. S. Yuan, unpublished data.

16. D. Ritter, R. A. Hamm, A. Feygenson, H. Temkin, M. B. Panish, and S. Chandrasekhar, "Bistable hot electron transport in InP/GaInAs composite collector heterojunction bipolar transistors," *Appl. Phys. Lett.*, **61**, 70 (1992).

17. G.-B Gao and H. Morkoç, "Material-based comparison for power heterojunction bipolar transistors," *IEEE Trans. Electron Devices*, **ED-38**, 2410 (1991).

18. N. O. Sokal and A. D. Sokal, "Class E, a new class of high-efficiency tuned single-ended switching power amplifiers," *IEEE J. Solid-State Circuits*, **SC-10**, 168 (1975).

19. J. A. Blanchard and J. S. Yuan, "Effect of collector current exponential decay on power efficiency for class E tuned power amplifier," *IEEE Trans. Circuits Systems*, **CS-41**, 69 (1994).

20. F. H. Rabb, "Effects of circuit variations on the class E tuned power amplifier," *IEEE J. Solid-State Circuits*, **SC-13**, 239 (1978).

21. T. Sowlati, A. T. Salama, J. Sitch, G. Rabjohn, and D. Smith, "Low voltage, high-efficiency GaAs class E power amplifiers for wireless transmitters," *IEEE J. Solid-State Circuits*, **SC-30**, 1074 (1995).

22. H. L. Krauss, C. W. Bostian, F. H. Raab, and H. A. Haus, *Solid State Radio Engineering*, Wiley, New York (1980).

23. S. A. Maas, B. L. Nelson, and D. L. Tait, "Intermodulation in heterojunction bipolar transistors," *IEEE Trans. Microwave Theory Tech.*, **MTT-40**, 442 (2992).

24. *SuperCompact User Manual*, Microwave Harmonica, Paterson, NJ (1992).

25. A. Samelis and D. Pavilidis, "Mechanisms determining third-order intermodulation distortion in AlGaAs/GaAs heterojunction bipolar transistors," *IEEE Trans. Microwave Theory Tech.*, **MTT-40**, 2374 (1992).

26. H. Yamada, S. Ohara, T. Iwai, Y. Yamagchi, K. Imanishi, and K. Joshin, "Self-linearizing technique for L-band HBT power amplifier: effect of source impedance on phase distortion," *IEEE Trans. Microwave Theory Tech.*, **MTT-44**, 2398 (1996).

27. G. Gonzalez, *Microwave Transistor Amplifiers: Analysis and Design*, Prentice Hall, Upper Saddle River, NJ (1984).

28. F. Ali and A. Gupta, *HEMTs and HBTs: Devices, Fabrication, and Circuits*, Artech House, Norwood, MA (1991).

29. K. W. Kobayashi, A. K. Oki, L. T. Tran, and D. C. Striet, "Ultra-low-dc power GaAs HBT S- and C-band low-noise amplifiers for portable wireless applications," *IEEE Trans. Microwave Theory Tech.*, **MTT-43**, 3055 (1995).

30. K. Maruhashi, M. Madihian, L. Desclos, K. Omda, and M. Kuzuhara, "A K-band monolithic oscillator integrated with a buffer amplifier using a device-circuit interaction design concept," *IEEE Trans. Microwave Theory Tech.*, **MTT-44**, 1424 (1996).

31. K. Takagi, S. Serikawa, and T. Kurita, "Phase noise and a low-frequency noise reduction method in bipolar transistors," *IEEE Trans. Electron Devices*, **ED-44**, 1180 (1997).

32. H. Rhodi, C. Su, and C. Stolte, "A study of the relation between device low-frequency noise and oscillator phase noise for GaAs MESFETs," *IEEE MTT-S Tech. Dig.*, 267 (1984).

33. K. Takagi, S. Serikawa, and T. Kurita, "Phase noise and a low-frequency noise reduction method in bipolar transistors," *IEEE Trans. Electron Devices*, **ED-44**, 1180 (1997).

34. H. J. Siweris and B. Schiek, "Analysis of noise up-conversion in microwave FET oscillators," *IEEE Trans. Microwave Theory Tech.*, **MTT-33**, 233 (1985).

35. N. Hayama, S. R. LeSage, M. Madihian, and K. Honjo, "A low-noise Ku-band AlGaAs/GaAs oscillator," *IEEE MTT-S. Tech. Dig.*, 679 (1988).

36. M. Tutt, D. Pavlidis, A. Khatibzadeh, and B. Bayraktaroglu, "The role of baseband noise and its up-conversion in HBT oscillator phase noise," *IEEE Trans. Microwave Theory Tech.*, **MTT-43**, 1461 (1995).

37. J. S. Yuan, "Modeling the bipolar oscillator phase noise," *Solid-State Electron.*, **37**, 1765 (1994).

38. G. Moulton, "Dig for the roots of oscillator noise," *Microwaves RF*, **4**, 65 (1986).

39. W. Liu and J. S. Harris, Jr., "Diode ideality factor for surface recombination current in AlGaAs/GaAs heterojunction bipolar transistors," *IEEE Trans. Electron Devices*, **ED-39**, 2726 (1992).

40. A. Katz, ed., *Indium Phosphide and Related Materials*: *Processing*, *Technology*, *and Devices*, Artech House, Norwood, MA (1992).

41. K. Osafune and Y. Yamauchi, "20-GHz 5-dB-gain analog multipliers with AlGaAs/GaAs HBTs," *IEEE Trans. Microwave Theory Tech.*, **MTT-42**, 518 (1994).

42. K.-C. Wang, P. M. Asbeck, M.-C. F. Chang, D. L. Miller, G. J. Sullivan, J. J. Corcoran, and T. Hornak, "Heating effects on the accuracy of HBT voltage comparators," *IEEE Trans. Electron Devices*, **ED-34**, 1729 (1987).

43. K. H Wang, S. M. Beccue, M.-C. F. Chang, R. B. Nubling, A. M. Cappon, C.-T. Tsen, D. M. Chen, P. M. Asbeck, and C. Y. Kwok, "Diode-HBT-logic circuits monolithically integrable with ECL/CML circuits," *IEEE J. Solid-State Circuits*, **SC-27**, 1372 (1992).

44. P. M. Enquist, D. B. Slater, Jr., and J. W. Swart, "Complementary AlGaAs/GaAs HBT I^2L (CHI^2L) technology," *IEEE Electron Device Lett.*, **EDL-13**, 180 (1992).

45. E. W. Greenwich, "An appropriate device figure of merit for bipolar CML," *IEEE Electron Device Lett.*, **EDL-12**, 18 (1991).

46. K. G. Ashar, "The method of estimating delay in switching circuits and the figure of merit of a switching transistor," *IEEE Trans. Electron Devices*, **ED-11**, 497 (1964).

47. D. D. Tang and P. M. Soloman, "Bipolar transistor design for optimized power–delay logic circuits," *IEEE J. Solid-State Circuits*, **SC-14**, 679 (1979).

48. W. Fang, "Accurate analytical delay expressions for ECL and CML circuits and their application to optimizing high-speed bipolar circuits," *IEEE J. Solid-State Circuits*, **SC-25**, 572 (1990).

49. M. Karlsteen and M. Willander, "Improved switch time of I^2L at low power consumption by using a SiGe heterojunction bipolar transistor," *Solid-State Electron.*, **38**, 1401 (1995).

50. J. J. Ebers and J. L. Moll, "Large-signal behavior of junction transistors," *Proc. IRE*, **42**, 1761 (1954).

51. Z. A. Shafi, P. Ashburn, and G. J. Parker, "Predicted propagation delay of Si/SiGe heterojunction bipolar ECL circuits," *IEEE J. Solid-State Circuits*, **SC-25**, 1268 (1990).

52. D. L. Harame, J. H. Comfort, J. D. Cressler, E. F. Crabbé, Y.-C. Sun, B. S. Meyerson, and T. Tice, "Si/SiGe epitaxial-base transistors: I. Materials, physics, and circuits," *IEEE Trans. Electron Devices*, **ED-42**, 455 (1995).

53. T. C. Lu and J. B. Kuo, "An analytical SiGe-base HBT model and its effects on a BiCMOS inverter circuit," *IEEE Trans. Electron Devices*, **ED-41**, 272 (1994).

54. T. C. Lu and J. B. Kuo, "An analytical pull-up transient model for BiCMOS inverters," *Solid State Electron.*, **35**, 1 (1992).

55. P. Bhattacharya, *Semiconductor Optoelectronic Devices*, Prentice Hall, Upper Saddle River, NJ (1994).

56. D. Fritzche, E. Kuphal, and R. Aulbach, "Fast response InP/InGaAsP heterojunction phototransistor," *Electron Lett.*, **17**, 178 (1981).

57. Y. Zhu, Y. Komatsu, S. Noda, Y. Takeda, and A. Sasaki, "Fabrication and characteristics of $Al_xGa_{1-x}As$ heterojunction phototransistors with wide-gap window," *IEEE Trans. Electron Devices*, **ED-38**, 1310 (1991).

58. S. Chandrasekhar, M. K. Hoppe, A. G. Dentai, C. H. Joyner, and G. J. Qua, "Demonstration of enhanced performance of an InP/InGaAs heterojunction phototransistor with a base terminal," *IEEE Electron Device Lett.*, **EDL-12**, 10 (1991).

59. S. A. Feld, F. R. Beytte, Jr., M. J. Hafich, H. Y. Lee, G. Y. Robinson, and C. W. Wilmsen, "Electrical and optical feedback in an InGaAs/InP light-amplifying optical switch (LAOS)," *IEEE Trans. Electron Devices*, **ED-38**, 2452 (1991).

60. A. Sasaki, S. Metavikul, M. Itoh, and Y. Takeda, "Light-to-light transducers with amplification," *IEEE Trans. Electron Devices*, **ED-35**, 780, (1988).

61. K. Hagimoto, Y. Miyamoto, T. Katoka, H. Ichino, and O. Nakajima, "Twenty-Gbit/s signal transmission using a simple high-sensitivity optical receiver," *Opt. Fiber Commun. Tech. Dig.*, 48 (1992).

62. E. Sano, M. Yoneyama, S. Yamahata, and Y. Matsuoka, "InP/InGaAs double-heterojunction bipolar transistors for high-speed optical receivers," *IEEE Trans. Electron Devices*, **ED-43**, 1826 (1996).

Index

Absorption efficiency, 447
Accelerated life testing, 327
Acoustic scattering, 18, 21
Activation energy, 338
Adjacent channel leakage power (ACP), 401
Alloy barrier effect, 63
Alloy scattering, 21
Anderson and Crowell method, 295
Auger generation, 282
Auger recombination, 248, 298, 330
Avalanche breakdown, 282, 284–285, 306–307, 309
Avalanche effect, 297
Avalanche multiplication, 287, 288, 290, 293–294, 300, 308

Backward Euler method, 259
Ballistic motion, 296, 352
Bandgap narrowing, 11–13, 93, 162, 298, 350
Base ballasting resistor, 319, 320
Base conductivity modulation, 71
Base current reversal, 298
Base pushout, 71–72, 89–90, 119, 122, 297
Base transit time, 76, 80, 111–112, 120, 122, 124, 136, 144–147, 152, 161, 163, 166, 170, 214, 303, 435, 442, 446
Base width modulation effect, 2, 69, 71, 80, 109, 204, 218
Boltzmann statistics, 255
Boltzmann transport equation, 228, 232, 250
Bulk recombination, 54, 56, 103, 205
Burn-in, 338–339

Carrier freeze-out, 80
Charge-control model, 191, 297
Charge-control relation, 191
Charge partitioning method, 265
Chemical vapor deposition (CVD), 14, 32, 34, 45, 158, 361
Coinwell–Weisskopf approximation, 234
Collapse loci, 312
Collapse phenomenon, 306–307
Collector breadown voltage, 39, 123, 142, 149, 296, 300, 389
Collector efficiency, 2, 323, 396
Collector offset voltage, 67–69, 205–206
Collector signal delay, 114–115, 243
Collector space-charge-layer transit time, 112, 120, 123, 152, 170, 303, 387
Collector space-charge-region widening, 90
Collector transit time, 243, 388
Complete ionization, 21
Common-base mode, 383
Common-emitter mode, 383
Conduction band barrier effect, 75, 212, 345
Conduction-band discontinuity, 6, 10, 58, 79, 101
Conduction-band energy gradient, 61
Constant-available-power-gain circles, 406
Constant-noise-figure circles, 406
Continuous-wave (CW), 381
Coulombic scattering, 18–19
Critical thickness, 14–15, 347
Cross-section transmission electron microscope (XTEM), 33
Current collapse, 318
Current crowding, 71–72, 80
Cutoff frequency, 2, 119–120, 122–123, 126, 163, 170, 262–264, 318, 322, 376, 396, 435–436, 455

dc-RF efficiency, 413–414
Depletion approximation, 110, 168
Diffusion effect, 24, 250
Dose-rate effect, 357
Drift-diffusion, 252
Drift-diffusion analysis, 291
Drift-diffusion equation, 251
Drift-diffusion model, 244
Drift-diffusion simulation, 253
Duty cycle 381
DX centers 44, 132
Dynamic hysteresis voltage, 426

Early voltage, 2, 69, 71, 76, 78–79, 82, 84, 86, 88–89, 149, 204, 215, 250, 252, 297, 345, 399–400, 451
Einstein relation, 249
Electromigration, 330–333
Electron cyclotron resonance (ECR) etching, 45
Emitter ballasting resistor, 314–315, 319, 383
Emitter delay, 161, 345, 442
Emitter thermal shunt, 317
Energy balance equation, 255
Energy balance model, 250–251, 253
Energy balance simulation, 252–253
Energy lagging effect, 293
Energy model, 293

Fan-out, 428
Fermi–Dirac statistics, 256
Figure of merit (FOM), 389
Finite element method, 256
Flicker (1/f) noise, 3, 129–131, 332, 349, 358–359, 362, 375, 414–415, 419, 421
Forward active mode, 57–58, 69, 205, 215, 318, 335, 341
Forward biased stress effect, 330
Forward collector method, 334
Forward collector stress and mesasure method (FCSAM), 355
Freeze-out effect, 247, 256

Gas-source MBE (GSBME), 27, 29–31, 43, 45–46, 350
Gain-bandwidth product, 2, 449
Gain compression, 404
Gummel method, 257–258
Gummel number, 11, 71 264, 345
Gummel–Poon model, 187, 204–206, 209, 215

h-y ration technique, 219–220
Harmonic balance method, 414
Heavy doping effect, 6, 12, 18, 77, 79
High injection barrier effect, 72–73, 75, 160
High injection effect, 2, 204, 445
Hot carrier, 238, 334, 351
Hot electron, 241, 330, 333, 336, 352–356, 388
Hot hole, 281, 354–356
Hydrodynamic model, 250

Impact ionization, 281–282, 286–288, 290–291, 294–295, 297–298, 308–309
Impurity scattering, 17–18, 20–21
Incomplete ionization, 247
Inelastic collision, 352
Infant mortality, 329
Injection efficiency, 40, 58–59, 61, 124, 166, 203
Input impedance circle method, 219
Input matching network, 397
Input return loss, 408–410
Integral charge-control relation (ICCR), 179–180
Interface recombination, 199–201
Interface traps, 337
Intermodulation distortion, 397, 399–400
Intermodulation intercept point, 399
Intermodulation response, 398–399
Intervalley scattering, 237
Inverse base width modulation effect, 71, 89–90, 152
Inverse Early voltage, 218
Inverse oscillator reflection coefficient, 413
Ionizing radiation, 358, 361

Kirk effect, 183, 243–244, 287, 388

Ledge passivation, 340
Lift-off, 37, 45
Limited reaction processing (LRP), 15, 32
Liquid-phase epitaxy, 1, 24
Local model, 291, 293
Low-dose-rate radiation, 363
Low-harmonic-distortion, 375

Maximum available gain, 383, 384–385
Maximum frequency of oscillation, 120, 123, 126, 149, 385, 387, 396, 455
Mean free path, 353
Mean-time-to-failure (MTTF), 327
Metal-organic chemical vapor deposition (MOCVD), 1, 25–27, 29, 36, 42–43, 46, 56, 103, 339, 350–351, 377
Metal-organic molecular beam epitaxy (MOMBE), 29, 31
Molecular beam epitaxy (MBE), 1, 15, 24–25, 27, 30, 32, 36–37, 40, 42–45, 103, 377
Moll–Ross current relation, 178
Monolithic microwave integrated circuit (MMIC), 376
Monte Carlos simulation, 137, 232, 234–235, 239, 250, 253, 291, 354
Multiquantum wells, 24

Negative different resistance (NDR), 102–103, 318
Newton method, 258
Neutral base recombination, 65, 82, 84, 86, 88–89, 99, 200
Noise figure, 2, 404–410
Noise power, 404–405
Nonequilibrium transport, 294
Nonlocal model, 291, 293

Offset voltage, 328–329, 335, 364
Offset current, 364
Open collector method, 221, 334
Open collector stress and measure (OCSAM), 355
Optical bistability, 450
Optical scattering, 21
Optoelectronic integrated circuits (OEICs), 445
Out-diffusion, 37, 59, 340–342, 345–346, 349, 351
Output return loss, 410

Phase cancellation technique, 219
Phase distortion, 401–402, 404
Phase noise, 2, 412, 415–416, 419–420, 422
Phonon scattering, 17–18
Power-added efficiency, 214, 322, 381–382, 389, 397
Power-delay product, 428, 433–434
Power efficiency, 380, 387, 390, 395–396

Quality factor, 391
Quantum efficiency, 447, 451
Quantum-mechanical effect, 166
Quantum-mechanical reflection, 187
Quantum-mechanical wavelengths, 25
Quantum well, 25–26
Quasi-Fermi level splitting, 65, 67
Quasi-neutral recombination, 97
Quasi-saturation region, 221

Radiation effect, 357–358
Radiation hardness, 2, 364, 366
Rapid thermal chemical vapor deposition (RTCVD), 333, 341
Reactive ion beam etching (RIBE), 38
Reactive ion etching (RIE) 38, 47, 50
Reflection high energy electron diffraction (RHEED), 42, 45
Regression characteristics, 321
Reverse active mode, 57–58, 205
Reverse base current effect, 187, 285, 287
Rutherford backscattering spectrometry (RBS), 23, 33

Scanning electron microscope (SEM), 331
Scattering parameters (*s*-parameters), 47, 113, 124, 133, 135–136, 214, 223–224, 383, 385, 456
Secondary ion mass spectrometry (SIMS), 21, 33–35
Self-heating effect, 92–93, 95–97, 170–171, 187, 202–204, 206, 208, 212, 251–252, 287, 289–290, 306, 318, 378
Setback layer, 59, 61, 166–167, 169–170, 203–204
Shot noise, 130
Sideband noise, 419
Signal-to-noise ratio, 450
Single event upset (SEU), 357–358, 369
Sticking coefficients, 30, 42–43
Shockley–Read–Hall (SRH) recombination rate, 199, 248, 298, 353
Space-charge recombination, 97, 99, 101, 193, 195, 197, 201–202, 349, 366
Spreading resistance profiling (SRP), 33
Superlattice, 336, 388
Surface passivation, 336, 422
Surface recombination, 3, 40, 64, 97–99, 102, 198, 200, 349–350, 362, 422, 427
Surface recombination velocity, 98
Surface scattering, 4

Thermal instability, 304, 310–311, 313–314, 317, 321, 322
Thermal noise, 130
Thermal runaway, 307
Thermal stability, 319, 323
Thermal shunt, 318–319
Thermionic emission, 55–56, 65, 101, 186, 201, 204, 352
Thermionic-field-emission, 185, 189
Third-order intermodulation, 401
Time-to-failure (TFF), 334, 356
Total dose irradiation, 365
Total dose-rate radiation, 364
Transconductance clipping, 143, 155
Transient collector current recovery time, 369
Transient radiation effect, 369
Transmission electron microscope (TEM), 14, 341
Tunneling, 55, 65, 101, 187, 204, 354, 357
Tunneling current, 183
Two-port network method, 219

Valence-band discontinuity, 3, 6, 10–11, 55, 72, 184, 200, 297, 341, 345
Valence-band offset, 4, 73
Vapor phase epitaxy, 24–25, 27
Velocity overshoot, 3, 8–10, 113–115, 119, 241, 243, 252–253, 255, 388
Velocity saturation, 71
Volterra series expansion method, 401

X-ray diffraction, 14